Pedology: Formation, Morphology and Classification of Soil

Pedology: Formation, Morphology and Classification of Soil

Editor: Katie Phillips

RCALLISTO
REFERENCE

www.callistoreference.com

Callisto Reference,
118-35 Queens Blvd., Suite 400,
Forest Hills, NY 11375, USA

Visit us on the World Wide Web at:
www.callistoreference.com

ISBN: 978-1-63239-797-3 (Hardback)

The publisher's policy is to use permanent paper from mills that operate a sustainable forestry policy. Furthermore, the publisher ensures that the text paper and cover boards used have met acceptable environmental accreditation standards.

Trademark Notice: Registered trademark of products or corporate names are used only for explanation and identification without intent to infringe.

Printed in the United States of America.

Cataloging-in-publication Data

Pedology : formation, morphology and classification of soil / edited by Katie Phillips.
   p. cm.
Includes bibliographical references and index.
ISBN 978-1-63239-797-3
   1. Soil science. 2. Soil formation. 3. Soil structure. 4. Soils--Classification. I. Phillips, Katie.
S591 .P43 2017
631.4--dc23

Table of Contents

Preface

Pedology is the science of the natural occurrence of various types of soil. Soil formation is the basis for planetary life and plays an important part in species evolution and propagation. This book brings together research on various aspects on Pedology and its relation to agriculture. Researchers and experts in the fields of soil science will find this book suitable for their study. This book discusses fundamentals as well as modern approaches of Pedology by unfolding the innovative aspects of this field.

Over the recent decade, advancements and applications have progressed exponentially. This has led to the increased interest in this field and projects are being conducted to enhance knowledge. The main objective of this book is to present some of the critical challenges and provide insights into possible solutions. This book will answer the varied questions that arise in the field and also provide an increased scope for furthering studies.

I hope that this book, with its visionary approach, will be a valuable addition and will promote interest among readers. Each of the authors has provided their extraordinary competence in their specific fields by providing different perspectives as they come from diverse nations and regions. I thank them for their contributions.

Editor

Influence of Vegetation Restoration on Topsoil Organic Carbon in a Small Catchment of the Loess Hilly Region, China

Yunbin Qin, Zhongbao Xin*, Xinxiao Yu, Yuling Xiao

Institute of Soil and Water Conservation, Beijing Forestry University, Beijing, China

Abstract

Understanding effects of land-use changes driven by the implementation of the "Grain for Green" project and the corresponding changes in soil organic carbon (SOC) storage is important in evaluating the environmental benefits of this ecological restoration project. The goals of this study were to quantify the current soil organic carbon density (SOCD) in different land-use types [cultivated land, abandoned land (cessation of farming), woodland, wild grassland and orchards] in a catchment of the loess hilly and gully region of China to evaluate the benefits of SOC sequestration achieved by vegetation restoration in the past 10 years as well as to discuss uncertain factors affecting future SOC sequestration. Based on soil surveys (N = 83) and laboratory analyses, the results show that the topsoil (0–20 cm) SOCD was 20.44 Mg/ha in this catchment. Using the SOCD in cultivated lands (19.08 Mg/ha) as a reference, the SOCD in woodlands and abandoned lands was significantly higher by 33.81% and 8.49%, respectively, whereas in orchards, it was lower by 10.80%. The correlation analysis showed that SOC and total nitrogen (TN) were strongly correlated ($R^2 = 0.98$) and that the average C:N (SOC:TN) ratio was 9.69. With increasing years since planting, the SOCD in woodlands showed a tendency to increase; however, no obvious difference was observed in orchards. A high positive correlation was found between SOCD and elevation ($R^2 = 0.395$), but a low positive correlation was found between slope and SOCD ($R^2 = 0.170$, $P = 0.127$). In the past 10 years of restoration, SOC storage did not increase significantly (2.74% or 3706.46 t) in the catchment where the conversion of cultivated land to orchards was the primary restoration pattern. However, the potential contribution of vegetation restoration to SOC sequestration in the next several decades would be massive if the woodland converted from the cropland is well managed and maintained.

Editor: Manuel Reigosa, University of Vigo, Spain

Funding: This study was supported by the Fundamental Research Funds for the Central Universities (NoTD2011-2), the Open Foundation of Key Laboratory of Soil and Water Loss Process and Control on the Loess Plateau of Ministry of Water Resources (201301), the National Natural Science Foundation of China (No. 41001362) and the College Student Scientific Research Training Project of Beijing Forest University (No. 201210022013). The funders had no role in study design, data collection and analysis, decision to publish, or preparation of the manuscript.

Competing Interests: The authors have declared that no competing interests exist.

* Email: xinzhongbao@126.com

Introduction

Afforestation and other vegetation restoration techniques have been considered effective practices for the sequestration of carbon (C) to mitigate carbon dioxide (CO_2) concentrations in the atmosphere [1–4]. Soil plays an important role in the global carbon cycle. The soil C pool to a one metre depth has been estimated to sequester approximately three times the amount of carbon sequestered by the atmospheric pool and about four times that in the biotic/vegetation pool [5]. Therefore, a relatively small change in the soil C pool can significantly mitigate or enhance CO_2 concentrations in the atmosphere [6–7].

Land-use change can significantly influence the accumulation and release of SOC. When an ecosystem is disturbed by land-use change, the original equilibrium of the soil carbon pool is broken and a new equilibrium is created. During this process, soil may act as either a source or a sink of carbon depending on the ratio between inflows and outflows [8]. Many studies have reviewed the effect of land-use change on SOC. For instance, deforestation for agricultural purposes is the primary reason for the SOC loss. It

was reported that approximately 25% of SOC was lost by conversion of primary forest into cropland [9]. Houghton (1999) estimated that 105 PgC was released into the atmosphere due to the conversion of forests to agricultural lands between 1850 and 1990 [10]. However, afforestation and reforestation on agricultural lands have been cited as effective methods for increasing SOC pool and reducing the atmospheric CO_2 concentration [3,11]. Morris et al. (2007) observed that placing agricultural soils in deciduous and conifer forests resulted in soil carbon accumulations of 0.35 and 0.26 $MgCha^{-1} yr^{-1}$, respectively. Based on a meta analysis of 33 recent publications, Laganière et al. (2010) reported that afforestation increased SOC stocks by 26% for croplands. Some studies found that croplands that were converted to abandoned lands or grasslands could also increase the SOC storage [9,13–14]. Therefore, understanding the influence of land-use changes on soil organic carbon is an important step in predicting climatic change and developing potential future CO_2 mitigation strategies.

To improve the carbon sink status of afforestation or other vegetation restoration methods on agricultural land, it is necessary to understand the control mechanisms of SOC dynamics to allow more carbon storage in soils [1]. A variety of factors will affect the quantity and quality of SOC after land-use changes. For example, climate variations have a significant effect on SOC, including temperature, precipitation, and potential evapotranspiration changes [11,15]. Laganière et al. (2010) reported that SOC restoration after afforestation was found to vary with the climate zone, with the temperate maritime zone having a higher SOC increase than others by approximately 17%. In addition, landscape and elevation also have a pronounced effect on SOC change at the catchment scale. Slope is an important topographical factor that affects soil erosion and also has an important influence on the soil nutrient loss of the slope surface [16]. Wang et al. (2012) reported that the SOCD was higher in shady slopes than in sunny slopes, and gentle slopes would generally sequester more SOC than steep slopes. Elevation differences can cause climatic and biological changes in the soil-forming environment and can influence the vertical distribution of SOC [18]. In natural ecosystems, it has been extensively documented that carbon and nitrogen are closely related [4,19]. Increased nitrogen retention may increase the carbon sequestration potential [1]. In Panama, Batterman et al. (2013) found that symbiotic N_2 fixation has potentially important implications for the ability of tropical forests to sequester CO_2 [20]. N_2-fixing tree species accumulated carbon up to nine times faster per individual than neighbouring non-fixing trees. The soil C/N ratio also has a significant effect on the rate of decomposition of organic compounds by soil microorganisms [21].

Soil erosion, as the most widespread form of soil degradation, has a large impact on the global C cycle, causing a severe depletion of SOC pools in the soil [22–24]. The total amount of C released by soil erosion each year has been estimated at approximately 4.0–6.0 Pg/year [23]. China's Loess Plateau, covering approximately 6.2×10^5 km^2, has the world's most severe soil erosion because of its unusual geographic landscape, soil and climatic conditions, and long history (over 5000 years) of human activity. Over 60–80% of the land in the Loess Plateau has been affected by soil erosion, with an average annual soil erosion of 2000 to 20000 t km^{-2} yr^{-1} [25–26]. Severe soil erosion has resulted in land degradation, which was manifested primarily in the thinning of the soil layer, nutrient loss and fertility reduction, which has directly caused decreases in the local farmers' income and has economically and socially hindered sustainable development [27–29]. Unreasonable human activities, such as deforestation and tillage on slopes, have further intensified soil erosion and land degradation in the Loess Plateau [30].

Since the 1950s, the Chinese government has launched many large-scale projects to attempt to control soil erosion and restore vegetation in the Loess Plateau, including large-scale afforestation in the 1970s and comprehensive control of soil erosion on the watershed scale in the 1980s and 1990s. Despite these efforts, there have not been significant increases in ecological benefits, which is largely due to the limitations and influences of bad natural conditions and unreasonable ecological restoration techniques. To control soil erosion and improve the quality of the local environment, "Grain for Green" was initiated by the government in the Loess Plateau in 1999. The government demanded that the agricultural lands with a slope of over 25 degrees be converted to forest, terrace orchards or grassland. To compensate famers for their economic loss, they will be given grain, cash and planting stocks by the government as subsidies and incentives for converting cultivated land back to forest, orchards or grassland. Currently, the Loess Plateau environment appears to be experiencing a recovery following more than 10 years of vegetation restoration [31–32].

Over the past decade, large-scale vegetation restoration efforts have brought obvious land-use changes to the Loess Plateau, and also have significantly influence on SOC sequestration. Therefore, understanding the effects of these dynamic changes and the corresponding changes in SOC storage caused by land-use change driven by the implementation of the Grain for Green project is important in evaluating the environmental benefits of this ecological restoration project. Recently, many studies have focused on SOC changes induced by the Grain for Green project in the Loess Plateau. These studies have mainly focused on the SOCD of different land-use types, the SOC pool, the rate of SOC change, the variation in SOC among different land-use conversions and factors which have influenced SOC sequestration in the Loess Plateau after vegetation restoration efforts [32–34]. However, for most of the researches, the study time period was shorter than 10 years as vegetation restoration was driven by the Grain for Green project. The study areas have been mostly located at the gully region of the Loess Plateau [2,17,35–36]. Little is known about the gully area of the Loess hilly region. To gain a complete understanding of the SOC change after restoration in this region, we selected the Luoyugou catchment of the Loess Plateau as the study area, which has the typical geomorphologic characteristics of the gully area of the Loess hilly region.

In this study, we hypothesised that the SOCD varied with land-use types in this catchment, and there was an increase in SOC storage of the total catchment after 10 years of restoration. Therefore, the objectives of this study were to (i) quantify the SOCD of different land-use types differ significantly in the study catchment, (ii) analyze those factors affecting SOCD, (iii) estimate the contribution of land-use conversions on SOC sequestration in the study area, in the past 10 years of restoration, and (iv) discuss the uncertainties in potential SOC sequestration during future ecological restoration efforts in the Loess Plateau.

Materials and Methods

Ethics statement

The administration of the Tianshui Experiment Station of Soil and Water Conservation, Yellow River Conservancy Commission and local farmers which are the owners of study lands gave permission for this research at each study site. We confirm that the field studies did not involve endangered or protected species.

Study area

Tianshui city (104°35′–106°44′E, 34°05′–35°10′N) of Gansu province, China, is located in the western side of the Qinling Mountains, which is the transitional zone between the Qinling Mountains and the Loess Plateau and also belongs to the third sub-region of the Loess hilly region in the middle portion of the Yellow River and the second class tributary of the Wei River. This region has the typical geomorphology of the gully area of the Loess hilly region, which includes earth-rocky mountainous areas, Loess ridges and hilly areas. In this area, the catchment landscape is an agroforestry landscape, and the climate has obvious transitional characteristics. Because of these special geomorphologic characteristics, Walter Lowermilk, deputy director of the US Department of Soil Conservation Service, chose this region in which to build an experimental station for soil and water conservation in 1941. Tianshui station, Suide station and Xifeng station now are three the best-known experimental stations for soil and water conservation in the Loess Plateau.

The Luoyugou catchment (105°30′–105°45′E, 34°34′–34°40′N), located in northern Tianshui, is a part of the observation area of the Tianshui soil and water conservation station. Its total area is 72.79 km^2, with a range from 1165–1895 m above sea level. The main topographic type is loess ridge landform, and the average slope is 19°. It has a typical continental monsoon climate with a mean annual precipitation of 548.9 mm (1986–2004). Approximately 78% of the rainfall is concentrated between July and September. The average annual temperature is 11.4°C (1986–2004),and the annual evaporation is 1293.3 mm. The main soil type in this catchment is mountain grey cinnamon soil, which, according to the Food and Agriculture Organization of the United Nations Educational, Scientific and Cultural Organization (FAO-UNESCO), belongs to the Cambisol soil group and accounts for 91.7% of the soil in the region [37–38] (Figure 1).

The main land-use types for the Luoyugou catchment include woodland, wild grassland, orchards, cultivated land, and abandoned land. In the woodland, the major tree species is black locust (*Robinia pseudoacacia L.*), most of which was artificially planted. Woodlands that are more than 30 years old have good water condition and low density because planting is located mainly beside the place beside water ditches and shady slope. They are the retained trees of the planting projects in the late 1950s to 1970s, without human management now. And most of woodland at the age of 10 to 30 years since planting is found in Fenghuang forest farm located in the gully head of this catchment, which has good management and protecting. Since 1999, the government of Tianshui city has widely implemented the "Grain for Green" project. A lot of cultivated land in the ridge top and slope top has been converted into woodland. Most of the woodland is less than 10 years since planting, and has a high density and bad management. Wild grassland is usually found on stone mountain of the northern catchment where the soil layer is thin and slope is steep. Human activities are restricted in there, only sometimes sheep may reach. The major species are dahurian bushclover (*Lespedeza dahurica (Laxm.) Schindl.*), russian wormwood (*Artemsia sacrorum Ledeb.*), digitate goldenbeard (*Bothriochloa ischaemun (L.) Keng*) and bunge needlegrass (*Stipa bungeana Trin.*), etc. The main fruit species under orchards are cherry (*Cerasus pseudocerasus (Lindl.) G. Don*), apple (*Malus pumila Mill.*) and apricot trees (*Armeniaca vulgaris Lam.*). The management system of orchards is the conventional tillage which is that the weeds beneath the trees are eliminated with herbicides and dead leaves, dried fruit and twigs are removed by manual blowers. Besides, there are not irrigation facilities and the tree water demand all comes from the rain. Fertiliser and pesticide are also applied in the orchards. This catchment is a rainfed agriculture region with a long cultivated history. The major species in the cultivated land are wheat (*Triticum aestivum L.*), maize (*Zea mays L.*) and edible rape (*Brassica campestris L.*), etc. Abandoned land is the cultivated land has been the ceased farming

activities, due to the demand of "Grain for Green" project, the bad environment conditions, the shortage of rural labour and the damage by wild animals, etc. After cultivated land abandoned, old field are spontaneously colonised by various plants, while a secondary succession process will gradually develop during different plant communities. Because of short time since abandoned, the main species in the abandoned land are annual and biennial herb and a bit of perennial herb, such as virgate wormwood (*Artemisia scoparia Waldst. et Kit.*) and green bristlegrass herb (*Utricularia australis R. Br.*).

Soil sampling and laboratory analysis

Based on the main land-use types and the percentage of each land-use-type area in the total catchment area (2008), we randomly selected 83 land-use blocks that included cultivated land (34 samples), abandoned cropland (9 samples), orchards (13 samples), wild grassland (8 samples) and woodland (19 samples) as investigation plots. Their latitudes and longitudes are shown in Table 1. Spatially separated plots were located at least 3 km from one another to help avoid pseudo-replication. In July 2012, the selected plots were surveyed and collected in the field using the Global Positioning System (GPS). The information recorded from the sampling plots in the field included geographic coordinates, slope, elevation, plant species, and water conservation measures. The years since planting or abandoned of the sampling plots were evaluated by recording tree diameters and heights, plant composition, the degree of decomposition of the topsoil crust and the crop's residual body, and by interviewing local farmers. In each sampling plot — woodland and orchards (10×10 m), cultivated land, abandoned land and wild grass land (5×5 m) — we randomly collected three soil samples using a stainless steel cutting ring 5.0 cm high and 5.0 cm in diameter to measure the soil bulk density (BD) of the topsoil layer (0–20 cm) and by taking five random soil samples between 0 and 20 cm depth with a 20 cm long soil auger. The five soil samples were manually homogenised to form a composite sample for each sampling plot, and a quarter of this sample was taken to the laboratory. These samples were air-dried and passed through a 2 mm sieve, while gravel and roots were removed from each soil sample. A quarter of each sub-sample was completely passed through a 0.25 mm sieve to determine SOC and TN. SOC was determined by the $K_2Cr_2O_7$-H_2SO_4 Walkey-Black oxidation method [39]. TN was measured using the micro-Kjeldahl procedure [40].

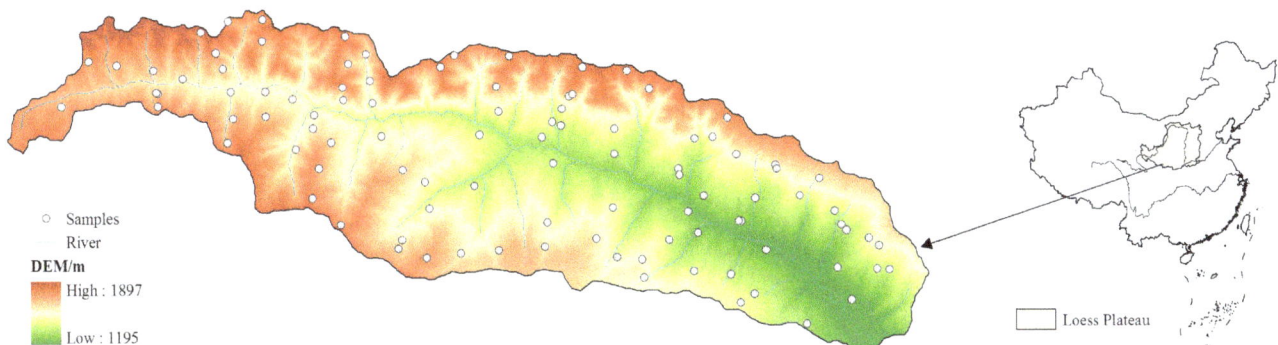

Figure 1. Location of the study area and distribution of the sampling points.

Table 1. Attributes of the studied sites.

Land-use	Sample sites	Years since planting (abandoned)/a	Elevation/m	Slope/(°)	Bulk density/(g/cm³)	Plants
Cultivated land	34	—	1536.94	4.85	1.29	Wheat, maize, rape
Abandoned cropland	9	8–10a	1559.40	4.22	1.35	*Artemisia scoparia, Leymus secalinus*
Orchards	13	<5a, 5–10a, >10a	1446.45	1.92	1.32	Cherry, apple, pear
Woodland	24	<10a, 10a–30a, >30a	1639.14	12.13	1.39	Robinia
Wild grassland	8	—	1526.55	13.00	1.44	*Artemisia sacrorum var. incana mattf, Stipa bungeana, Bothriochloa ischaemun*

Statistical analysis

The SOCD was calculated using the following equation:

$$SOCD = SOC \times D \times BD \qquad (1)$$

where SOCD was the soil organic carbon density (Mg ha^{-1}), SOC was the soil organic carbon content (%), D was the soil layer depth (cm), BD was the soil bulk density (g cm^{-3}).

The soil organic carbon storage (SOCS) was calculated using the following equation:

$$SOCS = SOCD_i \times A_i \times 100 \qquad (2)$$

where SOCS was the soil organic carbon storage (t), $SOCD_i$ was the soil organic carbon density in the i land-use (Mg ha^{-1}), and A_i was the area of the i land-use (km^2).

The increase in SOCS caused by land-use conversion was calculated using the following equation:

$$\text{Increase of SOCD (t)} = SOCD_i \times (A_{i2}\text{-}A_{i1}) \times 100 \qquad (3)$$

$$\text{Sequestration amount of SOCS (t)} = (SOCD_i - SOCD_c) \times (A_{i2} - A_{i1}) \times 100 \qquad (4)$$

where $SOCD_i$ was the SOCD in the i land-use (Mg ha^{-1}), A_{i1} and A_{i2} were the i land-use area in 2002 and 2008, respectively (km^2), and $SOCD_c$ was the SOCD of the cultivated land (Mg ha^{-1}). All SOCD values of the different land-use types in 2002 and 2008 were the average SOCD values of the soil data collected in 2012.

All data were analysed using Excel 2007 and SPSS (Statistical Package for the Social Sciences) 18.0 software. Analysis of variance (ANOVA) was used to determine the significance of the mean difference. Fisher's LSD test was used to compare the mean values of soil variables when the results of ANOVA were significant at $p < 0.05$. Pearson's test was used to analyse the correlation between SOCD and the soil variables. For all analyses, a $p < 0.05$ was used to test statistical significance. Finally, SOCD, land-use type, elevation and slope values were used to establish the multiple regression model. SOCD was the dependent variable, and others were the independent variables.

Satellite data acquisition and processing

In our study, two types of SPOT5 multispectral images were acquired on August 25, 2002 and May 5, 2008, which were multi-spectral images (resolution: 10 m) and panchromatic images (resolution: 2.5 m). To validate the final classification result, we selected 17 points located in the Luoyugou catchment to use as the observed data based on the land-use map and some thematic maps. In July 2012, these monitoring points were collected in the field to validate the panchromatic map, while we recorded 83 points with information of land-use type and longitude and latitude to validate the accuracy of the classification results. Using ArcGIS and ERDAS software, we chose the common pre-classification, which is the supervised classification with a Maximum Likelihood Decision (MLD), to classify the preprocessed images and acquire the data of different land-use types in 2002 and 2008. The overall accuracy and Kappa coefficient of the initial classification images were 87.50% and 0.85 in 2002, respectively, and 90.10% and 0.87 in 2008, respectively. The accuracy of the 88 typical types of land-use information used to validate the classification results was 95.74%, which implied that this classification result could meet the analysis demands.

Results

Topsoil organic carbon density under different land-uses

The one-way ANOVA indicated that land-use type had a significant effect on the SOCD ($F = 8.34$, $P < 0.01$). However, only the woodland was significantly different from the other land-uses ($P < 0.05$), while no significant differences were found between other land-use types. The average SOCD of the topsoil layer (0–20 cm) in the Luoyugou catchment was 20.44 Mg/ha, and was highest in the woodland and lowest in the orchards (Table 2). Using the SOCD in the cultivated land (19.08 Mg/ha) as a reference, the SOCD in the woodland and abandoned cropland significantly increased by 33.81% and 8.49%, respectively, while the SOCD in the orchards decreased by 10.80%. The SOCD in the wild grassland was close to that of cultivated land. In this area, the coefficient of variation of the SOCD was 26.46%, and was highest (27.08%) in the woodland and smallest (16.75%) in the orchards.

Change of topsoil SOC storage in the catchment

Based on the interpretation of the remote sensing data (2000), the results showed that the total area of the four main land-use types including woodland, orchards, cultivated land and wild grassland was 67.5 km^2, accounted for 92.09% of the total region of the Luoyugou catchment. The difference of 7.02% included river, residential points, and roads that were not taken into account in the calculation of the topsoil SOC storage. From 2002 to 2008, the net decrease in the cultivated land area was

Table 2. Topsoil (0–20 cm) SOCD of different land-uses.

Land-use	Total	Woodland	Orchards	Abandoned cropland	Wild grassland	Cultivated land
Sample sites/N	83	19	13	9	8	34
SOCD Mg/ha	20.44	25.53 a	17.02 b	20.70 b	19.35 b	19.08 b
Standard deviation	5.41	6.91	2.85	4.59	3.99	3.66
Minimum	10.47	16.45	11.03	16.55	12.59	10.47
Maximum	40.71	40.71	20.72	31.14	26.02	28.86
Coefficient of variation (%)	26.46	27.08	16.75	22.17	20.62	19.16

* A different letter means a difference significant at 0.05 level.

15.11 km^2, where 7.95 km^2 of this land was converted into the woodland, and 6.51 km^2 was converted into the orchards. The increased areas of land were accounted for 52.65% and 43.12% of the total conversion area, respectively. In 2008, the SOC storage in this catchment was 1.39×10^5 t and was higher than the SOC storage in 2002 (1.35×10^5 t) by 3706.46 t. When the cultivated land was converted into the woodland, the SOC sequestration contribution was 4484.83 t; however, when the cultivated land was converted into the orchards, the contribution was −723.02 t (Table 3).

Relationship of soil organic carbon and total nitrogen content

In this catchment, soil organic carbon and total nitrogen revealed a significant positive correlation ($R^2 = 0.978$, $p < 0.01$) that increased as total nitrogen content and soil organic carbon increased (Figure 2). A one-way ANOVA indicated that land-use had significant effect on the TN ($F = 3.07$, $P = 0.021$), and that had no significant effect on the soil C/N ratio ($F = 0.86$, $P = 0.48$). The average soil C/N was 9.69, with a range of 8.73–10.77. The soil C/N ratio in the woodland was the largest at 9.80 and that in the orchards was the smallest at 9.57. Other land-uses followed the order of abandoned cropland (9.76) > cultivated land (9.66) > wild grassland (9.59).

Relationship of years since planting and topsoil SOCD

For the woodland, years since planting had a significant effect on SOCD ($F = 13.00$, $P < 0.001$). SOCD increased as years since planting increased. However, SOC was lost initially after afforestation in the cultivated land. The SOCD of the woodland that was older than 30 years was higher than that of the woodland under 10 years old (17.45 Mg/ha) and the cultivated land (19.08 Mg/ha), improving by 74.44% and 59.54% respectively (Figure 3a). However, years since planting had no significant effect on SOCD in the orchards ($F = 2.01$, $P = 0.146$), where the SOCD at >10a was 18.04 Mg/ha and was 10.13% higher at <10a. But, they all were below that of the cultivated land (Figure 3b).

Relationship of elevation and topsoil SOCD

The correlation analysis showed that there was a significant positive correlation between elevation and SOCD ($R^2 = 0.395$ $P < 0.01$), using the land-use type as the covariate. The SOCD was calculated as the average value with each 100 m used as an elevation gradient. The results showed that with increasing elevation, SOCD had an increasing trend (Figure 4). At an elevation gradient of ≥1700 m, the SOCD was 26.57 Mg/ha,

which was higher than that at the elevation gradient of <1300 m by approximately 58.10%.

Relationship of slope and topsoil SOCD

A one-way ANOVA indicated that land-use had a significant effect on slope ($F = 2.87$, $P = 0.028$) where the average slopes of the woodland and wild grassland all were more than $11.00°$, and the average slopes of the cultivated land, abandoned cropland and orchards all were less than $5.00°$ (Table 4). The correlation analysis showed a low positive correlation between slope and SOCD ($R^2 = 0.170$, $P = 0.127$). A negative correlation was found between the slope and SOCD in the cultivated land, abandoned cropland and orchards ($R^2 = -0.210$, $F = 0.123$), and a positive correlation was found in the woodland and wild grassland ($R^2 = 0.250$, $F = 0.209$). For all correlations, the land-use type was used as the covariate.

Influence of multivariates on topsoil SOCD

Through normalisation processing of SOCD, elevation and slope and setting land-use dummy variables, the multiple regression model was established using SPSS software. The results were as follows:

$$y = 0.150 + 0.233x_1 + 0.054x_2 + 0.004s_1 - 0.025s_2 + 0.044s_3 + 0.168s_4 \,(R^2 = 0.329, \ p < 0.01) \tag{5}$$

where y was the topsoil SOCD, x_1 was elevation, x_2 was slope, s_1 was wild grassland, s_2 was orchards, s_3 was abandoned land, and s_4 was woodland. The cultivated land was used as the reference when the value of s_{1-4} was 0.

According to the multiple model, elevation, slope and land-use accounted for 32.9% of the SOCD variation using the cultivated land as the reference (Eqn. 5).

Discussion

SOC sequestration in different ecological restoration types

The different types of conversion result in different trends in SOC [8]. In this study, the highest SOCD in the woodland implied that the conversion from the cultivated land to the woodland may lead to the accumulation of more SOC than when cultivated land is converted into the orchards or abandoned cropland in this area (Table 2). Some previous research has also found that the conversion from cultivated land to woodland can increase SOC levels [3,17]. This is mainly because the conversion from cultivated land to woodland can increase topsoil SOCD

Table 3. Change in SOC storage from 2002 to 2008 in the Luoyugou catchment.

Land-use	2002			2008			SOC	
	Area/km²	SOC		Area/km²	SOC			
		Storage/t	Percentage/%		Storage/t	Percentage/%	Increased/t	Sequestration amount/t
Immature woodland	8.16	19764.05	14.62	15.48	37480.99	26.98	17716.94	3759.91
Mature woodland	2.35	7145.31	5.28	2.99	9087.78	6.54	1942.48	724.92
Orchards	4.11	7377.46	5.46	10.62	19082.52	13.74	11705.06	−723.02
Wild grassland	0.71	1374.26	1.02	1.31	2536.62	1.83	1162.36	16.22
Cultivated land	52.18	99557.47	76.63	37.07	70737.09	50.92	−28820.38	–
Total	67.50	135218.55	100.00	67.46	138925.00	100.00	3706.46	3778.03

* Because the area of this catchment in 2008 was less than that in 2002, by 0.04 km², so that about 71.58 t of SOC wasn't taken into the amount of SOC storage. Immature woodland refers to trees under 30 years of age since planting (SOCD = 24.22 Mg/ha), and mature woodland refers to trees more than 30 years of age since planting (SOCD = 30.44 Mg/ha).

through increasing biomass inputs into the soil, and reducing soil erosion [9,12]. In addition, compared to frequent human activity such as tillage and grazing, in the cultivated land and abandoned cropland, less human disturbance in the woodland also enhanced SOC accumulation.

Our research found that the SOCD in the immature woodland (<10 a) was lower than that in the cultivated land. However, the SOCD of the mature woodland (>30 a) was higher than that of the cultivated land, increasing by 59.54% (Figure 3a). Therefore, in the initial conversion from the cultivated land to the woodland, SOCD may decrease, and from 10a after afforestation SOCD increase rapidly and significantly. Other studies have found similar trends [4,15]. Degryze et al. (2004) in Michigan found no difference in topsoil (0–25 cm) soil C during the first ten years after afforesting in a cropland; however, soil C of the native forest (48.6 t/ha) was significantly greater than the cropland (31.9 t/ha) [41]. Lu et al. (2013), studying afforestation in the Loess Plateau, found that the time of the SOC source to sink transition was 3 to 8 years after afforestation. The decrease of SOC in the first few years following afforestation has been attributed to low net primary productivity of plants, decreased litter inputs and increased decomposition rates [12,29].

In this study, the topsoil SOCD in the orchards was less than that in the cultivated land, which implied that a decrease in SOC may occur when the cultivated land is converted into the orchards (Figure 3b). This result is similar to the results of Yang et al. in the gully region of the Loess Plateau [42]. However, Xue et al. (2011) reported that when the slope farmland was converted into the orchards, the SOC content increased slowly with increasing years since planting and reached its peak between 20 to 30 years [43]. Thirty years later, the SOC content of the converted orchards was 4.96 g/kg and was higher than that in the slope farmland by 97%. One difference between our study and that by Xue et al. (2011) is the different reference used. The reference in their study was slope cropland, whereas we used the average value of the slope and terrace cropland as the reference. Moreover, management measures in their study were better than ours. For example, they interplanted crops into the fruit trees at the seedling stage and used more fertiliser.

Management measures are one of the main factors affecting the loss of SOC in orchards. In this catchment, fruit, trimmed branches, and litter are removed from orchards, which can cause less organic compounds to enter the soil. Clear tillage in the orchards also causes more soil erosion. Therefore, in some farmland regions that are returning to their historic uses and thus need to develop fruit trees for economic purposes, methods for reducing the loss of SOC and enhancing SOC sequestration have become important. To manage this problem, some researchers have found that orchards managed with conservation practices such as growing grass and leguminous cover crop, mulching the ground, and frequently using organic and inorganic fertilisers could reduce soil erosion and improve the SOC content [44–45]. In addition, establishing new orchard management models oriented to SOC sequestration is also very necessary.

Many studies have found that the land-use change from the cultivated land to the abandoned land can increase the SOC stored in soil [13–14]. In our study, our results also showed that the SOCD may improve when the cultivated land is converted into the abandoned land (Table 2). With the termination of human disturbance, vegetation in the abandoned land may begin the process of self-succession [14], gradually forming original vegetation communities of their region, which are secondary wild grassland or secondary forest communities. However, this process of succession is slow, as the process that turns the abandoned land into the top secondary bunge needlegrass (*Stipa bungeana Trin.*) takes 40 to 50 years [46]. Further studies are needed to understand the methods needed to accelerate community succession and better sequester SOC in the abandoned land.

The research of Li et al. (2007) in the northern region of the Loess Plateau found that the SOCD of abandoned land at approximately 10 years of age was slightly lower than that in the secondary wild land where planted grassland (*Medicago sativa*) changed in the same number of years. After 6 to 10 years of restoration, degraded artificial grassland may form secondary bunge needlegrass (*Stipa bungeana Trin.*) [47–48]. Therefore, planting artificial grass in the abandoned land can shorten the time of succession, which ensures ecological benefits and increase economic benefits at the same time. However, in this study, the SOCD of the abandoned land at 8–10 years of age was higher

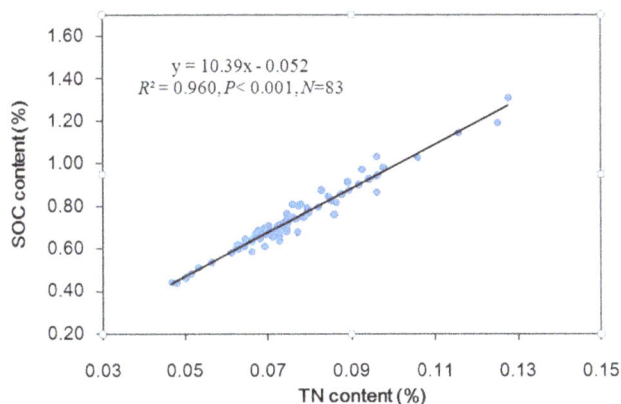

Figure 2. Relationship of topsoil soil organic carbon and total nitrogen content.

than in the wild grassland, which was different to the results of previous studies [47]. The main reason for this may be that these grasslands were distributed in the barren stony mountainous area. The average slope of these regions was 13°, and soil erosion was relatively severe. These factors led to low SOCD.

Benefits of SOC sequestration in the entire region

The SOC storage is determined by the dynamic equilibrium of SOC input and output. After approximately 10 years of ecological restoration, the area of cultivated land greatly decreased, while the area of woodland and orchards increased. From 2002 to 2008, the SOC storage of the entire area only improved by 2.74% (3706.46 t) (Table 3). Therefore, the ecological restoration has an increase in SOC storage, but not significant in the short-term. The main reason for this is that an increase in the SOC resulting from the woodland accumulation was offset by a loss in the SOC resulting from the orchards and the initial conversion stage. So, the SOC storage of the total area is in a relatively stable state. If all of the woodland values are calculated using the SOCD value of the woodland more than 30 years old (30.44 Mg/ha), then considering the loss of SOC by the orchards, the SOC storage of the total area can improve by 7.12% (9625.59 t). Therefore, SOC storage has great potential to increase in the future if the woodland converted from the cropland is well manage and maintain.

Although SOC storage has not increased in the past 10 years of restoration, with increasing ecological restoration, soil erosion has been controlled and the quality of the ecological environment, such as the air, water, vegetable coverage has gradually improved in this catchment.

Influencing factors on the SOC

Results from this study showed a strong correlation between SOC and TN (Table 2), which was consistent with earlier findings [33,36]. The trend of C:N ratio varying with the land-use types was similar to that of SOC. The woodland had a higher C:N ratio than other land-use types due to increased above and below ground biomass. The most prevalent tree in the woodland is the black locust, which has N-fixing capabilities, increased TN can promote tree growth, which results in an increase in SOC; it was found that the SOC increase was greater than the TN increase [33]. In catchments, elevation is a governing variable because of its effect on various environmental factors, such as temperature and precipitation. Our study showed a significant positive correlation between elevation and the SOCD (Figure 4). With rising elevation, topsoil SOC increased. This phenomenon was similar to results of other studies [18,49], which can be explained by the lower temperatures and increased precipitation with higher elevations. Increased precipitation can promote vegetation growth, which increases the accumulation of humus, and lower temperatures can limit the decomposition and turnover of SOC, leading to enhanced SOC storage.

The slope is one of the main topographical features affecting the soil erosion intensity as well as SOC loss and enrichment [16]. In the present study, land-use types significantly affected the correlation between the slope and SOCD, where there was a positive correlation in woodland and wild grassland and a negative correlation in cultivated land, abandoned land and orchards (Table 4). These outcomes were due mainly to the influence of vegetation type and coverage. Vegetation restoration on slopes such as planting trees and grass can increase surface vegetation coverage and effectively control soil erosion, further reducing the loss of SOC. These results have been found in other regions [17,30]. Thus, the slope land should be converted to woodland and grassland, or changed to terrace to maintain and increase SOC levels. The multiple regression analysis showed that land-use, slope and elevation accounted for 32.9% of the SOCD variation. The unexplained variation may be caused by the soil particle size,

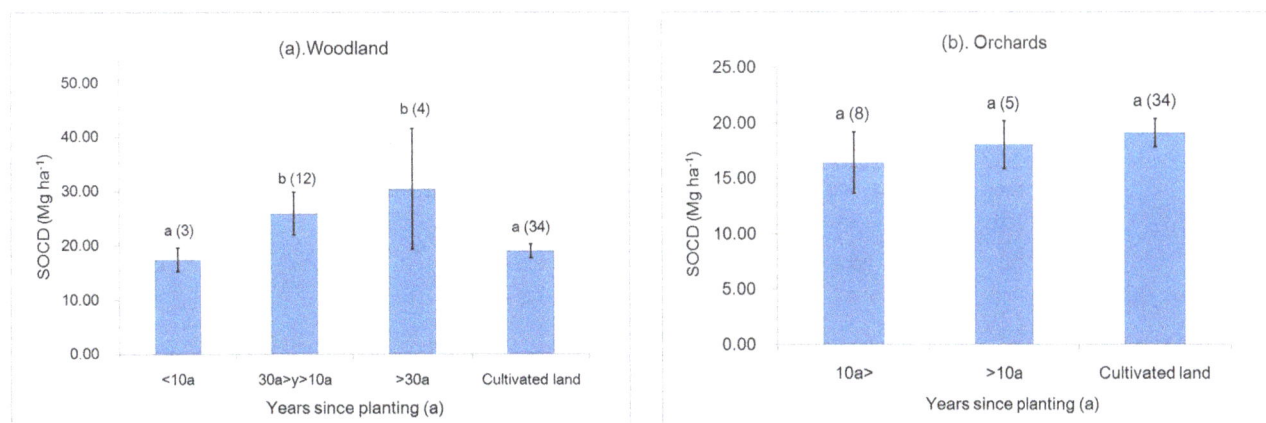

Figure 3. Relationship of years since planting and topsoil SOCD. * The error bars are the standard errors of the mean of SOCD and values above the bars is the number of observations (in parentheses). A different letter means a difference significant at P<0.05.

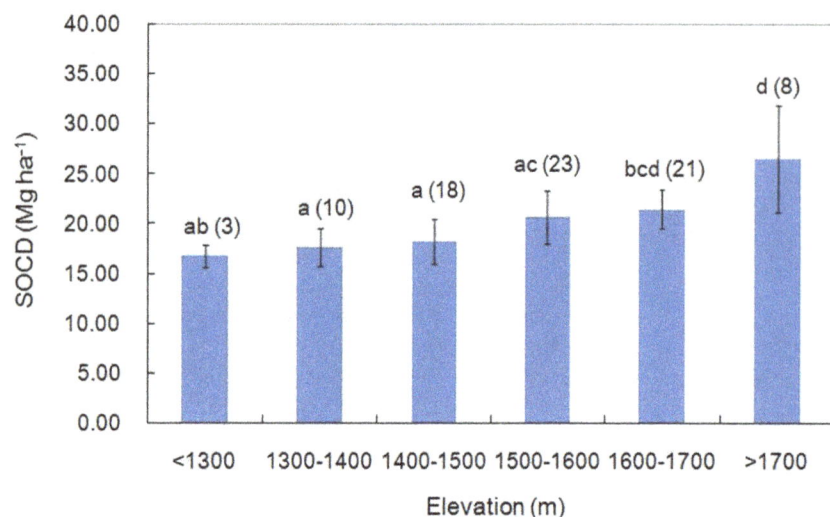

Figure 4. Relationship of elevation and topsoil SOCD. * The error bars are the standard errors of the mean of SOCD and values above the bars is the number of observations (in parentheses). A different letter means a difference significant at $P<0.05$.

soil aggregates, the aspect and the years of utilisation. Therefore, a benefits evaluation of SOC sequestration achieved through ecological restoration efforts should fully consider those influencing factors.

Uncertainties over SOC sequestration from future ecological restoration efforts

With ecological restoration efforts increasing, soil erosion in the Loess Plateau has significantly decreased, and ecological benefits such as vegetation coverage have gradually increased. However, in the long run, there are still uncertainties that will persist if we use the present SOC sequestration achievements from ecological restoration to evaluate possible future SOC sequestration benefits.

(I)   Influence of national policies: with the economy developing, the present subsidy standard for returning farmland to forests in the Loess Plateau has been increasingly unable to meet the basic living demands of local farmers. The competitiveness of this subsidy is also growing increasingly weak due to the influence of factors which are raising grain prices and creating policies which benefit farmers, including the abolition of the agricultural tax and the enhancement of subsidies for farming. Based

on the results of 2000 farmers' surveys, Cao et al. (2009) found that approximately 37.2% of farmers planned to return to cultivating forested areas and grasslands once the project's subsidies end in 2018. The smaller the subsidy amount to farmers, the fewer the farmers who are willing to participate in them. If the government is unable to improve farmers' income, much of the vegetation restored during the "Grain for Green" project is at risk of being re-converted into farmland and rangeland at the end of this project [50]. Therefore, it is very important and urgent to determine how to effectively consolidate the achievements of the "Grain for Green" project, especially once the project ends, while continually completing the task of returning farmland to forest. However, the government does not do well in this respect. Therefore, uncertainties exist concerning the benefits from SOC sequestration by ecological restoration in the future.

(II)   Influence of the environment: Most regions converted from cultivated land to forest are experiencing drought and soil impoverishment. There are no rich water resources in the Loess Plateau, and large-scale vegetation construction will certainly consume massive amounts of water. Lü et al. (2012) found that over half of the Loess

Table 4. Slope of different land-uses.

Land-use	Total	Woodland	Orchards	Abandoned cropland	Wild grassland	Cultivated land
Sample sites/N	83	19	13	9	8	32
Slope/°	6.65	11.58 a	1.92 b	4.22 ab	13.00 a	4.85 b
Standard deviation	10.85	14.76	4.80	8.39	12.10	8.93
Minimum	0	0	0	0	0	0
Maximum	45.00	45.00	15.00	20.00	30.00	30.00
Coefficient of variation %	163.16	127.46	250.00	198.82	93.08	184.12
Correlation between slope and SOCD/R^2	0.170	0.269	−0.169	−0.459	0.480	−0.229

* A different letter means a difference significant at 0.05 level. The signification of correlation between slope and SOCD all were more than 0.05 ($P>0.05$).

Plateau experienced a decrease in runoff (2–37 mm/year) with an average of 10.3 mm/year after the implementation of the "Grain for Green" project [51]. Therefore, with water consumption increases resulting from vegetation, some regions are likely to face a deficit of water resources or even a possible depletion. Drought, decreased rainfall and the excessive depletion of deep soil water by planted vegetation causes the formation of the dried soil layer [52]. The dried soil layer can lead to land degradation and heavily influences vegetation growth, which forms large-scale, low-efficiency production forests [53–54]. Once the dried soil layer appears, it is very difficult for recovery to occur. In the process of implementing "Grain for Green" some improper phenomena have occurred, including the improper selection of vegetation restoration types and excessive close planting conducted in the blind pursuit of economic benefits and political achievement [55], and this has accelerated the formation of the dried soil layer. Therefore, in the long run, ecological restoration in the Loess Plateau is likely to face water storage restrictions. Thus, the benefits of SOC sequestration may be uncertain in the future.

(III) Influence of private and local government interests: Many regions would prefer that cultivated land be converted into orchards when the cultivated land needs to be converted due to this conversion type may generate a higher land value. However, the SOC sequestration benefits resulting from this type of conversion are lower than in the conversion from cultivated land to woodland, and this conversion type may even cause SOC loss. Over time, farmers may remove old trees in orchards converted from cultivated land, or abandoned land. In addition, according to changes in market demand, farmers may replace other tree species before the original trees have matured. The non-ecological woodland converted from cultivated land also can be harvested under reasonable conditions. However, the SOC sequestration effects of these changes have not been studied. In the future, many woodlands and orchards converted from cultivated land will almost certainly undergo cutting and regeneration. The benefits of SOC sequestration are uncertain in the future unless we can understand the SOC variation under these changes.

(IV) The influence of urbanization and economic interests: since the reforms and open policies were established, labour exports from rural areas have increased year after year. Lack of labour has caused the abandonment of large areas of rural cultivated lands [56]. Recent research has shown that the SOC content could increase when cultivated land is converted into abandoned land. However, given the increasing speed of abandonment, it is unclear how long this phenomenon can continue. In addition, these labours may choose to go home to engage in farming in the future, affecting by the economic crisis of the original working regions and other reasons such as older age and homesickness. This transform may bring new pressure on the local ecological environment, which may cause the loss of SOC. Therefore, the future benefits of SOC sequestration are uncertain.

(V) Influence of climate change: Here, climate change mainly refers to precipitation decline, temperature increase and elevated carbon dioxide levels. According to the results of

Xin et al. (2011), the drying trend of the Loess Plateau was highly significant, and annual rainfall showed an obvious decreasing trend over the past five decades (1956–2008) by approximately -1.4 mm/a [57]. The reduced precipitation aggravated the water resource shortage situation in the Loess Plateau, which led to the poor tree growth. This outcome may affect SOC sequestration by vegetation restoration, because the temperature increased and elevated carbon dioxide levels have opposite effects on SOC sequestration. Increasing temperatures may promote SOC decomposition; however, elevated carbon dioxide levels may also improve the net primary productivity of plants and increase both the residual body of vegetation and SOC storage [58–59]. Over the last 50 years (1961–2010), the annual mean temperature has significantly increased by $1.91\,^{\circ}C$ in the Loess Plateau [60]. The atmospheric concentration of CO_2 has increased from 280 p.p.m. to 385 p.p.m. between the pre-industrial era and 2008 [61–62]. There is no clear research concerning what functions of the increasing temperatures and elevated carbon dioxide levels for SOC storage are more pronounced. Therefore, some uncertainties exist regarding the possible benefits of SOC sequestration in the future.

Generally speaking, there are many factors that may affect SOC sequestration in the future and may increase or decrease SOC storage. Factors that may be detrimental for SOC storage include: (i) the low national subsidy standard for returning farmland into forest or grasslands and the low income of local farmers; (ii) the persistent, deteriorating environment situation and; (iii) land-use changes caused by private and local government interests. Because of these disadvantageous factors, it is very important to determine how to effectively increase SOC storage. Creating new agriculture products by the implementation of more modern agriculture techniques and offering more work opportunities in urban areas for farmers to increase their income could partially solve these problems. Additionally, the large-scale afforestation that may be exacerbating the soil water shortage in the Loess Plateau should be controlled, especially in vulnerable arid and semi-arid regions, and should fully consider the affordability of such environmental efforts before vegetation restoration is conducted.

Conclusions

In this study, land-use type had a significant influence on the topsoil SOCD. Compared with the average SOCD of cultivated land, the average SOCD in woodland was significantly larger by 33.81%, while the average SOCD in the orchards decreased by 10.80%. Therefore, the SOCD may improve when cultivated land is converted into woodland and may decline when cultivated land is converted into orchards. Over time, there was an increasing trend of SOCD in woodland, but only a small change of SOCD in the orchards. Based on the remote sensing data on land-use change, we found that the "Grain for Green" project did not significantly increase SOC storage in the Luoyugou catchment in the past 10 years of restoration. This is mainly because the increase of SOC storage by woodland was offset by a SOC storage loss in the orchards and decreases in SOC storage in the initial stage of land-use conversion. If all of the woodland area was calculated as 30.44 t/km^2 of SOCD, the potential contribution to SOC sequestration in this catchment would increase by 9625.59 t, improving the SOC storage ratio by 7.12%. Therefore, it has a huge potential for future environmental restoration efforts if we could manage and maintain the woodland converted from the cropland well. However, we cannot ignore other factors that affect SOC sequestration, including climate change and national policies.

References

1. Morris SJ, Bohm S, Haile-Mariam S, Paul EA (2007) Evaluation of carbon accrual in afforested agricultural soils. Global Change Biology, 13: 1145–1156.
2. Wang YF, Fu BJ, Lü YH, Chen LD (2011) Effects of vegetation restoration on soil organic carbon sequestration at multiple scales in semi-arid Loess Plateau, China. Catena, 85: 58–66.
3. Sauer TJ, James DE, Cambardella CA, Hernandez-Ramirez G (2012) Soil properties following reforestation or afforestation of marginal cropland. Plant Soil, 360: 375–390.
4. Li DJ, Niu SL, Luo YQ (2012) Global patterns of the dynamics of soil carbon and nitrogen stocks following afforestation: a meta-analysis. New Phytologist, 195: 172–181.
5. Lal R (2004) Agricultural activities and the global carbon cycle. Nutrient Cycling in Agroecosystems, 70: 103–116.
6. Powlson D (2005) Will soil amplify climate change? Nature, 433: 204–205.
7. Smith P, Martino D, Cai ZC, Gwary D, Janzen H, et al (2008) Greenhouse gas mitigation in agriculture. Philosophical Transactions of the Royal Society, Series B, 363: 789–813.
8. Guo LB, Gifford RM (2002) Soil carbon stocks and land-use change: a meta analysis. Global Change Biology, 8: 345–360.
9. Don A, Schumacher J, Freibauer A (2011) Impact of tropical land-use change on soil organic carbon stocks- a meta-analysis. Global Change Biology, 17: 1658–1670.
10. Houghton RA (1999) The annual net flux of carbon to the atmosphere from changes in land-use 1850–1990. Tellus, 51B: 298–313.
11. Lal R (2005) Forest soils and carbon sequestration. Forest Ecology and Management, 220: 242–258.
12. Laganière J, Angers AD, Paré D (2010) Carbon accumulation in agricultural soils after afforestation: a meta-analysis. Global Change Biology, 16: 439–453.
13. Raiesi F (2012) Soil properties and C dynamics in abandoned and cultivated farmlands in a semi-arid ecosystem. Plant Soil, 351: 161–175.
14. Novara A, Gristina L, Mantia TL, Rühl J (2013) Carbon dynamics of soil organic matter in bulk soil and aggregate fraction during secondary succession in a Mediterranean environment. Geoderma, 193–194: 213–221.
15. Paul KI, Ploglase PJ, Nyakuengama JG, Khanna PK (2002) Change in soil carbon following afforestation. Forest Ecology and Management, 168: 241–257.
16. Wang BQ, Liu GB (1999) Effects of relief on soil nutrient losses in sloping fields in hilly region of Loess Plateau. Soil Erosion and Soil and Water Conservation, 5: 18–22 (in Chinese).
17. Wang Z, Liu GB, Xu MX, Zhang J, Wang Y, et al (2012) Temporal and spatial variations in soil organic carbon sequestration following revegetation in the hilly Loess Plateau, China. Catena, 99: 26–33.
18. Zhou Y, Xu XG, Ruan HH, Wang JS, Fang YH, et al (2008) Mineralization rates of soil organic carbon along an elevation gradient in Wuyi Mountain of Southeast China. Ecology, 27: 1901–1907 (in Chinese).
19. Luo YQ, Hui DF, Zhang DQ (2006) Elevated CO$_2$ stimulates net accumulations of carbon and nitrogen in land ecosystems: a meta-analysis, Ecology, 87: 53–63.
20. Batterman SA, Hedin LO, Breugel Mv, Ransijn J, Craven DJ, et al (2013) Key role of symbiotic dinitrogen fixation in tropical forest secondary succession. Nature, 502: 224–227.
21. Huang CY (2000) Soil Science. Beijing: China Agriculture Press, 311p.
22. Gregorich EG, Greer KJ, Anderson DW, Liang BC (1998) Carbon distribution and losses: erosion and deposition effects. Soil & Tillage Research, 47: 291–302.
23. Lal R (2003) Soil erosion and the global carbon budget. Environment International, 29: 437–450.
24. Chartier MP, Rostagno CM, Videla LS (2013) Selective erosion of clay, organic carbon and total nitrogen in grazed semiarid rangelands of northeastern Patagonia, Argentina. Journal of Arid Environments, 88: 43–49.
25. Fu BJ (1989) Soil erosion and its control in the Loess Plateau of China. Soil Use and Management, 5: 76–81.
26. Shi H, Shao MA (2000) Soil and water loss from the Loess Plateau in China. Journal of Arid Environments, 45: 9–20.
27. Fu BJ, Chen LX, Qiu Y (2002) Land-use structure and ecological processes in the Loess Plateau. Beijing: The Commercial Press, pp: 1–12 (in Chinese).
28. Feng XM, Wang YF, Chen LD, Fu BJ, Bai GS (2010) Modeling soil erosion and its response to land-use change in hilly catchments of the Chinese Loess Plateau. Geomorphology, 118: 239–248.
29. Lu N, Liski J, Chang RY, Akujärvi A, Wu X, et al (2013) Soil organic carbon dynamics following afforestation in the Loess Plateau of China. Biogeosciences Discuss, 10: 11181–11211.
30. Zheng FL (2006) Effect of Vegetation Changes on soil erosion on the Loess Plateau. Pedosphere, 16: 420–427.
31. Gong J, Chen LD, Fu BJ, Huang Y, Huang Z, et al (2006) Effect of land-use on soil nutrients in the loess hilly area of the Loess Plateau, China. Land Degradation & Development, 17: 453–465.
32. Chen LD, Gong J, Fu BJ, Huang ZL, Huang YL, et al (2007) Effect of land-use conversion on soil organic carbon sequestration in the loess hilly area, loess plateau of China. Ecological Research, 22: 641–648.
33. Fu XL, Shao MA, Wei XR, Horton R (2010) Soil organic carbon and total nitrogen as affected by vegetation types in Northern Loess Plateau of China. Geoderma, 155: 31–35.
34. Chang RY, Fu BJ, Liu GH, Liu SG (2011) Soil carbon sequestration potential for "Grain for Green" project in Loess Plateau, China. Environment Management, 48: 1158–1172.
35. Wei J, Cheng JM, Li WJ, Liu WG (2012) Comparing the effect of naturally restored forest and grassland on carbon sequestration and its vertical distribution in the Chinese Loess Plateau. PLoS ONE, 7: e40123. doi:10.1371/journal.pone.0040123
36. Lei D, ShangGuan ZP, Sweeney S (2013) Changes in soil carbon and nitrogen following land abandonment of farmland on the Loess Plateau, China. PLoS ONE 8: e71923. doi:10.1371/journal.pone.0071923
37. Yu XX, Zhang XM, Niu LL, Yue YJ, Wu SH, et al (2009) Dynamic evolution and driving force analysis of land-use/cover change on loess plateau catchment. Transaction of the CSAE, 25: 219–225 (in Chinese).
38. Zhao Y, Yu XX (2013) Effects of climate variation and land-use change on runoff-sediment yield in typical watershed of loess hilly-gully region. Journal of Beijing Forestry University, 35: 39–45 (in Chinese).
39. Nelson DW, Sommers LE (1982) Total carbon, organic carbon, and organic matter. In: Page AL, Miller RH, Keeney DR (eds) Methods of soil analysis, Part 2, Chemical and microbial properties. Agronomy Society of America, Agronomy Monograph 9, Madison, Wisconsin, pp 539–552.
40. Institute of Soil Sciences, Chinese Academy of Sciences (ISSCAS) (1978) Physical and chemical analysis methods of soil. Shanghai: Shanghai Science Technology Press, pp 7–15.
41. Degryze S, Six J, Paustian K, Morris SJ, Paul EA, et al (2004) Soil organic carbon pool changes following land-use conversions. Global Change Biology, 10: 1120–1132.
42. Yang YL, Guo SL, Ma YH, Chen SG, Sun WY (2008) Changes of orchard soil carbon, nitrogen and phosphorus in gully region of Loess Plateau. Plant Nutrition and Fertilizer Science, 14: 685–691 (in Chinese).
43. Xue S, Liu GB, Zhang C, Zhang CS (2011) Analysis of effect of soil quality after orchard established in hilly Loess Plateau. Scientia Agricultura Sinica, 44: 3154–3161 (in Chinese).
44. Umali BP, Oliver DP, Forrester S, Chittleborough DJ, Hutson JL, et al (2012) The effect of terrain and management on the spatial variability of soil properties in an apple orchard. Catena, 93: 38–48.
45. Guimarães DV, Gonzaga MIS, Silva TOd, Silva TLd, Dias NdS, et al (2013) Soil organic matter pools and carbon fractions in soil under different land-uses. Soil & Tillage Research, 126: 177–182.
46. Zou HY, Cheng JM, Zhou L (1998) Natural recoverage succession and regulation of the prairie vegetation on the Loess Plateau. Research of Soil and Water Conservation, 5: 126–138 (in Chinese).
47. Li YY, Shao MA, Shang Guan ZP, Fan J, Wang LM (2006) Study on the degrading process and vegetation succession of Medicago sativa grassland in North Loess Plateau, China. Acta Prataculturae Sinica, 15: 85–92 (in Chinese).
48. Li YY, Shao MA, Zhang JY, Li QF (2007) Impact of grassland recovery and reconstruction on soil organic carbon in the northern Loess Plateau. Acta Ecologica Sinica, 27: 2279–2287 (in Chinese).
49. Leifeld J, Bassin S, Fuhrer J (2005) Carbon stocks in Swiss agricultural soils predicted by land-use, soil characteristics, and elevation. Agriculture, Ecosystems and Environment, 105: 255–266.
50. Cao SX, Xu CG, Chen L, Wang XQ (2009) Attitudes of farmers in China's northern Shaanxi Province towards the land-use changes required under the Grain for Green Project, and implications for the project's success. Land-use Policy, 26: 1182–1194.
51. Lü YH, Fu BJ, Feng XM, Zeng Y, Liu Y, et al (2012) A policy-driven large scale ecological restoration: quantifying ecosystem services changes in the Loess Plateau of China. PLoS ONE, 7: e31782, doi:10.1371/journal.pone.0031782
52. Pan ZB, Zhang L, Yang R, Li SB, Dong LG, et al (2012) Overview on research progress of soil drought in semiarid regions of the Loess Plateau. Research of Soil and Water Conservation, 19: 287–291, 298 (in Chinese).
53. Wang L, Shao MA (2004) Soil desiccation under the returning farms to forest on the Loess Plateau. World Forestry Research, 17: 57–60 (in Chinese).
54. Chen HS, Shao MA, Li YY (2008) Soil desiccation in the Loess Plateau of China. Geoderma, 143: 91–100.
55. Niu JJ, Zhao JB, Wang SY (2007) A study on plantation soil desiccation in the upper reaches of the Fenhe River basin based on deep soil experiments. Geographical Research, 26: 773–781 (in Chinese).

Author Contributions

Conceived and designed the experiments: YBQ ZBX XXY YLX. Performed the experiments: YBQ ZBX. Analyzed the data: YBQ ZBX. Contributed reagents/materials/analysis tools: YBQ ZBX XXY YLX. Wrote the paper: YBQ ZBX XXY YLX.

56. Duan FL, Lin Z, Xiong YQ (2007) Analysis on the phenomenon of farmland abandoned by the reason of rural laborers moving out for work. Rural Economy, 16–19 (in Chinese).

57. Xin ZB, Yu XX, Li QY, Lu XX (2011) Spatiotemporal variation in rainfall erosivity on the Chinese Loess Plateau during the period 1956–2008. Regional Environmental Change, 11: 149–159.

58. William HS (1999) Carbon and agriculture: carbon sequestration in soils. Science, 284: 2095.

59. Guo GF, Zhang CY, Xu Y (2006) Effects of climate change on soil organic carbon storage in terrestrial ecosystem. Ecology, 25: 435–442 (in Chinese).

60. Wang QX, Fan XH, Qin ZD, Wang MB (2012) Change trends of temperature and precipitation in the Loess Plateau Region of China, 1961–2010. Global and Planetary Change, 92–93: 138–147.

61. Intergovernmental Panel on Climate Change (2007) Climate Change 2007: The Science of Climate Change. Cambridge University Press, Cambridge.

62. Lal R (2009) Challenges and opportunities in soil organic matter research. European Journal of Soil Science, 60: 158–169.

2

Vertical Profiles of Soil Water Content as Influenced by Environmental Factors in a Small Catchment on the Hilly-Gully Loess Plateau

Bing Wang[1], Fenxiang Wen[1], Jiangtao Wu[1], Xiaojun Wang[1]*, Yani Hu[2]

1 College of Environmental Science and Resources, Shanxi University, Taiyuan, China, 2 Library, Hebei University of Science and Technology, Shijiazhuang, China

Abstract

Characterization of soil water content (SWC) profiles at catchment scale has profound implications for understanding hydrological processes of the terrestrial water cycle, thereby contributing to sustainable water management and ecological restoration in arid and semi-arid regions. This study described the vertical profiles of SWC at the small catchment scale on the hilly and gully Loess Plateau in Northeast China, and evaluated the influences of selected environmental factors (land-use type, topography and landform) on average SWC within 300 cm depth. Soils were sampled from 101 points across a small catchment before and after the rainy season. Cluster analysis showed that soil profiles with high-level SWC in a stable trend (from top to bottom) were most commonly present in the catchment, especially in the gully related to terrace. Woodland soil profiles had low-level SWC with vertical variations in a descending or stable trend. Most abandoned farmland and grassland soil profiles had medium-level SWC with vertical variations in varying trends. No soil profiles had low-level SWC with vertical variations in an ascending trend. Multi-regression analysis showed that average SWC was significantly affected by land-use type in different soil layers (0–20, 20–160, and 160–300 cm), generally in descending order of terrace, abandoned farmland, grassland, and woodland. There was a significant negative correlation between average SWC and gradient along the whole profile ($P<0.05$). Landform significantly affected SWC in the surface soil layer (0–20 cm) before the rainy season but throughout the whole profile after the rainy season, with lower levels on the ridge than in the gully. Altitude only strongly affected SWC after the rainy season. The results indicated that land-use type, gradient, landform, and altitude should be considered in spatial SWC estimation and sustainable water management in these small catchments on the Loess Plateau as well as in other complex terrains with similar settings.

Editor: Andrew C. Singer, NERC Centre for Ecology & Hydrology, United Kingdom

Funding: This work was supported by the Natural Science Foundation of China (No. 41201277 and 41101025) and the Natural Science Foundation of Shanxi Province of China (No. 2014011034-2). The funders had no role in study design, data collection and analysis, decision to publish, or preparation of the manuscript.

Competing Interests: The authors have declared that no competing interests exist.

* Email: xjwang@sxu.edu.cn

Introduction

Soil water content (SWC) is a critical factor for plant growth and a determinant of plant distribution in arid and semiarid areas such as China's Loess Plateau [1, 2]. Vertical distribution of SWC can greatly affect soil water movement [3], thereby greatly affecting the biomass production and water use efficiency of plants (e.g., switchgrass) under water stress [4]. Plant-available water stored in the soil profile has a buffering capacity, which, in deep layers, prolongs or alleviates the effects of seasonal or inter-annual drought on plant growth and soil water flux to the atmosphere [5–7]. Research has provided strong evidence that deep soil water depletion plays a key role in sustainable agriculture, ecological restoration, and terrestrial water cycling on the Loess Plateau [8–10]. However, measurement of SWC profiles has been frequently conducted at different spatial scales. The results thus need to be converted before comparison analysis or practical uses. The SWC profile in small catchment is considered to be at a moderate scale for data exchanging. In particular, small catchment is thought to be the basic unit for integrated soil and water loss management in

complicated terrain of the Loess Plateau [11, 12]. Characterization of SWC profiles and evaluation of relevant influencing factors at the small catchment scale have implications for hydrological modeling of soil water dynamics and sustainable management of soil water resources in similar areas.

Classical statistics is frequently used to analyze the variability of SWC profiles at the small catchment scale, which involves the estimation of descriptive parameters such as average (mean), variance, standard deviation (STD), and coefficient of variation (CV). Average SWC at individual soil depth intervals or across the whole soil profile is extensively determined. The CV of SWC is also routinely calculated as the temporal variable in a certain period of time or the spatial variable across a specific area. The SWC profile can be divided into distinct intervals by considering its average and CV which exhibit complex spatial-temporal relationships in several plots or watersheds [13–16]. Additionally, ranking method, clustering method, and semivariogram model have been applied for the division of SWC profile [3, 17–20]. However, the above-mentioned methods cannot clearly reflect the variation trend in SWC profiles. Thus, great effort has been made

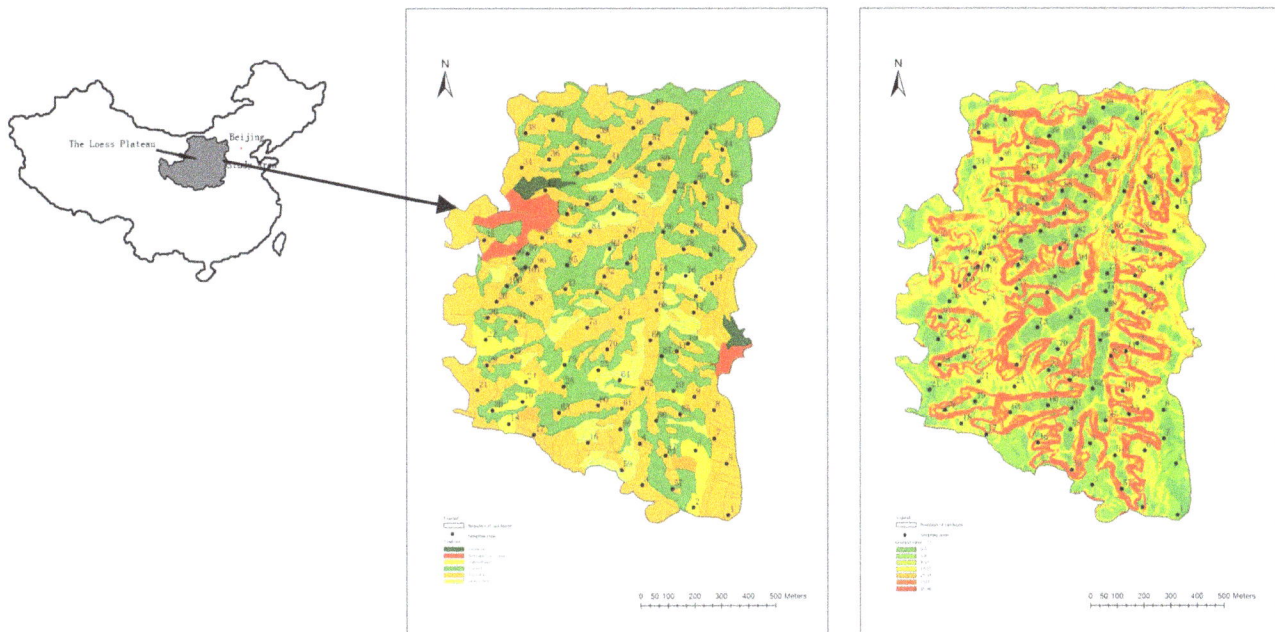

Figure 1. Location of Sanyanjing catchment and distribution of 101 sampling points in the Sanyanjing catchment.

to describe the vertical profiles of SWC through comparing variation curves or variation ranges, in small watersheds related to different land-use types, vegetation species, and/or terrain factors [3, 17, 18, 20–22]. If a massive sample size is involved, however, it becomes difficult to distinguish the vertical profiles and major influencing factors of SWC by direct comparisons.

In recent decades, a great number of studies have been conducted on the spatiotemporal variability of SWC and related influencing factors worldwide. Canton [23] pointed out that wasteland-scale spatial variability of SWC is mainly controlled by surface cover and soil properties in a semi-arid region of Spain, where surface cover counteracts the influence of terrain factors (including gradient, aspect, topographic wetness index, and distance from the river) on SWC distribution. Burnt [24] reported a topographic index which can simulate changes in high-level SWC in a humid climate zone of Devin County, UK. O'loughlin [25] estimated the spatial pattern of SWC distribution in a small catchment using humidity index model based on digital terrain dataset. Hawley [15] discovered that topography is the major factor responsible for the spatial distribution of SWC in an agricultural region, where resultant SWC variation is diminished by vegetation in a moist climate zone in Chickasha, USA. In arid and semi-arid areas, catchment-scale distribution of SWC is strongly affected by land-use/vegetation and topographic indices, e.g., land-use type, soil organic matter content, tillage, soil physical properties, gradient, and aspects [6, 17–19, 21, 26].

Many researchers have focused on the quantification of environmental parameters such as topographic factor, vegetation type, soil texture, and land-use type, in attempt to evaluate their impacts on the variability of SWC. At the small catchment scale, little information is available on the major factors affecting vertical profiles of SWC in cinnamon soil (Haplic Lixisols, FAO) zone on the hilly and gully Loess Plateau [3, 19]. As a regional water reservoir experiencing depletion, the plateau region requires measurements and characterization of deep SWC profiles for the thick soil layer. However, soil sampling for SWC profile analysis at the catchment scale has been commonly conducted at <200 cm

depth [18–20, 23]. Wang et al. [21] exceptionally examined SWC along the 0–21 m soil profile on the Loess Plateau, but the reliability of their tests might be affected by a small sample size (11 sites). Deep soil sampling at a larger number of sites and statistical analysis of parameters involving soil depth information will contribute to better understanding of the vertical profiles and influencing factors of SWC.

In the present study, we characterized the vertical profiles of SWC in a small catchment on the Loess Plateau by cluster analysis of two descriptive parameters (mean and regression gradient). Sampling was carried out in the 0–300 cm profile at 101 points throughout the catchment before and after the rainy season, to meet the demand for deep depth, spatial representativeness and temporal comparability. The influences of selected environmental factors (land-use type, topographic factors, and landform) on average SWC were examined by multi-linear regression [6, 26–32]. The results were discussed in order to provide new insights to the vertical profiles and influencing factors of SWC on the Loess Plateau, further providing reference data for sustainable management of water resource in small catchment areas in the semi-arid region with complex terrain.

Material and Methods

1 Site description

This study was conducted in Sanyanjing catchment ($112°2'13''$, $37°46'23''$), which is located on the east margin of the Loess Plateau in Shouyang county, mid-east Shanxi province, China (Figure 1). The catchment has a total area of 1.32 km^2 and the elevation ranges from 1001 to 1160 m. It is a hilly and gully area with mostly deep gully erosion slopes. The landform consists of ridge and gully.

The catchment area has a semi-arid continental climate (Cwa by Koppen Climate Classification) with an average annual precipitation of 474.2 mm (1967–1999). Snow in the winter accounts for ~8% and rainfall in July to September for ~73% of annual precipitation. Monthly average precipitation, potential evapotranspiration and precipitation in 2013 are shown in Figure 2. Annual mean temperature in this area is 8.1°C, with a maximum of 34.7°C and a minimum −20.6°C. The soil type is cinnamon soil (Haplic lixisols, FAO), which consists of 54–62% silt and 10.95–30.15% sand with the bulk density of 1.3–1.4 g/cm^3. Soil texture was measured using a particle size analyzer (SEDIMAT 4–12, UGT, Germany). Soil bulk density was determined through sampling with cutting rings (inner diameter 5.0 cm, volume 100 cm^3) and drying in an oven (105°C, 24 h). The profile of soil texture related to different landforms is listed in Table 1. Maximum soil depth is mostly down to 300 cm on the ridge and bare rock could be rarely seen only on the northern margin of the gully area.

The distribution of land-use types across the catchment is shown in Figure 1. Terrace is the dominant land-use type, accounting for about 60% of the study area. Few terraces had been abandoned for natural restoration of vegetation because of the Grain for Green project since 2000. Grassland is mainly covered with herbs and semi-shrubs, which had never been reclaimed for several decades. About 80% of the woodland is covered with semi-shrubs at steep slopes unsuitable for sampling.

2 Soil sampling

Ethics statement. Sampling activities at the farmland were allowed by the owners. No specific permissions were required at other locations because they were not privately-owned or protected in any way and the field activities did not involve any endangered or protected species.

A total of 101 sampling points were designed in a 150 m×150 m grid throughout the catchment area by considering major land-use types, including terrace (83), abandoned farmland (9), grassland (3), and woodland (6). Soil sampling was carried out during two periods in 2013, from April 29 to May 4 (before the rainy season) and from October 28 to November 1 (after the rainy season). No precipitation occurred during the two sampling periods or a week before sampling. Each sample was taken at 20 cm intervals along the 0–300 cm soil profile using an auger (inner diameter 5.0 cm). The samples were kept in capped aluminum boxes for transportation. Measurement of SWC was conducted using an oven-drying method (105°C, 24 h). At the

majority of the sampling points, soils were collected along a vertical profile over 300 cm. A few exceptions were in the north of the gully at the lowest altitude where weathered rock was occasionally encountered. Background information of the 101 sampling points is summarized in Table 2.

3 Data analysis

To identify the variability of SWC profiles at the catchment scale, descriptive parameters were calculated for each profile. Further, we calculated the linear regression coefficient (K value) between SWC and soil depth to represent the variation trend of SWC vertical profiles and the mean value to describe the average level of SWC along the 0–300 cm profile.

The SWC profiles were classified using a combined cluster analysis of the K and mean values. Cluster analysis is the process of grouping a set to data objects into multiple groups (or clusters), so that objects within a cluster share high similarity but are dissimilar to those in other clusters. In this approach, dissimilarities and similarities are assessed based on the attribute values describing the objects and often involve distance measures. Cluster analysis is a statistical classification method for discovering whether the individuals of a population gall into different groups by making quantitative comparisons of multiple characteristics [34]. Here a combined cluster analysis was conducted in three steps: 1) cluster of the mean to three groups, which present the average level of SWC along the vertical soil profile, 2) cluster of K to three groups, which reflect the variation trend of SWC profiles (top to bottom), and 3) combination of the two sets of groups into nine new groups using the between-groups linkage method with squared Euclidean distance criteria [34].

For group comparisons, SWC profiles (0–300 cm) of the same group were averaged and re-plotted. The average curves of SWC were compared between groups to identify the major factors influencing SWC in individual soil layers (0–20, 20–160, and 160–300 cm). On the basis of cluster analysis, the influences of land-use type, topography and landform on average SWC in individual soil layers were examined by multi-regression analysis. The independent variables were land-use type, landform type, Sin(gradient), Sin(aspect), flow accumulation (calculated cell numbers to a grid cell from surrounding cells with the ArcGIS hydrology analysis

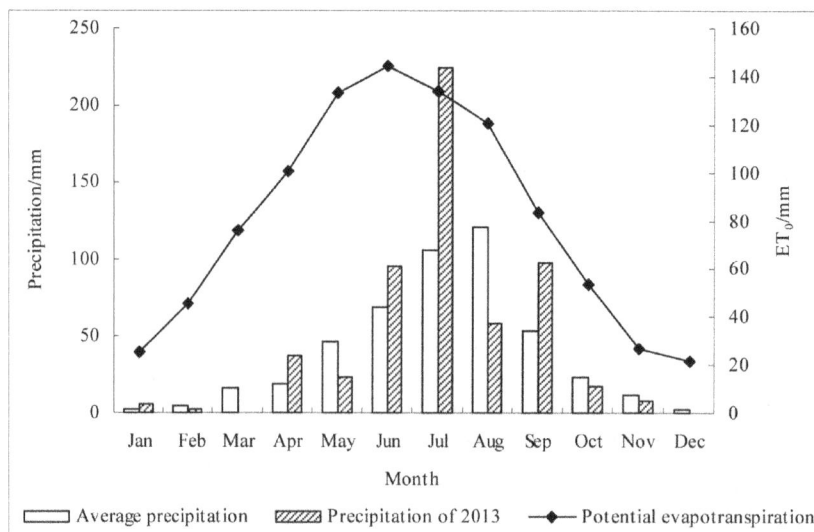

Figure 2. Average annual precipitation and potential evapotranspiration of 1967–1999 and precipitation in 2013 in the Sanyanjing catchment, Shanxi province, China.

Table 1. Soil texture in vertical profiles related to different landforms in the Sanyanjing catchment in Shanxi province, China.

Point description	Soil depth (cm)	Sand (%)		Silt (%)		Clay (%)
		>0.05 mm (%)	0.05–0.02 mm (%)	0.02–0.0063 mm (%)	0.0063–0.002 mm (%)	<0.002 mm (%)
	0–20	26.05	30.40	20.60	7.00	15.95
	20–40	21.35	34.80	20.80	6.90	16.15
	40–60	30.25	25.90	21.60	4.50	17.75
	60–80	28.25	28.80	21.70	2.80	18.45
	80–100	18.45	38.00	20.80	6.70	16.05
	100–120	26.45	30.10	20.40	5.20	17.85
Terrace at	120–140	26.95	27.90	22.00	4.40	18.75
ridge	140–160	16.35	32.60	24.20	8.10	18.75
(Point 33)	160–180	27.75	22.40	22.70	7.90	19.25
	180–200	36.55	17.80	20.60	8.00	17.05
	200–220	18.45	29.00	24.30	9.00	19.25
	220–240	20.45	28.50	23.20	8.60	19.25
	240–260	22.05	27.30	23.90	8.80	17.95
	260–280	31.35	20.30	22.20	9.40	16.75
	280–300	24.65	29.80	20.70	5.90	18.95
	0–20	30.15	20.00	23.20	6.40	20.25
	20–40	27.75	21.80	21.90	6.10	22.45
	40–60	25.95	26.80	20.50	4.50	22.25
	60–80	23.85	27.09	21.20	1.20	26.65
	80–100	0.95	31.00	20.10	20.10	27.85
	100–120	10.95	23.40	18.40	18.40	28.85
Terrace at	120–140	7.45	27.30	19.20	19.20	26.85
ridge	140–160	17.45	19.50	18.90	18.90	25.25
(Point 39)	160–180	10.45	25.80	18.20	18.20	27.35
	180–200	15.75	24.70	17.00	17.00	25.55
	200–220	13.35	21.40	18.60	18.60	28.05
	220–240	20.85	8.80	22.00	22.00	26.35
	240–260	20.25	12.40	20.00	20.00	27.35
	260–280	12.15	23.70	19.10	19.10	25.95
	280–300	24.95	29.00	20.50	0.50	25.05

module), and elevation. The former two factors were categorical variables converted into dummy variables before introduced into the regression analysis; and the latter four factors were continuous variables produced using digital elevation model at 1-m resolution.

SWC data were statistically analyzed in SPSS13.0 (SPSS Inc., Chicago, IL, USA), and topographic features were analyzed in ArcGIS 10.0 (ESRI, Redlands, CA, USA). SWC profiles were drawn in Microsoft Excel 2010 (Microsoft Corp., Redmond, WA, USA) and then clustered in SPSS 13.0 by considering descriptive parameters (maximum, minimum, mean, CV, STD, and K). Multi-regression analysis was performed in SPSS 13.0, with a *P*-value less than 0.05 considered statistically significant.

Results

1 Vertical profiles and descriptive parameters of SWC

The vertical profiles (0–300 cm) of SWC at 101 sampling points before the rainy season were drawn (Figure 3). These SWC profiles showed dynamic variations across the catchment study area, with substantial differences in the soil layers below 100 cm. At a few sampling points, there were obvious soil water depletion (e.g., 67, 75, 85, and 94) and an increasing trend (top to bottom) of SWC (e.g., 12 and 36). High degrees of soil desiccation were rarely detected in the lower soil layers, and low SWC was mainly found in the lower soil layers of woodland.

Descriptive parameters such as maximum, minimum, mean, and CV, STD are commonly used to reveal the spatial-temporal variability of SWC. However, these parameters cannot reflect the variation trend of SWC vertical profiles. To this end, the K value of SWC to profile depth was introduced for quantification of variation trend of SWC vertical profiles (Figure 4). Results showed that before the rainy season, SWC substantially varied between 5.87% and 34.72%, whereas the mean, STD, CV, and K values respectively ranged from 10.57% to 21.76%, 0.47 to 4.53, 3% to 24%, and −0.0405 to 0.0274 along the vertical soil profile (0–

Table 2. Background information of 101 soil sampling points in the Sanyanjing catchment study area in Shanxi province, China.

Land-use type	Vegetation	Landform type	Soil profile/cm	Sampling points
Terrace (n = 83)	Maize	Ridge	300	1, 3–5, 7–9, 12–18, 21–22, 25–26, 28, 32–35, 37, 51, 53, 91, 95, 96, 98
			260	99
		Gully	300	29, 30, 38–46, 48–50, 54–60, 62, 63, 65, 66, 68–74, 77–84, 87–90
			280	47
			260	31, 61
			220	52
			160	86
	Millet	Ridge	300	11
	Maize +five-year-walnut	Gully	300	27
Abandoned farmland	Subshrubs + herbs	Ridge	300	6, 10, 23
(n = 9)			240	2
	Subshrubs + herbs + few ulmus pumila	Gully	220	19
	Robinia peseudoacacia + subshrubs + herbs	Ridge	300	24
	Robinia peseudoacacia+ subshrubs + herbs	Gully	300	64
	Herbs + few almond-apricot	Gully	300	20
	Poplar + subshrubs +herbs	Gully	300	75
Grassland	Subshrubs +herbs	Ridge	300	92, 93
(n = 3)		Gully	300	76
Woodland	Poplar	Gully	300	97
(n = 6)			280	85
	Poplar + herbs	Gully	140	101
	Poplar + subshrubs	Gully	280	100
	Poplar + subshrubs + herbs	Gully	300	67, 94

300 cm). The ranges of the parameters after the rainy season were generally similar with those before the rainy season.

According to the division criteria of Nielson [24], CV in the range of 10–100% indicates moderate variability. Thus, the vertical variability of SWC at all sampling points in Sanyanjing catchment (Figure 4) can be classified to the medium degree. K is the linear regression coefficient between SWC and soil depth. A positive value of K indicates that SWC increases with increasing soil depth. Inversely, a negative value of K indicates that SWC decreases with increasing soil depth. The positive and negative K values of SWC data (Figure 4) are indicative of different variation trends of SWC vertical profiles in the catchment.

2 Clustering of SWC profiles

The vertical profile of SWC across the catchment can be described more clearly using cluster analysis. The 101 SWC profiles before the rainy season were classified into the first three groups by considering the mean value of SWC (Figure 5a), and the second three groups by considering the K value of SWC to soil depth (Figure 5b). The mean value of SWC ranged from 10.57% to 13.13% (low level), 14.15% to 16.86% (medium level), and 17.13% to 21.76% (high level) in the first three groups, whereas the K value of SWC to soil depth ranged from −0.0405 to 0.0106 (decreasing trend), 0.0144 to 0.0163 (stable trend), and 0.0194 to 0.0274 (increasing trend) in the second three groups. By

combining the two cluster series, we obtained nine groups of SWC profiles (Table 3).

Before the rainy season, vertical profiles of SWC in groups 1–3 featured low-level SWC (Table 3). In group 1, SWC decreased along the vertical profile (0–300 cm) in woodland (2) and abandoned farmland (1) located in the gully area. In group 2, SWC remained stable along the vertical profile in terrace (2) and grassland (1) located on the ridge as well as woodland (1) located in the gully (Table 4). No sampling points were classified into group 3 with increasing SWC along the vertical profile.

Vertical profiles of SWC in groups 4–6 featured medium-level SWC (Table 3). In group 4, SWC decreased along the vertical profile on the ridge related to terrace (4) as well as in the gully related to abandoned farmland (1) and woodland (1) located in the gully. In group 5, SWC remained stable along the vertical profile in terrace (7) mostly located in the gully, few terrace (3) and abandoned farmland (1) located on the ridge, and grassland (1) and woodland (1) located in the gully. In group 6, SWC increased along the vertical profile in terrace (2) located on the ridge (Table 4).

Vertical profiles of SWC in groups 7–9 featured high-level SWC (Table 3). In group 7, SWC decreased along the vertical profile in terrace (5), grassland (1), and abandoned farmland (1) located on the ridge, and terrace (2), abandoned farmland (1) and woodland (1) located in the gully. In group 8, SWC remained stable along

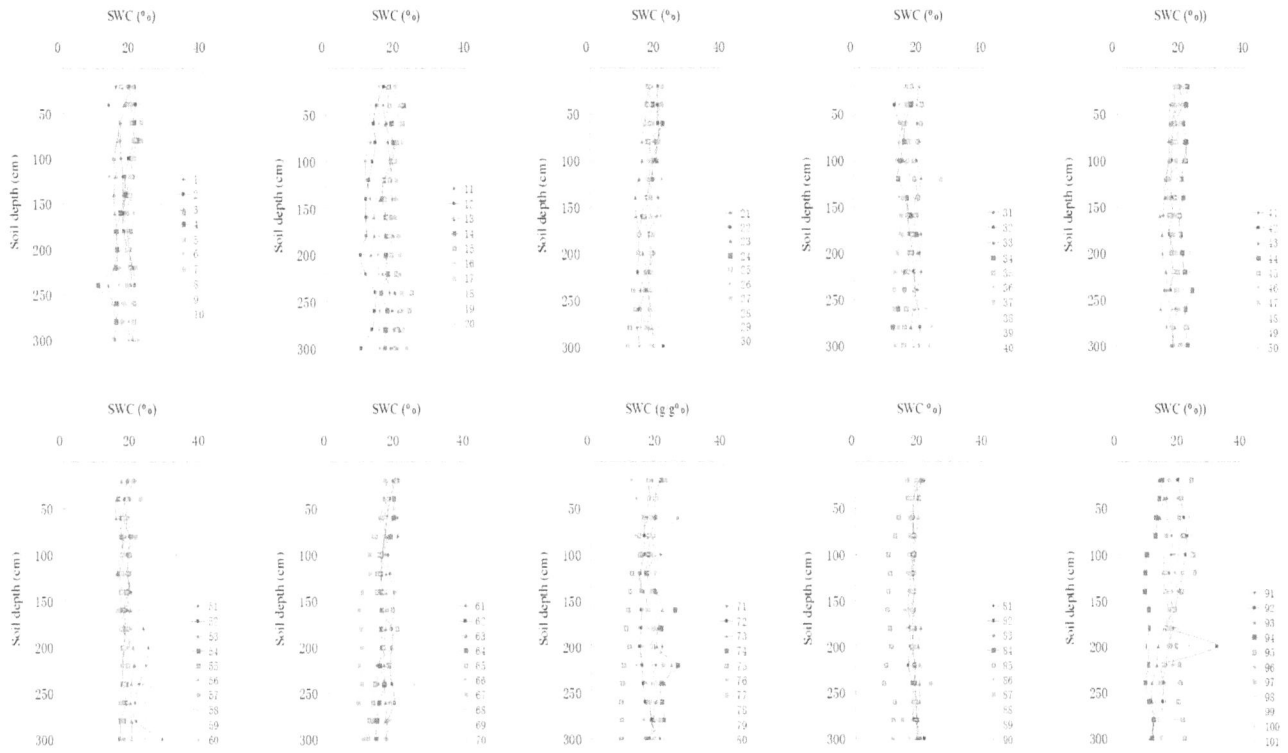

Figure 3. Vertical profiles of soil water at 101 sampling points in the Sanyanjing catchment.

the vertical profile at up to 58 of 101 sampling points, far more than other groups. Most sampling points of group 2 were located in the gully related to terrace (41), and few were on the ridge related to terrace (14) and abandoned farmland (3). In group 9, SWC increased along the vertical profile in terrace located on the ridge (3) and in the gully (1) (Table 4).

Similar grouping of SWC profiles was obtained with data collected after the rainy season. Overall, soil profiles of group 8 with high-level SWC in a stable trend were most commonly present in the catchment, more after the rainy season than before the rainy season. Group 3 of SWC profiles with low level and increasing trend was absent in the study area.

3 The relationships between average SWC and selected environmental factors

According to cluster analysis, there were nine combinations of SWC profiles in terms of average level and variation trend. We averaged SWC profiles of the same group and plotted the average curves (Figure 6), to examine differences of SWC profiles among various types. From Figure 4, we divided the whole soil profile (0–30 cm) into three layers (0–20, 20–160, and 160–300 cm) for multiple linear regression analysis. The results showed that selected environmental factors had significant linear correlations with average SWC at individual layers of 0–20 cm ($P<0.001$, $R^2=0.30$; $P<0.001$, $R^2=0.37$), 20–160 cm ($P=0.01$, $R^2=0.19$; $P<0.001$, $R^2=0.39$), and 160–300 cm ($P<0.001$, $R^2=0.32$; $P<0.001$, $R^2=0.43$; Table 5).

Before the rainy season, average SWC in the lower soil layer (10–20 cm) was significantly lower in grassland and woodland than in terrace, with no significant difference between abandoned farmland and terrace ($P_{D51}=0.109$, $P_{D52}=0.003$, $P_{D53}=0.047$, $P_{X1}=0.013$, and $P_{D61}=0.005$; Table 5). Additionally, average

SWC decreased with increasing gradient, with higher levels on the ridge than in the gully.

In the lower soil layer (20–160 cm), average SWC decreased significantly with increasing gradient ($P=0.026$; Table 5), and was significantly lower in woodland than in the other three types of land-use types, with no significant differences among the latter three types. Other environmental factors had no significant linear correlation with average SWC ($P>0.05$).

In the deeper soil layer (160–300 cm), there also existed a significantly negative correlation between Sin(gradient) and average SWC ($P=0.001$; Table 5). Average SWC obviously increased with increasing gradient and was significantly higher in terrace than in abandoned farmland, grassland, and woodland (in descending order).

Similar results can be seen in the data collected after the rainy season. That is, land-use type was the major environmental factor affecting average SWC, whereas landform and altitude strongly affected average SWC only in specific periods and soil layers (Table 5). In the whole vertical profile (0–300 cm), SWC occurred at high levels from upper to deeper layers in terrace, with the lowest level in woodland. Compared with data of terrace, average SWC was relatively low in grassland and woodland in the top (0–20 cm) and deeper soil layers (160–300 cm), with significantly low levels in abandoned farmland soils in the deeper layer only. In the deeper soil layer (160–300 cm), average SWC varied with different land-use types in descending order of terrace > abandoned farmland > grassland > woodland.

Discussion

1 Vertical profiles of SWC at the catchment scale

According to Wang [21], the variability of SWC (as indicated by the CV) varies notably across the whole Loess Plateau, i.e., 15% in

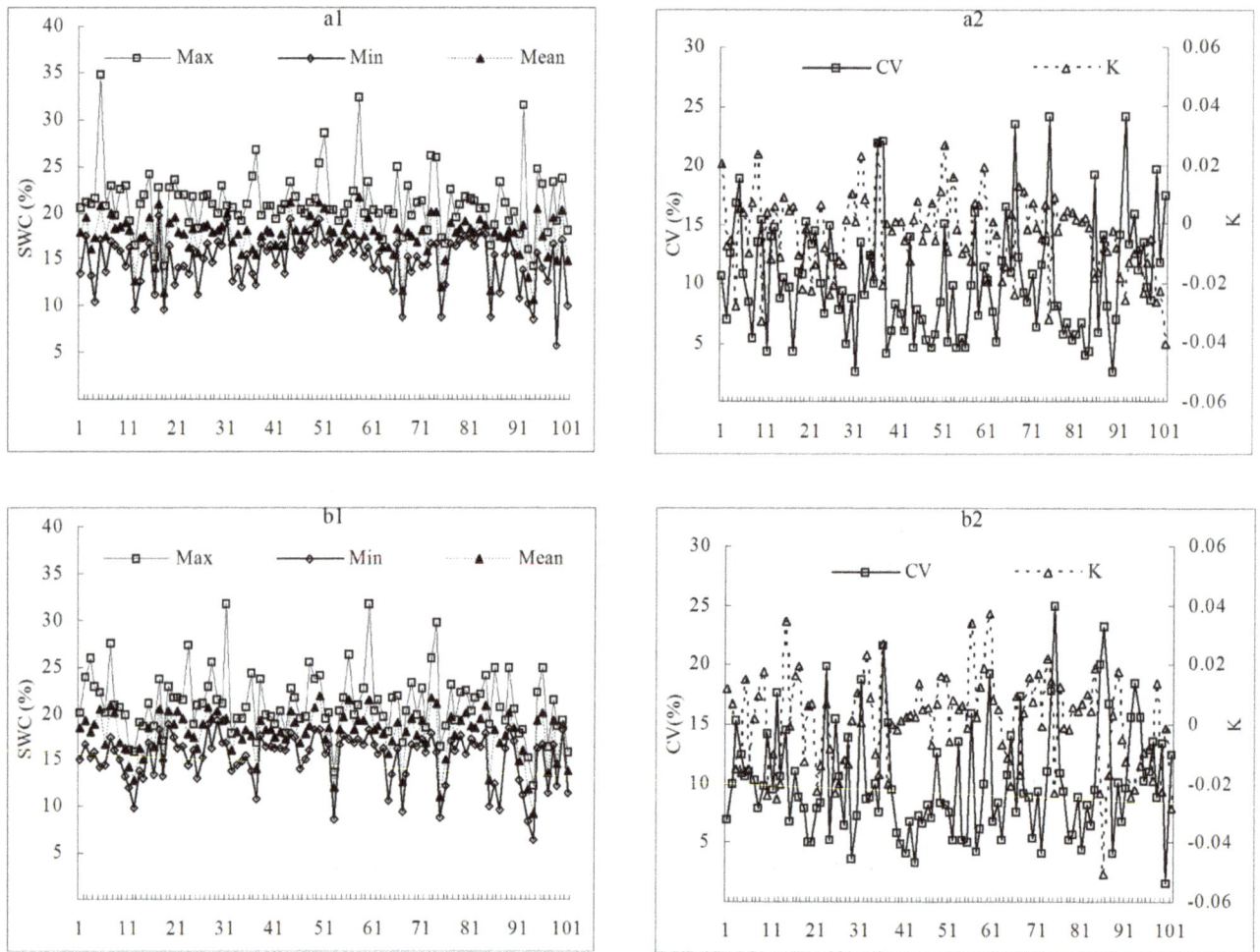

Figure 4. Statistical parameters of soil water content at 101 sampling points across the Sanyanjing catchment. (a. before the rainy season; and b. after the rainy season.)

Changwu and 55% in Shenmu. In the small catchment of Sanyanjing, SWC profiles exhibited weak and medium degrees of variability at 0–300 cm depth [33], with CV in the range of 3–24% (Figure 2). The lower variability of SWC profiles in our study area may be related to the higher SWC levels across the catchment (Pearson correlation coefficient between average SWC and CV,

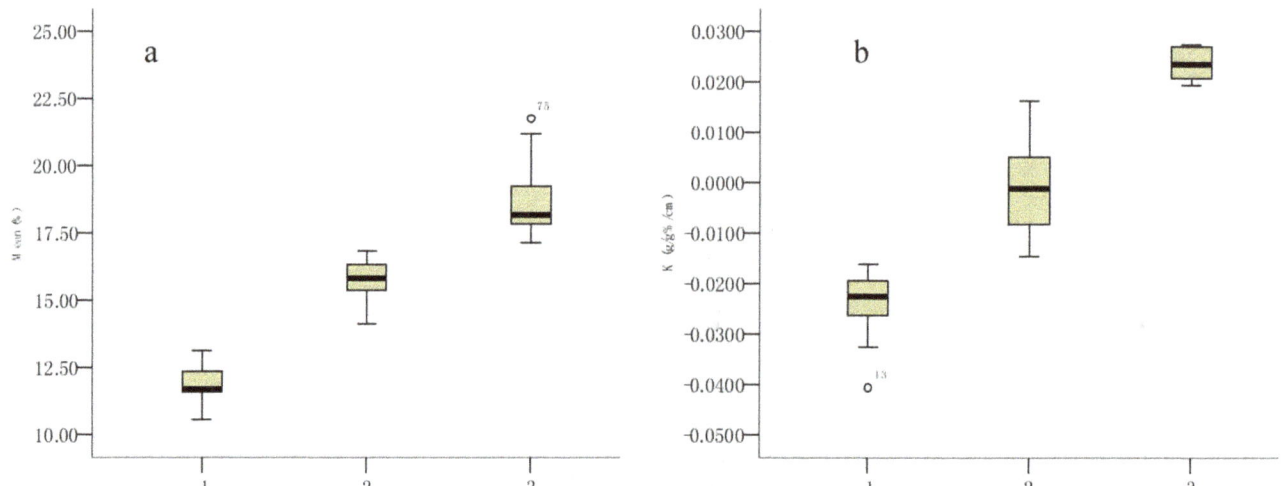

Figure 5. Grouping of 101 vertical soil water profiles in the Sanyanjing catchment before the rainy season by cluster analysis of the mean value (a) and regression gradient (K, b).

Table 3. Combined grouping of 101 vertical profiles of soil water content (0–300 cm) in Sanyanjing catchment by cluster analysis of the mean value and regression gradient.

Cluster by mean K	Combined grouping	Quantity of points — Before the rainy season	Quantity of points — After the rainy season	Point Nos. — Before the rainy season	Point Nos. — After the rainy season
1	1	3	0	65,67,85	–
2	2	4	1	12,18,93,94	94
3	3	0	0	–	–
1	4	6	15	25,37,64,91,99,101	11,12,18,37,53,64,67,75,76,85,91,93,97,99,101
2	5	13	0	3,16,23,34,41,43,56,63,65,70,72,76,97	3,16,23,34,41, 43,56,63,65,70,72,76,97
3	6	2	1	32,36	14
1	7	11	1	4,10,19,21,26,35,61,86,92,96,100	86
2	8	58	80	2,5–8,11,13–15,17, 20,22,24,27–31,33, 38–40,42,44–50,52, 53,55,56–59,62,66, 68,69,71,73,74, 77–84,87–90,95,98	2–10,13,15,16, 17,19–36,38–52,54,55,57–59, 61–63,65,66,68,69–74, 77–84, 87–90,92,95,96,98,100
3	9	4	3	1,9,51,60	1, 51,60

Table 4. Grouping richness of 101 vertical profiles of SWC (0–300 cm) in relation to different land-use types in Sanyanjing catchment before the rainy season.

Land-use type	Grouping								
	1	2	3	4	5	6	7	8	9
Terrace	0	2	0	4	10	2	7	54	4
Abandoned	1	0	0	1	1	0	2	0	0
Grassland	0	1	0	0	1	0	1	0	0
Woodland	2	1	0	1	1	0	1	0	0

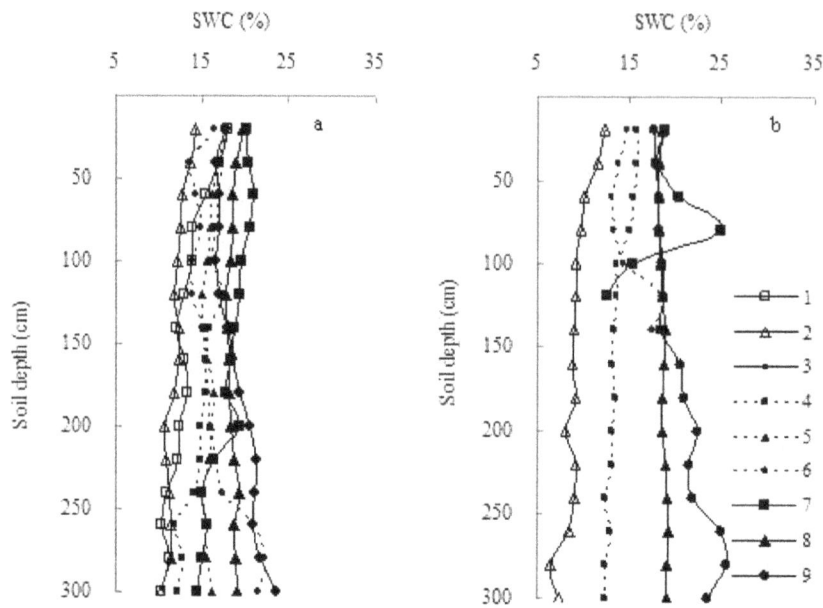

Figure 6. Vertical soil water profiles in relation to different groups in the Sanyanjing catchment study area before and after the rainy season (a. before the rainy season; and b. after the rainy season).

−0.40; $P<0.01$). Qiu [17] found that wetter soil with greater vertical variations in an increasing trend along the SWC profile (mean 13.03%; and STD, 2.3%) is representative in a dry year in Danangou catchment on the Loess Plateau, where the land-use pattern (including slope farmland, terrace, and orchard) differs from that in our study area.

Cluster analysis of the mean and K values provides a clear description for the overall variability of SWC in the vertical profile. Based on combined grouping, the 101 vertical SWC profiles were classified into nine groups with high, medium, and low levels associated with increasing, stable, and decreasing trends (Table 3). More than half of the SWC profiles were obtained from terrace soils in the gully and classified into group 8 (58/101 before the rainy season and 80/101 after the rainy season) with high-level SWC in a stable trend (Tables 3, 4). Despite that all sampling points of woodland were also located in the gully, their average SWC remained the lowest among different land-use types and mostly descended along the vertical profile (Table 4). The above differences can be attributed to the lower soil water consumption by maize crop in the terrace, which generally has shallower root distribution and less above-ground biomass than trees in the woodland. Our observations coincide with previous findings on the Loess Plateau that soil water conditions of terrace, gully farmland, and dam land are better than that of artificial woodland. The latter land-use type is associated with soil desiccation, especially in deep soil layers [2, 22, 35, 36].

Although the cluster analysis divided vertical SWC profiles into nine groups, only eight types were present in the Sanyanjing catchment and no sampling points were classified into group 3 (i.e., low-level SWC with an increasing trend from top to bottom). According to previous research in semi-arid regions, if SWC occurs at low level in the upper soil layer, deep-root crops, shrubs, and trees will consume more soil water in the deeper soil layers through root extraction [37–40]. Additionally, it is hard to achieve soil water recharge in the deeper soil layers by precipitation infiltration because the depth of soil water infiltration is shallow. Therefore, soil desiccation exists in the lower soil layer in case of

no groundwater recharge [8–10]. These mechanisms explain the absence of high-level SWC with an increasing trend along the vertical profile in the small catchment of Sanyanjing (Table 3).

2 Effects of environmental factors on average SWC at the catchment scale

Consistent with cluster analysis (Table 3), multiple regression analysis showed that land-use type had a significant effect on soil water status in the small catchment of Sanyanjing (Table 4, 5). This result coincides with the data previously reported in small catchments on the Loess Plateau [17, 18, 22, 23, 29]. For example, Zhang [23] concluded that average SWC (20–200 cm) descends with different land-use types (farmland > grassland > shrub land > and woodland, n = 80) in the small catchment of Zhifanggou. Bai [42] found that average SWC (0–500 cm) ranges from of 9% to 16% in orchard, gradient farmland, terrace, and grassland, but remains less than 10% in shrub land and most woodland (n = 91) in Nangou catchment in the central area of Loess Plateau, Ansai, Shaanxi. The consistency of the data demonstrates that cluster analysis is a reliable method for characterization of SWC profiles.

The effect of land-use type on SWC can be related to the differences existing in anthropogenic activity and vegetation type [22]. Average SWC was found significantly higher in terrace and abandoned farmland than in grassland and woodland along the 0–300 cm profile (Table 5). Abandoned farmland and terrace are associated with artificial tillage in the surface soil layer, which improves soil porosity and loosens soil structure, further enhancing soil water infiltration [22, 43]. Additionally, soil water consumption by crops is less than that in grassland and woodland due to lower leaf area index [21], contributing to the accumulation of SWC. The above mechanisms account for the greater average of SWC profiles with a stable trend to soil depth in terrace and abandoned farmland.

Difference in root distribution is another factor contributing the effect of land–use type on SWC [40]. In the Sanyanjing catchment, average SWC of woodland was higher in the 0–20 cm soil layer but lower in the 20–160 and 160–300 cm soil

Table 5. Multi-linear regression analysis of soil water content and selected environmental factors in three layers (0–20, 20–160, and 160–300 cm) of the vertical soil profile in the Sanyanjing catchment study area.

| Model | Y$_1$ | | | Before the | rainy season | | | | | |
| | | | | Y$_2$ | | | Y$_3$ | | |
	Unstandardized coefficients B	Standardized coefficients (Beta)	Sig.	Unstandardized coefficients B	Standardized coefficients (Beta)	Sig.	Unstandardized coefficients B	Standardized coefficients (Beta)	Sig.
Constant	14.714		0.066	6.031		0.498	18.514		0.069
X$_1$	−2.743	−0.23	0.013	−2.751	−0.22	0.026	−4.471	−0.296	0.001
X$_2$	−0.112	−0.037	0.68	0.042	0.013	0.889	−0.261	−0.068	0.449
X$_3$	4.86E−06	0.025	0.783	−7.80E−06	−0.038	0.694	1.20E−05	−0.05	0.578
X$_4$	0.005	0.081	0.507	0.011	0.177	0.175	0	0.005	0.97
D$_{51}$	1.09	0.146	0.109	0.019	0.002	0.98	−0.788	−0.084	0.354
D$_{52}$	−3.472	−0.278	0.003	−0.845	−0.064	0.503	−2.553	−0.162	0.073
D$_{53}$	−1.715	−0.191	0.047	−3.216	−0.342	0.001	−4.745	−0.384	0
D$_{61}$	−1.512	−0.348	0.005	−0.607	−0.134	0.307	−0.833	−0.151	0.212
	$R^2 = 0.30$	(P<0.001)		$R^2 = 0.19$	(P = 0.001)		$R^2 = 0.32$	(P<0.001)	

| Model | Y$_1$ | | | After the | rainy season | | | | | |
| | | | | Y$_2$ | | | Y$_3$ | | |
	Unstandardized coefficients B	Standardized coefficients (Beta)	Sig.	Unstandardized coefficients B	Standardized coefficients (Beta)	Sig.	Unstandardized coefficients B	Standardized coefficients (Beta)	Sig.
Constant	11.426		0.111	−5.811		0.452	−9.045		0.41
X$_1$	−2.01	−0.178	0.041	−2.763	−0.221	0.01	−3.688	−0.206	0.015
X$_2$	0.313	0.11	0.201	−0.132	−0.042	0.618	0.294	0.063	0.44
X$_3$	6.30E−06	−0.034	0.692	1.80E−05	−0.089	0.285	1.60E−05	−0.054	0.508
X$_4$	0.007	0.125	0.278	0.023	0.359	0.002	0.027	0.291	0.009
D$_{51}$	−0.622	−0.088	0.307	−1.161	−0.149	0.08	−1.81	0.162	0.053
D$_{52}$	−2.374	−0.2	0.021	−2.372	−0.181	0.032	−4.166	−0.222	0.008
D$_{53}$	−2.599	−0.305	0.001	−4.807	−0.51	0	−6.862	−0.467	0
D$_{61}$	−2.082	−0.506	0	−1.512	−0.332	0.004	−3.14	−0.476	0
	$R^2 = 0.37$	(P<0.001)		$R^2 = 0.39$	(P = 0.001)		$R^2 = 0.43$	(P<0.001)	

Dependent Variable: Y$_1$ (soil water content of 0–20 cm layer).
Y$_2$ (average soil water content of 20–160 cm layer).
Y$_3$ (average soil water content of 160–300 cm layer).
Independent Variables: X$_1$ = Sin(gradient), X$_2$ = Sin(aspect), X$_3$ = flowaccu, X$_4$ = elevation.
Dummy Variables: X$_5$ = terrace, (D$_{51}$, D$_{52}$, D$_{53}$) = (0,0,0); X$_5$ = abandoned farmland, (D$_{51}$, D$_{52}$, D$_{53}$) = (1,0,0).
X$_5$ = grassland, (D$_{51}$, D$_{52}$, D$_{53}$) = (0,1,0); X$_5$ = woodland, (D$_{51}$, D$_{52}$, D$_{53}$) = (0,0,1).
X$_6$ = ridge, (D$_{61}$) = 1; X$_6$ = gully, (D$_{61}$) = 0.
D represents sub-variable; binary variables 0 and 1 for the absence and presence of some land-use type or landform, respectively.

layers than data of grassland (Table 5). The varying trends of SWC profiles between grassland and woodland can be related to different distribution of root system in individual soil layers and stratified root extraction of soil water. The rooting depth of maize crop is reported to be approximately 100 cm and most maize roots are distributed in the soil layer of 0–20 cm, shallower than average rooting depths in grassland (20–60 cm) and woodland (20–100 cm) [40]. Diverse root distribution patterns can lead to different levels of soil water consumption by plants, contributing to great variability of SWC level.

In addition to land-use type, topographic factors strongly affected SWC in the study area (Table 4). This is because the distribution of wind and solar radiation varies with different topographic conditions, leading to different levels of soil evaporation, runoff on gradient, and soil water infiltration [41]. Gradient negatively affected SWC in the soil layers of 0–20, 20–160, and 160–300 cm (Table 5), possibly due to the increased runoff with increasing gradient and resultant reduction of precipitation infiltration [17, 27, 29, 41]. Other topographic factors including aspect and flow accumulation had no significant effects on average SWC in the three soil layers (Table 5). Similarly, Gómez [29] referred that aspect has no obvious influence on SWC in burned and unburned areas. Shi [19] suggested that aspect and catchment area significantly affect SWC during the wet period only, whereas elevation has a significant effect on SWC in arid and humid periods but not in semi-arid and semi-humid periods. In the present study, we found the effect of elevation on SWC of the three layers varying with the period of time and being significant after the rainy season only.

As for the landform type, location of sampling points significantly affected SWC only in the surface layer (0–20 cm) before the rainy season and throughout all the three layers (0–20, 20–160, and 160–300 cm) after the rainy season, with greater values in the gully than on the ridge (Table 5). The effect of landform type on SWC can be related to different levels of soil evaporation as affected by wind strength and solar radiation and soil physical properties. Similarly, Zhang [22] suggested that average SWC descends with different landforms as gully > terrace > slop land > hill top.

Overall, land-use type is the most significant factor affecting SWC while topographic factors and landform type are interacting jointly at the catchment-scale. Because the impact of environ-mental factors on SWC varies in different periods, it is necessary to increase the observation frequency, in order to better understand the spatiotemporal distribution and influencing factors of SWC in the small catchment. Such work will provide reference data for selecting reasonable environmental parameters in catchment scale SWC simulation over different periods of time.

Conclusions

In this study, cluster analysis enables catchment-scale characterization of soil water profiles in terms of average level and variation trend along the vertical profile, allowing for simple and clear interpretation of the results. A total of nine groups of soil water profiles are recognized but those with low-level soil water content and a decreasing trend are not present in the Sanyanjing catchment. Land-use type, gradient, landform type, and altitude are the major environmental factors significantly influencing average soil water content in the hilly and gully catchment with complex terrain. The former two factors strongly affect soil water content along the 0–300 cm soil profile, whereas effects exerted by the latter two factors vary by soil layer and season.

Understanding the vertical profile of soil water content and evaluation of related major influencing factors in individual soil layers can help with sustainable land use and water management in catchment areas on the hilly and gully Loess Plateau as well as in arid and semi-arid areas with complex terrain. For better estimation of soil water profiles in small catchments, other factors such as fertilization, coverage, and soil physical properties may be considered with respect to specific soil layers.

Acknowledgments

We give our thanks to Prof./Dr. Wenzhao Liu (Institute of Soil and Water Conservation, Northwest A & F University, Yangling) for suggestions on experimental arrangement.

Author Contributions

Conceived and designed the experiments: BW FW JW XW YH. Performed the experiments: BW FW JW XW YH. Analyzed the data: BW FW JW XW YH. Contributed reagents/materials/analysis tools: BW FW JW XW YH. Wrote the paper: BW FW JW XW YH.

References

1. Engelbrecht BMJ, Comita LS, Condit R, Kursar TA, Tyree MT, et al. (2007) Drought sensitivity shapes species distribution patterns in tropical forests. Nature 447: 80–82.

2. Yang L, Wei W, Mo BR, Chen LD (2011) Soil water under different artificial vegetation restoration in the semi-hilly region of the Loess Plateau. Acta Ecologica Sinica 31: 3060–3068 (in Chinese with English abstract).

3. Xing G, Zhang XM, Fei XL, Wu YX (2012) Study on soil moisture content under different land use types in Sunjiacha basin. Agricultural Research in the Arid Areas 30: 225–229 (in Chinese with English abstract).

4. Li JW, Zuo HT, Li QF, Fan XF, Hou XC (2011) Effect of soil water spatial distribution pattern on switchgrass during first growing season. Acta Agrista Sinica 19: 43–50 (in Chinese with English abstract).

5. Jipp PH, Nepstad DC, Cassel DK, Carvalho C (1998) Deep soil moisture storage and transpiration in forests and pastures of seasonally-dry Amazonia. Climatic Change 39: 395–412.

6. Grassini P, You JS, Hubbard KG, Cassman KG (2010) Soil water recharge in a semi-arid temperate climate in the central US Great Plains. Agricultural Water Management 97: 1063–1069.

7. Markewitz D, Devine S, Davidson EA, Brando P, Nepstad DC (2010) Soil moisture depletion under simulated drought in the Amazon: impacts on deep root uptake. New Phytologist 187: 592–607.

8. Li YS (2001) Fluctuation of yield on high-yield field and desiccation of the soil on dryland. Acta Pedologica Sinica 38: 353–356 (in Chinese with English abstract).

9. Huang MB, Dang TH, Gallichand J, Goulet M (2003) Effect of increased fertilizer applications to wheat crop on soil-water depletion in the Loess Plateau, China. Agricultural Water Management 58: 267–278.

10. Liu WZ, Zhang XC, Dang TH, Zhu OY, Li Z, et al. (2010) Soil water dynamics and deep soil recharge in a record wet year in the southern Loess Plateau of China. Agricultural Water Management 97: 1133–1138.

11. Beldring S, Gottschalk L, Seibert J, Tallaksen LM (1999) Distribution of soil moisture and groundwater levels at patch and catchment scales. Agricultural and Forest Meteorology 98–99: 305–324.

12. Li B, Rodell M (2013) Spatial variability and its scale dependency of observed and modeled soil moisture over different climate regions. Hydrology and Earth System Sciences 17: 1177–1188.

13. Henninger DL, Petersen GW, Engman ET (1976) Surface soil moisture within a watershed: Variations, factors influencing, and relationship to surface runoff. Soil Science Society of American Journal 40: 773–776.

14. Jones EB, Owe M, Schmugge TJ (1982) Soil moisture variation patterns observed in Hand county, South Dakota. Water Recources Bulletin 18: 949–954.

15. Hawley ME, Jackson TJ, Mccuen RH (1983) Surface soil moisture variation on small agricultural watersheds. Journal of Hydrology 62: 179–200.

16. Robinson M, Dean TJ (1993) Measurement of near surface soil water content using a capacitance probe. Hydrological Processes 7: 77–86.

17. Qiu Y, Fu BJ, Wang J, Chen LD (2000) Quantitative analysis of relationships between spatial and temporal variation of soil moisture content and

environmental factors at a gully catchment. Acta Ecologica Sinica 20: 741–747 (in Chinese with English abstract).

18. Zeng C, Shao MA, Wang QJ, Zhang J (2011) Effects of land use on temporal-spatial variability of soil water and soil-water conservation. Acta Agriculturae Scandinavica Section B-Soil and Plant Science 61: 1–13.

19. Shi ZH, Zhu HD, Chen J, Fang NF, Ai L (2012) Spatial heterogeneity of soil moisture and its relationships with environmental factors at small catchment level. Chinese Journal of Applied Ecology 23: 889–895 (in Chinese with English abstract).

20. Chen LD, Huang ZL, Gong J, Fu BJ, Huang YL (2007) The effect of land cover/vegetation on soil water dynamic in the hilly area of the loess plateau, China. Catena 70: 200–208.

21. Wang YQ, Shao MA, Liu ZP, Orton R (2013) Regional-scale variation and distribution patterns of soil saturated hydraulic conductivities in surface and subsurface layers in the loessial soils of China. Journal of Hydrology 487: 13–23.

22. Zhang R, Cao H, Wang YQ, Huang CQ, Tan WF (2012) spatial variability of soil moisture and its influence factors in watershed of gully region on the loess plateau. Research of Soil and Water Conservation 19: 52–58 (in Chinese with English abstract).

23. Canton Y, Sole-benet A, Domingo F (2004) Temporal and spatial patterns of soil moisture in semiarid badlands of SE Spain. Journal of Hydrology 285: 199–214.

24. Burnt TP, Butcher DP (1985) Topographic controls of soil moisture distributions. Journal of Soil Science 36: 469–486.

25. O'loughlin EM (1981) Saturation regions in catchments and their relations to soil and topographic properties. Journal of Hydrology 53: 229–246.

26. Huang J, WU P, Zhao XN (2012) Effects of rainfall intensity, underlying surface and slope gradient on soil infiltration under simulated rainfall experiments. Catena 104: 93–102

27. Qiu Y, FU BJ, Wang J, Chen LD (2003) Spatiotemporal prediction of soil moisture content using multiple-linear regression in a small catchment of the Loess Plateau, China. Catena 54: 173–195.

28. Qiu Y, Fu B, Wang J, Chen L, Meng Q, Zhang Y (2010) Spatial prediction of soil moisture content using multiple-linear regressions in a gully catchment of the Loess Plateau, China. Journal of Arid Environments 74: 208–220.

29. Gómez-Plaza A, Martínez-Mena M, Albaladejo J, Castillo VM (2001) Factors regulating spatial distribution of soil water content in small semiarid catchments. Journal of Hydrology 253: 211–226.

30. Dripps WR, Bradbury KR (2007) A simple daily soil-water balance model for estimating the spatial and temporal distribution of groundwater recharge in temperate humid areas. Hydrogeology Journal 15: 433–444.

31. Yao XL, Fu BJ, Lu YH, Sun FX, Wang S, et al. (2013) Comparison of four spatial interpolation methods for estimating soil moisture in a complex terrain catchment. PLoS One 8(1): e54660.

32. Wang MB, Li HJ (1995) Quantitative study on the soil water dynamics of various forest plantations in the loess plateau region in northwestern Shanxi. Acta Ecologica Sinica 15: 172–184 (in Chinese with English abstract).

33. Wang YQ, Zhang XC, Han FP (2008) Profile variability of soil properties in check dam on the Loess Plateau and its functions. Environmental Science 29: 1020–1026 (in Chinese with English abstract).

34. Jain AK (2010) Data clustering: 50 years beyond K-means. Pattern Recognition Letters 31: 651–666.

35. Huang YL, Chen LD, Fu BJ, Wang YL (2005) Spatial pattern of soil water and its influencing factors in gully catchment of the Loess Plateau. Journal of Natural Resources 20: 483–492 (in Chinese with English abstract).

36. Zou JL, Shao MA, Gong SH (2011) Effects of different vegetation and soil types on profile variability of soil moisture. Research of Soil and Water Conservation 18: 12–17 (in Chinese with English abstract).

37. Kizito F, Dragila M, Se'ne M, Lufafa A, Diedhiou I, et al. (2006) Seasonal soil water variation and root patterns between two semi-arid shrubs co-existing with Pearl millet in Senegal, West Africa. Journal of Arid Environments 67: 436–455.

38. Li J, Chen B, Li XF, Zhao YJ, Ciren YJ, et al. (2008) Effects of deep soil desiccation on artificial forestlands in different vegetation zones on the Loess Plateau, China. Acta Ecologica Sinica 28: 1429–1445 (in Chinese with English abstract).

39. Cheng LP, Liu WZ (2013) Long term effects of farming system on soil water content and dry soil layer in deep loess profile of Loess Tableland in China. Journal of Integrative Agriculture. 13(6): 1382–1392.

40. Wang XZ, Jiao F (2011) Partition of soil moisture profiles based on sequential clustering method. Journal of Northwest A &F University 39: 191–201,196 (in Chinese with English abstract).

41. Fu XL, Shao MA, Wei XR, Wang HM, Zeng C (2013) Effects of monovegetation restoration types on soil water distribution and balance on a hillslope in northern Loess Plateau of China. Journal of Hydrologic Engineering 18: 413–421.

42. Bai TL, Yang QK, Shen J (2009) Soil variability of soil moisture vertical distribution and related affecting factors in hilly and gully watershed region of Loess Plateau. Chinese Journal of Ecology 28: 2508–2514 (in Chinese with English abstract).

43. Lian G, Guo XD, Fu BJ, Hu CX (2006) Spatial variability of bulk density and soil water in a small catchment of the Loess Plateau. Acta Ecologica Sinica 26: 647–654 (in Chinese with English abstract).

3

Land Suitability Assessment on a Watershed of Loess Plateau Using the Analytic Hierarchy Process

Xiaobo Yi[1,2], Li Wang[1,2]*

1 College of Resources and Environment, Northwest A&F University, Yangling, Shaanxi, China, **2** State Key Laboratory of Soil Erosion and Dryland Farming on the Loess Plateau, Northwest A&F University, Yangling, Shannxi, China

Abstract

In order to reduce soil erosion and desertification, the Sloping Land Conversion Program has been conducted in China for more than 15 years, and large areas of farmland have been converted to forest and grassland. However, this large-scale vegetation-restoration project has faced some key problems (e.g. soil drying) that have limited the successful development of the current ecological-recovery policy. Therefore, it is necessary to know about the land use, vegetation, and soil, and their inter-relationships in order to identify the suitability of vegetation restoration. This study was conducted at the watershed level in the ecologically vulnerable region of the Loess Plateau, to evaluate the land suitability using the analytic hierarchy process (AHP). The results showed that (1) the area unsuitable for crops accounted for 73.3% of the watershed, and the main factors restricting cropland development were soil physical properties and soil nutrients; (2) the area suitable for grassland was about 86.7% of the watershed, with the remaining 13.3% being unsuitable; (3) an area of 3.95 km^2, accounting for 66.7% of the watershed, was unsuitable for forest. Overall, the grassland was found to be the most suitable land-use to support the aims of the Sloping Land Conversion Program in the Liudaogou watershed. Under the constraints of soil water shortage and nutrient deficits, crops and forests were considered to be inappropriate land uses in the study area, especially on sloping land. When selecting species for re-vegetation, non-native grass species with high water requirements should be avoided so as to guarantee the sustainable development of grassland and effective ecological functioning. Our study provides local land managers and farmers with valuable information about the inappropriateness of growing trees in the study area along with some information on species selection for planting in the semi-arid area of the Loess Plateau.

Editor: Matteo Convertino, University of Florida, United States of America

Funding: This work was financially supported by the National Natural Science Foundation of China (Nos. 51239009 and 41271239) (http://www.nsfc.gov.cn), and the CAS Action-plan for Western Development (KZCX2-XB3-13) (http://www.cas.cn/). The funders had no role in study design, data collection and analysis, decision to publish, or preparation of the manuscript.

Competing Interests: The authors have declared that no competing interests exist.

* E-mail: wangli5208@nwsuaf.edu.cn

Introduction

Sandy desertification and soil erosion are two of the most serious problems affecting China's water and land resources. The World Bank has suggested that potentially more than 331 million hectares of land are susceptible to desertification (about one third of the area of China) while about 262 million hectares are actually affected [1]. Previous studies have shown that soil erosion affects about 360 million hectares of land in China, which is about 38% of its total area, and this proportion is more than three times the global average [2–3]; land in China is considered to be amongst the most severely eroded in the world. For example, on the Loess Plateau, erosion rates are about 8000–25,000 tonnes km^{-2} year^{-1} in the gully areas of Shanxi and Shaanxi Provinces. Intensive cultivation of the steep hillsides has resulted in the loss of an estimated 1.6 billion tonnes of soil annually to the Yellow River [4]. In addition, dry conditions combined with the fine texture of the loess soil make the area very susceptible to dust storms. Due to the severe soil erosion and sandy desertification, the eco-environment of the Loess Plateau has been severely degraded, which has seriously affected sustainable development in the region [5]. With the environmental goals of reducing soil erosion and desertification, the Chinese Central Government launched the

Sloping Land Conversion Program (SLCP, also known as Grain for Green or Grain to Green) in the late 1990s, with the intension of increasing the country's forest and grassland cover by retiring steeply sloping and marginal land from agricultural production [6–9]. This program aimed to restore degraded ecosystems, reduce poverty and assist rural households to move towards more sustainable economic activities [2,10].

However, after more than 10 years, this large-scale ecological-recovery project has faced some key problems that have limited the successful development of the current SLCP policy [11]. Surveys and case studies of SLCP have consistently identified insufficient technical support and arid conditions as being the key constraints to achieving program goals [3]. For example, in the north part of the Loess Plateau, the potential for converting cropland into forests has been over estimated: tree species have been planted in areas that are better suited to growing shrubs or grass, and species with high water requirements have been planted in areas where drought-tolerance is required. The result has been soil desiccation and the development of dry soil layers because the planted trees exploited water stored in the deeper soil layers and prevented them from being recharged by rainwater [12,7]. Consequently, once the stored water was exhausted, the limited precipitation was insufficient to maintain normal growth in the revegetated areas [13], leading to the vegetation dying or the production of stunted trees (colloquially referred to as "little old man trees", which are only about 20% of the normal height for their age) [14–17]. Thus, implementing the SLCP policy on the Loess Plateau requires careful consideration of several factors that have a significant effect on land use change and land suitability associated with carrying out the re-vegetation program. Currently, China is facing increased environmental pressures with shortages of water potentially limiting development, especially in its dryer northern and western regions, including the Loess Plateau [12]. In order to reduce soil water depletion and implement a successful re-vegetation program for environmental improvement on the Loess Plateau it is essential to identify appropriate species to plant. It is important to evaluate the land suitability and to develop land use plans that maximize reduction in soil erosion while at the same time minimizing water yield reduction, thus ensuring the sustainable growth and succession of vegetation. Only the proper implementation of the SLCP and successful re-vegetation can make a real contribution to efforts being made to combat the urgent environmental problems of soil erosion and desertification, as well as of climate change and loss of biodiversity, currently confronting the Loess Plateau, in particular, and China as a whole. In this study, the analytic hierarchy process (AHP) was used to assess the suitability of cropland, grassland and forestland in the Liudaogou watershed, and the main limiting factors for different land use options were quantitatively analyzed. The AHP has been widely used b**y** decision-makers and researchers. It is a mathematical method, developed by Saaty in 1977 [18] and improved by the same author in 1980 [19], for analyzing complex decisions involving many criteria [20]. It has been widely used in site selection, suitability analysis, regional planning, and land consolidation analysis [21–22].The success of the AHP as a practical and reliable method is highlighted by its extensive application in the past two decades [23–24]. Furthermore, its simplicity in relation to its power was a significant factor in the choice for its use in the presented study, and all the mentioned factors ensured that the study objectives would be successfully achieved. The objectives of the study are to determine which land use is best suited to the re-vegetation program by assessing the land suitability, to identify the constraints for future land conservation, and to provide a scientific basis for decision-making in the successful implementation of the Sloping Land Conversion Program, not only for the study's watershed but across the whole Loess Plateau.

Materials and Methods

Study area

This study was conducted in the Liudaogou watershed, located in Shenmu County, Shaanxi Province, China ($110°21'–110°23'$E, $38°46'–38°51'$N; Fig. 1). The Liudaogou watershed covers 6.9 km^2 and is located at the center of the wind-water erosion crisscross region in the north part of the Loess Plateau. This area suffers its most serious water erosion in summer and autumn and its most serious wind erosion in winter and spring [25–26]. The watershed is representative of the ecotone between the grass–pastoral and the agricultural areas, in the transitional zone between the desert aeolian deflation zone (the Mu Us Desert) and the loess hilly area (Loess Plateau), as well as being between the arid and the semi-arid regions. For these reasons, the Shenmu Erosion and Environmental Research Station (SEERS) of the Chinese Academy of Sciences was built within the watershed [27]. The study area is located at altitudes ranging from 1080 to 1270 m above mean sea level and has a semiarid continental monsoon climate, with a mean annual temperature of 8.4°C. The monthly

mean temperature ranges from 9.7°C in January to 23.7°C in July. The mean annual precipitation is 437 mm, 77% of which occurs from June to September. Mean annual potential evapotranspiration can be as high as 1800 mm, which would result in a water deficit of 1350 mm year^{-1}. The area has a deep (up to 100 m) loess layer, which originated during the Quaternary period. The dominant soil type (cultivated loessial soil), is a Ust-Sandic Entisol, which is loess-derived and is consequently easily eroded by both water and wind.

Analytical hierarchy process (AHP)

The application of AHP for making a decision about land suitability in this study involved four main steps, as follows:

The first step was to decompose the decision problem into a hierarchical structure where the attributes and plans were present as inter-related elements. Based on a qualitative analysis of the environment in the study area, the final hierarchical structures were separated into four levels (Fig. 2). The first level was the overall land suitability. The second level was composed of subsystems: geological and topographical conditions; nutrient status; and soil physical properties. The third level consisted of the specific factors that affected the land suitability. The fourth level comprised each assessment unit (cell).

Elevation (C_1) and slope gradient (C_2) were selected to represent geological and topographical conditions (B_1). Slope is a crucial factor affecting vegetation structure and soil erosion on the Loess Plateau. Variation in elevation has an impact on soils, microclimatic effects, and other processes that could affect land suitability. Soil organic matter (C_3), total nitrogen (C_4), available phosphorous (C_5), and available potassium (C_6) were selected as factors that represented nutrient status (B_2). These four nutrition factors are closely related to land use and cover change, and understanding the effects of land use change on soil organic matter and nitrogen is important to sustainable management of land resources and associated watershed processes, as well as regional responses to global climatic change [28]. Three factors were selected to represent the soil physical properties. Soil bulk density (C_7), soil texture (C_8) and soil water content (C_9) strongly influence plant growth and land use, and soil properties and plant recovery processes have distinct characteristics in the typical desertified sandy land of the Loess Plateau [17]. Socio-economic factors were not considered in this study because it was limited to a small watershed where such factors are generally less important than the physical ones.

The second step involved a pair-wise comparison of the elements based on a nine point weighting scale; this generated the input data (Table 1). The comparison was carried out for each decision element at 1–$(n–1)$ levels, where n was the matrix size. A matrix was generated as a result of the pair-wise comparisons and weights for the criteria were derived from these calculations.

A matrix of scores could be developed from the comparisons, given by

$$R = \begin{vmatrix} a_{11} & a_{12} & \dots & a_{1n} \\ a_{21} & a_{22} & \dots & a_{2n} \\ \dots & \dots & \dots & \dots \\ a_{n1} & a_{n2} & \dots & a_{nn} \end{vmatrix}$$

where a_{ij} indicated how much more important the ith objective was than the jth objective, while making a suitable material handling/equipment selection decision. For all i and j, it was necessary that $a_{ii}=1$ and $a_{ij}=1/a_{ji}$. The possible assessment

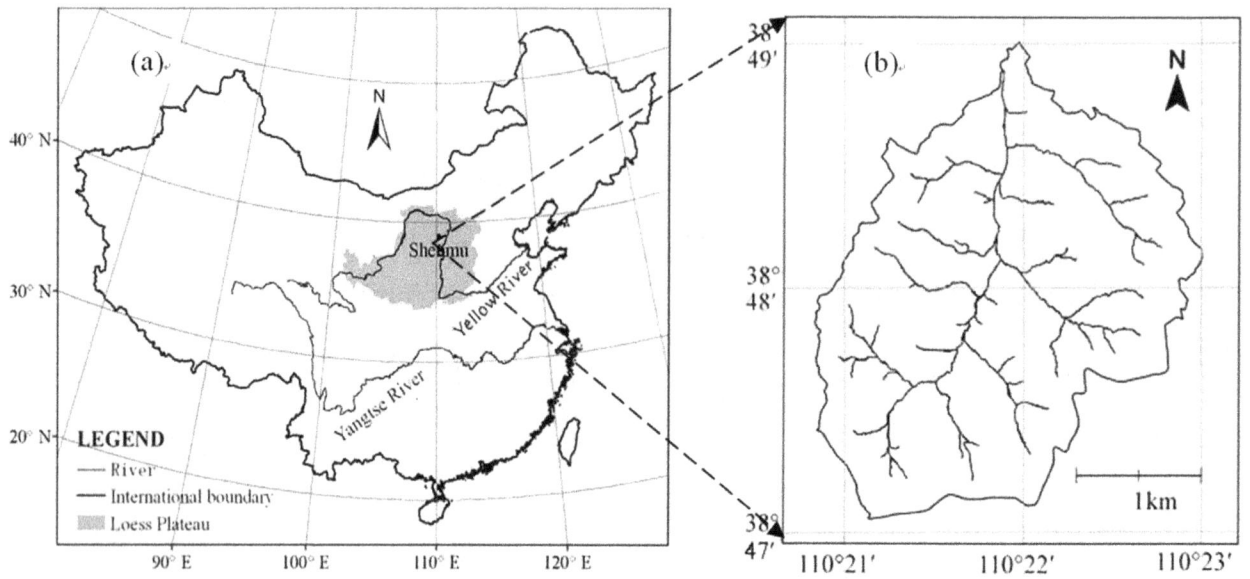

Figure 1. Location of the study site. (a) China; (b) the Liudaogou watershed.

values of a_{ij} in the pair-wise comparison matrix, along with their corresponding interpretations, are shown in Table 1.

The scores were normalized by dividing each element within the matrix by the sum of the column j, to create a normalized matrix, Rw:

$$Rw = \begin{vmatrix} \dfrac{a_{11}}{\sum a_{i1}} & \dfrac{a_{12}}{\sum a_{i2}} & \cdots & \dfrac{a_{1n}}{\sum a_{in}} \\ \dfrac{a_{21}}{\sum a_{i1}} & \dfrac{a_{22}}{\sum a_{i2}} & \cdots & \dfrac{a_{2n}}{\sum a_{in}} \\ \cdots & \cdots & \cdots & \cdots \\ \dfrac{a_{n1}}{\sum a_{i1}} & \dfrac{a_{n2}}{\sum a_{i2}} & \cdots & \dfrac{a_{nn}}{\sum a_{in}} \end{vmatrix}$$

The weight associated with each objective (c_i) could be estimated as the mean of the normalized scores in row i of the Rw matrix. Hence, c_i was calculated to give a matrix of weights, C:

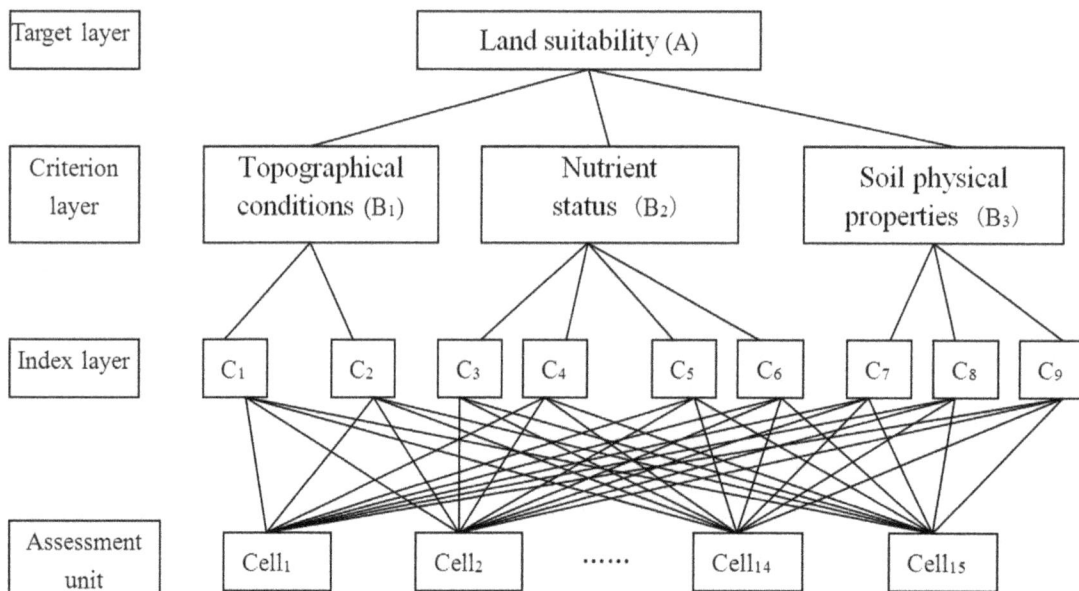

Figure 2. Hierarchical structure of land suitability.

Table 1. AHP pair-wise comparison scale for variables *i and j*.

Intensity of the relative importance	Definition
1	Equal importance of *i* and *j*
3	Moderate importance of *i* over *j*
5	Strong importance of *i* over *j*
7	Very strong importance of *i* over *j*
9	Extremely strong importance of *i* over *j*
2, 4, 6, 8	Intermediate values

$$
C = \begin{vmatrix} c_1 \\ c_2 \\ \cdots \\ c_n \end{vmatrix} = \begin{vmatrix} \dfrac{a_{11}}{\sum a_{i1}} + \dfrac{a_{12}}{\sum a_{i2}} + \cdots + \dfrac{a_{1n}}{\sum a_{in}} \\[2mm] \dfrac{a_{21}}{\sum a_{i1}} + \dfrac{a_{22}}{\sum a_{i2}} + \cdots + \dfrac{a_{2n}}{\sum a_{in}} \\[2mm] \cdots \quad \cdots \quad \cdots \quad \cdots \\[2mm] \dfrac{a_{n1}}{\sum a_{i1}} + \dfrac{a_{n1}}{\sum a_{i2}} + \cdots + \dfrac{a_{nn}}{\sum a_{in}} \end{vmatrix}
$$

where n is the number of objectives being compared, and the c_i value indicates the relative degree of importance (weight) of the *ith* objective.

The third step was to check for the consistency of the weight values (c_i) underlying the theoretical validity of the comparison matrix. In order to determine consistency, the consistency vector $(R \times C$ matrix) and x_i were calculated as:

$$
R \times C = \begin{vmatrix} a_{11} & a_{12} & \cdots & a_{1n} \\ a_{21} & a_{22} & \cdots & a_{2n} \\ \cdots & \cdots & \cdots & \cdots \\ a_{n1} & a_{n2} & \cdots & a_{nn} \end{vmatrix} \times \begin{vmatrix} c_1 \\ c_2 \\ \cdots \\ c_n \end{vmatrix} = \begin{vmatrix} x_1 \\ x_2 \\ \cdots \\ x_n \end{vmatrix}
$$

The eigenvalue of the pair-wise comparison matrix, $\lambda_{\max}$, was then estimated using the following equation:

$$
\lambda_{\max} = \frac{1}{n} \sum_{i=1}^{n} \frac{x_i}{c_i}
$$

An approximation to the consistency index (CI) was calculated and the consistency judgment was checked for the appropriate value of n by CR. CI and CR were calculated as [18]:

$$
CI = \frac{\lambda_{\max} - n}{n - 1}
$$

$$
CR = \frac{CI}{RI}
$$

where RI is the random consistency index. The RI values for different numbers of n are listed in Table 2.

The rule commonly used in the AHP was applied, whereby if CR was less than 0.10 (i.e. 10%), the degree of consistency was acceptable; and if it was greater than 0.10 it was considered that there were significant inconsistencies and, in such a case, the AHP

would not produce meaningful results [29], and it would be necessary to review and improve the judgments.

The fourth step was a complete evaluation based on the rating of the final weights in the decision plan. Matrixes for layers B and C and for C and D were generated using the same method as that used for layers A and B. By comparing the final values by simple rankings, the weights of all elements in each level of the hierarchy relative to the entire level could be obtained. These in turn were all ranked, and were carried from the upper layer to the lower layer. The combined weight (W_i) of each assessment factor was determined for the integrated assessment of land suitability in the Liudaogou watershed.

Membership value standardization for assessment factors

In the process of land suitability assessment, a primary step was to ensure a standardized measurement system for all the factors considered. Since those factors have different standards of measurements, they had to be standardized to a uniform rating scale; in this study the scale was between 1 and 4 for ease of analysis. Assigning values to specific factors required specific decision rules in the form of thresholds for each factor. Various statistical and empirical guidelines from the related national codes and literature were used to determine the boundary values. As a general guideline, a positive correlation between the value awarded and suitability was employed. The class boundaries and standardized measurements employed for each factor are given for arable land in Section 3.1.1 and for forestland and grassland in Section 3.2.1. The integer numbers ranging from 1 to 4 were assigned to high, moderate, marginal and unsuitable classes, respectively. The next step involved assigning a new value for the degree of membership of every attribute to each index at each level. All these new values ranged from 0 to 1 where 0 indicated a poor fit and 1 indicated a perfect fit for membership of an attribute to a given index.. The membership degree function (F_i) of each assessment factor is given as follows:

The membership degree of soil organic matter, total nitrogen, available phosphorus and available potassium to the soil nutrient status criterion were calculated using the following equations:

$$
F_i = \begin{cases} 0.1 & M_i \leq M_{i1} \\ 0.1 + 0.9 \times \dfrac{M_i - M_{i1}}{M_{i2} - M_{i1}} & M_{i1} < M_i < M_{i2} \\ 1 & M_i > M_{i2} \end{cases}
$$

where M_{i1} and M_{i2} are the lower and upper limits of the value range of a factor.

Table 2. Random consistency index for AHP.

n	1	2	3	4	5	6	7	8	9
RI	0.00	0.00	0.58	0.90	1.12	1.24	1.32	1.41	1.45

The membership degree of slope gradient was given by:

$$F_i = \begin{cases} 1.0 & M_i < M_{i1} \\ 1 - 0.9 \times \dfrac{M_i - M_{i1}}{M_{i2} - M_{i1}} & M_{i1} \leq M_i < M_{i2} \\ 0.1 & M_i \geq M_{i2} \end{cases}$$

The membership degree of physical clay content was given by:

$$F_i = \begin{cases} 0 & M_i \leq M_{i1}, M_i \geq M_{i2} \\ \dfrac{M_i - M_{i1}}{M_{i0} - M_{i1}} & M_{i1} < M_i < M_{i0} \\ \dfrac{M_{i2} - M_i}{M_{i2} - M_{i0}} & M_{i0} < M_i < M_{i2} \\ 1 & M_i = M_{i0} \end{cases}$$

where M_{i0} means the best physical clay content in the study area.
The membership degree of altitude was given by:

$$F_i = \begin{cases} 1 & M_i \leq M_{i1} \\ \dfrac{M_{i2} - M_i}{M_{i2} - M_{i1}} & M_{i1} < M_i < M_{i2} \end{cases}$$

The degree of membership of soil water content and bulk density was given by:

$$F_i = \begin{cases} 1 & M_i \geq M_{i2} \\ \dfrac{M_i}{M_{i2}} & M_i < M_{i2} \end{cases}$$

Assessment of land suitability

The integrated classification of land suitability for the study area was obtained by integrating the combined weight (W_i) and the total degree of membership for different factors (F_i) as follows:

$$E(j) = \sum_{i=1}^{n} F_i W_i$$

The maximum membership principle was adopted to define the comprehensive classification of land suitability. If $E_{i0} = \max E_i (1 \leq i \leq n)$, the comprehensive classification was E_{i0}.

Data

Field survey and soil sampling were conducted during August and September 2011. The land use status in 1993 and 2011 is presented in Table 3. When compared with 1993, there was significantly less arable land in 2011, whilst there had been increases in the extent of grassland, land that had been build on

and land that was not in use. The main reason was the construction of four collieries during this period; this resulted in the transfer of the work force (local farmers) from traditional cultivation to the coal industry [30]. Based on land use and topography, we selected 15 assessment units: three for representing arable land, five for forestland and seven for grassland. Details of the assessment units are presented in Table 4.

Altitude was measured in the field by double frequency RTK-GPS. Slope gradient was determined using a compass incorporating an inclinometer. Sampling transects were placed along the slope in each assessment unit. For each transect, 3 plots (15×15 m) were established for the field investigation and sampling. In each sampling plot, a 1.0 m long×0.7 m wide×0.5 m deep pit was dug to allow measurements of soil bulk density. Samples were collected using a 5.0 cm diameter by 5.0 cm long stainless cutting ring with which samples were collected from the top 20 cm of the soil profile. Soil samples in the ring were extracted and the roots were carefully removed by hand. The soils were dried at 105°C to constant weight and bulk density was calculated by dividing the dry mass by the known volume. Soils were also collected with a soil auger (4 cm-diameter) to estimate profile water content in 10 cm layers to a depth of 40 cm. Soil water content was determined by oven-drying samples at 105°C.

Composite samples of about 1 kg from each assessment unit were collected and then air-dried and ground to pass through 1.00 mm and 0.25 mm nylon screens prior to laboratory analysis. The sieved soil was used to measure soil particle composition (physical clay content) and soil organic matter (SOM), respectively. Particle composition was measured by the laser diffraction technique using a MasterSizer 2000 (Malvern Instruments, Malvern, England), equipped with a low-power (2 mW) Helium-Neon laser with a wavelength of 633 nm as the light source. Soil organic carbon (SOC) was measured by an Elementar Vario EL element analyzer and soil organic matter (SOM) was obtained by multiplying SOC content values by 1.723.

Soil total nitrogen (TN) was measured following the Kjeldahl digestion method [31]. Total phosphorus (TP) was determined colorimetrically after wet digestion with sulfuric acid and perchloric acid, and available phosphorus (AP) was extracted following the Olsen bicarbonate extractable P method [32]. Available K was determined by the ammonium acetate extraction method [33].

Table 3. Land use in the Liudaogou watershed (%) in 1993 and 2011.

Year	Arable land	Forestland	Grassland	Others[2]
1993[1]	31.3	25.7	39.6	3.4
2011	16	26	44	14

Note:
[1] Data cited are from Yang et al. (1994).
[2] Refers to built-up and unused land.

Table 4. Detailed information about the assessment units.

Assessment unit	Land use	Slope gradient (°)	Altitude (m)	Soil organic matter (g/kg)	Total nitrogen (g/kg)	Available phosphorus (mg/kg)	Available potassium (mg/kg)	Bulk density (g/cm³)	Soil texture (physical clay content ≤0.01 mm) (%)	Soil water content (%)	Main species
1	Arable land	7	1170	14.29	0.41	4.64	203.78	1.60	17.52	9.45	*Solanum tuberosum*
2	Arable land	5	1127	22.26	0.39	15.83	420.90	1.57	12.99	12.16	*Setaria italica*
3	Arable land	0	1155	16.02	0.24	7.36	244.80	1.59	11.37	12.09	*Zea mays*
4	Forestland	15	1195	9.54	0.22	5.12	269.09	1.53	6.01	3.71	*Populus simonii*
5	Forestland	14	1201	13.84	0.35	5.87	279.82	1.60	7.07	4.46	*Populus simonii*
6	Forestland	10	1170	12.50	0.26	5.17	176.56	1.54	5.14	2.00	*Pinus tabullaeformis*
7	Forestland	12	1154	5.80	0.07	3.77	72.55	1.55	6.59	3.78	*Populus simonii*
8	Forestland	17	1175	3.08	0.05	3.14	133.91	1.56	6.86	3.91	*Populus simonii*
9	Grassland	17	1160	2.90	0.12	3.06	119.29	1.60	9.21	3.90	*Stipa bungeana*
10	Grassland	15	1196	1.99	0.08	2.46	92.97	1.58	19.13	3.51	*Medicago sativa*
11	Grassland	20	1189	8.14	0.16	4.58	116.34	1.58	2.27	3.64	*Medicago sativa*
12	Grassland	30	1204	6.23	0.13	3.95	297.40	1.61	12.45	8.60	*Lespedeza daurica*
13	Grassland	10	1237	11.02	0.22	7.28	157.24	1.59	39.50	9.54	*Stipa bungeana*
14	Grassland	9	1152	6.38	0.14	4.11	85.04	1.59	18.86	8.17	*Stipa bungeana*
15	Grassland	18	1197	2.46	0.08	2.98	186.94	1.57	20.52	8.50	*Stipa bungeana*

Results and Discussion

Suitability assessment for arable land

Standardized evaluation factors for arable land. Arable land suitability was divided into 4 classes, i.e. highly suitable (S1), moderately suitable (S2), marginally suitable (S3) and unsuitable (N). All factors were standardized to a uniform rating scale for ease of analysis. Standardized values for specific factors required specific decision rules in the form of thresholds for each factor. Various statistical and empirical guidelines from the related national–provincial codes and literature were used to determine the boundary values. In addition, twelve experts were invited to act as the decision makers to guarantee the reliability of boundary values. The experts involved were ecologists, pedologists, geographers, forest and grass managers, land resource experts and environmental protection experts; all have undertaken related studies of vegetation restoration and ecological– environmental issues in the study area and all are familiar with the Loess Plateau. Thus, they were able to suggest reasonable values for the thresholds. As a general guideline, there was a positive correlation between the value assigned and suitability. The class boundaries and standardized measurements applied for each factor are shown in Table 5.

Distribution of suitable cropland and associated statistics. Based on the method described above, suitable cropland classes were calculated and the main limiting factors were identified for each assessment unit (Table 6). Assessment units 1 and 3 in arable land and unit 14 in grassland were considered to be of marginal suitability, assessment unit 2 in arable land was classified as being of moderate suitability, and the other units from forest and grasslands were classified as being unsuitable for crops. Generally, an area of 0.40 km^2, accounting for 6.7% of the total area of the watershed (not including the built-up and unused lands), was classified as being moderately suitable for crops, and 1.19 km^2 (20.0%) as marginally suitable for crops (Table 7). In total, the area suitable for crops comprised about 26.7% of the watershed, so the area unsuitable for crops accounted for 73.3% of the watershed. The main factors restricting cropland development were soil physical properties and soil nutrients. First, the soil organic matter content and total nitrogen were generally low. This was particularly true in forest and grasslands, where the total nitrogen ranged between 0.05 and 0.35 g/kg (on average 0.17±0.09 g/kg), and was significantly lower than the generally reported values for soils (0.48–0.91 g/kg; on average 0.71) in the center of the Loess Plateau [34]. Soil organic matter was in the

range 1.99–13.84 g/kg (on average 6.99±4.05 g/kg), and lower than the generally reported values for soils (7.0–15.0 g/kg; on average 11.0) in the center of the Loess Plateau [35]. Secondly, soil texture and soil water content obviously affected the land's suitability to grow crops in the watershed. In forestland in particular, the clay content was significantly lower than that of crop and grasslands, suggesting that planting trees adversely affected the soil structure in the study area. Our data were consistent with those presented in previous studies [26,30] indicating that species such as *Populus simonii* can cause soil degradation. The watershed is located in the ecotone between arid and the semi-arid regions, and soil water is always a limiting factor for plant growth because of the low levels of precipitation (about 437 mm) and high potential evapotranspiration (1800 mm). In forestland, soil water content ranged between 2.00% and 4.46%, which was close to the wilting point. Grasslands also exhibited soil water deficits compared with the croplands, although soil water content was higher than that in forestlands. Previous studies have shown that planting *P. simonii* (a tree species) and *Medicago sativa* (alfalfa, a forage legume) could cause large reductions in soil water content due to their high water consumption, and this could result in the formation of dry soil layers on land where they have been planted [36–37]. Usually, soil drying due to high plant water use adversely affects soil physical properties, for example, reducing aggregate stability and soil surface roughness [38–39]. Thus, soil water deficit and soil physical properties may have a compounding effect.

It is noteworthy that available phosphorus was in the range 2.46–5.87 mg/kg (on average 4.29±1.39 mg/kg) in forest and grass lands, and was, therefore, very much higher than generally reported for soils (0.91–1.76 g/kg; on average 1.10) in the center of the Loess Plateau. Available potassium was in the range 85.04–297.40 mg/kg (on average 165.60±78.49 mg/kg), which is also significantly higher than generally reported for soils (77.2–170.0 mg/kg) in the center of the Loess Plateau [35]. Therefore, these two nutrients were not the limiting factors for cropland development in the watershed.

Suitability assessment for forest and grassland

Standardized evaluation factors for forest and grassland. As for the arable land suitability assessment, forest and grassland suitability was also divided into 4 categories: highly suitable (S1), moderately suitable (S2), marginally suitable (S3) and unsuitable (N). The class boundaries and standardized measure-

Table 5. Land characteristics, thresholds and degree of suitability for arable land.

Land characteristics	Suitability			
	High (S1)	**Moderate (S2)**	**Marginal (S3)**	**Unsuitable (N)**
Slope gradient (°)	≤5	5~15	15~25	>25
Altitude (m)	≤1170	1170~1195	1195~1220	1220~1273.9
Soil organic matter (g/kg)	≥20	13~20	6~13	<6
Total nitrogen (g/kg)	≥0.28	0.16~0.28	0.05~0.16	<0.05
Available phosphorus (mg/kg)	≥15	9~15	3~9	<3
Available potassium (mg/kg)	≥250	150~250	50~150	<50
Bulk density (g/cm³)	-	-	-	>1.60
Soil texture (physical clay content) (%)	15	-	-	>30 or <10
Soil water content (%)	≥12	10~12	8~10	<8

Table 6. Cropland suitability and limiting factors for each assessment unit.

Assessment unit	Combined weight	Grade	Limiting factors
1	0.709	S3	-
2	0.989	S2	-
3	0.795	S3	-
4	0.466	N	Soil water and texture
5	0.619	N	Soil water and texture
6	0.501	N	Soil water and texture
7	0.259	N	Soil water, texture and organic matter
8	0.244	N	Soil water, texture and organic matter
9	0.287	N	Soil water, texture and organic matter
10	0.260	N	Soil water, organic matter
11	0.347	N	Soil water and texture
12	0.390	N	Bulk density
13	0.607	N	Soil texture and slope gradient
14	0.427	S2	-
15	0.371	N	Soil organic matter and available P

ments applied for each factor are shown in Table 8. Altitude is not a limiting factor for forest and grassland development.

Distribution of forest and grassland suitability. Only assessment unit 6 in the forestland and unit 11 in the grassland were classified as being unsuitable for grassland (Table 9). The limiting factor for unit 6 was the low soil water content resulting from the high water consumption by tree species, and the limiting factor for unit 11 was soil texture because of its very low clay content (Table 4). Overall, an area of 0.79 km², accounting for 13.3% of the total area of the watershed, was classified as highly suitable for grassland, 1.18 km² (20.0%) was classified as being moderately suitable for grassland, and 3.16 km² (53.4%) as being marginally suitable for grassland (Table 10). In total, the area suitable for grassland was about 86.7% of the watershed, with the remaining 13.3% being unsuitable. This result is consistent with a previous study suggesting that development of vegetation in the watershed should focus on grassland [30]. Hou et al. [27] reported that the watershed is located within the boundary of the 400 mm rainfall isoline, representing the demarcation between cropping and pastoral regions. From the perspective of the climate, the watershed is more suitable for the growth and development of herbaceous plants than for trees. It is also notable that the area is not entirely appropriate for grassland, with 53.4% being classified as only marginally suitable. Besides soil water content and soil

Table 7. Areas of land suitable for growing crops in the Liudaogou watershed.

Grade	Area (km²)	Percentage (%)
S1	0	0
S2	0.40	6.7
S3	1.19	20.0
N	4.34	73.3

texture, total nitrogen may be a subsidiary factor limiting the suitability of this land for use as grassland – it is present at low levels in forest and grasslands (Table 11).

With respect to forestland suitability, assessment units 2 and 3 in the arable land were classified as being moderately suitable, assessment unit 1 in the arable land and units 13 and 14 in the grassland were considered marginally suitable, and other units in the forest and grasslands were unsuitable for use as forest. Overall, an area of 0.79 km², accounting for 13.3% of the total area of the watershed, was classified as moderately suitable for forestland and 1.19 km² (20.0%) as marginally suitable for forestland; the remaining 3.95 km², accounting for 66.7% of the watershed was unsuitable for forest (Table 12).

Like cropland, the main factors restricting the establishment of forest were soil water content, organic matter and total nitrogen. This means soil water and nutrition conditions were not suitable for widespread afforestation. The slope gradients of the arable and grassland assessment units that were suitable for forest were all below 10°, indicating that slope gradient may be a subsidiary limiting factor with respect to forests. Usually, steep slopes are associated with low soil water content, with increasing soil erosion and decreasing infiltration [40–42].

Implications for the Sloping Land Conversion Program

On the Loess Plateau, rapid population growth and an economic boom, coupled with severe soil erosion and desertification, have led to deterioration of the natural environment and a reduction in biodiversity [43,4]. The Sloping Land Conversion Program is, therefore, considered to be a necessary step for increasing vegetation cover and restoring ecosystem service functions, thus promoting sustainable environmental and economic development. However, during implementation of the SLCP, tree planting was overemphasized, resulting in inappropriate planting schemes [34]. Lü et al. [8] reported that the large areas converted from farmland to woodland have resulted in decreased regional water yield as the climate warms and dries on the Loess Plateau. Successful ecological rehabilitation programs have, thus, been largely dependent on innovative ecosystem management systems and technical support. Land suitability evaluation indicates that the Liudaogou watershed is suited to grass growth and is unsuitable for forest growth on the sloping land; this is due mainly to water and nitrogen deficits. In fact, there is still 1.79 km² (26% of the total area) of forestland in the watershed, with *Populus simonii* and *Pinus tabulaeformis* that were mainly planted in the late 1970s and after 1999 [27,30]. Hou [27] and Yang et al. [44] reported that the forest in the Liudaogou watershed could not function as a stable forest ecosystem. The height of the 20-year-old *Populus simonii* trees averaged about 4–6 m (the smallest just 2 m), the diameter at breast height (DBH) was about 5–6 cm, and the volume of timber was just 0.0031 m³ per individual. In general, the growth of *Populus simonii* stopped after about 15 years. Under normal circumstances, the height of a 20-year-old *Populus simonii* tree would be more than 20 m with DBH ≥15 cm. Due to limited cover and poor growth, the *Populus simonii* forest is very restricted in its ecological function. It was originally planted as a part of a shelterbelt in the "Three North" Protective Forest Program, aiming to combat desertification and soil erosion in the Northwest of China [45–46]. However, according to a survey conducted by Hou et al. [47], the wind erosion depth in the *Populus simonii* forestland amounts to 1.91–4.68 cm/a, corresponding to 1.9×10^4–4.7×10^4 m³ soil loss per year. Compared with the natural grass vegetation, it does not prevent wind erosion, but greatly accelerates it. The main reasons for the stunted tree growth are: (1) there is insufficient soil water available to maintain normal

Table 8. Land characteristics, thresholds and degree of suitability for forest and grasslands.

Land characteristics	Forest or grassland	Suitability High (S1)	Moderate (S2)	Marginal (S3)	Unsuitable (N)
Slope gradient (°)	Both	≤15	15~25	25~35	>35
Soil organic matter (g/kg)	Forest	≥7.97	5.5~7.97	3.08~5.5	<3.08
	Grass	≥5.41	3.50~5.41	1.99~3.5	<1.99
Total nitrogen (g/kg)	Forest	≥0.19	0.12~0.19	0.05~0.12	<0.05
	Grass	≥0.11	0.08~0.11	0.05~0.08	<0.05
Available phosphorus (mg/kg)	Forest	≥4.62	3.88~4.62	3.14~3.88	<3.14
	Grass	≥3.66	2.52~3.66	2.46~2.52	<2.46
Available potassium (mg/kg)	Forest	≥250	150~250	50~150	<50
	Grass	≥250	150~250	50~150	<50
Bulk density (g/cm³)	Both	-	-	-	>1.65
Soil texture (physical clay content) (%)	Both	15	-	-	>40 or <5
Soil water content (%)	Forest	≥15	11.5~15	8~11.5	<8
	Grass	≥4.78	4.14~4.78	3.50~4.14	<3.50

growth rates; (2) there is insufficient soil fertility (e.g. besides low total nitrogen, the Soil Organic Carbon Density is about 1.18–2.81 kg/m², significantly lower than the national mean values for soils of 11.52–12.04 kg/m²) [48]; and (3) there is insufficient management because of the very limited economic value of forest – local farmers are more willing to devote themselves to coal-mining which generates a much higher income.

Although our forestland suitability assessment indicated that about 33.3% of total area is suitable for forest growth (Table 12), such areas are currently arable land and some grassland with gentle gradients (Table 4). Compared to tree species, crop and grass species have shallow roots and consume relatively little water [30], so do not deplete soil water to any great depth; however, the

current arable lands are dammed, terraced fields that collect valuable rainfall because of the low soil erosion rates. Therefore, those arable lands and grasslands with gentle gradients were classified as being moderately or marginally suitable for forest development. We predict that planting trees in these areas would cause soil water depletion because of high water consumption combined with the low precipitation rates in the watershed.

The grassland suitability assessment indicated that about 86.7% of the area of the watershed is suitable for grass growth, thus the SLCP in the watershed should focus on conversion of cropland to grassland. Wang et al. [17] reported that soil physical properties such as bulk density, hydraulic conductivity, mean weight diameter, and the stability of >1 mm macro-aggregates have been significantly ameliorated in the 0–20 cm soil layer under secondary natural grasslands, implying that natural grass (*Stipa bungeana* Trin.) restoration is an appropriate and sensible approach to re-vegetation in the wind–water erosion region of the northern Loess Plateau of China. Planting alfalfa (*Medicago sativa*) and korshinsk peashrub (*Caragana korshinskii*) has no effect on the soil physical conditions, and may even reduce bulk density and soil permeability because of their very high water consumption with deep and vigorous root systems. Therefore, care is required when selecting species for re-vegetation to promote shifts from arable land to grasslands in a more environmentally compatible manner. Non-native grassland species, such as the korshinsk peashrub and alfalfa, have high water requirements and should be avoided

Table 9. Grassland suitability and limiting factors for each assessment unit.

Assessment unit	Combined weight	Grade	Limiting factors
1	0.984	S2	-
2	0.995	S1	-
3	0.989	S2	-
4	0.926	S3	-
5	0.965	S1	-
6	0.821	N	Soil water
7	0.751	S3	-
8	0.538	S3	-
9	0.683	S3	-
10	0.441	S3	-
11	0.865	N	Soil texture
12	0.945	S3	-
13	0.948	S2	-
14	0.948	S3	-
15	0.632	S3	-

Table 10. Areas suitable for use as grassland in the Liudaogou watershed.

Grade	Area (km²)	Percentage (%)
S1	0.79	13.3
S2	1.18	20.0
S3	3.16	53.4
N	0.79	13.3

Table 11. Forestland suitability and limiting factors for each assessment unit.

Assessment unit	Combined weight	Grade	Limiting factors
1	0.900	S3	-
2	0.952	S2	-
3	0.945	S2	-
4	0.806	N	-
5	0.820	N	-
6	0.756	N	Soil water
7	0.423	N	Organic matter and soil water
8	0.249	N	Organic matter and soil water
9	0.507	N	Organic matter, soil water and total nitrogen
10	0.458	N	Organic matter, soil water and total nitrogen
11	0.708	N	Soil water
12	0.603	N	Organic matter and soil texture
13	0.865	S3	-
14	0.658	S3	-
15	0.556	N	Organic matter and total nitrogen

Table 12. Areas of land suitable for forest growth in the Liudaogou watershed.

Grade	Area (km²)	Percentage (%)
S1	0	0
S2	0.79	13.3
S3	1.18	20.0
N	3.95	66.7

when choosing species to plant in the watershed.

Conclusion

Land use management involves complex decision-making that requires an understanding of many factors. This paper presents the AHP as a decision support tool for use when selecting an appropriate land use, which is an important issue for the sustainable development of the Liudaogou watershed. Based on the AHP, grassland was found to be the most suitable land-use type to support the aims of the Sloping Land Conversion Program in the watershed. Under the constraints of soil water shortage and nutrient deficit, crops and forests were considered to be

inappropriate land uses in the study area, especially on sloping land. All forest should be converted to grassland because continued growth of trees will damage the soil water environment and increase desertification problems. When selecting species for re-vegetation, non-native grass species with high water requirements should be avoided so as to guarantee the sustainable development of grassland and effective ecological functioning. In the future, the area of abandoned cropland is likely to increase rapidly due to the government policy embodied in the Sloping Land Conversion Program. Our study provides local land managers and farmers with valuable information about the inappropriateness of growing trees in the study area along with some information on species selection for planting in the semi-arid area of the Loess Plateau.

Acknowledgments

We wish to thank graduate students Wang Jianguo, Shi Zhanfei and Wang Mei for their help in field survey and data analysis in the Laboratory. We also wish to thank Dr. David Warrington for his invaluable comments and suggestions on this paper.

Author Contributions

Conceived and designed the experiments: LW XBY. Performed the experiments: XBY LW. Analyzed the data: XBY. Contributed reagents/materials/analysis tools: XBY LW. Wrote the paper: LW XBY.

References

1. World Bank (2001) China: air, land and water, environmental priorities for a New Millennium. World Bank, Washington, DC.
2. SFA (2003) Sloping land conversion program plan. (2001–2010) (in Chinese).
3. Michael T, Bennett MT (2008) China's sloping land conversion program: Institutional innovation or business as usual? Ecological economics 65: 699–711.
4. Wang L, Shao MA, Wang QJ, Gale WJ (2006) Historical changes in the environment of the Chinese Loess Plateau. Environmental science & policy 9: 675–684.
5. Ding CR (2003) Land policy reform in China: assessment and prospects. Land Use Policy 20: 109–120.
6. Uchida E, Xu JT, Rozelle S (2005) Grain for Green: cost-effectiveness and sustainability of China's conservation set-aside programme. Land Economics 81: 247–264.
7. Wang L, Wang QJ, Wei SP, Shao M, Li Y (2008) Soil desiccation for Loess soils on natural and regrown areas. Forest Ecology and Management 255(7): 2467–2477.
8. Lü Y, Fu B, Feng X, Zeng Y, Liu Y, et al. (2012) A policy-driven large scale ecological restoration: quantifying ecosystem services changes in the Loess Plateau of China. PLoS ONE 7(2): e31782. doi:10.1371/journal.pone.0031782.
9. Wei J, Cheng J, Li W, Liu W (2012) Comparing the effect of naturally restored forest and grassland on carbon sequestration and its vertical distribution in the Chinese Loess Plateau. PLoS ONE 7(7): e40123. doi:10.1371/journal.pone.0040123.
10. Yin RS, Zhao MJ (2012) Ecological restoration programs and payments for ecosystem services as integrated biophysical and socioeconomic processes—China's experience as an example. Ecological Economics 73: 56–65.

11. Wang XH, Lu CH, Fang JF, Shen YC (2007) Implications for development of grain-for-green policy based on cropland suitability evaluation in desertification-affected north China. Land Use Policy 24: 417–424.

12. McVicar TR, Li LT, Van Nie TG, Zhang L, Li R, et al. (2007) Developing a decision support tool for China's revegetation program: Simulating regional impacts of afforestation on average annual streamflow in the Loess Plateau. Forest Ecology and Management 251: 65–81.

13. Yang RJ, Fu BJ, Liu GH, Ma KM (2004) Research on the relationship between water and eco-environment construction in Loess Hilly and Gully Region. Chinese Journal of Environmental Science 25(2): 37–42 (in Chinese with an English abstract).

14. Yang WZ, Tian JL (2004) Essential exploration of soil aridization in the Loess Plateau. Acta Pedologica Sinica 41(1): 1–6 (in Chinese with English abstract).

15. McVicar TR, Van Niel TG, Li LT, Wen ZM, Yang QK, et al. (2010) Parsimoniously modelling perennial vegetation suitability and identifying priority areas to support China's re-vegetation program in the Loess Plateau: Matching model complexity to data availability. Forest Ecology and Management 259: 1277–1290.

16. Wang L, Wei SP, Horton R, Shao M A (2011) Effects of vegetation and slope aspect on water budget in the hill and gully region of the Loess Plateau of China. Catena 87: 90–100.

17. Wang L, Wei SP, Shao HB, Wu YJ, Wang QJ (2012) Simulated water balance of forest and farmland in the hill and gully region of the Loess Plateau of China. Plant Biosystems, dx.doi.org/10.1080/11263504.2012.709198.

18. Saaty TL (1977) A scaling method for priorities in hierarchical structures. Journal of Mathematical Psychology 15: 234–281.

19. Saaty TL (1980) The analytical hierarchy process. McGraw Hill, New York.

20. Kurttila M, Pesonen M, Kangas J, Kajanus M (2000) Utilizing the analytic hierarchy process (AHP) in SWOT analysis-a hybrid method and its application to a forest-certification case. Forest Policy and Economics 1 (1): 41–52.

21. Ayalew L, Yamagishi H, Marui H, Kanno T (2005) Landslides in Sado Island of Japan: Part II. GIS-based susceptibility mapping with comparisons of results from two methods and verifications. Engineering Geology 81: 432–445.

22. Cay T, Uyan M (2012) Evaluation of reallocation criteria in land consolidation studies using the Analytic Hierarchy Process (AHP). Land Use Policy 30: 541–548.

23. Mardle S, Pascoe S (2004) Management objective importance in fisheries: an evaluation using the Analytic Hierarchy Process (AHP). Environmental Management 33(1):1–11.

24. Pascoe S, Dichmont CM, Brooks K, Pears R, Jebreen E (2013) Management objectives of Queensland fisheries: Putting the horse before the cart. Marine Policy 37: 115–122.

25. Li M, Li ZB, Liu PL, Yao WY (2005) Using cesium-137 technique to study the characteristics of different aspect of soil erosion in the Wind–Water Erosion Crisscross Region on Loess Plateau of China. Applied Radiation and Isotopes 62: 109–113.

26. Wang L, Mu Y, Zhang QF, Jia Z K (2012) Effects of vegetation restoration on soil physical properties in the wind-water erosion region of the northern Loess Plateau of China. CLEAN - Soil, Air, Water 40(1): 7–15.

27. Hou QC (1994) Comprehensive analysis on natural conditions and environmental harnessing in the experimental Area. Memoir of NISWC, Academia Sinica and Ministry of Water Resources, 18: 136–143 (in Chinese with English abstract).

28. Fahey B, Jackson R (1997) Hydrological impacts of converting native forests and grasslands to pine plantations, South Island, New Zealand. Agricultural and Forest Meteorology 84: 69–82.

29. Chakraborty S, Banik D (2006) Design of a material handling equipment selection model using analytic hierarchy process. The International Journal of Advanced Manufacturing Technology 28: 1237–1245.

30. Wang L, Zhang QF, Wei SP, Wang QJ (2009) Vegetation restoration model in a watershed of a coal mining area in the water and wind erosion crossing zone of the Loess Plateau. Journal of Beijing Forestry University 31 (2): 36–43.

31. Bremmer JM, Mulvaney CS (1982) Nitrogen—total. In: Page AL, Miller RH, Keeney DR (Eds.), Methods of soil analysis, Part 2—Chemical and microbiological properties. ASA-SSSA, Madison, WI, pp. 595–624.

32. Olsen SR, Sommers LE (1982) Phosphorous. In: Page AL, Miller RH, Keeney DR (eds) Methods of soil analysis Part 2, Chemical and microbial properties. Agronomy Society of America, Agronomy Monograph 9, Madison, Wisconsin, pp 403–430.

33. Pratt PF (1965) Potassium. In: Black, C.A. (Ed.), Methods of Soil Analysis. Part 2. Chemical and Microbiological Properties. Am. Soc. of Agron, Inc., Madison, p. 1022–1030.

34. Wang XL, Guo SL, Ma YH, Huang DY, Wu JS (2007) Effects of land use type on soil organic C and total N in a small watershed in loess hilly-gully region. Chinese Journal of Applied Ecology 18 (6): 1281–1285 (in Chinese with English abstract).

35. Bai WJ, Jiao JY, Ma XH, Jiao F (2005) Soil environmental effects of artificial woods in abandoned croplands in the Loess Hilly-gullied region. Journal of Arid Land Resources and Environment 19(7): 135–141 (in Chinese with English abstract).

36. Jiang N, Shao MA, Lei TW (2007) Soil water characteristics of different typical land use patterns in Water–Wind Erosion Interlaced Region. Journal of Beijing Forestry University 29: 134–137 (in Chinese with English abstract).

37. Huo Z, Shao MA, Horton R (2008) Impact of gully on soil moisture of shrubland in Wind–Water Erosion Crisscross Region of the Loess Plateau. Pedosphere 18: 674–680.

38. Reid JB, Goss MJ (1982) Suppression of decomposition of C-labeled plant roots in the presence of living roots of maize and perennial ryegrass. Journal of Soil Science 33: 387–395.

39. Grant CD, Dexter AR (1987) Generation of microcracks in moulded soils by rapid wetting. Australian Journal of Soil Research 27(1): 169–182.

40. Tang KL, Zhang KL, Lei AL (1998) Research on up-limit slope of "grain for green" in the plowlands of Loess Hilly and Gully Region. Chinese Science Bulletin 43(2): 200–203 (in Chinese).

41. Luk SH, Cai Q, Wang GP (1993) Effects of surface crusting and slope gradient on soil and water losses in the hilly Loess region, North China. Catena 24: 29–45.

42. Ng CWW, Shi Q (1997) A numerical investigation of the stability of unsaturated soil slopes subjected to transient seepage. Computers and Geotechnics 22 (1): 1–28.

43. Li WH (2004) Degradation and restoration of forest ecosystems in China. Forest Ecology and Management 201: 33–41.

44. Yang G, Song YX (1993) Analysis on the current situation of forestry and forestry developing strategy in Liudaogou Watershed of Shenmu Experimental Area. Memoir of Northwestern Institute of Soil and Water Conservation, 18: 106–112.

45. Fang JY, Chen AP, Peng CH, Zhao SQ, Ci LJ (2001) Changes in forest biomass carbon storage in China between 1949 and 1998. Science 292: 2320–2322.

46. Wang XM, Zhang CX, Hasi E, Dong ZB (2010) Has the Three Norths Forest Shelterbelt Program solved the desertification and dust storm problems in arid and semiarid China? Journal of Arid Environments 74(1): 13–22.

47. Hou QC, Wang YK, Yang G (1996) Several problems of vegetation construction in criss2cross belt of bind2water erosion. Bulletin of Soil and Water Conservation 16 (5): 36–40.

48. Li YY, Shao MA, Zheng JY, Li QF (2007) Impact of grassland recovery and reconstruction on soil organic carbon in the northern Loess Plateau. Acta Ecologica Sinica 27 (6): 1–9.

4

Climate and Land Use Controls on Soil Organic Carbon in the Loess Plateau Region of China

Yaai Dang[1,2,3 ♪], Wei Ren[2 ♪], Bo Tao[2], Guangsheng Chen[2], Chaoqun Lu[2], Jia Yang[2], Shufen Pan[2], Guodong Wang[3], Shiqing Li[1], Hanqin Tian[2]*

1 State Key Laboratory of Soil Erosion and Dryland Farming on the Loess Plateau, Institute of Soil and Water Conservation, Northwest A&F University, Yangling, Shaanxi, China, 2 International Center for Climate and Global Change Research, School of Forestry & Wildlife Sciences, Auburn University, Auburn, Alabama, United States of America, 3 College of Science, Northwest A&F University, Yangling, Shaanxi, China

Abstract

The Loess Plateau of China has the highest soil erosion rate in the world where billion tons of soil is annually washed into Yellow River. In recent decades this region has experienced significant climate change and policy-driven land conversion. However, it has not yet been well investigated how these changes in climate and land use have affected soil organic carbon (SOC) storage on the Loess Plateau. By using the Dynamic Land Ecosystem Model (DLEM), we quantified the effects of climate and land use on SOC storage on the Loess Plateau in the context of multiple environmental factors during the period of 1961–2005. Our results show that SOC storage increased by 0.27 Pg C on the Loess Plateau as a result of multiple environmental factors during the study period. About 55% (0.14 Pg C) of the SOC increase was caused by land conversion from cropland to grassland/forest owing to the government efforts to reduce soil erosion and improve the ecological conditions in the region. Historical climate change reduced SOC by 0.05 Pg C (approximately 19% of the total change) primarily due to a significant climate warming and a slight reduction in precipitation. Our results imply that the implementation of "Grain for Green" policy may effectively enhance regional soil carbon storage and hence starve off further soil erosion on the Loess Plateau.

Editor: Xiujun Wang, University of Maryland, United States of America

Funding: This study was supported by NASA Land Cover and Land Use Change Program (NNX08AL73G), NASA Interdisciplinary Science Program (NNG04GM39C), Chinese Universities Scientific Fund (z10921007), US National Science Foundation Grants (AGS-1243220, CNS-1059376), State Key Laboratory of Soil Erosion and Dryland Farming on the Loess Plateau Foundation (K318009902-1410), and Shaanxi Administration of Foreign Expert Affairs Science and Technology Activities Fundation (201327). Study design was supported by NASA Land Cover and Land Use Change Program (NNX08AL73G), NASA Interdisciplinary Science Program (NNG04GM39C) and Chinese Universities Scientific Fund (z10921007). Data collection and analysis was supported by NASA Land Cover and Land Use Change Program (NNX08AL73G), NASA Interdisciplinary Science Program (NNG04GM39C), and State Key Laboratory of Soil Erosion and Dryland Farming on the Loess Plateau Foundation (K318009902-1410). Preparation of the manuscript was supported by NASA Land Cover and Land Use Change Program (NNX08AL73G), US National Science Foundation Grants (AGS-1243220, CNS-1059376), and State Key Laboratory of Soil Erosion and Dryland Farming on the Loess Plateau Foundation (K318009902-1410).

Competing Interests: The authors have declared that no competing interests exist.

* E-mail: tianhan@auburn.edu

♪ These authors contributed equally to this work.

Introduction

Soil organic carbon (SOC), the major component of soil organic matter, plays a key role in the terrestrial carbon cycle and thus has drawn great attention from scientific community. It is a dynamic component of terrestrial systems, affecting carbon exchange between terrestrial ecosystem and the atmosphere [1,2]. SOC storage is nearly three times as large as carbon storage in vegetation and twice as large as global atmospheric carbon storage [3]. Soil has higher potential to sequester more carbon (such as converting the type of land use) in the future [2,4], therefore, increasing soil carbon storage is one of the most economical and effective ways to alleviate the greenhouse effect, which has become a hot scientific and political issue during the past decades.

Changes in climate and land use, caused by both natural and anthropogenic processes, have greatly influenced the terrestrial carbon balance during the past decades [5,6,7,8]. It was reported that about one fourth of anthropogenic CO_2 emissions were due to land cover and land use change (LCLUC), especially deforestation

[9]. Long-term experimental studies have confirmed that SOC is highly sensitive to land conversion from natural ecosystems, such as forest or grassland, to agricultural land, resulting in substantial SOC loss [6,10]. In addition, LCLUC may also cause carbon depletion by influencing soil respiration [11]. It was estimated that global carbon release from SOC mineralization owing to agricultural activities was approximately 0.80 Pg C/year (1 Pg = 10^{15} g) [12]. Globally, land use change resulted in a carbon release of (1.6±0.8) Pg C per year to the atmosphere during the period of 1990s [13]. However, the effects of conversions from cropland to grassland/forest on the SOC storage have not been fully understood and there still remains large uncertainty.

The Loess Plateau of China (Figure 1), located in the geographic center of China (33°43'N 100°54'E to 41°16'N 114°33'E), covers a total area of 628,000 km^2, which is about 6.5% of China's total land area. The Loess Plateau is characterized by highly erodible soils, steep slopes, being subjected to heavy rain, and low vegetation coverage due to excess exploitation of land resource and improper land use [4,14]. During the past decades, serious soil erosions caused by natural and anthropogenic disturbances (e.g., climate change, natural disasters, LCLUC etc.) occurred in the area of the Loess Plateau. Previous reports also indicated that the warming and drying climate in this region has significantly aggravated soil erosion [15]. As a result, the large amount of fine surface soil eroded from the loess area is transported into the Yellow River and acts as the main source of sediment of this river, which runs through the Loess Plateau and is considered to be the most turbid river in the world. Due to these disturbances, soil carbon storage on the Loess Plateau is much lower compared to other regions in China [16]. In general, adjusting the land use pattern so as to restore the degraded ecosystems and to modify the local rural income structure is regarded as the main measures to control soil and water erosion and to improve farmers' living conditions on the Loess Plateau. Since the 1950s, a series of conservation policies have been implemented in this region, such as extensive tree planting since the 1970s, integrated soil erosion controls on the watershed scale in the 1980s and the 1990s [17,18], and the government-funded project "Grain for Green" in 1999, aiming at transforming the low-yield slope cropland into grassland/forest. The implementation of these policies improved vegetation coverage, altered land use patterns, and changed the SOC storage. Although many field experiments have been performed to explore the impacts of both drying and warming climate and LCLUC on soil carbon storage on the Loess Plateau [19,20], little attention has been paid to the regional impacts of these factors and their interactions.

Over recent decades, many field observations and control experiments have been conducted to explore the effects of climate and land use change on the SOC in this region and make it possible to study the regional effects of climate change and LCLUC on SOC. In addition, many approaches, including eddy covariance flux tower, inventory, remote sensing techniques, forward and inversion models, have been used to examine the regional carbon budget on the Loess Plateau [21,22,23,24,25]. Among them, process-based ecosystem modeling is one of the most effective approaches to estimate regional SOC storage and fluxes in different terrestrial ecosystems driven by multiple global changes factors [7,11,24,26,27]. To address the effects of changes in climate, land use, and other environmental factors on SOC storage in this region, the Dynamic Land Ecosystem Model (DLEM), a highly integrated process-based model [28], was

applied to evaluate the spatial and temporal patterns of SOC storage on the Loess Plateau during 1961–2005. The objectives of this study are: 1) to investigate the temporal and spatial patterns of SOC storage on the Loess Plateau; and 2) to identify the relative contribution of climate and land use changes to the SOC storage changes.

Methods

Model Description

The DLEM is a process-based terrestrial ecosystem model, which aims at simulating the impacts of natural and anthropogenic disturbances on the structure and functions of terrestrial ecosystems over the spatial and temporal contexts. The DLEM has been widely used to simulate the effects of climate variability and change, elevated atmospheric CO_2, tropospheric ozone pollution, land use change, and increasing nitrogen deposition, etc. on terrestrial carbon storage and fluxes in China and other regions across the globe [15,29,30,31,32,33].

In this study, DLEM simulates two kinds of LCLUC: land conversion from natural ecosystem to cropland, and cropland abandonment. In the DLEM model, the balance of soil organic matter depends on the transformation of litter (LIT) to soil organic matter, the fractions of conversion from gross primary production (GPP) to dissolved organic carbon (DOC), the returned organic matter from production decay (PRD) (e.g., manure), the growth of microbe, the methane production from dissolved organic carbon, and the carbon loss from soil organic matter (SOM) decomposition.

$$\frac{dC_{som}}{dt} = k_{tr}LITC_{loss} + k_{gppdoc}GPP + k_{prd}PRD_{docom} - k_{rh}SOMC_{docom} - k_{Lucc}C_{som} - DOC_{loss,methane}$$

where k_{tr} is the transfer rate of decomposed LIT to SOM; k_{gppdoc} is the fraction of GPP converted to soil DOC; k_{prd} is the returned rate of decomposed (or consumed) PRD to SOM pools as manure; k_{rh} is the fraction of decomposed SOM that is converted to CO_2 through heterotrophic respiration; k_{Lucc} is coefficient for quick carbon loss from SOM due to land use conversion; $DOC_{loss,methane}$ is DOC consumed for the growth of production of methane. More

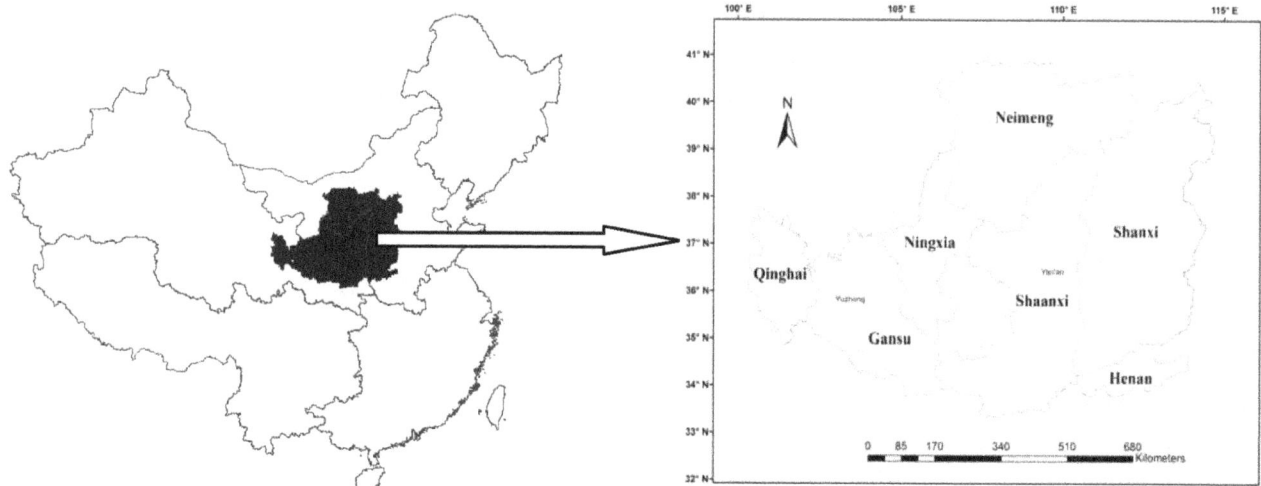

Figure 1. Location of the Loess Plateau, China.

detailed processes were described in our previous papers [15,29,30].

Input Data Description

The major input data in the DLEM include: (1) daily climatic data (i.e. maximum, minimum, average temperature, precipitation, relative humidity, and radiation) and atmospheric chemistry data (i.e. tropospheric O_3, atmospheric CO_2 and nitrogen deposition); (2) soil properties (including soil type, bulk density, depth, pH, soil texture) which are derived from the 1:1 million soil map based on the Second National Soil Survey of China [34,35,36]; (3) contemporary vegetation map for 2000 which was developed from Landsat Enhanced Thematic Mapper (ETM) imagery [37]; (4) long-term land use history which was developed on the basis of three recent (1990, 1995 and 2000) land cover maps and historical census datasets [38,39]. All the input datasets were developed at the spatial resolution of 10 km×0 km. Detailed information about other input data were described in our previous studies [15,29,30,40].

Climate change. Average air temperature on the Loess Plateau increased at a rate of 0.030°C/year from 1961 to 2005 (Figure 2a), higher than those reported for the entire China (about 0.029°C/year) and the global average level (about 0.010°C/year) [41] in the same period. The most rapid increase in temperature occurred during the 1990s. Air temperature increased from the north to the south of the Loess Plateau (Figure 2b), with the greatest increase in the northern Loess Plateau (e.g., the north of Shanxi and the northwest of Inner Mongolia).

The Loess Plateau can be mainly divided into three climate zones according to the precipitation: the northern Loess Plateau with precipitation below 400 mm, the central Loess Plateau with precipitation between 400 and 500 mm, and the southern Loess Plateau with precipitation above 500 mm [42,43]. The precipitation less than 550 mm/year occurred across most areas of the Loess Plateau. Over the past 45 years, the mean annual precipitation was approximately 423 mm, with the lowest of 288 mm in 1997 and the highest of 661 mm in 1964 (Figure 2c). A slightly decreasing trend at a rate of 1.27 mm/year in precipitation was found on the Loess Plateau from 1961 to 2005. This temporal trend of precipitation was consistent with a previous study based on the meteorological observations according to 99 stations for the period 1956–2005 on the Loess Plateau [44]. Figure 2d further indicated that decreases in precipitation occurred in the most areas of the central and southern Loess Plateau from 1961–1990 to 1991–2005. The regions with the most obvious drying trend were located in the central and southern Loess Plateau, especially in the north of Shaanxi and the center of Shanxi Province with a reduction of more than 80 mm in the recent 15 years (1991–2005) compared to the 1961–1990 average.

LCLUC. Expansion of cropland and pasture was driven by social-economic factors on the Loess Plateau during the past decades, though to some extent, the soil erosion and frequent drought events limited the massive expansion of cropland. Since the 1950s, many measures have been implemented to alleviate and control serious soil erosion on the Loess Plateau. For example, "Grain for Green" project, which was launched in 1999, recommended that cropland with slopes greater than 15° should be converted back to natural vegetation. Driven by such governmental policies, the Loess Plateau experienced a remarkable change in land use, characterizing by a large area of

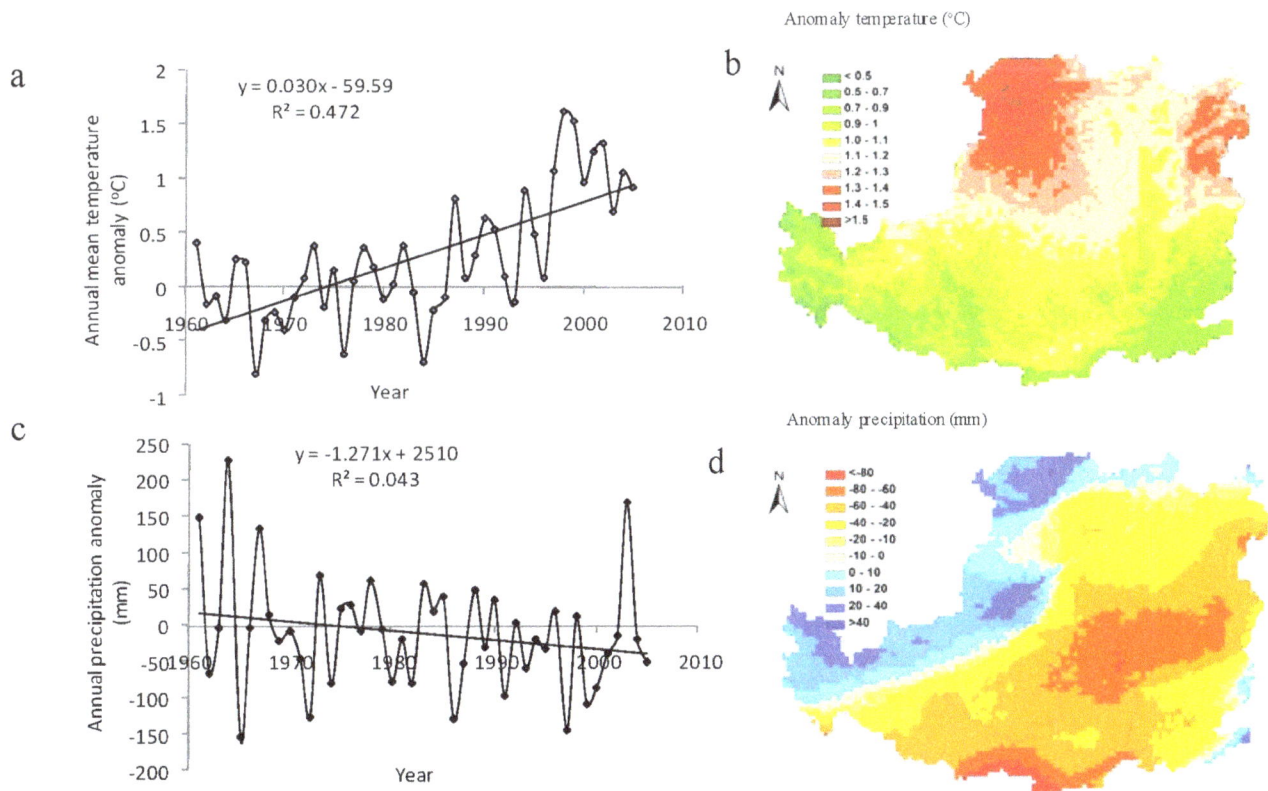

Figure 2. Anomaly of annual mean temperature (a), and precipitation (c) on the Loess Plateau from 1961 to 2005; and spatial distribution of temperature anomaly (b) and precipitation anomaly (d) during 1991–2005 (relative to 1961–1990 average);

conversion from cropland to grassland/forest. In order to well understand the LCLUC history on the Loess Plateau, we analyzed the spatial and temporal variations from 1961 to 2005 (Figure 3–4). Figure 3 showed that cropland area on the Loess Plateau decreased slowly until 1999, and followed by a sharp decrease. However, the grassland area showed an opposite changing trend in the same periods. Since 1961, the area of cropland decreased by 19.61%, while grassland and forest increased by 7.31% and 6.75%, respectively.

Land use patterns exhibited large spatial variations on the Loess Plateau (Figure 4). Grassland was the dominant vegetation type in the northern Loess Plateau. Cropland, grassland and shrubland were the main vegetation type in the middle Loess Plateau. In contrast, cropland and forest occupied large area of the southern Loess Plateau. Compared to 1960, the coverage of cropland decreased mainly due to land conversion from cropland to grassland, especially in the middle and southern part of Loess Plateau. During 1999–2005, cropland largely shrank in some areas of the middle Loess Plateau, especially in the northern Shaanxi and middle of Shanxi Province. These results were consistent with previous studies [18,45].

Model Parameterization and Evaluation

DLEM has been well calibrated and intensively validated against the site-level observed carbon fluxes and pool sizes from Chinese Ecosystem Research Network (CERN) and other previous studies [29,38,46]. In this study, we further compared our simulated results with field observation data and survey data from the Second National Soil Survey (1979–1983) to evaluate the model performance in simulating the SOC storage on the Loess Plateau. The simulated SOC storage as influenced by multiple environmental factors significantly correlated with the observed data ($R^2 = 0.738$, $p<0.01$, Figure 5), indicating that DLEM could capture the spatial and temporal patterns of SOC storage on the Loess Plateau. Several factors may contribute to the difference between model results and field observations. First, the input datasets were developed at a spatial resolution of $10 \text{ km} \times 10 \text{ km}$ for driving DLEM simulations. Each grid was assumed to have the uniform climate, land use type, and vegetation cover in the model. As a well-known climate-sensitive zone and a fragile ecological belt, some subtle changes in climate, land use or other factors on the Loess Plateau might cause large difference in the SOC storage even within one grid cell. This difference in the same grid will be reflected in the field experiments but neglected in the model

simulations. Second, the shortage of field observation data may weaken the capability of model to realistically capture the magnitude and patterns in SOC changes, which has long been identified as one of the biases in the large-scale model development. In addition, model simplification and neglecting microbial biomass might be another potential reason [13,30].

Simulation Experimental Design and Model Run

In this study, four main simulation experiments were designed to analyze the effects of climate change alone, LCLUC alone, the interaction of climate and LCLUC, and the combined effects of all environmental factors on the SOC storage on the Loess Plateau (Table 1). The experiment I was designed to provide a 'best-estimate' of spatial and temporal patterns of SOC driven by major environmental factors changes including climate variability, LCLUC, elevated CO_2, N fertilizer, N deposition, and O_3 pollution, etc. In experiment II and III, we simulated the contributions of climate variability alone and the LCLUC alone, respectively. In the experiment IV, we tried to understand the interactive effect between climate change and LCLUC on the SOC storage. The model simulations began with an equilibrium run to obtain the baseline carbon pools for each grid. A spin-up of about 100 years was applied if the climate change was included in the simulation experiments. Finally, the model was run in transient mode driven by the daily climate data and other time-variant or invariant input data.

Results and Analysis

Temporal Changes in SOC on the Loess Plateau

Model simulation indicated that the SOC storage over the entire Loess Plateau displayed substantial temporal and spatial variations in the context of multiple environmental factors changes (Figure 6a, b). As a whole, the combination of all these environmental factors considered in this study (e.g. climate, LCLUC and others environment factors) caused a net increase of about 0.27 Pg C in SOC storage from 1961 to 2005. The SOC storage kept relatively stable before the 1970s, and then gradually increased in the following two decades, and rapidly increased since 2000 (Figure 6a). Our further analysis found that the decrease of cropland area was relatively slower from the 1960s to 1999, and became rapid after then. Most of the abandoned cropland were replaced by grassland (Figure 3). Our results implied that temporal pattern in SOC change was partly related to the land use change on the Loess Plateau.

Spatial Variation in SOC on the Loess Plateau

We found that the spatial patterns of SOC storage were primarily controlled by the precipitation distribution. Spatially, the SOC storage increased gradually from the north to the south along an increasing precipitation gradient on the Loess Plateau. Large SOC increases were found in the southern and central Loess Plateau (south of Shaanxi and Shanxi Province in particular); and some other areas show a slight increase of SOC storage in the past decades. Due to changes in multiple environmental factors, the SOC storage increased throughout the majority of the Loess Plateau over the past 45 years (Figure 6b). Some areas showed a significant increase in SOC storage of more than 200 g C/m^2. However, a significant decrease occurred in the middle of central Loess Plateau (especially in the Northern Shaanxi Province), where experienced the most obvious warming and drying tendency in the past decades (Figure 2 b,d), releasing more than 100 g C/m^2 during the study period. We further found that the SOC increase

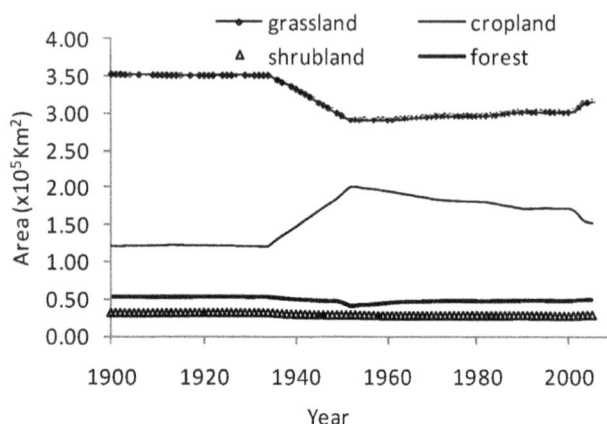

Figure 3. Area of major land use cover on the Loess Plateau during 1900–2005.

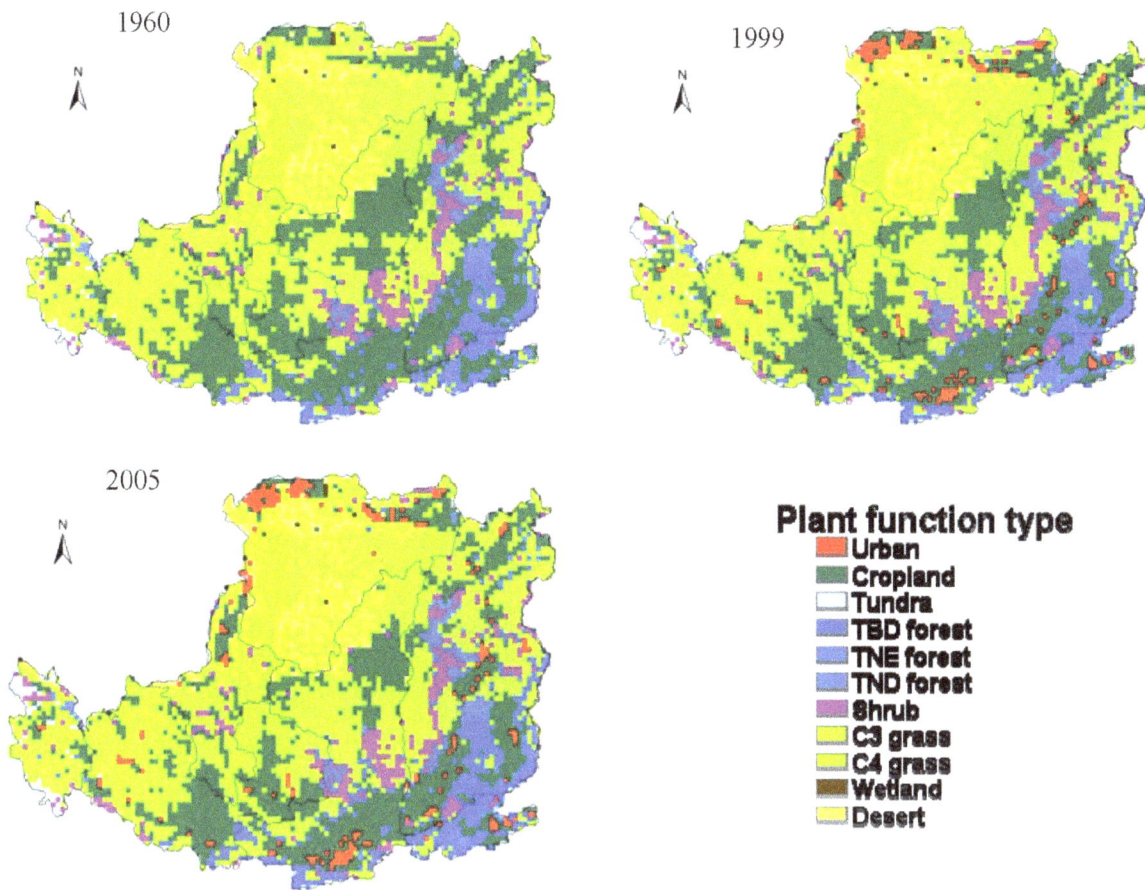

Figure 4. Spatial variations in LCLUC on the Loess Plateau in different year.

in the northern Loess Plateau was lower than that in the southern Loess Plateau during the past decades.

Individual Factorial Contributions to Changes in SOC Storage

To well understand the influence of climate on the SOC storage, we simulated the SOC storage on the Loess Plateau under the influence of climate change alone. DLEM simulated results

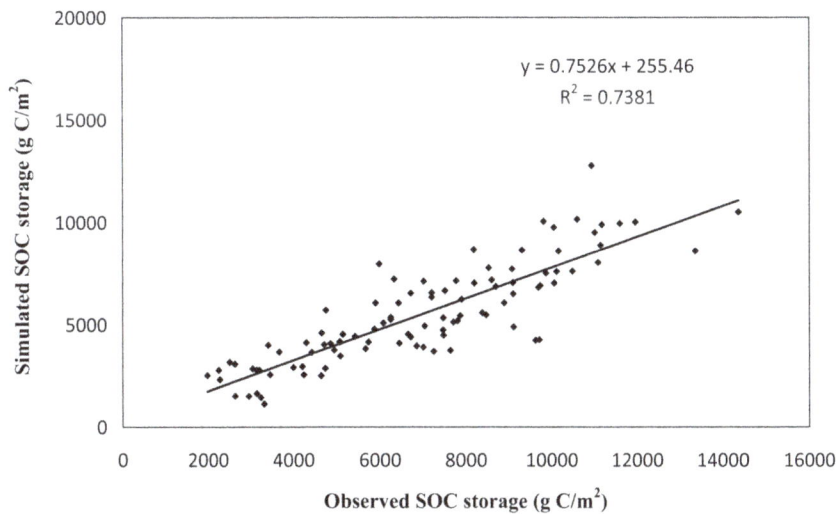

Figure 5. Correlation between simulated and observed SOC storage based on 91 soil samples.

Table 1. Simulation experiments.

Simulation Experiment	Environmental factors		
	Climate	LCLUC	Others
I. All	1960–2005	1960–2005	1960–2005
II. Climate only	1960–2005	1960	1960
III. LCLUC only	1960	1960–2005	1960
IV. Climate-LCLUC	1960–2005	1960–2005	1960

Notes: Climate-LCLUC means the combination effects of climate and LCLUC. Others include the effects of atmospheric CO_2, ozone pollution (AOT 40 index), nitrogen deposition, nitrogen fertilizer application, etc. All means the simulation experiment which includes all above environmental factors.

showed that SOC storage decreased by about 0.05 Pg C with substantial inter-annual fluctuations throughout the Loess Plateau from 1961 to 2005 (Figure 6c). SOC storage showed a slowly decreasing trend before the late 1990s, followed by a sharp decrease until the early 21st century. Since 1961, SOC storage decreased in most areas of the Loess Plateau under the influences of the climate, with a maximum carbon release of 200 g C/m^2 in some areas of the southern and the central Loess Plateau (Figure 6d). Figure 6d also indicated that SOC storage decreased in the southeastern humid monsoon climatic regions, and to a lesser extent in the continental dry climatic regions in the northern Loess Plateau during 1961–2005.

Considering the single effect of LCLUC, our results showed that the SOC storage increased by 0.14 Pg C during 1961–2005 (Figure 6e). We found that the most rapid increase in the SOC storage occurred after the late 1970s (Figure 6e). DLEM simulation results also showed that obvious increases of SOC storage were found in most areas of the central and the southern Loess Plateau (Figure 6f).

Relative Contributions of Climate, LCLUC, and their Interactions

Our simulated results indicated that with multiple environmental changes, SOC continuously increased from 1961 to 2005 on the Loess Plateau, resulting in a net increase in SOC storage by 0.27 Pg C (5.97 Tg C/year) (Figure 7). Among these factors, LCLUC was obviously the major factor affecting the magnitude of SOC change on the Loess Plateau, leading to a significant increase (0.14 Pg C) in SOC storage, accounting for 55% of the net increase in SOC in the past 45 years. However, in the same period, warming and drying climate greatly reduced SOC storage by 0.05 Pg C, approximately 19% of the total SOC storage change. The interaction between climate and LCLUC contributed to the net SOC increases by 3%.

Discussion

SOC Storage Change on the Loess Plateau

In this study, the combination of all environmental factors (climate, LCLUC and others environment factors) caused a net increase of about 0.27 Pg C in SOC storage, indicating that soil acted as a weak carbon sink on the Loess Plateau during the past 45 year. The temporal pattern of SOC storage was largely influenced by land use change and relevant land use policies.

Except the central Loess Plateau, the SOC storage increased throughout the entire region, with a smaller sink in the northern Loess Plateau and a larger sink in the southern Loess Plateau during 1961–2005, which is consistent with previous reports [43,47,48]. The less increase of SOC storage in the northern Loess

Plateau might be due to more sandy soil, lower soil carbon input from plant biomass, and larger soil carbon loss from erosion. This is consistent with previous reports, indicating the soil in the northern Loess Plateau contained more sand which could accumulate less carbon than the soil with more clay [49,50,51,52]. On the other hand, the area of forest and shrubland, characterized by higher aboveground biomass and productivity than cropland and grassland [43,53], decreased from the south to north over the Loess Plateau (Figure 4). This also contributed to realtively less SOC increase in the northern Loess Plateau. In addition, the extensive soil erosions in the northern Loess Plateau was suggested to further explain the lower contents of SOC storage [2,17,54].

Comparisons with Field Observations

Our estimation of SOC storage and its change on the Loess Plateau were comparable to other studies. Based on the 0–100 cm SOC data collected from 382 sampling sites across the entire Loess Plateau in 2008, Liu et al. [55] indicated that mean SOC storage was 7.70 kg C/m^2. Using data gathered by the Second National Soil Survey of China (1979–1983), Xu et al. [56] estimated that SOC storage was 1.07 Pg in 0–20 cm soils layers on the Loess Plateau region (with a land area of 429,800 km^2). A soil survey conducted throughout the Loess Plateau region indicated that the SOC storage amounted to 1.23 Pg C in the 0–20 cm soil layers with land area of 592,900 km^2 during 1985–1988, [55]. Converting SOC storage from different soil depth to the same soil depth according to the summary of the vertical distribution of soil organic in the 0–100 cm soil [31], SOC storage were 6.22 kg C/m^2 and 5.19 kg C/m^2 in Xu et al.'s [57] and Liu et al.'s [55] reports (be mentioned but not published) respectively. In the same soil depth, DLEM-estimated average SOC storage was 6.45 kg C/ m^2 in 2005, which falls in the range of previous studies.

Tian et al. [7] estimated an average SOC sink of 94 Tg C/year in the terrestrial ecosystem of China as influenced by multiple global change factors during 1961–2005. Our results showed that soil on the Loess Plateau acted as a weak sink of 5.97 Tg C/year under the combined experiment during the same period. Considering that the Loess Plateau accounts for approximately 6.5% of the country's land surface, implying that the SOC storage on the Loess Plateau was close to the country-level average. At national level, Wang et al. [35] suggested that SOC storage in China decreased by about 5.69 g C/m^2 per year during the 1960s–1980s, while our study showed that the SOC storage on the Loess Plateau significantly increased by 4.30 g C/m^2 per year during the same period. This indicated that successful implementation of land conservation measures was beneficial and effective to reduce soil erosion and improve soil properties, thus enhance the soil carbon sequestration over the Loess Plateau.

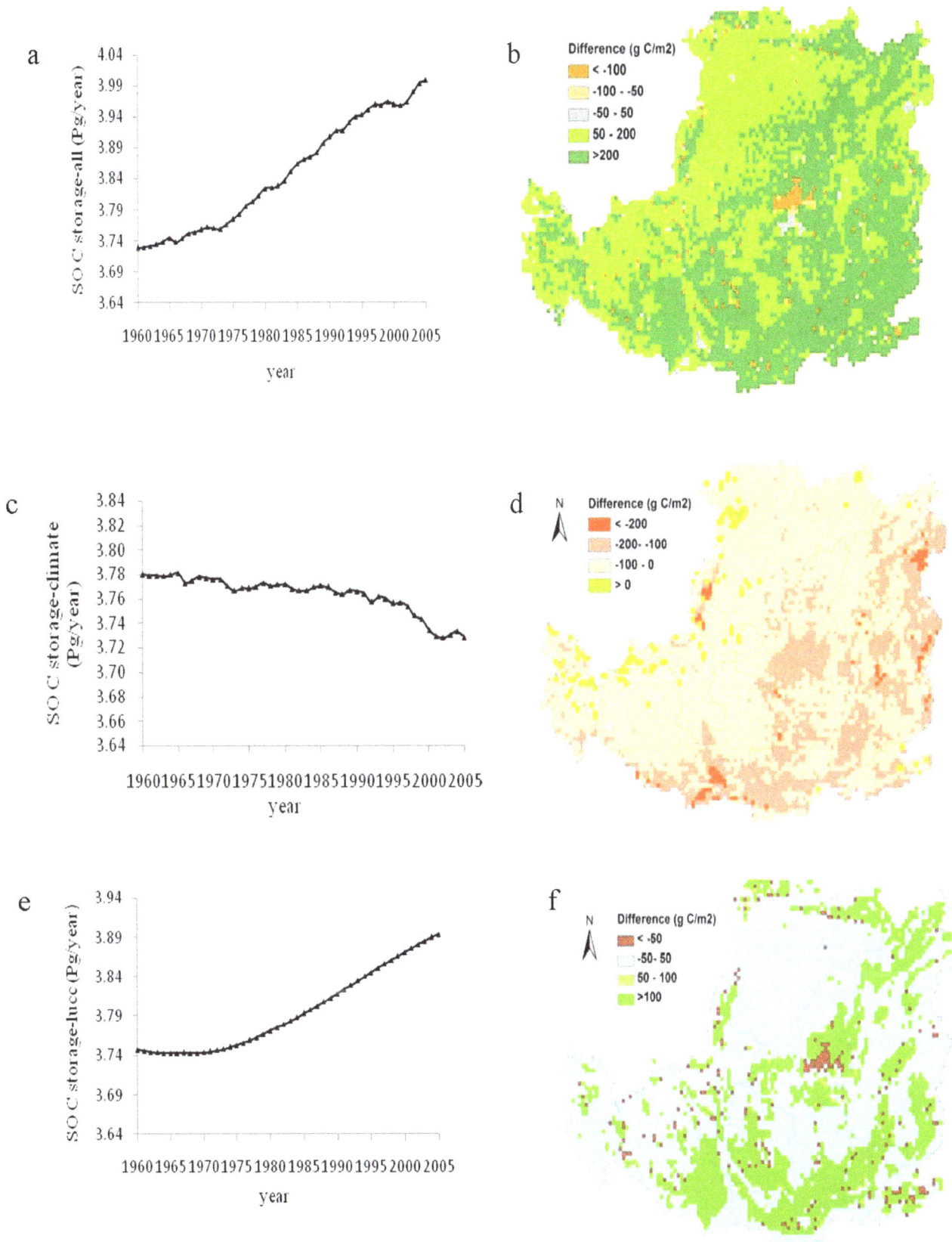

Figure 6. Spatiotemporal variations of SOC storage under experiments I, II, and III on the Loess Plateau from 1960–2005. Note: The "difference" means change in SOC storage between 2005 and 1960.

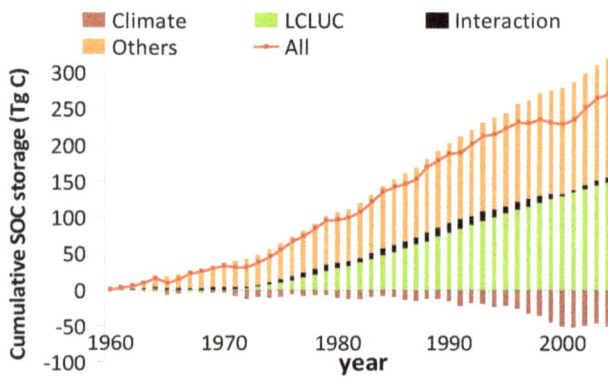

Figure 7. Contributions of multi-factor global changes to accumulative SOC change during 1961–2005. Note: "interaction" referrers to the interactive effects between climate and LCLUC.

Climate Controls on the Spatiaotemporal Patterns of SOC Storage

Climatic factors, especially precipitation and temperature, play an important role in long-term variations of SOC due to their effects on the quantity and quality of organic residue inputs and on the rates of soil organic matter and litter decomposition [1,12,58]. DLEM-simulated results showed that SOC storage decreased by about 0.05 Pg C with a significant inter-annual fluctuations throughout the Loess Plateau from 1961 to 2005 as influenced by climate change alone (Figure 6c). As shown in Figure 2, precipitation decreased at a rate of 1.27 (mm/year) during 1961–2005, which was a notable factor that influenced the change of SOC on the Loess Plateau, especially after 1990. Meanwhile, increased air temperature (0.030°C/year) might accelerate the evapotranspiration and potentially aggravate the water deficiency, thus cause the formation of drying soil layer and suppress the growth of vegetation. Due to the warming and drying climate, dried soil layer was widely distributed in the hilly and gully areas of the Loess Plateau. The development of drying soil layer has been

regarded as a key cause for the decrease of SOC storage in some areas on the Loess Plateau in the past decades [59].

Climate also had a substantial effect on the spatial distribution of SOC storage. Our results suggested that SOC storage decreased in most areas of the Loess Plateau under the influences of the climate change since 1961 (Figure 6d). This result could be explained partly by the distribution of temperature and precipitation, which is well-known to have a positive relationship with SOC decomposition [60]. During 1961–2005, both temperature and precipitation were higher in the central and southern Loess Plateau, compared to the northern Loess Plateau. Therefore, the higher loss of soil carbon from decomposition might be an important cause for lower SOC accumulation in the central and southern Loess Plateau. This result was supported by previous studies [43,50,57].

Land Use Controls on the Spatiaotemporal Patterns of SOC Storage

During the past decades, the Loess Plateau has experienced a complex change in land use pattern (Figure 3). Considering the LCLUC alone, DLEM-simulated results showed that the SOC storage increased by 0.14 Pg C during 1961–2005 (Figure 6e), which was significantly larger than the influence of climate change alone. However, in the first few years, the SOC changed slowly, even had a decreasing tendency in the 1960s, and then increased gradually since the 1980s.

Since the 1950s, various soil and water conservation measures including afforestation, cropland abandonment, and terrace construction etc., have been implemented in this region [17,18]. All of these measures directly or indirectly affected the vegetation cover and further influenced the SOC storage on the Loess Plateau. Previous studies demonstrated that soil can lose up to 20–40 percent of organic carbon into the atmosphere when perennial vegetation land was converted into cultivation land [61,62]. On the contrary, conversion from cropland to perennial vegetation land or shrubland was found to accumulate SOC by increasing carbon derived from new vegetation and decreasing carbon loss from decomposition and erosion [11,43,51,63]. Liu et al. [47] reported that SOC in shrubland, which was converted from

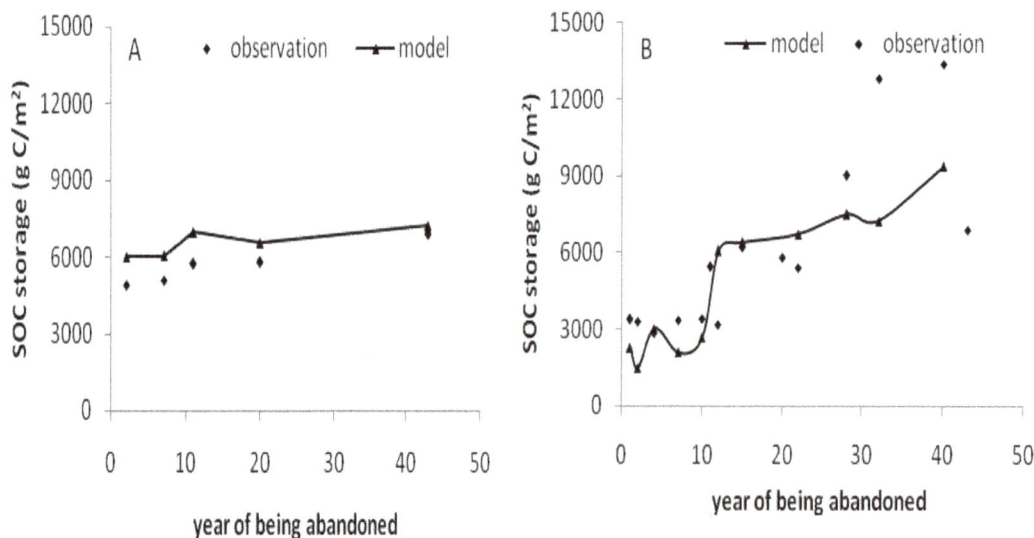

Figure 8. Comparison between simulated (DLEM) and observed SOC storage. Notes: A: cropland abandoned inYuzhong (Gansu Province), the observation data come from Jinping Jiang et al. [67]; B: cropland abandoned in Yan'an (Shaanxi Province), the observation data come from Junmin Wang et al. [2].

cropland in 1985, were 27.7%–34.8% higher than that of the cropland in 2010 on the Loess Plateau. Fu et al. [64] suggested that cropland abandonment significantly increased the density and stock of SOC in 0–100 cm soil profiles on the Loess Plateau. Feng et al. [48] found a total of 96.1 Tg of additional carbon had been sequestered on the Loess Plateau since China's "Grain for Green" program during 2000–2008, by using remote sensing techniques and ecosystem modelling. Our results are comparable with those previous findings.

LCLUC might induce an immediate change in vegetation coverage but a lagged effect on the change of SOC storage [65]. In this study, the SOC changed slowly, and even had a decreasing trend in the 1960s. We also found that the most rapid increase in the SOC storage occurred after the late 1970s, which might be partly due to extensive and conducted tree planting projects for alleviating soil erosion. The large-scale cropland abandonment in response to the recently implemented "Grain for Green" policy also contributed to the increase of SOC storage on the Loess Plateau. The plantation from "Grain for Green" Project would keep a large proportion of carbon in wood which need long time to return to soil. During the initial period of land conversion from cropland to forests, leaf biomass of trees were very low, so the litters on the ground decreased and resulted in a slight change (even decrease) in SOC. During the period of implementing "Grain for Green" project, litters on the ground accumulated gradually with the growth of planted trees, resulting in increasing SOC storage after a certain time period. Compared with simulated results in the context of multiple environmental factors, a similar temporal pattern for the SOC change was found when considering LCLUC alone from 1961 to 2005 (Figure 6a,e). This further implied that LCLUC was the dominant factor controlling temporal variations of SOC storage.

DLEM simulated results also showed obvious increases of SOC storage in most areas of the central and southern Loess Plateau (Figure 6f). Further analysis found that changes in SOC storage were smaller in the LCLUC alone simulation experiment than that with the combination of all environmental change (Figure 6b). However, SOC storage changed slightly in most areas of the northern Loess Plateau from 1961 to 2005, which were also suggested by other research results (e.g., [18,43]).

Interactive Effects of LCLUC and Climate on SOC Storage

Based on DLEM simulations, Tian et al. [7] found that LCLUC accounted for 17% of the net carbon increase, but climate change reduced it by 4% in China in the past decades. However, we found that LCLUC increased SOC storage by 0.14 Pg C, accounting for 55% of the net increase in SOC in the past 45 years over the Loess Plateau. In the same period, warming and drying climate greatly reduced SOC storage by 0.05 Pg C, approximately 19% of the total SOC storage change. These results implied that the SOC was more sensitive to LCLUC and climate change on the Loess Plateau comparing to other regions in China. In the past decades, LCLUC, particularly in the southern Loess Plateau, enhanced the vegetation coverage and reduced the anthropogenic disturbance, which further enhanced the SOC storage.

The interaction between climate and LCLUC contributed to the net SOC increases by 3%. Our results also showed that the interactive effects among environmental factors can't be neglected in attributing the changes of SOC storage in response to environmental factors on the Loess Plateau. Although the interaction among environmental factors has been recognized long before [66], most of the field experiments still overlook it. This study further demonstrated that the modeling approach may

serve as one complementary tool for the field experiments in addressing interactive effects among multiple environmental factors.

Effects of Cropland Abandonment on SOC Storage

Cropland abandonment and natural vegetation recovery were important implementations to mitigate soil loss on the Loess Plateau. Previous studies demonstrated that former land use types, soil property, climate change, and soil management practices were crucial to changes of SOC storage in the establishment of perennial vegetation type [49,50]. Other factors, such as the abandonment age, have also been found to play a significant role in SOC accumulation and should not be ignored [4,50].

In this study, we chose two sites (Yuzhong City and Yan'an City) (Marked in Figure 1) located on the Loess Plateau to explore SOC storage pattern at different abandonment stages (or restoration age) and to further evaluate the DLEM simulated results against the observations (Figure 8). It showed that DLEM results could well capture the distribution characteristics of the SOC storage in different abandonment stages. Both model results and field observations demonstrated that the restoration age played a key role in SOC accumulation. Jiang et al. [67] indicated that SOC storage changed little in early abandonment stage in a semiarid hilly area of Yuzhong City, followed by an obvious increase after 9–12 years, and then increased stably (Figure 8A). Wang et al. [2] found the similar temporal pattern through studying the change of SOC storage during different successional stages of rehabilitated grassland in Yan'an City. Compared to cropland, the rehabilitated grassland had a lower SOC storage at the early stage (1–12 years). However, SOC storage was higher than that in cropland after 15 years and then increased steadily (Figure 8B). Generally, long-term abandonment (>10 years) and the following colonization of natural vegetation could lead to substantial increase of SOC storage. The difference of the SOC storage between two sites over the early abandonment stage might be attributed to difference in environmental factors such as climate and soil property. Yan'an City is located in a semi-arid and warm temperate zone with less precipitation and more obvious drying and warming tendency, compared with Yuzhong City during the past years. This also partly explained the lower SOC storage in cropland in this climate zone than others over the Loess Plateau.

Our results indicated that cropland abandonment could increase SOC storage, improve soil quality and promote ecosystem restoration, especially in the warm and dry climate zone. However, these effects will emerge after a relative longer time period (e.g., >10 years) under local environmental conditions.

Uncertainties

This study examined temporal and spatial patterns of SOC storage and attributed these patterns to multiple environmental factors on the Loess Plateau during 1961–2005. There were several uncertainties which need to be addressed in our future work. First, impacts of some ecological processes, such as soil erosion, were not separated from other environmental factors in this study. The soil erosion area could be as high as 45.4×10^4 km^2 (72.3% of the land area) over the Loess Plateau. Soil erosion has long been identified as one of key factors controlling the change of SOC storage [47]. Further efforts should be put on interactions among soil erosion and other environmental factors. Second, the input datasets were developed at the spatial resolution of 10 km × 10 km which was the finest dataset for the Loess Plateau. However, it is still difficult to capture subtle change of SOC storage due to high heterogeneity in land surface processes. In the long run, finer gidded datasets would be greatly helpful for further quantifying temporal and spatial changes in SOC storage on the Loess Plateau. In addition, the uncertainties from other input data, model structure, and parameterization need to be further specified in the furture efforts.

Conclusions

This study examined effects of climate and land use changes on SOC storage on the Loess Plateau of China in the context of multiple global changes by using an integrated ecosystem model. The results showed that temperature on the Loess Plateau has significantly increased, while precipitation slightly decreased during 1961–2005. Meanwhile, this region experienced a remarkable change in land cover and land use, characterized by conversions from cropland to grassland/forest owing to the government policies to alleviate soil and water losses during the past decades. The overall change in SOC storage due to multiple environmental factors was estimated to be a net increase of SOC storage by 0.27 Pg C during 1961–2005, indicating that soil on the Loess Plateau acted as a carbon sink in this period. Among multiple factors, LCLUC led to a significant increase in SOC storage of 0.14 Pg C, accounting for 55% of the net increase in SOC. In contrast, climate change reduced SOC storage by 0.05 Pg C (approximately 19% of the total SOC change). The interaction of climate and LCLUC accounted for 3% of the net increase in SOC. Our results were consistent with field observation data and both of them suggested that SOC storage could be enhanced significantly by the conversion of cropland to grassland along with the increasing abandonment age on the Loess Plateau.

However, the magnitude could be influenced by local environmental conditions.

This study provides the first attempt to quantify relative effects of multiple environmental factors (climate and LCLUC in particular) on regional SOC storage on the Loess Plateau over the past decades. The results drawn from this study provide insight for land management as well as policy-making to enhance carbon sequestration and alleviate the serious soil erosion conditions on the Loess Plateau. To reduce uncertainties in estimating effects of climate and land use changes on SOC storage, it is needed to put further efforts in developing more reliable and fine-resolution input data and improve model representation of some other processes relevant to SOC, such as soil erosion.

Acknowledgments

This study was supported by NASA Land Cover and Land Use Change Program (NNX08AL73G), NASA Interdisciplinary Science Program (NNG04GM39C), Chinese Universities Scientific Fund (z10921007), US National Science Foundation Grants (AGS-1243220, CNS-1059376), State Key Laboratory of Soil Erosion and Dryland Farming on the Loess Plateau Foundation (K318009902-1410), and Shaanxi Administration of Foreign Expert Affairs Science and Technology Activities Fundation (201327). Thank Mingliang Liu, Xiaofeng Xu, and Jiyuan Liu for their contributions in the development of spatial data and the DLEM model.

Author Contributions

Conceived and designed the experiments: HT YD WR. Performed the experiments: YD WR. Analyzed the data: YD WR BT GC. Contributed reagents/materials/analysis tools: GC BT JY SP. Wrote the paper: YD WR HT BT GC CL JY SP GW SL.

References

1. Lal R (2004) Soil carbon sequestration impacts on global climate change and food security. Science 304: 1623–1627.
2. Wang Y, Fu B, Lü Y, Chen L (2011) Effects of vegetation restoration on soil organic carbon sequestration at multiple scales in semi-arid Loess Plateau, China. Catena 85: 58–66.
3. Post WM, Emanuel WR, Zinke PJ, Stangenberger AG (1982) Soil carbon pools and world life zones.
4. Zhang K, Dang H, Tan S, Cheng X, Zhang Q (2010) Change in soil organic carbon following the 'Grain-for-Green' programme in China. Land Degradation & Development 21: 13–23.
5. Eglin T, Ciais P, Piao S, Barre P, Bellassen V, et al. (2010) Historical and future perspectives of global soil carbon response to climate and land-use changes. Tellus B 62: 700–718.
6. Martin D, Lal T, Sachdev C, Sharma J (2010) Soil organic carbon storage changes with climate change, landform and land use conditions in Garhwal hills of the Indian Himalayan mountains. Agriculture, Ecosystems & Environment 138: 64–73.
7. Tian H, Melillo J, Lu C, Kicklighter D, Liu M, et al. (2011) China's terrestrial carbon balance: Contributions from multiple global change factors. Global Biogeochemical Cycles 25.
8. Wang X, Feng Z, Ouyang Z (2001) The impact of human disturbance on vegetative carbon storage in forest ecosystems in China. Forest ecology and management 148: 117–123.
9. Barnett TP, Adam JC, Lettenmaier DP (2005) Potential impacts of a warming climate on water availability in snow-dominated regions. Nature 438: 303–309.
10. Paul EA (1997) Soil organic matter in temperate agroecosystems: long term experiments in North America: CRC PressI Llc.
11. Post WM, Kwon KC (2000) Soil carbon sequestration and land-use change: processes and potential. Global change biology 6: 317–327.
12. Lal R, Bruce J (1999) The potential of world cropland soils to sequester C and mitigate the greenhouse effect. Environmental Science & Policy 2: 177–185.
13. Schimel J (2001) 1.13-Biogeochemical Models: Implicit versus Explicit Microbiology. Global Biogeochemical Cycles in the Climate System: 177–183.
14. Liu S, Bliss N, Sundquist E, Huntington TG (2003) Modeling carbon dynamics in vegetation and soil under the impact of soil erosion and deposition. Global Biogeochemical Cycles 17.
15. Zhang C, Tian H, Pan S, Liu M, Lockaby G, et al. (2008) Effects of forest regrowth and urbanization on ecosystem carbon storage in a rural–urban gradient in the southeastern United States. Ecosystems 11: 1211–1222.
16. Li Z, Liu W-z, Zhang X-c, Zheng F-l (2009) Impacts of land use change and climate variability on hydrology in an agricultural catchment on the Loess Plateau of China. Journal of hydrology 377: 35–42.
17. Chen L, Gong J, Fu B, Huang Z, Huang Y, et al. (2007) Effect of land use conversion on soil organic carbon sequestration in the loess hilly area, loess plateau of China. Ecological Research 22: 641–648.
18. Xin Z, Xu J, Zheng W (2008) Spatiotemporal variations of vegetation cover on the Chinese Loess Plateau (1981–2006): impacts of climate changes and human activities. Science in China Series D: Earth Sciences 51: 67–78.
19. Fu B-J, Wang Y-F, Lu Y-H, He C-S, Chen L-D, et al. (2009) The effects of land-use combinations on soil erosion: a case study in the Loess Plateau of China. Progress in Physical Geography 33: 793–804.
20. Wang G, Huang J, Guo W, Zuo J, Wang J, et al. (2010) Observation analysis of land-atmosphere interactions over the Loess Plateau of northwest China. Journal of Geophysical Research: Atmospheres (1984–2012) 115.
21. Zhao M, Heinsch FA, Nemani RR, Running SW (2005) Improvements of the MODIS terrestrial gross and net primary production global data set. Remote Sensing of Environment 95: 164–176.
22. Fang J, Guo Z, Piao S, Chen A (2007) Terrestrial vegetation carbon sinks in China, 1981–2000. Science in China Series D: Earth Sciences 50: 1341–1350.
23. Peters W, Jacobson AR, Sweeney C, Andrews AE, Conway TJ, et al. (2007) An atmospheric perspective on North American carbon dioxide exchange: CarbonTracker. Proceedings of the National Academy of Sciences 104: 18925–18930.
24. Piao S, Ciais P, Friedlingstein P, de Noblet-Ducoudré N, Cadule P, et al. (2009) Spatiotemporal patterns of terrestrial carbon cycle during the 20th century. Global Biogeochemical Cycles 23.
25. Li K, Wang S, Cao M (2004) Vegetation and soil carbon storage in China. SCIENCE IN CHINA SERIES D EARTH SCIENCES-ENGLISH EDITION- 47: 49–57.
26. King J, Bradley R, Harrison R, Carter A (2004) Carbon sequestration and saving potential associated with changes to the management of agricultural soils in England. Soil Use and Management 20: 394–402.
27. Yan H, Cao M, Liu J, Tao B (2007) Potential and sustainability for carbon sequestration with improved soil management in agricultural soils of China. Agriculture, Ecosystems & Environment 121: 325–335.
28. Tian H, Chen G, Liu M, Zhang C, Sun G, et al. (2010) Model estimates of net primary productivity, evapotranspiration, and water use efficiency in the terrestrial ecosystems of the southern United States during 1895–2007. Forest ecology and management 259: 1311–1327.
29. Ren W, Tian H, Liu M, Zhang C, Chen G, et al. (2007) Effects of tropospheric ozone pollution on net primary productivity and carbon storage in terrestrial ecosystems of China. Journal of Geophysical Research: Atmospheres (1984–2012) 112.

30. Xu X, Tian H, Zhang C, Liu M, Ren W, et al. (2010) Attribution of spatial and temporal variations in terrestrial methane flux over North America. Biogeosciences Discussions 7: 5383–5428.

31. Tian H, Xu X, Lu C, Liu M, Ren W, et al. (2011) Net exchanges of CO2, CH4, and N2O between China's terrestrial ecosystems and the atmosphere and their contributions to global climate warming. Journal of Geophysical Research: Biogeosciences (2005–2012) 116.

32. Ren W, Tian H, Tao B, Huang Y, Pan S (2012) China's crop productivity and soil carbon storage as influenced by multifactor global change. Global Change Biology 18: 2945–2957.

33. Tian H, Chen G, Zhang C, Liu M, Sun G, et al. (2012) Century-scale responses of ecosystem carbon storage and flux to multiple environmental changes in the southern United States. Ecosystems 15: 674–694.

34. Shi H, Shao M (2000) Soil and water loss from the Loess Plateau in China. Journal of Arid Environments 45: 9–20.

35. Wang S, Tian H, Liu J, Pan S (2003) Pattern and change of soil organic carbon storage in China: 1960s–1980s. Tellus B 55: 416–427.

36. Zhang XP, Zhang L, McVicar TR, Van Niel TG, Li LT, et al. (2008) Modelling the impact of afforestation on average annual streamflow in the Loess Plateau, China. Hydrological Processes 22: 1996–2004.

37. Liu J, Tian H, Liu M, Zhuang D, Melillo JM, et al. (2005) China's changing landscape during the 1990s: Large-scale land transformations estimated with satellite data. Geophysical Research Letters 32.

38. Liu M, Tian H (2010) China's land cover and land use change from 1700 to 2005: Estimations from high-resolution satellite data and historical archives. Global Biogeochemical Cycles 24.

39. Tian H, Chen G, Zhang C, Melillo JM, Hall CA (2010) Pattern and variation of C: N: P ratios in China's soils: a synthesis of observational data. Biogeochemistry 98: 139–151.

40. Chen H, Tian H, Liu M, Melillo J, Pan S, et al. (2006) Effect of land-cover change on terrestrial carbon dynamics in the southern United States. Journal of environmental quality 35: 1533–1547.

41. Trenberth K, Jones P, Ambenje P, Bojariu R, Easterling D, et al. (2007) Observations: Surface and Atmospheric Climate Change, chap. 3 of Climate Change 2007: The Physical Science Basis. Contribution of Working Group I to the Fourth Assessment Report of the Intergovernmental Panel on Climate Change [Solomon, S., Qin, D., Manning, M., Marquis, M., Averyt, KB, Tignor, M., Miller, HL and Chen, Z.(eds.)]., 235–336. Cambridge University Press, Cambridge, UK and New York, NY, USA.

42. Li R, Yang W, Li B (2008) Research and future prospects for the Loess Plateau of China. SciencePress, Beijing (in Chinese).

43. Chang R, Fu B, Liu G, Liu S (2011) Soil carbon sequestration potential for "Grain for Green" project in Loess Plateau, China. Environmental management 48: 1158–1172.

44. Xin Z, Xu J, Ma Y (2009) Spatio-temporal variation of erosive precipitation in Loess Plateau during past 50 years. Scientia Geographica Sinica 29: 89–104.

45. Song Y, Ma M-g (2007) Study on vegetation cover change in Northwest China based on SPOT VEGETATION data. Journal of Desert Research 27: 89–93.

46. Ren W, Tian H, Xu X, Liu M, Lu C, et al. (2011) Spatial and temporal patterns of CO2 and CH4 fluxes in China's croplands in response to multifactor environmental changes. Tellus B 63: 222–240.

47. Liu X, Li F-M, Liu D-Q, Sun G-J (2010) Soil organic carbon, carbon fractions and nutrients as affected by land use in semi-arid region of Loess Plateau of China. Pedosphere 20: 146–152.

48. Feng X, Fu B, Lu N, Zeng Y, Wu B (2013) How ecological restoration alters ecosystem services: an analysis of carbon sequestration in China's Loess Plateau. Scientific reports 3.

49. Guo Z, Ruddiman WF, Hao Q, Wu H, Qiao Y, et al. (2002) Onset of Asian desertification by 22 Myr ago inferred from loess deposits in China. Nature 416: 159–163.

50. Paul K, Polglase P, Nyakuengama J, Khanna P (2002) Change in soil carbon following afforestation. Forest ecology and management 168: 241–257.

51. LAGANIÈRE J, ANGERS DA, PARÉ D (2010) Carbon accumulation in agricultural soils after afforestation: a meta-analysis. Global Change Biology 16: 439–453.

52. Cheng L, Zhao X (2011) Soil mineralized nutrients changes and soil conservation benefit evaluation on 'green project grain' in ecologically fragile areas in the south of Yulin city, Loess Plateau. African Journal of Biotechnology 10: 2230–2237.

53. Xiao Y (1990) Comparative studies on biomass and productivity of Pinus tabulaeformis plantations in different climatic zones in Shaanxi Province [China]. Acta Phytoecologica et Geobotanica Sinica 14.

54. Fu X, Shao M, Wei X, Horton R (2010) Soil organic carbon and total nitrogen as affected by vegetation types in Northern Loess Plateau of China. Geoderma 155: 31–35.

55. Liu Z, Shao Ma, Wang Y (2011) Effect of environmental factors on regional soil organic carbon stocks across the Loess Plateau region, China. Agriculture, Ecosystems & Environment 142: 184–194.

56. Xu X, Zhang K, Peng W (2003) Spatial distribution and estimating of soil organic carbon on Loess Plateau. J Soil Water Conserv 17: 13–15.

57. Xu X, Liu W, Kiely G (2011) Modeling the change in soil organic carbon of grassland in response to climate change: effects of measured versus modelled carbon pools for initializing the Rothamsted Carbon model. Agriculture, Ecosystems & Environment 140: 372–381.

58. Schlesinger WH (1990) Evidence from chronosequence studies for a low carbon-storage potential of soils. Nature 348: 232–234.

59. Han F, Hu W, Zheng J, Du F, Zhang X (2010) Estimating soil organic carbon storage and distribution in a catchment of Loess Plateau, China. Geoderma 154: 261–266.

60. Lehmann J, Skjemstad J, Sohi S, Carter J, Barson M, et al. (2008) Australian climate–carbon cycle feedback reduced by soil black carbon. Nature Geoscience 1: 832–835.

61. Houghton R, Hackler J, Lawrence K (1999) The US carbon budget: contributions from land-use change. Science 285: 574–578.

62. van der Werf GR, Morton DC, DeFries RS, Olivier JG, Kasibhatla PS, et al. (2009) CO2 emissions from forest loss. Nature Geoscience 2: 737–738.

63. Richter DD, Markewitz D, Trumbore SE, Wells CG (1999) Rapid accumulation and turnover of soil carbon in a re-establishing forest. Nature 400: 56–58.

64. Fu B-J, Zhang Q-J, Chen L-D, Zhao W-W, Gulinck H, et al. (2006) Temporal change in land use and its relationship to slope degree and soil type in a small catchment on the Loess Plateau of China. Catena 65: 41–48.

65. Kuzyakov Y, Gavrichkova O (2010) Review: Time lag between photosynthesis and carbon dioxide efflux from soil: a review of mechanisms and controls. Global Change Biology 16: 3386–3406.

66. Dermody O (2006) Mucking through multifactor experiments; design and analysis of multifactor studies in global change research. New Phytologist 172: 598–600.

67. Jiang J-P, Xiong Y-C, Jiang H-M, Ye D-Y, Song Y-J, et al. (2009) Soil microbial activity during secondary vegetation succession in semiarid abandoned lands of Loess Plateau. Pedosphere 19: 735–747.

5

Spatial and Temporal Variations of Crop Fertilization and Soil Fertility in the Loess Plateau in China from the 1970s to the 2000s

Xiaoying Wang[1,2], Yanan Tong[1,2]*, Yimin Gao[1], Pengcheng Gao[1], Fen Liu[1], Zuoping Zhao[1], Yan Pang[1]

1 College of Natural Resources and Environment, Northwest A&F University, Yangling, China, 2 Key Laboratory of Plant Nutrition and the Agri-environment in Northwest China, Ministry of Agriculture, Yangling, China

Abstract

Increased fertilizer input in agricultural systems during the last few decades has resulted in large yield increases, but also in environmental problems. We used data from published papers and a soil testing and fertilization project in Shaanxi province during the years 2005 to 2009 to analyze chemical fertilizer inputs and yields of wheat (*Triticum aestivum* L.) and maize (*Zea mays* L.) on the farmers' level, and soil fertility change from the 1970s to the 2000s in the Loess Plateau in China. The results showed that in different regions of the province, chemical fertilizer NPK inputs and yields of wheat and maize increased. With regard to soil nutrient balance, N and P gradually changed from deficit to surplus levels, while K deficiency became more severe. In addition, soil organic matter, total nitrogen, alkali-hydrolysis nitrogen, available phosphorus and available potassium increased during the same period. The PFP of N, NP and NPK on wheat and maize all decreased from the 1970s to the 2000s as a whole. With the increase in N fertilizer inputs, both soil total nitrogen and alkali-hydrolysis nitrogen increased; P fertilizer increased soil available phosphorus and K fertilizer increased soil available potassium. At the same time, soil organic matter, total nitrogen, alkali-hydrolysis nitrogen, available phosphorus and available potassium all had positive impacts on crop yields. In order to promote food safety and environmental protection, fertilizer requirements should be assessed at the farmers' level. In many cases, farmers should be encouraged to reduce nitrogen and phosphate fertilizer inputs significantly, but increase potassium fertilizer and organic manure on cereal crops as a whole.

Editor: Cheng–Sen Li, Institute of Botany, China

Funding: The authors thank the Special Fund for Agro-scientific Research in the Public Interest of China (201103003) and the Soil Quality Foundation of China (2012BAD05B03) for their financial support. The funders had no role in study design, data collection and analysis, decision to publish, or preparation of the manuscript.

Competing Interests: The authors have declared that no competing interests exist.

* Email: tongyanan@nwsuaf.edu.cn

Introduction

China has only 9% of the world's arable land and feeds nearly 22% of the world population [1–2]. This depends heavily on increasing grain production with the use of chemical fertilizers. Before the 1970s, farmers maintained the original agricultural practices, such as crop rotation, diversified plantation, manure application and legume crop integration, for soil fertility maintenance and pest and disease control. Since the late 1980s, the practice of applying organic manure in arable cropping systems has nearly come to an end [2–6]. From then on, almost all available organic manure has been used on vegetables and fruit trees, while the nutrients for cereal crops have been mainly in the form of chemical fertilizers. From 1970 to 2010, total annual grain production in China increased from 240 to 546 million tons (a 128% increase). However, inorganic fertilizer application increased from 3.51 to 55.62 million tons (a 1485% increase) over the same period [7].

Soil quality indicators are measurable soil properties that benefit food production or other specific functions, including physical, chemical and biological characteristics [8]. The increase or decrease in single soil index values, such as soil organic matter, total nitrogen and available nutrients, amplitude of variation and variation in time, can be used as a monitoring index for agricultural land management [9–11]. Given the spatial and temporal variation in characteristics of soil quality, it is necessary to compare or analyze two or more phase changes to understand the nature and mechanisms of soil quality [12].

Farmland fertilization is one of the most effective ways to maintain soil fertility and increase crop yields [13–15]. For this reason, information on household fertilization levels is of great value. In addition, wheat and maize are two of the most important food crops throughout the world, and they account for 51.7% of the total area for food crops and 53.5% of the total food production in 2010 in China [7]. Chemical fertilizer consumption data from official Chinese statistics do not contain information on usage for each kind of crop. It is imprecise to analyze and evaluate fertilizer efficiency using total amounts, because the distribution and application of fertilizer on specific crops are ambiguous [16].

Thus, the objectives of this study were to: (1) reveal the spatial and temporal variations of chemical fertilization and yields of wheat and maize at the farmers' level from the 1970s to the 2000s in the Loess Plateau in China; (2) reveal the spatial and temporal variations of soil fertility over the same period; and (3) reveal the relationships among fertilizer inputs, crop yields and soil fertility.

Materials and Methods

Ethics Statement

This study has been approved by the Agricultural Technology Extension Center of Shaanxi province, which is responsible for fertilization and soil fertility in Shaanxi province. All data in this study can be published and shared.

Study area

Shaanxi province (Figure 1) is located in the middle reaches of the Yellow River and the upper reaches of the Yangtze River of the eastern part of northwest China, and it falls between latitudes $31°42'$ and $39°35'N$, and longitudes $105°29'$ and $111°15'E$. The area is $2.058×10^5$ km^2, extending about 880 km from north to south and 160 to 490 km from east to west. The whole province from north to south can be divided into four agro-ecological zones, which include the Loess Plateau area of northern Shaanxi, the Weibei dry plateau, the Guanzhong irrigated area and the Qin-Ba mountain area of southern Shaanxi; the previous three regions belong to the Loess Plateau and in this study they are abbreviated as North, Weibei, and Guanzhong, respectively. The Loess Plateau region in China, covers five provinces (including Shaanxi province), stretches over an area of 0.62 million km^2, and consists of typical semiarid and arid areas with rainfed farming [17–18]. Winter wheat is planted in the regions of Weibei and Guanzhong, while summer maize is planted in the Guanzhong region and spring maize in the North and Weibei regions. Main soil types and climatic conditions in the different regions are shown in Table 1.

Data sources

The data from the 1970s to the 1990s was extracted from 380 published papers reporting household fertilization and soil fertility in the study area; the screening process and results are shown in Figure 2. Data from the 2000s was collected from the project "soil testing and formulated fertilization in Shaanxi province during the years 2005 to 2009."

Statistics

The data were analyzed by EXCEL software. In this study, we used the following equations to analyze the soil nutrient balance and partial factor productivity (PFP) of fertilizer:

$$Soil\ nutrient\ balance = nutrient\ input\ rate \\ - nutrient\ output\ rate \quad (Eq\ 1)$$

where the nutrient input rate represents chemical fertilizer input, and the nutrient output rate represents amounts extracted in crop products and above ground biomass;

$$PFP = Y/F \quad (Eq\ 2)$$

where Y represents crop yields, and F represents chemical fertilizer input.

Results

Spatial and temporal variations of chemical fertilization and yields of wheat and maize at the farmers' level in different regions of Shaanxi province

The average chemical fertilizer NPK inputs for both wheat and maize at the farmers' level increased for decades in the different regions (Figure 3). In the Weibei and Guanzhong regions,

chemical fertilizer N inputs for wheat in the 1970s were 45 kg ha^{-1} and 52 kg ha^{-1}, respectively, and in the 2000s they increased to 185 kg ha^{-1} and 195 kg ha^{-1}, respectively. In these two regions, chemical fertilizer P$_2$O$_5$ inputs were 45 kg ha^{-1} and 46 kg ha^{-1} in the 1970s and they increased to 112 kg ha^{-1} and 115 kg ha^{-1} in the 2000s. In the 1980s, farmers started to use the chemical fertilizer K$_2$O for wheat, which was increased from 0.5 kg ha^{-1} and 2.3 kg ha^{-1} to 22.8 kg ha^{-1} and 22.5 kg ha^{-1}, respectively, during the 1980s to the 2000s in the two regions. For maize in the North, Weibei and Guanzhong regions, chemical fertilizer N inputs were 48 kg ha^{-1}, 89 kg ha^{-1} and 36 kg ha^{-1} and they increased to 237 kg ha^{-1}, 223 kg ha^{-1} and 244 kg ha^{-1}, respectively, from the 1970s to the 2000s. Unlike wheat, from the 1980s onward farmers were awarded for using the chemical fertilizers P$_2$O$_5$ and K$_2$O for maize, and their use has increased greatly.

In accordance with increased chemical fertilizer NPK inputs (Figure 3), the average yields of wheat and maize showed increasing trends in the different regions over the four decades (Figure 4). In the Weibei and Guanzhong regions, from the 1970s to the 2000s, yields of wheat changed from 1883 kg ha^{-1} and 3377 kg ha^{-1} to 4269 kg ha^{-1} and 6437 kg ha^{-1}, with increase rates of 127% and 91%, respectively. In the North, Weibei and Guanzhong regions, yields of maize changed from 3636 kg ha^{-1}, 2519 kg ha^{-1} and 4232 kg ha^{-1} to 7867 kg ha^{-1}, 7077 kg ha^{-1} and 6886 kg ha^{-1}, with increase rates of 116%, 181% and 63%, respectively, for the same period.

Spatial and temporal variations of soil nutrient balance from the inputs and uptake on wheat and maize plots in different regions of Shaanxi province

Because the farmers tended not to use organic manure for cereal crops, especially from the 1980s onward, the soil nutrient inputs only include chemical fertilizers, and the nutrient uptakes include those extracted in crop products and above ground biomass. The nutrient balance was calculated as the difference between the average input and uptake (Eq. 1). Other losses, from leakage and gaseous loss, were not included in these calculations. In the 1970s, N was deficient on wheat and maize plots in the different regions (except for maize plots in the Weibei region). Then from the 1980s N was consistently at surplus levels, and it displayed an upward trend with time. In the 2000s, N surpluses on wheat plots were 74 kg ha^{-1} and 29 kg ha^{-1} in the Weibei and Guanzhong regions, respectively; meanwhile N surpluses on maize plots were 64 kg ha^{-1}, 67 kg ha^{-1} and 93 kg ha^{-1} in the North, Weibei and Guanzhong regions, respectively (Figure 5).

In the Weibei and Guanzhong regions, the amount of surplus P$_2$O$_5$ on wheat plots increased each year from the 1970s to the 2000s, and surplus amounts increased from 24 kg ha^{-1} and 9 kg ha^{-1} to 65 kg ha^{-1} and 44 kg ha^{-1}, respectively. In the North, Weibei and Guanzhong regions, P$_2$O$_5$ was deficient on maize plots in the 1980s; then it gradually reached surplus levels until the 2000s with the increased application of chemical fertilizer phosphorus. The balance of P$_2$O$_5$ on maize plots increased from -34 kg ha^{-1}, -7 kg ha^{-1} and -35 kg ha^{-1} to 29 kg ha^{-1}, 28 kg ha^{-1} and -11 kg ha^{-1}, respectively, in the three regions from the 1980s to the 2000s. It is worth noting, that winter wheat and summer maize were in a rotation system in the Guanzhong region, so total P$_2$O$_5$ was in surplus in this region in the 2000s and the amount was 33 kg ha^{-1} (Figure 5).

Although farmers have been awarded for using K$_2$O chemical fertilizer in recent years, the amount used was still small (Figure 3), and it was usually from compound fertilizers. So K$_2$O deficiency has become more serious (Figure 5). In the 2000s, K$_2$O deficiency

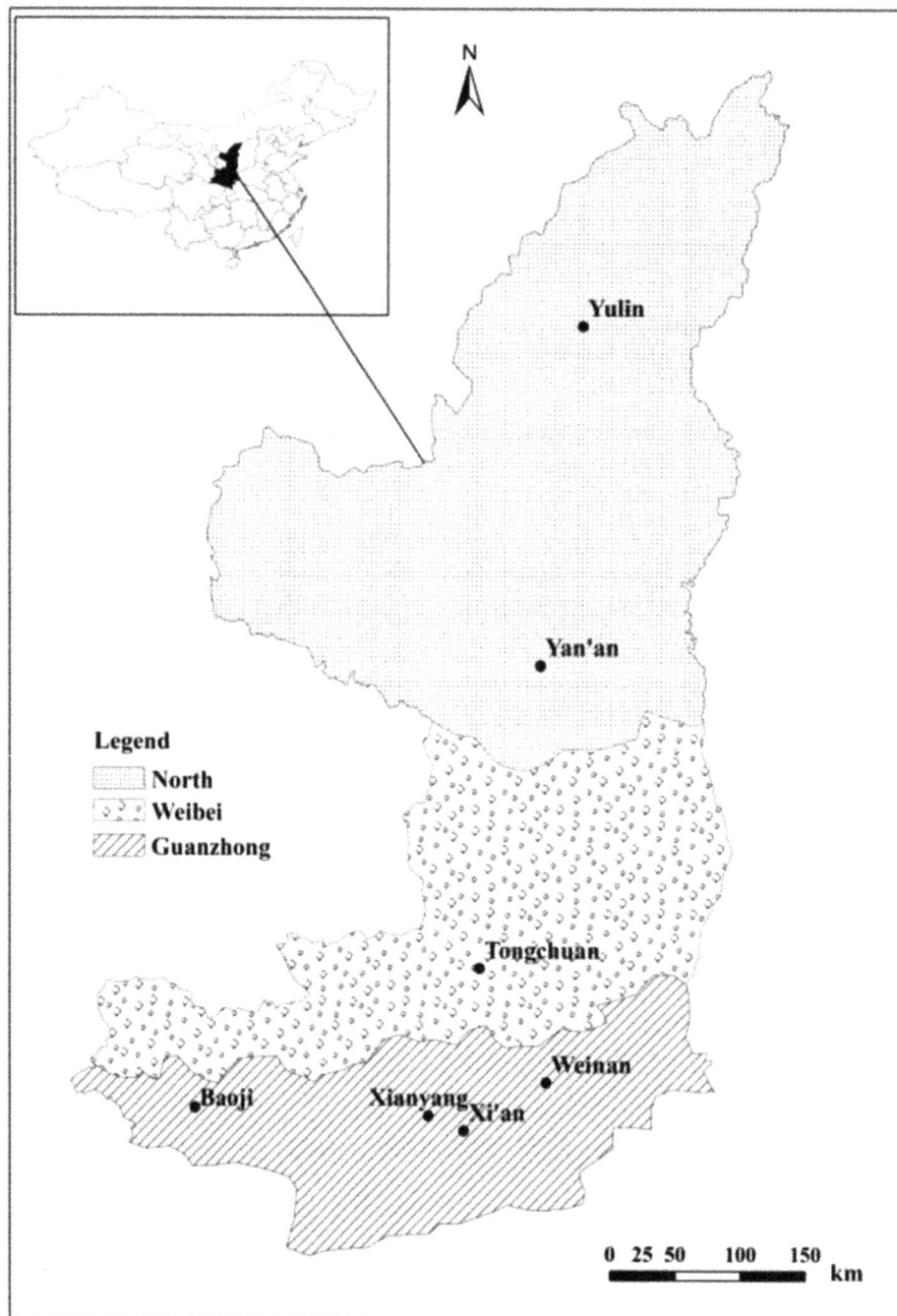

Figure 1. Map of the study area.

Table 1. Main soil types and climatic conditions in the different regions.

Region	Main soil types	Annual mean temperature (°C)	Annual precipitation (mm)
North	Castanozems, Sierozems, Loess soils	8~11	275~590
Weibei	Black loess soils, Loess soils	9~13	530~630
Guanzhong	Cinnamon soils	10~14	600~720

Identification

Literature search
Databases: CJFD, Vip and Wan-fang
(n=1595)
Limts: Chinese-language articles only

Identified through other sources
(n=3)

Screening

After duplicates removed (n=1361)

Articles screened on basis of title
and abstract (n=967)

Excluded (n=394)
    Multiple publication (n=15)
    Results of other provinces
    and regions (n=379)

Eligibility

Manuscript review and application
of inclusion criteria (n=380)

Excluded (n=587)
    No data (n=398)
    The data do not belong to
    1970s to 1990s (n=88)
    Uncertain data units (n=101)

Included in qualitative synthesis
(n=380)

Included

Included in quantitative synthesis
(meta-analysis) (n=380)

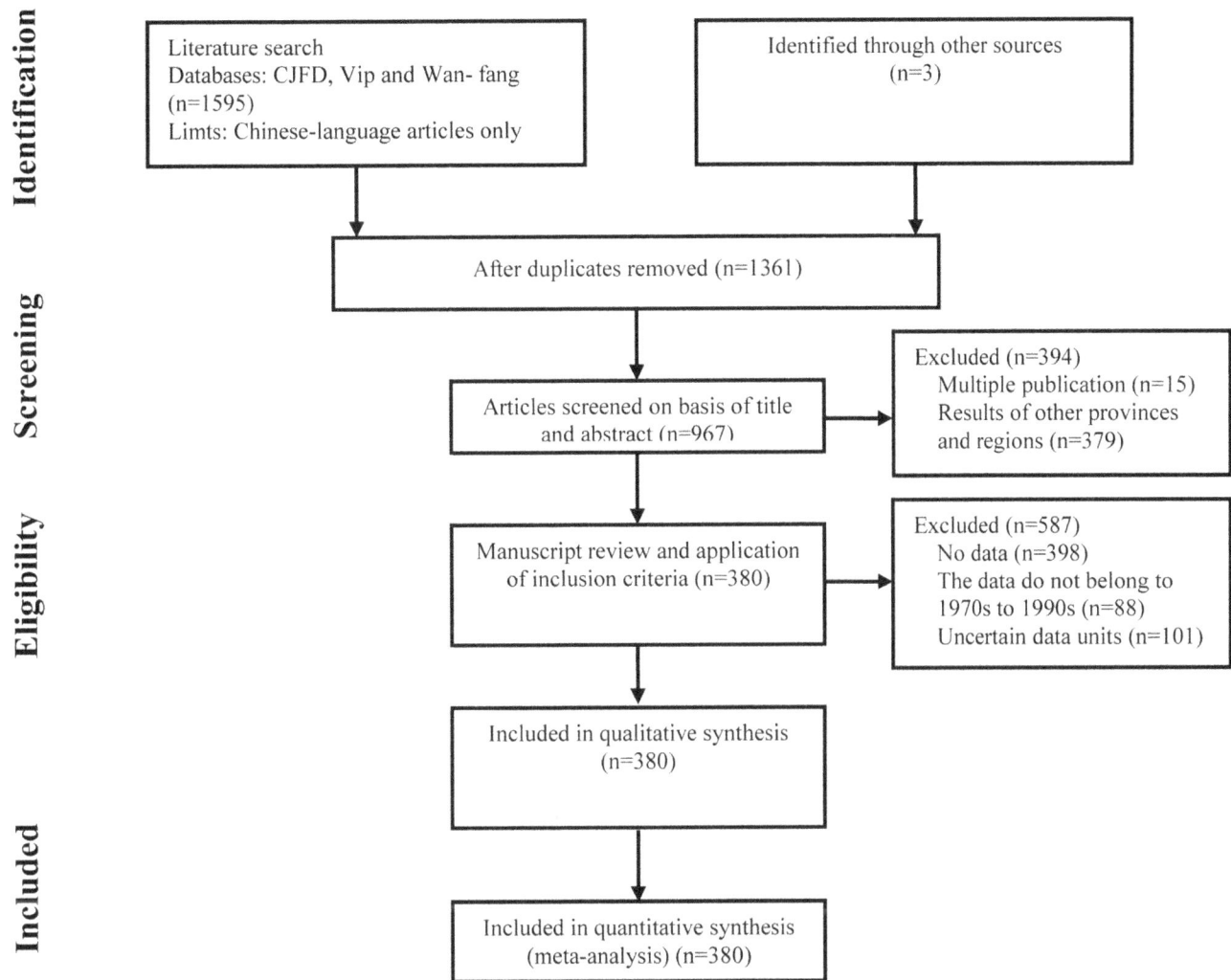

Figure 2. The screening process and results for literature from the 1970s to the 1990s.

levels on wheat plots were -102 kg ha^{-1} and -165 kg ha^{-1} in the Weibei and Guanzhong regions, respectively; meanwhile K_2O deficiency levels on maize plots were -179 kg ha^{-1}, -137 kg ha^{-1} and -147 kg ha^{-1} in the North, Weibei and Guanzhong regions, respectively.

Spatial and temporal variations of soil fertility in different regions of Shaanxi province

In the different regions of Shaanxi province, soil fertility indexes, including organic matter, total nitrogen, alkali-hydrolysis nitrogen, available phosphorus and available potassium, all increased from the 1970s to the 2000s. Simultaneously, each of these five indicators increased from the north to the south during the same period (North<Weibei<Guanzhong) (Figure 6). In the North, Weibei and Guanzhong regions from the 1970s to the 2000s, organic matter varied from 0.57%, 1.01% and 1.12% to 0.83%, 1.26% and 1.50%, with increase rates of 46%, 26% and 43%, respectively; total nitrogen varied from 0.04%, 0.07% and 0.07% to 0.05%, 0.08% and 0.09%, with increase rates of 42%, 9% and 14%, respectively; alkali-hydrolysis nitrogen varied from 29.95 mg kg^{-1}, 20.43 mg kg^{-1} and 30.81 mg kg^{-1} to 35.20 mg kg^{-1}, 58.70 mg kg^{-1} and 68.40 mg kg^{-1}, with increase rates of 18%, 187% and 122%, respectively; available phosphorus varied

from 4.98 mg kg^{-1}, 7.13 mg kg^{-1} and 9.90 mg kg^{-1} to 8.10 mg kg^{-1}, 14.60 mg kg^{-1} and 26.40 mg kg^{-1}, with increase rates of 63%, 105% and 167%, respectively; available potassium varied from 85.60 mg kg^{-1}, 56.78 mg kg^{-1} and 111.75 mg kg^{-1} to 99.60 mg kg^{-1}, 160.70 mg kg^{-1} and 170.40 mg kg^{-1}, with increase rates of 16%, 183% and 52%, respectively.

Relationships among fertilizer inputs, crop yields and soil fertility in different regions of Shaanxi province

Because farmers used little P and K fertilizers in the 1970s and 1980s (Figure 3), only PFP of N, NP and NPK were calculated in the study (Eq. 2). The PFP of N, NP and NPK on wheat and maize decreased from the 1970s to the 2000s as a whole in the different regions (Table 2). The PFP of N on wheat in the Weibei and Guanzhong regions were 42 kg kg^{-1} and 65 kg kg^{-1}, respectively, in the 1970s, which decreased to 23 kg kg^{-1} and 33 kg kg^{-1}, respectively, in the 2000s. Meanwhile the PFP of N on maize in the North and Guanzhong regions were 76 kg kg^{-1} and 118 kg kg^{-1}, respectively, and they decreased to 33 kg kg^{-1} and 28 kg kg^{-1} from the 1970s to the 2000s. In the Weibei region, the PFP of N on maize changed slightly from 28 kg kg^{-1} to 32 kg kg^{-1}, which resulted from the use of high N inputs relative to the other two regions (up to 89 kg ha^{-1}) in the 1970s (Figure 3). This led to

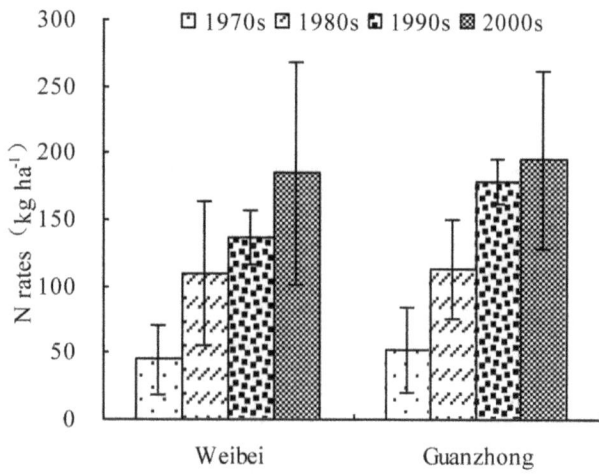

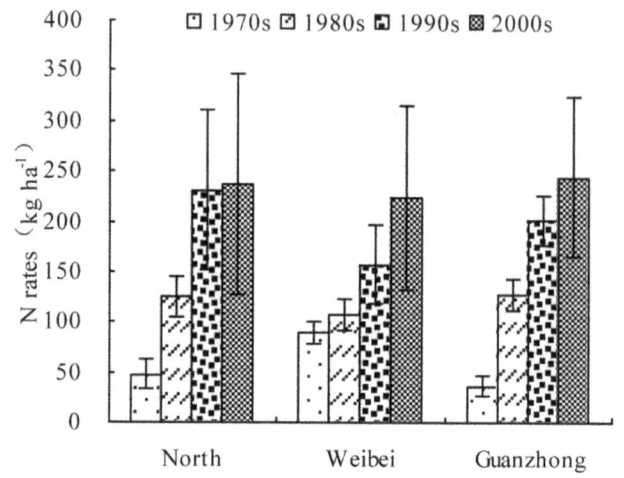

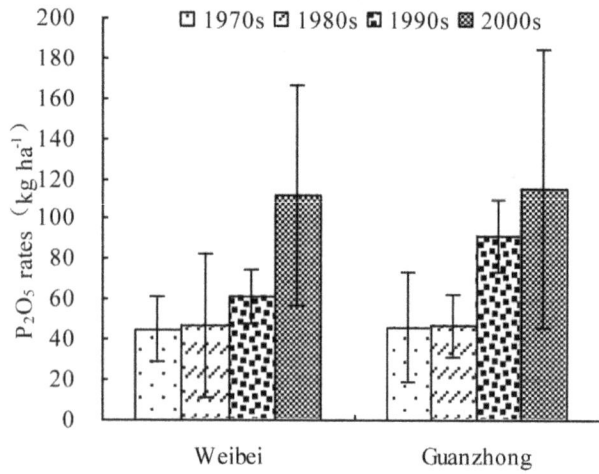

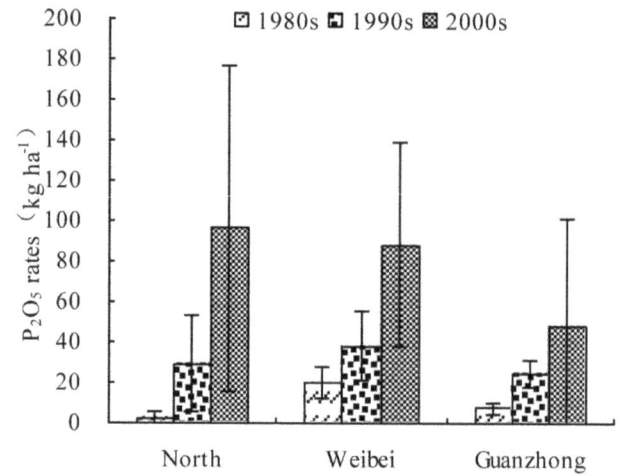

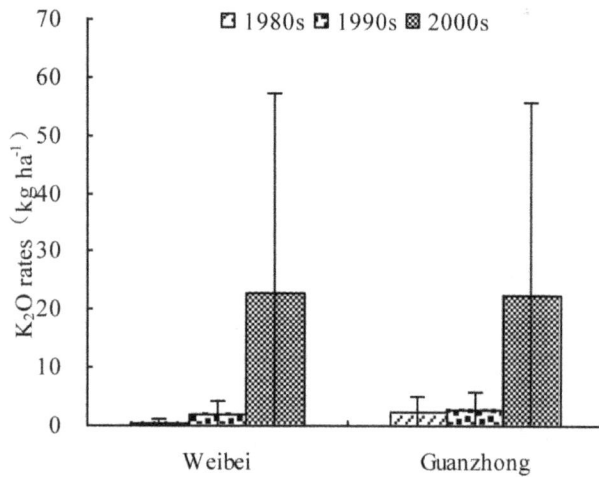

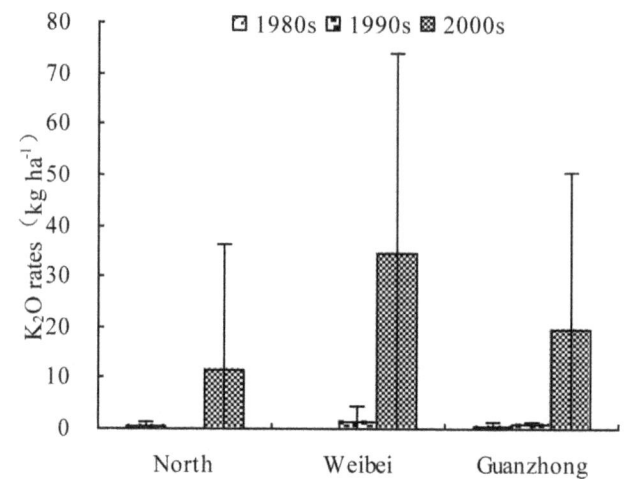

Wheat   **Maize**

Figure 3. Variations of chemical fertilization for wheat and maize at the farmers' level in different regions of Shaanxi province (error bars show standard deviations).

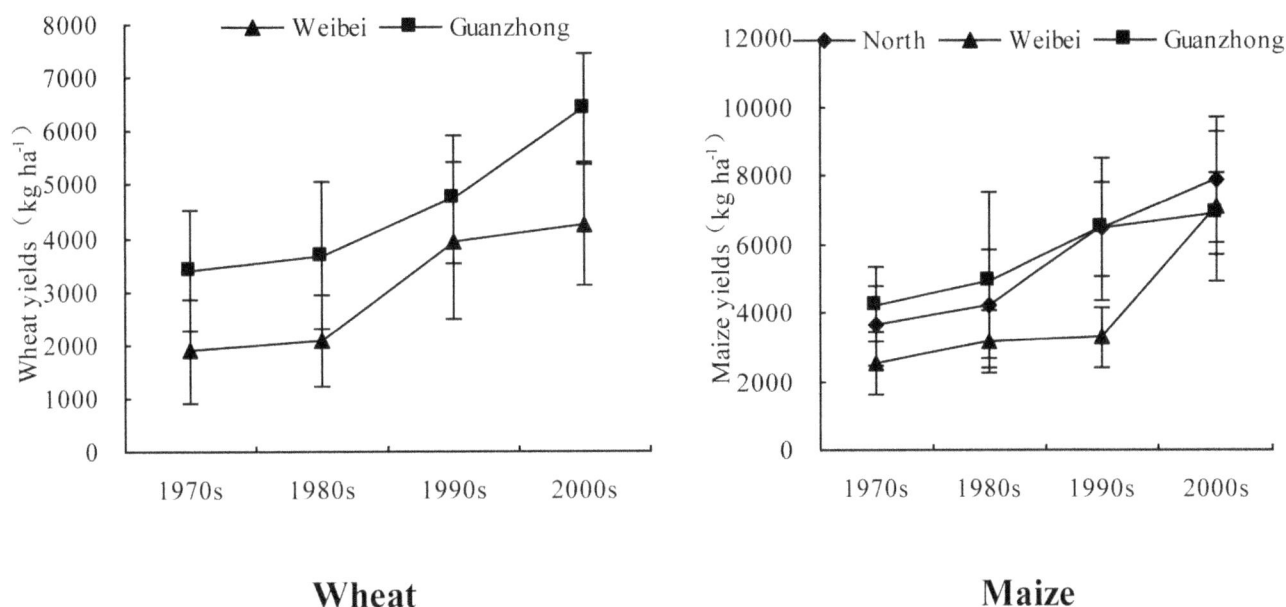

Wheat

Maize

Figure 4. Variations of yields for wheat and maize at the farmers' level in different regions of Shaanxi province (error bars show standard deviations).

a low PFP of N in that period. The PFP of NP on wheat decreased to 14 kg kg^{-1} and 21 kg kg^{-1} in the Weibei and Guanzhong regions, respectively; in maize it decreased to 24 kg kg^{-1}, 23 kg kg^{-1} and 24 kg kg^{-1} in the North, Weibei and Guanzhong regions, respectively. Similar to N and NP, the PFP of NPK on wheat decreased to 13 kg kg^{-1} and 19 kg kg^{-1} in the Weibei and Guanzhong regions, respectively; in maize it decreased to 23 kg kg^{-1}, 20 kg kg^{-1} and 22 kg kg^{-1} in the North, Weibei and Guanzhong regions, respectively (Table 2).

In order to find relationships among soil fertility, crop yields and fertilizer rates, we used the Weibei region as an example. The values of fertilization, crop yields and soil fertility did not have one to one correspondence from the 1970s to the 1990s, so their mean value from each period was examined (Figures 7 and 8). Although the sample size was small and some relationships did not reach significant levels, with the increase in N fertilizer inputs, soil total nitrogen and alkali-hydrolysis nitrogen both increased. P fertilizer increased soil available phosphorus and K fertilizer increased soil available potassium significantly (Figure 7). At the same time, soil organic matter, total nitrogen, alkali-hydrolysis nitrogen, available phosphorus and available potassium all had positive impacts on wheat yields (Figure 8).

Discussion

Fertilizer use efficiency of both wheat and maize decreased from the 1970s to the 2000s as a whole in the Loess Plateau of Shaanxi (Table 2), which was consistent with national trends. Nitrogen fertilizer, phosphorus fertilizer and potassium fertilizer use efficiencies were 30–35%, 15–20% and 35–50%, respectively, from 1981 to 1983, and the average values decreased to 28%, 12% and 32% on cereal crops by 2001 to 2005 in China [19]. This suggested that the effect of chemical fertilizers on increasing grain production had diminished. The PFP of N on wheat in the Weibei and Guanzhong regions decreased to 23 kg kg^{-1} and 33 kg kg^{-1}, respectively, and the PFP of N on maize in the North, Weibei and Guanzhong regions were 33 kg kg^{-1}, 32 kg kg^{-1} and 28 kg kg^{-1}, respectively, in the 2000s (Table 2). Zhang et al. [19] reported

average PFP values of N for wheat and maize of 43 kg kg^{-1} and 52 kg kg^{-1}, respectively, in China. Dobermann and Cassman [20] reported a global average PFP of N for cereals of 44 kg kg^{-1}. This indicated that nitrogen use efficiency on wheat and maize in the Loess Plateau of Shaanxi was much lower than the current national and global levels. Excessive fertilization has been the main reason for low fertilizer use efficiency in China [19]. In addition, Liu et al. [21] reported that in agro-ecosystems, surplus N increased from 1978 to 2005 throughout the country, and our findings on the Loess Plateau were consistent with this trend. For example, in the 2000s, chemical fertilizer N inputs on maize were 237 kg ha^{-1}, 223 kg ha^{-1} and 244 kg ha^{-1} in the North, Weibei and Guanzhong regions, respectively (Figure 3); meanwhile N surpluses on maize plots were 64 kg ha^{-1}, 67 kg ha^{-1} and 93 kg ha^{-1}, respectively, in the three regions (Figure 5). This indicated that excessive N fertilization was a serious problem in the Loess Plateau, and the same phenomenon has been reported many times in China, for example, in Beijing [16,22], Shandong [1,23–25], and Jiangsu [26–27]. Excessive N fertilization not only wastes resources, but also leads to many serious environmental problems [28–31] including nitrate pollution of groundwater [32–37], eutrophication of surface water [38–39], greenhouse gas emissions and other forms of air pollution [40–42], acid rain [43–46], soil acidification [36,47–50] and so on. On the other hand, a lower fertilization rate does not necessarily reduce crop yields [51]. Many studies have shown that reducing the current N application rates by 30 to 60% could increase N fertilizer efficiency, while still maintaining crop yields and substantially reducing N losses to the environment [31,52–53].

Like nitrogen, phosphate fertilizer inputs (Figure 3), P surpluses (Figure 5) and soil available phosphorus levels (Figure 6) all increased in the last 40 years on the Loess Plateau in Shaanxi. Similar results have been noted in north China and all over the country [25,54]. Yang et al. [55] reported that maintaining soil available phosphorus at a relatively high level requires a P application rate of about 80 kg ha^{-1} yr^{-1} in winter wheat/ summer maize rotation systems in the Guanzhong region. Our

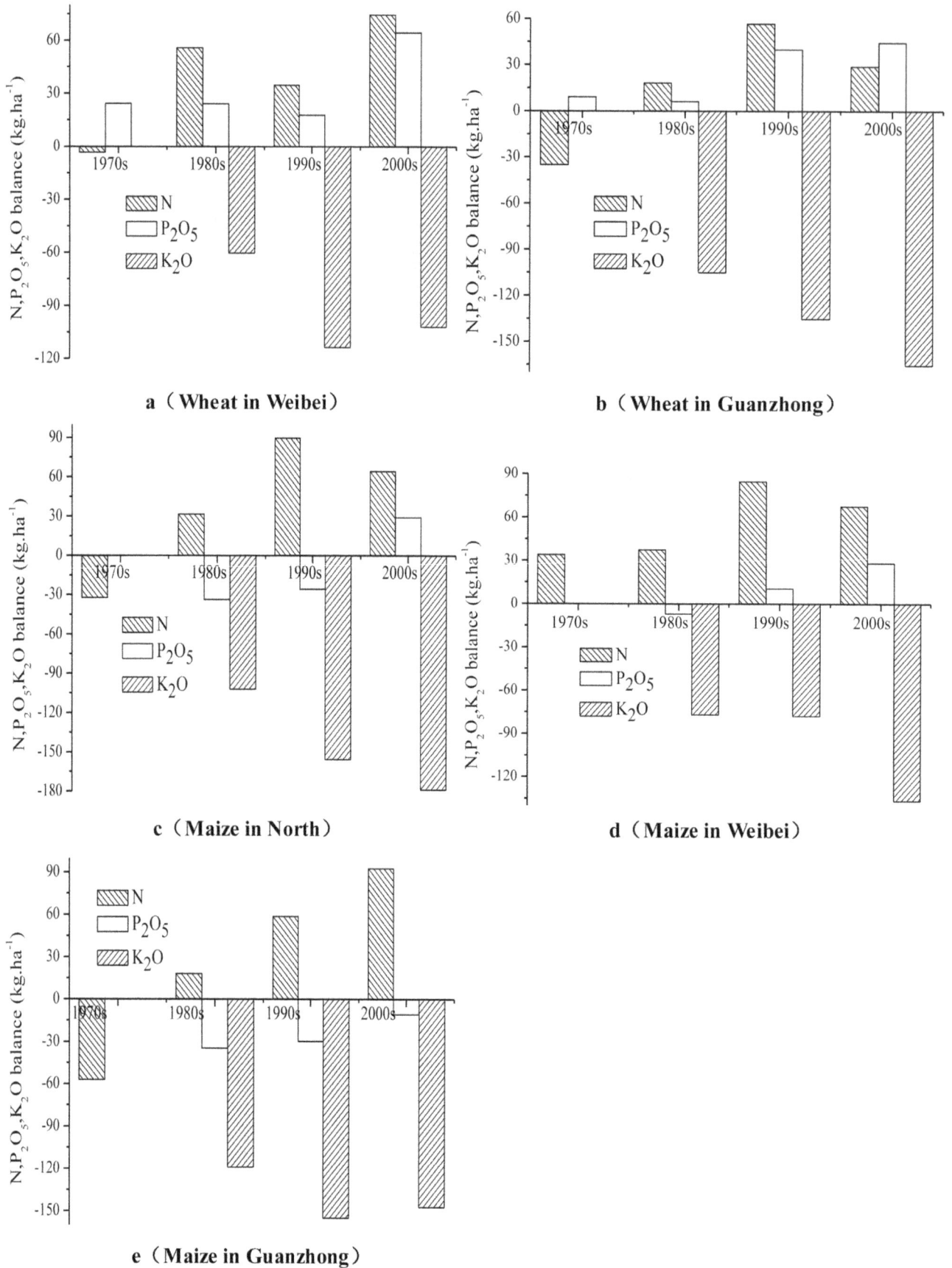

a（Wheat in Weibei）

b（Wheat in Guanzhong）

c（Maize in North）

d（Maize in Weibei）

e（Maize in Guanzhong）

Figure 5. Variations of soil nutrient balance on wheat and maize plots in different regions of Shaanxi province.

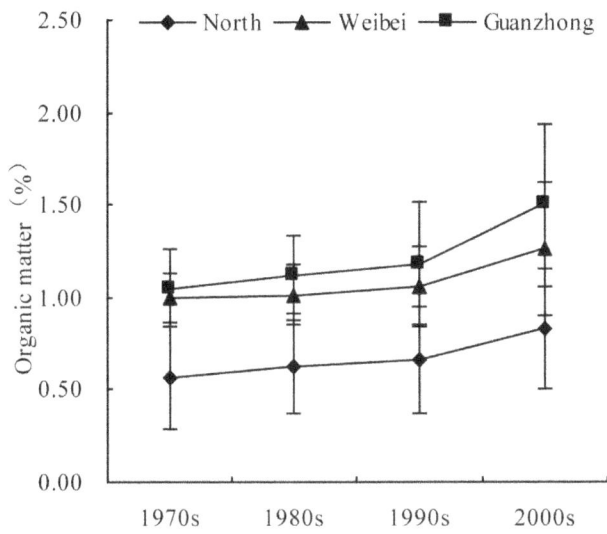

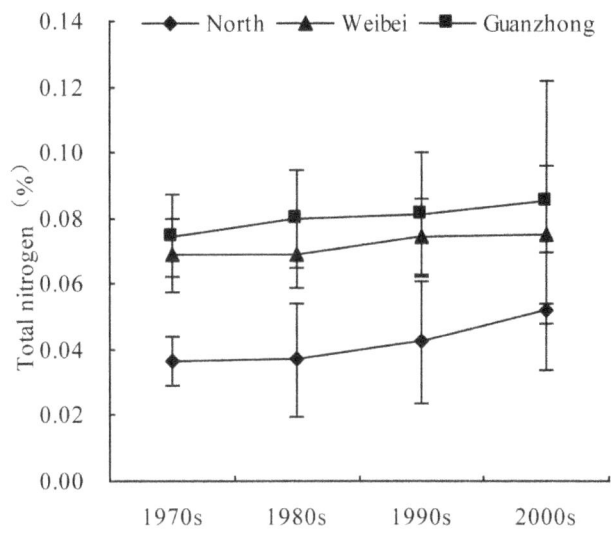

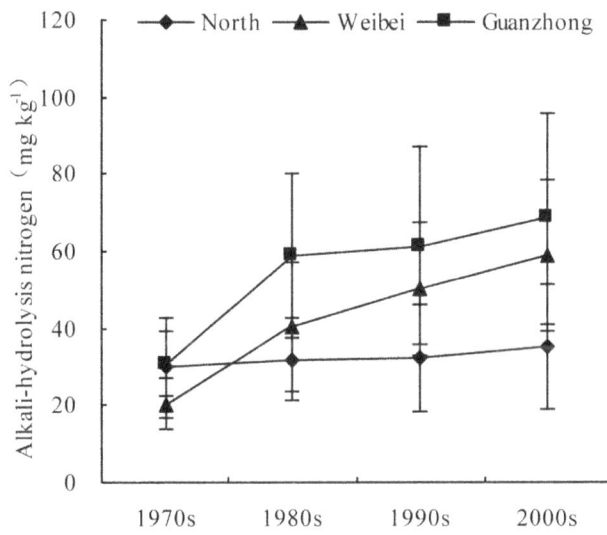

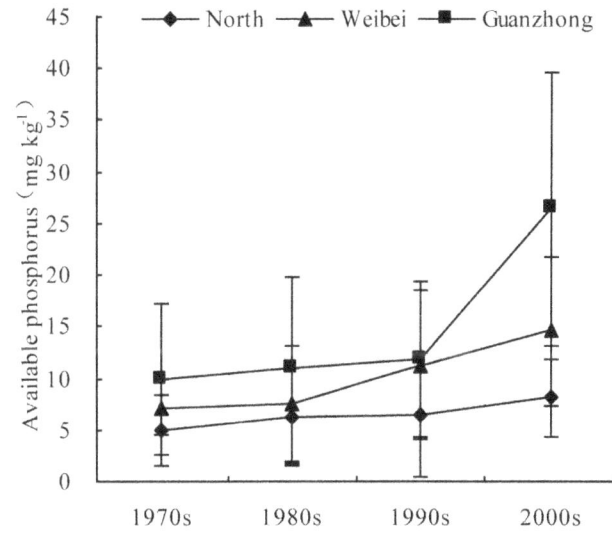

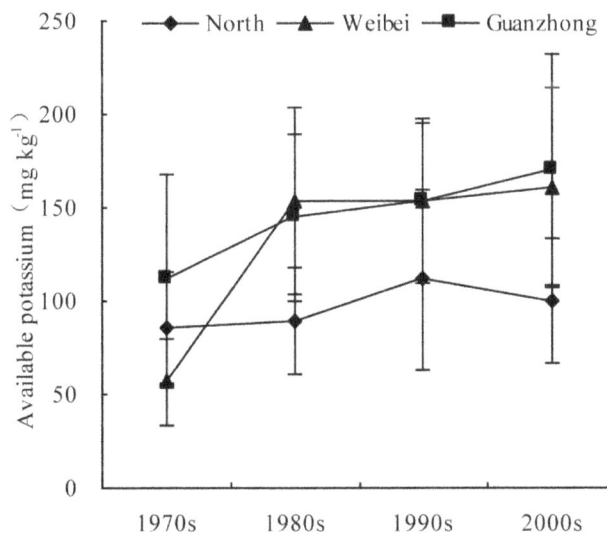

Figure 6. Variations of soil organic matter, total nitrogen, alkali-hydrolysis nitrogen, available phosphorus and available potassium in different regions of Shaanxi province (error bars show standard deviations).

results showed phosphate fertilizer inputs of up to 163 kg ha^{-1} in winter wheat/summer maize rotation systems in this region in the 2000s (Figure 3). This indicated that P fertilization was also excessive, which not only wasted resources but also led to many serious environmental problems [28–31]. Phosphate fertilizer production consumes more than 80% of the phosphate rock resources [56], but phosphate rock resources are limited and high grade material is in short supply [57]. In addition, the phosphate fertilization utilization ratio of the main crops ranged from 7% to 20%. It averages 12% in China [19], which has led to phosphorus accumulation in the soil, increasing the risk of non-point source pollution from surface runoff [58]. Agricultural non-point source pollution has become an increasingly serious problem in China, primarily because it leads to eutrophication.

In spite of increased K fertilizer inputs on wheat and maize in recent years (Figure 3), the soil K balance has become increasingly negative (Figure 5) and soil available potassium has increased (Figure 6) in the last 40 years. This phenomenon was previously reported in northwest and north China [55,59–60]. Evidently, K fertilizer application was not the only source of K absorbed by crops. The primary sources of K for crops were weathering of parent materials [60–61], release of K into the soil from increased soil organic matter and changes in soil pH [61]. Yang et al. [55] found that soil organic matter content in all treatments (including those without fertilizer) significantly increased over time and soil pH dropped from the initial value of 8.65 to 8.58 from 1991 to 2010 during long-term field trials in the Guanzhong region. Our results showed that in the North, Weibei and Guanzhong regions soil organic matter increased from 0.57%, 1.01% and 1.12% to 0.83%, 1.26% and 1.50%, respectively, from the 1970s to the 2000s (Figure 6). The average soil pH has declined 0.5 units with the overuse of N fertilizer in the past two decades in China [62]. Li et al. [63] reported that the soil pH decreased from the initial value of 8.76 to 8.56 from 1992 to 2008 during long-term field trials in

the North region. There may be other mechanisms involved, for example, crops might draw on K in the deeper soil layers or from the non-exchangeable pool. The contribution of K from the subsoil could be considerable [64]. Witter and Johansson [65] found that 41–47% of the K was from the subsoil for green manure crops. Many studies have shown that crops use non-exchangeable K [66–67]. Decreases in the abundance of non-exchangeable K with simultaneous increases in exchangeable and water-soluble K concentrations suggest that much of the K taken up by crops comes from non-exchangeable species via solution and exchangeable phases in a way that establishes and maintains the equilibrium between various forms of K in the soil [66].

Fertilizer rates had a large effect on soil fertility. With the increase in N fertilizer inputs, both soil total nitrogen and alkali-hydrolysis nitrogen increased; P fertilizer increased soil available phosphorus and K fertilizer increased soil available potassium significantly in the Weibei region (Figure 7). It has been reported that after 25 years of N fertilization, soil organic carbon and total nitrogen had increased by 18% and 26%, respectively, from 1984 to 2009 in the Weibei region [18]. Cai and Hao [68] also found that accumulation of soil nitrogen initially increased and then decreased with increasing nitrogen, and total nitrogen and alkali-hydrolysis nitrogen content reached the highest value or the second highest value of 135 kg ha^{-1} on wheat plots in the Weibei region, which was in accordance with findings in northwest and north China by Li et al. [63] and Lin et al. [69]. Through long-term field experimentation on the Loess Plateau in Shaanxi, Li et al. [63] and Hao et al. [70] found that with increases in P fertilizer inputs, soil available P increased significantly. Similar results have been obtained in northeast and northwest China by Geng et al. [71] and Zhao et al. [72], and also in America by Griffin et al. [73]. In addition, Li et al. [74] found that with increased K fertilizer inputs, soil available K increased significantly in a long-term field experiment on the Loess Plateau. Further-

Table 2. Variations of PFP of fertilizer on wheat and maize in the different regions (kg kg^{-1}).

Crop	Fertilizer type	Region	1970s	1980s	1990s	2000s
Wheat	N	Weibei	42	19	29	23
		Guanzhong	65	33	26	33
	N+P$_2$O$_5$	Weibei	21	13	20	14
		Guanzhong	34	23	17	21
	N+P$_2$O$_5$+K$_2$O	Weibei	21	13	20	13
		Guanzhong	34	23	17	19
Maize	N	North	76	34	28	33
		Weibei	28	30	21	32
		Guanzhong	118	39	32	28
	N+P$_2$O$_5$	North	76	33	25	24
		Weibei	28	25	17	23
		Guanzhong	118	37	29	24
	N+P$_2$O$_5$+K$_2$O	North	76	33	25	23
		Weibei	28	25	17	20
		Guanzhong	118	37	29	22

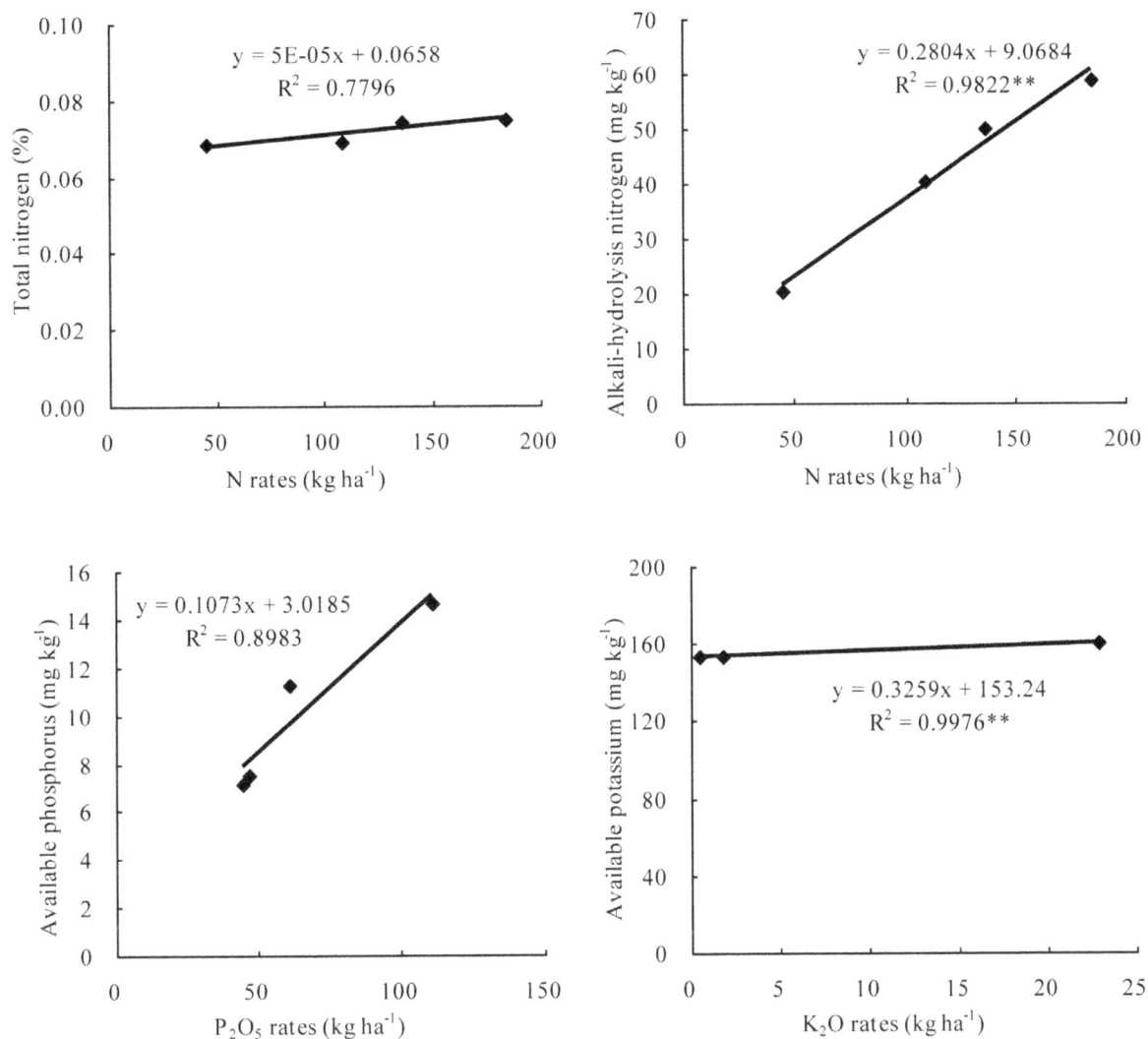

Figure 7. Relationships between N rates and total nitrogen, N rates and alkali-hydrolysis nitrogen, P₂O₅ rates and available phosphorus and K₂O rates and available potassium on wheat plots in the Weibei region of Shaanxi province. **Significance level: P< 0.01.

more, many studies in this area have shown that on the basis of N and P fertilizer application, long-term K fertilizer application can increase soil available K and grain yields [75–76].

Our research also found that soil fertility had a positive impact on crop yields (Figure 8). Zhou et al. [77] revealed that soil organic carbon and total nitrogen concentrations had a significant effect on crop yields in the semi-arid Loess Plateau by long-term experimentation. Higher yields without fertilizer were generally obtained in soils with higher average soil organic matter concentrations. For example, yields without fertilizer <4000 kg ha^{-1} were obtained with average soil organic matter concentrations of 1.41% for winter wheat and 1.46 for summer maize. In contrast, average soil organic matter concentrations were 1.69% for winter wheat and 1.61% for summer maize for plots with yields>6000 kg ha^{-1} without fertilizer in north China [78]. Gong et al. [79] also found that the contribution percentage of basic soil productivity to wheat yield was significantly correlated with soil organic carbon, total nitrogen, available nitrogen, available phosphorus and available potassium in long-term soil fertility experiments in north China. Similar results have been obtained in

other parts of mainland China [80], indicating that inherent soil productivity contributed to the substantial increase in China's crop yields.

In addition, although the use of chemical fertilizers to supplement NPK nutrients in the soil is important, many researchers at home and abroad reported that the application of chemical fertilizer in combination with organic manure is helpful in maintaining soil fertility (especially soil organic carbon) and buffering capacity, and in reducing NO_3-N accumulation in the soil, while maintaining high soil productivity [4,81–87].

Conclusions

From the 1970s to the 2000s in the North, Weibei and Guanzhong regions of the Loess Plateau in Shaanxi province, chemical fertilizer NPK inputs and yields of wheat and maize increased at the farmers' level. In the 1970s, N was deficient on wheat and maize plots in the different regions; thereafter N was in surplus. In the same way, P gradually changed from deficit to surplus levels. In addition, soil organic matter, total nitrogen, alkali-hydrolysis nitrogen, available phosphorus and available

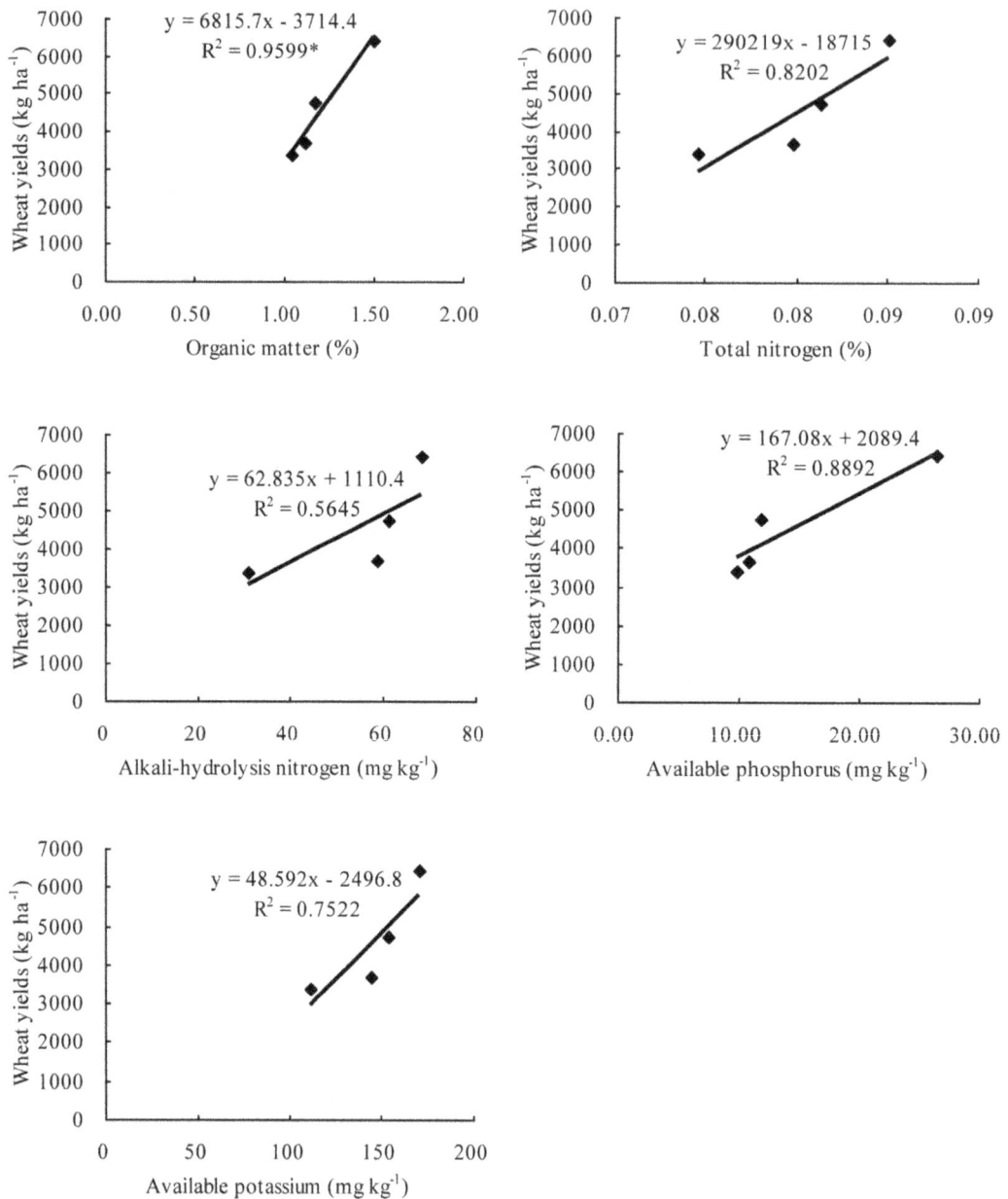

$y = 6815.7x - 3714.4$
$R^2 = 0.9599*$

$y = 290219x - 18715$
$R^2 = 0.8202$

$y = 62.835x + 1110.4$
$R^2 = 0.5645$

$y = 167.08x + 2089.4$
$R^2 = 0.8892$

$y = 48.592x - 2496.8$
$R^2 = 0.7522$

Figure 8. Relationships between wheat yield and soil organic matter, total nitrogen, alkali-hydrolysis nitrogen, available phosphorus and available potassium in the Weibei region of Shaanxi province. *Significance level: P<0.05.

potassium increased over the same period. However, K deficiencies became more and more severe. The PFP of N, NP and NPK on wheat and maize all decreased from the 1970s to the 2000s as a whole. With the increase in N fertilizer inputs, both soil total nitrogen and alkali-hydrolysis nitrogen increased; P fertilizer increased soil available phosphorus and K fertilizer increased soil available potassium significantly. At the same time, soil organic matter, total nitrogen, alkali-hydrolysis nitrogen, available phosphorus and available potassium all had positive impacts on crop yields. In order to promote food safety and environmental protection, farmers should be encouraged to assess their fertilizer needs carefully. Many can reduce nitrogen and phosphate fertilizer inputs significantly and increase potassium fertilizer and organic manure on cereal crops.

Acknowledgments

We are grateful to Harald Grip and Lars Lövdahl for their help in writing this paper. The authors would also like to thank the Agricultural Technology Extension Center of Shaanxi province for the help with data collection.

Author Contributions

Conceived and designed the experiments: YT PG. Analyzed the data: XW YG. Contributed reagents/materials/analysis tools: YT YG PG. Wrote the paper: XW. Collected the data: XW FL ZZ YP.

References

1. Cui ZL, Chen XP, Zhang FS (2010) Current nitrogen management status and measures to improve the intensive wheat–maize system in China. AMBIO 39: 376–384.

2. Gao C, Sun B, Zhang TL (2006) Sustainable nutrient management in Chinese agriculture: challenges and perspective. Pedosphere 16(2): 253–263.

3. Zhu ZL, Chen DL (2002) Nitrogen fertilizer use in China-Contributions to food production, impacts on the environment and best management strategies. Nutrient Cycling in Agroecosystems 63: 117–127.

4. Jiang D, Hengsdijk H, Dai TB, de Boer W, Qi J, et al. (2006) Long-term effects of manure and inorganic fertilizers on yield and soil fertility for a winter wheat-maize system in Jiangsu, China. Pedosphere 16(1): 25–32.

5. Luo SM (2007) To discover the secret of traditional agriculture and serve the modern ecoagriculture. Geographical Research 26(3): 609–615. (in Chinese).

6. Zhang F, Qiao Y, Wang F, Zhang W (2007) A perspective on organic agriculture in China: Opportunities and challenges. Proceedings of 9th German Scientific Conference on Organic Agriculture.

7. Department of Rural Surveys, National Bureau of Statistics (1971–2011) China Rural Statistical Yearbook. China Statistics Press. Beijing, China. (in Chinese).

8. Karlen DL, Mausbach MJ, Doran JW, Cline RG, Harris RF, et al. (1997) Soil quality: A concept, definition and framework for evaluation. Soil Science Society of America Journal 61: 4–10.

9. Wang XJ, Gong ZT (1998) Assessment and analysis of soil quality changes after eleven years of reclamation in subtropical China. Geoderma 81: 339–355.

10. Arshad MA, Martin S (2002) Identifying critical limits for soil quality indicators in agro-ecosystems. Agriculture, Ecosystems & Environment 88: 153–160.

11. Huang B, Sun WX, Zhao YC, Zhu J, Yang RQ, et al. (2007) Temporal and spatial variability of soil organic matter and total nitrogen in an agricultural ecosystem as affected by farming practices. Geoderma 139: 336–345.

12. Hoosbeek MR, Bryant RB (1992) Towards the quantitative modeling of pedogenesis: A review. Geoderma 55: 183–210.

13. Smil V (2001) Enriching the Earth: Fritz Haber, Carl Bosch, and the Transformation of World Food Production. MIT Press. Cambridge, UK.

14. Gong W, Yan XY, Wang JY (2011) Effect of long-term fertilization on soil fertility. Soils 43: 336–342. (in Chinese).

15. Bierman PM, Rosen CJ, Venterea RT, Lamb JA (2012) Survey of nitrogen fertilizer use on corn in Minnesota. Agricultural Systems 109: 43–52.

16. Wang SR (2002) Current status and evaluation of crop fertilization in Shaanxi province and Beijing city, Ph.D. thesis, China Agricultural University, Beijing, China. (in Chinese).

17. Liu GB (1999) Soil conservation and sustainable agriculture on the Loess Plateau: Challenges and prospects. AMBIO 28: 663–668.

18. Guo SL, Zhu HH, Dang TH, Wu JS, Liu WZ, et al. (2012) Winter wheat grain yield associated with precipitation distribution under long-term nitrogen fertilization in the semiarid Loess Plateau in China. Geoderma 189: 442–450.

19. Zhang FS, Wang JQ, Zhang WF, Cui ZL, Ma WQ, et al. (2008) Nutrient use efficiencies of major cereal crops in China and measures for improvement. Acta Pedologica Sinica 45: 915–924. (in Chinese).

20. Dobermann A, Cassman KG (2005) Cereal area and nitrogen use efficiency are drivers of future nitrogen fertilizer consumption. Science in China (Series C: Life Sciences) 48: 745–758.

21. Liu Z, Li BG, Fu J (2009) Nitrogen balance in agro-ecosystem in China from 1978 to 2005 based on DSS. Transactions of the CSAE 25(4): 168–175. (in Chinese).

22. Zhao JR, Guo Q, Guo JL, Wei DM, Wang CW, et al. (1997) The chemical fertilizer inputs and yields of grain fields in the suburbs of Beijing. Beijing Agricultural Sciences 15(2): 36–38. (in Chinese).

23. Ma WQ (1999) Current status and evaluation of crop fertilization in Shandong province, Ph.D. thesis, China Agricultural University, Beijing, China. (in Chinese).

24. Li JL, Cui DJ, Meng XX, Li XL, Zhang FS (2002) The study of fertilization condition and question in protectorate vegetable in Shouguang Shandong. Chinese Journal of Soil Science 33: 126–128. (in Chinese).

25. Zhen L, Zoebisch MA, Chen GB, Feng ZM (2006) Sustainability of farmers' soil fertility management practices: A case study in the North China Plain. Journal of Environmental Management 79: 409–419.

26. Richter J, Roelcke M (2000) The N-cycle as determined by intensive agriculture-examples from central Europe and China. Nutrient Cycling in Agroecosystems 57: 33–46.

27. Ma LH, Zhang Y, Sui B, Liu CL, Wang P, et al. (2011) The impact factors of excessive fertilization in Jiangsu province. Journal of Yangzhou University 32(2): 48–52, 80. (in Chinese).

28. Gao XZ, Ma WQ, Du S, Zhang FS, Mao DR (2001) Current status and problems of fertilization in China. Chinese Journal of Soil Science 32(6): 258–261. (in Chinese).

29. Cui ZL, Chen XP, Miao YX, Zhang FS, Sun QP, et al. (2008) On-farm evaluation of the improved soil N_{min}-based nitrogen management for summer maize in North China Plain. Agronomy Journal 100: 517–525.

30. Cui ZL, Zhang FS, Chen XP, Miao YX, Li JL, et al. (2008) On-farm evaluation of an in-season nitrogen management strategy based on soil N_{min} test. Field Crops Research 105: 48–55.

31. Ju XT, Xing GX, Chen XP, Zhang SL, Zhang LJ, et al. (2009) Reducing environmental risk by improving N management in intensive Chinese agricultural systems. Proceedings of the National Academy of Sciences 106: 3041–3046.

32. Tong YA, Emteryd O, Lu DQ, Grip H (1997) Effect of organic manure and chemical fertilizer on nitrogen uptake and nitrate leaching in a Eum-orthic anthrosols profile. Nutrient Cycling in Agroecosystems 48: 225–229.

33. Ju XT, Kou CL, Zhang FS, Christie P (2006) Nitrogen balance and groundwater nitrate contamination: Comparison among three intensive cropping systems on the North China Plain. Environmental Pollution 143: 117–125.

34. Ju XT, Liu XJ, Zhang FS, Roelcke M (2004) Nitrogen fertilization, soil nitrate accumulation, and policy recommendations in several agricultural regions of China. AMBIO 33: 300–305.

35. Yan X, Jin JY, He P, Liang MZ (2008) Recent advances on the technologies to increase fertilizer use efficiency. Agricultural Sciences in China 7(4): 469–479.

36. Guo SL, Wu JS, Dang TH, Liu WZ, Li Y, et al. (2010) Impacts of fertilizer practices on environmental risk of nitrate in semiarid farmlands in the Loess Plateau of China. Plant and Soil 330: 1–13.

37. Gao Y, Yu G, Luo C, Zhou P (2012) Groundwater nitrogen pollution and assessment of its health risks: A case study of a typical village in rural-urban continuum, China. PloS ONE 7(4): e33982.

38. Tilman D, Fargione J, Wolff B, D'Antonio C, Dobson A, et al. (2001) Forecasting agriculturally driven global environmental change. Science 292: 281–284.

39. Huang GQ, Wang XX, Qian HY, Zhang TL, Zhao QG (2004) Negative impact of inorganic fertilizer application on agricultural environment and its countermeasures. Ecology and Environment 13(4): 656–660. (in Chinese).

40. Mosier AR, Duxbury JM, Freney JR, Heinemeyer O, Minami K (1996) Nitrous oxide emissions from agricultural fields: Assessment, measurement and mitigation. Plant and Soil 181: 95–108.

41. Zhang JF, Han XG (2008) N_2O emission from the semi-arid ecosystem under mineral fertilizer (urea and superphosphate) and increased precipitation in northern China. Atmospheric Environment 42: 291–302.

42. Li H, Qiu JJ, Wang LG, Tang HJ, Li CS, et al. (2010) Modelling impacts of alternative farming management practices on greenhouse gas emissions from a winter wheat–maize rotation system in China. Agriculture, Ecosystems and Environment 135: 24–33.

43. Krusche AV, de Camargo PB, Cerri CE, Ballester MV, Lara LBLS, et al. (2003) Acid rain and nitrogen deposition in a sub-tropical watershed (Piracicaba): ecosystem consequences. Environmental Pollution 121: 389–399.

44. Menz FC, Seip HM (2004) Acid rain in Europe and the United States: an update. Environmental Science & Policy 7: 253–265.

45. Wu D, Wang SG, Shang KZ (2006) Progress in research of acid rain in China. Arid Meteorology 24(2): 70–77. (in Chinese).

46. Huang DY, Xu YG, Peng PA, Zhang HH, Lan JB (2009) Chemical composition and seasonal variation of acid deposition in Guangzhou, South China: Comparison with precipitation in other major Chinese cities. Environmental Pollution 157: 35–41.

47. Dai ZH, Liu YX, Wang XJ, Zhao DW (1998) Changes in pH, CEC, and exchangeable acidity of some forest soils in southern China during the last 32–35 years. Water, Air, and Soil Pollution 108: 377–390.

48. Zhang HM, Wang BR, Xu MG, Fan TL (2009) Crop yield and soil responses to long-term fertilization on a red soil in southern China. Pedosphere 19(2): 199–207.

49. Zhao X, Xing GX (2009) Variation in the relationship between nitrification and acidification of subtropical soils as affected by the addition of urea or ammonium sulfate. Soil Biology & Biochemistry 41: 2584–2587.

50. Huang S, Zhang WJ, Yu XC, Huang QR (2010) Effects of long-term fertilization on corn productivity and its sustainability in an Ultisol of southern China. Agriculture, Ecosystems and Environment 138: 44–50.

51. Ma WQ, István S (2008) Can sharp decrease of fertilizer input lead obvious reduction of crop yield?. Ecology and Environment 17: 1296–1301. (in Chinese).

52. Peng SB, Buresh RJ, Huang JL, Yang JC, Zou YB, et al. (2006) Strategies for overcoming low agronomic nitrogen use efficiency in irrigated rice systems in China. Field Crops Research 96: 37–47.

53. Yi Q, Zhang XZ, He P, Yang L, Xiong GY (2010) Effects of reducing N application on crop N uptake, utilization, and soil N balance in rice-wheat rotation system. Plant Nutrition and Fertilizer Science 16: 1069–1077. (in Chinese).

54. Cao N, Zhang YB, Chen XP (2009) Spatial-temporal change of phosphorus balance and the driving factors for agroecosystems in China. Chinese Agricultural Science Bulletin 25: 220–225. (in Chinese).

55. Yang XY, Sun BH, Zhang SL (2014) Trends of yield and soil fertility in a long-term wheat-maize system. Journal of Integrative Agriculture 13: 402–414.

56. Zhang WX (2011) Development and utilization trend of phosphate resources in China. Journal of Wuhan Institute of Technology 33: 1–5. (in Chinese).

57. Zhang WF, Ma WQ, Zhang FS, Ma J (2005) Comparative analysis of the superiority of China's phosphate rock and development strategies with that of the United States and Morocco. Journal of Natural Resources 20: 378–386. (in Chinese).

58. van Bochove E, Thériault G, Dechmi F, Leclerc ML, Goussard N (2007) Indicator of risk of water contamination by phosphorus: Temporal trends for the Province of Quebec from 1981 to 2001. Canadian Journal of Soil Science 87: 121–128.

59. Liu EK, Yan CR, Mei XR, He WQ, Bing SH, et al. (2010) Long-term effect of chemical fertilizer, straw, and manure on soil chemical and biological properties in northwest China. Geoderma 158: 173–180.

60. Tan DS, Jin J Y, Jiang LH, Huang SW, Liu ZH (2012) Potassium assessment of grain producing soils in North China. Agriculture, Ecosystems and Environment 148: 65–71.

61. Munson RD (1985) Potassium in Agriculture. Soil Science Society of America Madison, Wisconsin, USA.

62. Guo JH, Liu XJ, Zhang Y, Shen JL, Han WX, et al. (2010) Significant acidification in major Chinese croplands. Science 327: 1008–1010.

63. Li Q, Xu MX, Liu GB, ZhaoYG, Tuo DF (2013) Cumulative effects of a 17-year chemical fertilization on the soil quality of cropping system in the Loess Hilly Region, China. Journal of Plant Nutrition and Soil Science 176: 249–259.

64. Kautz T, Amelung W, Ewert F, Gaiser T, Horn R, et al. (2013) Nutrient acquisition from arable subsoils in temperate climates: A review. Soil Biology & Biochemistry 57: 1003–1022.

65. Witter E, Johansson G (2001) Potassium uptake from the subsoil by green manure crops. Biological Agriculture & Horticulture 19: 127–141.

66. Singh M, Singh VP, Reddy DD (2002) Potassium balance and release kinetics under continuous rice-wheat cropping system in Vertisol. Field Crops Research 77: 81–91.

67. Sharma A, Jalali VK, Arora S (2010) Non-exchangeable potassium release and its removal in foot-hill soils of North-west Himalayas. Catena 82: 112–117.

68. Cai Y, Hao MD (2013) Effects of long-term nitrogen fertilization on wheat in Loess Plateau. Journal of Triticeae Crops 33: 983–987. (in Chinese).

69. Lin ZA, Zhao BQ, Yuan L, Bing-So H (2009) Effects of organic manure and fertilizers long-term located application on soil fertility and crop yield. Scientia Agricultura Sinica 42: 2809–2819. (in Chinese).

70. Hao MD, Fan J, Wei XR, Pen LF, Lu L (2005) Effect of fertilization on soil fertility and wheat yield of dryland in the Loess Plateau. Pedosphere 15(2): 189–195.

71. Geng YH, Cao GJ, Ye Q, Qi QG, Wu P, et al. (2013) Effects of different phosphorus applications on soil available phosphorus, phosphorus absorption and yield of spring maize. Journal of South China Agricultural University 34: 470–474. (in Chinese).

72. Zhao J, Hou ZA, Li SX, Liu LP, Huang T, et al. (2014) Effects of P rate on soil available P, yield and nutrient uptake of maize. Journal of Maize Sciences 22: 123–128. (in Chinese).

73. Griffin TS, Honeycutt CW, He Z (2003) Changes in soil phosphorus from manure application. Soil Science Society of America Journal 67: 645–653.

74. Li LF, Hao MD, Li YM, Gao CQ (2009) Research on characteristics of spatial distribution and availability of soil potassium forms under long-term fertilization in the dryland of the Loess Plateau. Agricultural Research in the Arid Areas 27: 127–131, 142. (in Chinese).

75. Wang HT, Jin JY, Wang B, Zhao PP (2010) Effects of long-term potassium application and wheat straw return to cinnamon soil on wheat yields and soil potassium balance in Shanxi. Plant Nutrition and Fertilizer Science 16: 801–808. (in Chinese).

76. Zhang YL, Lu JL, Jin JY, Li ST, Chen ZQ, et al. (2012) Effects of chemical fertilizer and straw return on soil fertility and spring wheat quality. Plant Nutrition and Fertilizer Science 18: 307–314. (in Chinese).

77. Zhou ZC, Gan ZT, Shangguan ZP, Zhang FP (2013) Effects of long-term repeated mineral and organic fertilizer applications on soil organic carbon and total nitrogen in a semi-arid cropland. European Journal of Agronomy 45: 20–26.

78. Fan MS, Lai R, Cao J, Qiao L, Su YS, et al. (2013) Plant-based assessment of inherent soil productivity and contributions to China's cereal crop yield increase since 1980. PloS ONE, 8(9): e74617.

79. Gong FF, Zha Y, Wu XP, Huang SM, Xu MG, et al. (2013) Analysis on basic soil productivity change of winter wheat in fluvo-aquic soil under long-term fertilization. Transactions of the Chinese Society of Agricultural Engineering 29(12): 120–129. (in Chinese).

80. Tang YH, Huang Y (2009) Spatial distribution characteristics of the percentage of soil fertility contribution and its associated basic crop yield in mainland China. Journal of Agro-Environment Science 28: 1070–1078. (in Chinese).

81. Gami SK, Ladha JK, Pathak H, Shah MP, Pasuquin E, et al. (2001) Long-term changes in yield and soil fertility in a twenty-year rice-wheat experiment in Nepal. Biology and Fertility of Soils 34: 73–78.

82. Yang SM, Li FM, Malhi SS, Wang P, Suo DR, et al. (2004) Long-term fertilization effects on crop yield and nitrate nitrogen accumulation in soil in northwestern China. Agronomy Journal 96: 1039–1049.

83. Mando A, Ouattara B, Somado AE, Wopereis MCS, Stroosnijder L, et al. (2005) Long-term effects of fallow, tillage and manure application on soil organic matter and nitrogen fractions and on sorghum yield under Sudano-Sahelian conditions. Soil Use and Management 21: 25–31.

84. Li J, Zhao BQ, Li XY, Jiang RB, Bing SH (2008) Effects of long-term combined application of organic and mineral fertilizers on microbial biomass, soil enzyme activities and soil fertility. Agricultural Sciences in China 7(3): 336–343.

85. Banger K, Kukal SS, Toor G, Sudhir K, Hanumanthraju TH (2009) Impact of long-term additions of chemical fertilizers and farmyard manure on carbon and nitrogen sequestration under rice–cowpea cropping system in semi-arid tropics. Plant and Soil 318: 27–35.

86. Majumder B, Mandal B, Bandyopadhyay PK (2008) Soil organic carbon pools and productivity in relation to nutrient management in a 20-year-old rice-berseem agroecosystem. Biology and Fertility of Soils 44: 451–561.

87. Moharana PC, Sharma BM, Biswas DR, Dwivedi BS, Singh RV (2012) Long-term effect of nutrient management on soil fertility and soil organic carbon pools under a 6-year-old pearl millet–wheat cropping system in an Inceptisol of subtropical India. Field Crops Research 136: 32–41.

6

Evapotranspiration Measurement and Crop Coefficient Estimation over a Spring Wheat Farmland Ecosystem in the Loess Plateau

Fulin Yang[1], Qiang Zhang[1]*, Runyuan Wang[1], Jing Zhou[2]

1 Key Laboratory of Arid Climatic Change and Reducing Disaster of Gansu Province, Key Open Laboratory of Arid Climatic Change and Disaster Reduction of China Meteorological Administration (CMA), Institute of Arid Meteorology, CMA, Lanzhou, China, **2** State Key Laboratory of Grassland Agro-ecosystems, College of Pastoral Agriculture Science and Technology, Lanzhou University, Lanzhou, China

Abstract

Evapotranspiration (ET) is an important component of the surface energy balance and hydrological cycle. In this study, the eddy covariance technique was used to measure ET of the semi-arid farmland ecosystem in the Loess Plateau during 2010 growing season (April to September). The characteristics and environmental regulations of ET and crop coefficient (Kc) were investigated. The results showed that the diurnal variation of latent heat flux (LE) was similar to single-peak shape for each month, with the largest peak value of LE occurring in August (151.4 W m^{-2}). The daily ET rate of the semi-arid farmland in the Loess Plateau also showed clear seasonal variation, with the maximum daily ET rate of 4.69 mm day^{-1}. Cumulative ET during 2010 growing season was 252.4 mm, and lower than precipitation. Radiation was the main driver of farmland ET in the Loess Plateau, which explained 88% of the variances in daily ET (p<0.001). The farmland Kc values showed the obvious seasonal fluctuation, with the average of 0.46. The correlation analysis between daily Kc and its major environmental factors indicated that wind speed (Ws), relative humidity (RH), soil water content (SWC), and atmospheric vapor pressure deficit (VPD) were the major environmental regulations of daily Kc. The regression analysis results showed that Kc exponentially decreased with Ws increase, an exponentially increased with RH, SWC increase, and a linearly decreased with VPD increase. An experiential Kc model for the semi-arid farmland in the Loess Plateau, driven by Ws, RH, SWC and VPD, was developed, showing a good consistency between the simulated and the measured Kc values.

Editor: Wen-Xiong Lin, Agroecological Institute, China

Funding: This research was jointly supported by National Basic Research Program of China (2012CB955304, 2013CB430206), National Natural Science Foundation of China (31300376, 40830957, 41275118, 41305134), Natural Science Foundation of Gansu province (1208RJYA025), China Postdoctoral Science Foundation (2012M512044), Meteorological Research Program of Gansu Provincial Meteorological Service (2012-15), and Science Research Foundation for Drought Meteorology (IAM201312). The funders had no role in study design, data collection and analysis, decision to publish, or preparation of the manuscript.

Competing Interests: The authors have declared that no competing interests exist.

* E-mail: zhangqiang@cma.gov.cn

Introduction

Water cycle is the key process in the multi-layers interaction of earth system. Land surface evapotranspiration (ET), as the important segment of water cycle, is the main way for water consumption of earth system, playing an important role in regional and global climate [1,2]. Farmland ET refers to the overall water flux sent to the air by vegetation and earth surface, which is the important component of water balance. About 60% rainfall and 99% water in farmland system are consumed by ET around and the world [3]. Furthermore, as a component of energy balance, ET is also the important consumption of surface available energy [1]. Farmland irrigation and ET can occupy 60~80% of the net radiation during the growing season [4].

Recently, a series studies were carried out to investigate the characteristics of farmland ET, suing process-based models, remote sensing [5], calculation, and direct observation [6,7]. By applying eddy covariance technology, Zhang *et al.* [8] found that most of net radiation was consumed by crop latent heat during the growing period, and higher ratio between ET and net radiation for summer corn (*Zea mays* L.) than that of winter wheat (*Triticum*

aestivum L.) in Licheng area. Yang *et al.* [9] and Li *et al.* [10] investigated the effects of net radiation (Rn), soil water content (SWC), leaf area index (LAI) and rainfall seasonal distributions to the daily ET changes in corn ecosystem. On the other hand, the empirical models, such as Priestley–Taylor [11] and crop coefficient (Kc) methods [12], were mostly used to estimate the crop actual ET.

Loess Plateau is the unique land type and ecological environmental region in China, has a total area of 0.63 million km^2 with nearly 70% covered with thick loess. Most regions of the Loess Plateau receive less annual precipitation with large precipitation variation and suffer serious spring drought. Furthermore, the Loess Plateau is the climate demarcation area in China and the main distribution area of ecotone between agriculture and animal husbandry, an important and unique region whose water cycle will exert significant impact on the atmospheric circulation and agriculture development in East Asia. However, water resources shortage becomes the current bottleneck that restricts the agricultural sustainable development. Studies indicated that the Loess Plateau is the sensitive area to climate change as climate warming will aggravate the soil drought by accelerating the water exchange process [13]. ET, as the critical process of water circulation, had become an important scientific issue in agricultural sustainable development and impacts of climate change. Knowledge of the farmland ET process in the Loess Plateau is an important research content in water and energy balance of the terrestrial ecosystem, and promoting the reasonable utilization of limited water resources in the Loess Plateau.

Several measurement researches on the farmland ET in the Loess Plateau by lysimeter [14] or eddy correlation system [15] are become available. Moreover, Li [16,17] compared the adaptability of different ETo estimation methods in the Loess Plateau area and analyzed the time-space variation characteristics of crop reference evapotranspiration (ETo) in the Loess Plateau based on Penman-Monteith model. Zhang *et al.* [18] compared the differences of observing methods for land surface ET in the Loess Plateau, showing that the actual ET estimated based on Penman-Monteith model and crop coefficient is significantly lower than that observed through eddy covariance method and lysimeter.

The spring wheat is the main grain crop in the semi-arid rain-fed region of Loess Plateau. Precipitation and ET are critical for spring wheat, which is the main water-demande crop of region, because the irrigation and infiltration were unavailable completely. An accurate estimate of crop ET is required for appropriate water management in order to increase water uses efficiency in the water-limited region [12]. Crop coefficient method was recommended by Food and Agriculture Organization of the United Nations to estimate the ET amount in the different terrestrial ecosystems, which is widely applied in various ecological systems, such as crops, grassland ecosystem [19,20]. In terms of Kc method, it is the key to obtain specific reliable Kc value to apply Kc method for ET estimation accurately. Some studies revealed that Kc not only has significant differences in different types of ecological system, but also is influenced by various environment factors [20,21]. However, most current researches regard Kc as a constant or suppose the Kc during the specific crop growing period as a constant value, ignoring its daily variation [18]. The approximate treatment on Kc values will influence the accuracy of ET estimation. Therefore, it is of much significance for ET dynamic simulation to investigate Kc characteristics of different underlying surfaces.

In this study, we carried out an successive ET observation in semi-arid farmland ecosystem in the Loess Plateau during 2010 growing season (from 1st April to 30th September) by eddy covariance system, analyzed the dynamics of farmland ET and Kc, and then developed an empirical Kc model.

Materials and Methods

1.1 Ethics Statement

No specific permits were required for the described field studies. The location is not privately-owned or protected in any way during the study period, and the field studies did not involve endangered or protected species in these areas.

1.2 Study Site

The study site carried out at Dingxi Arid Meteorology and Ecological Environment Experimental Station (35°33′ N,104°35′ E, 1896.7 m a.s.l), where located in the west of Dingxi city in Gansu Province, China. This area belongs to semi-arid temperate continent climate with an annual average temperature of 7.1°C. The lowest average monthly temperature is 7.0°C (January) and the highest average monthly temperature is 19.0°C (July). It has an average annual precipitation of 382.5 mm (1979 to 2008, from the Dingxi meteorological station), with most of the precipitation

(80%) falling between May and September. It has flat and homogenous underlying surface, and the soil type is loess-like loam with an average bulk density of 1.38 g cm^{-3}. The dominate crops was the spring wheat (*Triticum aestivum* L. cv. 'Dingxi 24').

1.3 Water and Heat Fluxes Measurements

Although there are several methods in ET measurements currently, the eddy covariance technology has become one of prior measure approach in determining the water exchange between atmosphere and land ecological system boundary layer due to its advantages of less theoretical assumptions and high measurement accuracy, and the measure data are often used to examine the simulation accuracy of model [7,22]. The farmland ET in semi-arid area of the Loess Plateau was measured by eddy covariance in this study. The eddy covariance system, composed of three-dimensional sonic anemometer (CSAT-3, Campbell Scientific, USA) and an open path infrared gas analyzer (IRGA, Li-7500, LI-COR, USA), is mainly applied for measuring the exchange of latent heat flux and sensible heat flux between surface and atmosphere. The sampling frequency was set as 10 Hz and the real-time observed data was recorded by the data logger (CR5000, Campbell Scientific, USA).

The positive values of latent heat flux and sensible heat flux represents that energy transferred from land surface to atmosphere, while the negative values mean opposite. Before data calculation, the latent heat flux and sensible heat flux were treated with spike detection and removal and coordinate rotation were performed. In addition, sonic temperature fluctuations were taken into account to correct the fluxes of sensible heat; the Webb-Pearman-Leuning correction was used to adjust density changes resulting from fluctuations in latent heat [23]. Furthermore, all anomalous values of latent heat flux and sensible heat flux were deleted through following the criteria [20]: (1) the incomplete measuring data caused by power failure, instrument calibration, and so on; (2) precipitation events; (3) anomalous data detected by Papale *et al.* [24]. About 15.9% of EC flux data within the observation period in 2010 were deleted. Falge *et al.* [25] method was applied to fill the data gaps.

1.4 Environmental Factors Measurements

The experimental site was equipped with an automatic weather system (Campbell Scientific, USA) to measure environmental factors on the semi-arid farmland ecosystem in the Loess Plateau, including precipitation (PPT) (52203, RM Young, USA), net radiation (Rn) (CNR-1, Kipp & Zonen, Netherlands), air temperature (Ta), air relative humidity (RH) (HMP45C, Vaisala, Finland), wind speed (Ws) at height of 2 m, and soil water content (SWC) at 10 cm depth (CS616, Campbell Scientific, USA). The sampling frequency of both common meteorological and soil environmental factors was set as 0.1 Hz, and the data acquisition unit (CR1000, Campbell Scientific, USA) was applied for data storage. The half-hourly soil heat fluxes were calculated through Ts and SWC data, which were expressed in Yang *et al.* [26] in detail.

1.5 Reference Evapotranspiration and Crop Coefficient

The daily ETo of the semi-arid farmland in the Loess Plateau was calculated by Penman-Monteith model, as following [19]:

$$ET_o = \frac{0.408\Delta(Rn-G)+\gamma\frac{900}{T+273}u_2(e_s-e_a)}{\Delta+\gamma(1+0.34u_2)} \quad (1)$$

where ETo is the reference ET, Rn is net radiation; G is soil heat flux; γ is the psychrometric constant (kPa°C^{-1}); Ta is air temperature (°C); u is wind speed; e_s is saturation vapor pressure; e_a is actual vapor pressure; e_s-e_a represents the differential saturation vapor pressure; and Δ is the slope of saturation vapor pressure curve (kPa°C^{-1}). Crop actual water consumption (ET) can be calculated from referential evapotranspiration (ETo) and Kc [27]. Numerically, crop coefficient is the ratio of ET to ETo [19]:

$$Kc = ET/ET_o \qquad (2)$$

ET measured by eddy covariance directly was regarded as ET (mm), while the calculation result obtained through Penman-Monteith model was regarded as ETo (mm).

1.6 Statistical Parameters

A quantitative evaluation on the fitting effect of ET model was conducted by using root-mean-square error, index of agreement, coefficient of determination and regression coefficient through the origin. The more the root-mean-square error close to 0, and the more the index of agreement, coefficient of determination and regression coefficient through the origin close to 1, that indicates the better the fitting effect of the model [20].

Results and Discussion

2.1 Characteristics of Environmental Factors

The seasonal variation of PPT, SWC, and Rn from April to September are presented in Figure 1. There was totally 285.9 mm precipitation in the whole observation period, with the maximum daily precipitation of 26.9 mm (DOY 180). There had a little rainfall during early to middle of May as well as middle and late June, appearing obvious periodic drought. SWC was very sensitive to rainfall as a daily precipitation, especially the over 10 mm rainfall events. The seasonal variation of SWC showed significant fluctuation during the observation period, and SWC ranged from 7.2% to 19.9%, with a mean value of 12.0%. An obvious drought was appeared in the farmland ecosystem during the middle and late June. There was no rainfall record for 18 days successively, from 10th to 27th June, and the SWC on 27th June, with only valued 8.5%. This serious drought decreased the RH significantly, but greatly increased atmospheric vapor pressure deficit (VPD) during this period. RH and VPD produced approximate opposite variation trend within the observation period. Furthermore, Rn and Ta showed obvious seasonal variations. Although the Rn in July was relative higher, it produced relative greater daily fluctuation during the whole observation period. Ta was relative stable, with a mean value of 15.4°C, compared to Rn. However, Ta in April was relatively low, increased gradually later, reached the peak in the beginning of August.

2.2 Characteristics of Farmland ET in the Semi-arid Area of the Loess Plateau

According to Figure 2, the diurnal variation of monthly latent heat flux during the study period was character as "single-peak" curve, without the "ET highland" phenomenon around noon that reported by Guo et al. [28] in the winter wheat farmland. For the nighttime, latent heat flux was lower with relative stable, but it increased gradually after sunrises, reaching the peak around noon, which then decreased again and became stable till sunset. The latent heat flux peaks appeared between 12:00 and 14:00. The latent heat flux peaks appeared about 12:00 in April and May, while it appeared about one and half hour later (about 13:30) in

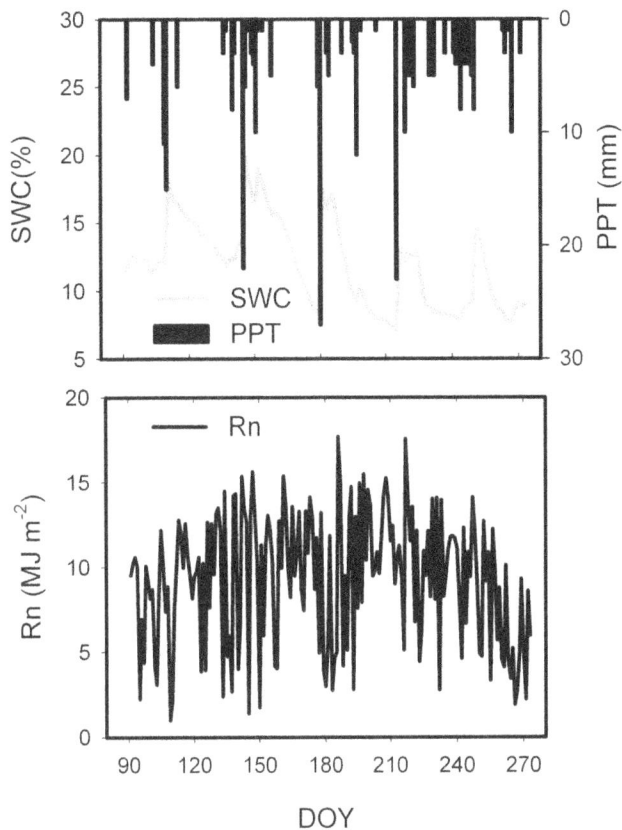

Figure 1. Seasonal variation of environmental factors over the semi-arid farmland ecosystem in the Loess Plateau.

June, July, August and September. Besides, there were significant differences between daily peaks of latent heat flux in each month, with the highest daily peak appeared in August while the lowest daily peak appeared in May. The latent heat flux peak values from April to September were 121.2, 92.3, 105.2, 135.2, 151.4 and 106.9 W m^{-2} respectively. The peak value of latent heat flux and H lagged about one hour later that of Rn. The daily variations of three components (Rn, latent heat flux and sensible heat flux) of energy balance didn't lie in same phase, which may be related with their different ways of energy transmission. The main energy resources of Rn (solar radiation, atmospheric radiation and earth surface radiation) were transmitted through electromagnetic wave, while that of latent heat flux and sensible heat flux was mainly transmitted through atmospheric turbulence with the later transmission far slower than the former one. Moreover, the different physical measure plane of radiation sensor and EC system may be another important reason causing the phase differences of energy components [29].

There was significant seasonal variation in daily ET rate of the semi-arid farmland ecosystem in the Loess Plateau (Figure 3), with the maximum daily ET rate of 4.69 mm day^{-1} (DOY 115), minimum daily ET rate of 0.24 mm day^{-1} (DOY 180) and the mean daily ET rate of 1.38±0.75 mm day^{-1} during the whole growing season. Furthermore, there was obvious difference in monthly ET, with relatively high in July and August, but low in May and September. The accumulative ET during the study period was 252.4 mm, lower than 11.7% precipitation and significantly lower than the averaged precipitation over years in the same period (329.3 mm, 1979 to 2008 observation data from Dingxi weather station), which indicated that the farmland ET

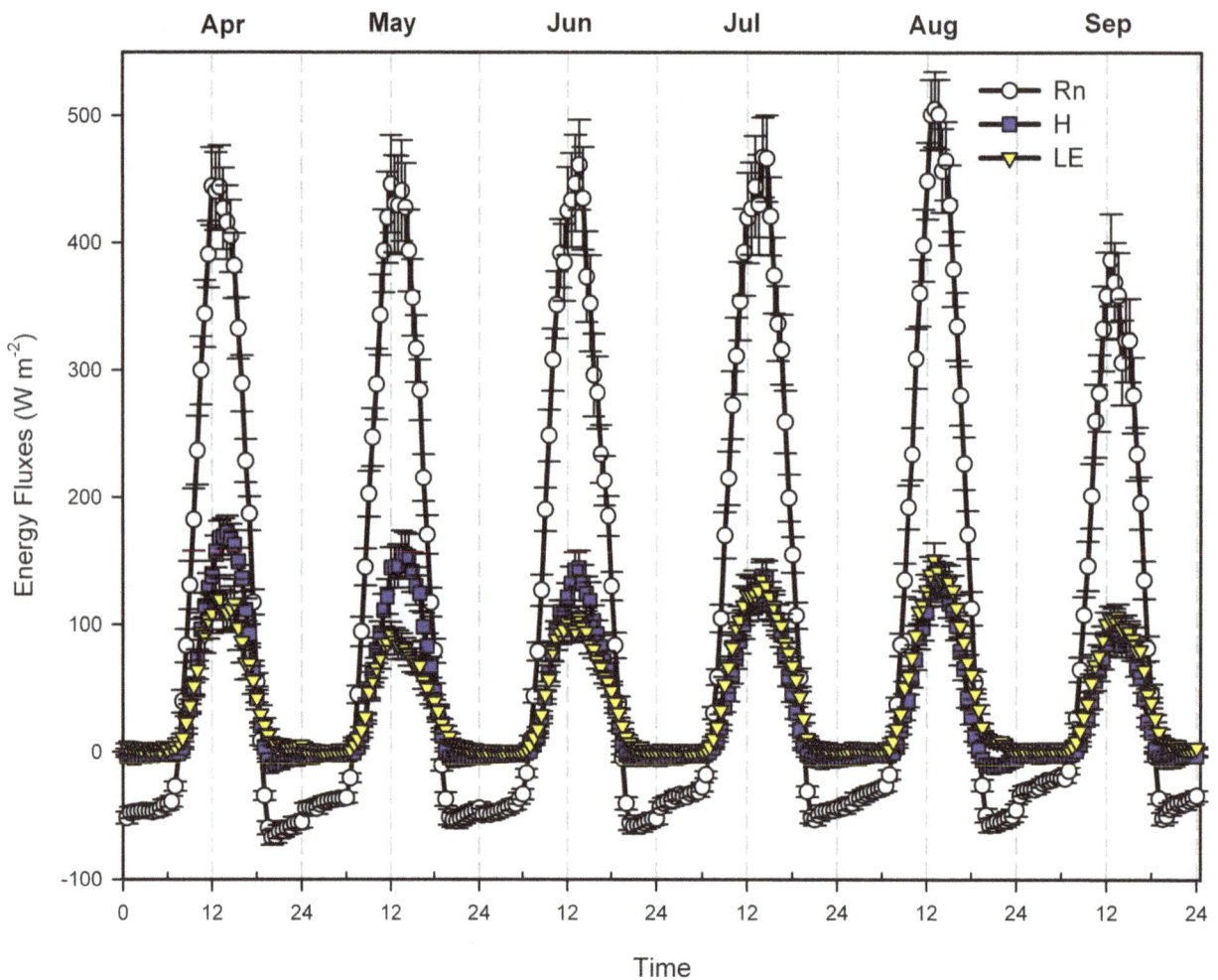

Figure 2. Monthly averaged diurnal variations of energy fluxes over the semi-arid farmland ecosystem in the Loess Plateau. Error bars represent one standard error. Rn, net radiation (W m^{-2}); H, sensible heat flux (W m^{-2}); LE, latent heat flux (W m^{-2}).

mainly derived from natural rainfall during 2010 growing season in the semi-arid area of the Loess Plateau.

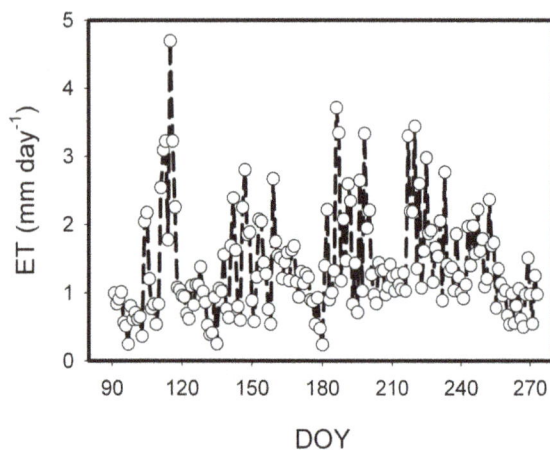

Figure 3. Seasonal variations of evapotranspiration (ET) over the semi-arid farmland ecosystem in the Loess Plateau.

Seasonal variations in abiotic variables (Rn, SWC, Ta, and VPD) and biotic variables (LAI) exert regulations on water exchange between atmosphere and terrestrial ecosystems [30,31]. Rn was the main environmental factor of ET in the semi-arid farmland of the Loess Plateau, followed by SWC (Figure 1, 3). ET was highly linked with net radiation (Rn), and there was a significant linear relationship between ET and Rn. Rn explained 88% of the variances in daily ET (regression analysis, p<0.001, Figure 4a). Similar environmental controls were found by Yu *et al.* [32] with Rn as the main driver of ET in a wetland and seasonal patterns of ET closely following radiation. Similarly, a linear increase of ET with SWC was observed (Figure 4b). Nevertheless, SWC explained only 44% of the variance in ET.

2.3 Characteristics of Farmland Crop Coefficient in the Semi-arid Area of the Loess Plateau

The farmland Kc values during the growing season in the semi-arid area of the Loess Plateau showed the obvious seasonal fluctuation (Figure 5), with relatively higher in the middle and late April. There were six days with over 1.0 of Kc value, and the Kc value on 25th April even reached 1.70. However, Kc decreased dramatically in the beginning of May, which valued less than 1 and mainly fluctuated near 0.50 in the following months (from June to September). Therefore, rainfall events will exert obvious

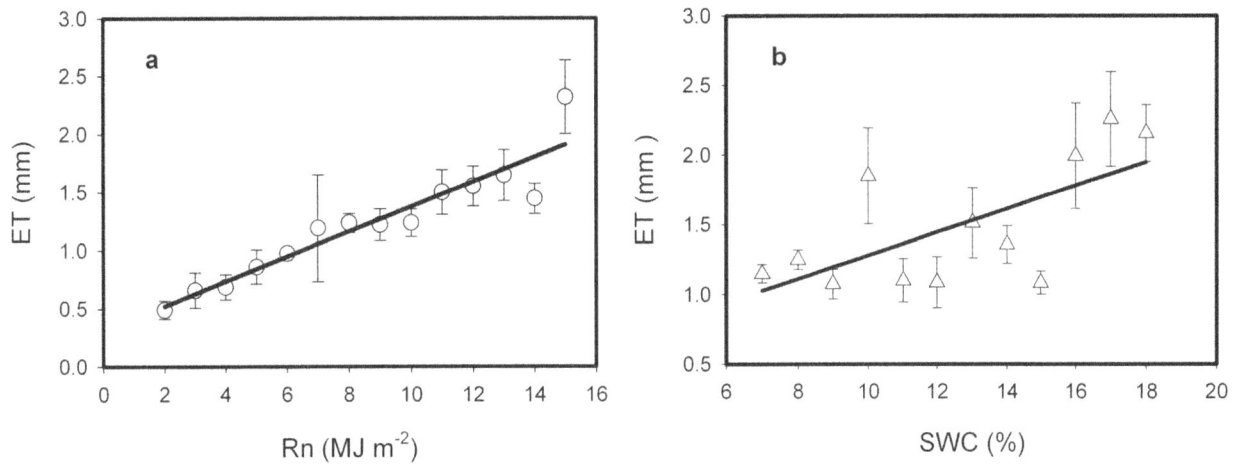

Figure 4. The response of evapotranspiration (ET) to net radiation (Rn) and soil water content (SWC) over the semi-arid farmland ecosystem in the Loess Plateau. ET data were averaged with Rn bins of 1 MJ m^{-2} (a), and SWC bins of 1% (b), respectively, during the growing season over the semi-arid farmland ecosystem in the Loess Plateau. Error bars represent one standard error.

impact on the seasonal variation of Kc. The Kc reached the lowest value (0.10) in 12th May due to the less rainfall and drought weather in the middle of May. The averaged Kc during the whole observation period was 0.46±0.25 (mean ± standard deviation).

2.4 Relationship between Kc and Environmental Factors

The Pearson correlation analysis between Kc and major environment factors showed that Ws was the most important in controlling Kc of semi-arid farmland (p<0.01). RH and VPD also showed very significant correlation to Kc (p<0.01), so as SWC (p<0.05) at the 95% confidence level. However, Kc were no significant correlation with Rn and Ta (p>0.05) (Table S1). Based on the correlation analysis results, Kc can be expressed by its four significant correlated environmental factors statistically, Ws, RH, SWC and VPD (Eq. 3):

$$Kc = f(Ws, RH, SWC, VPD) \qquad (3)$$

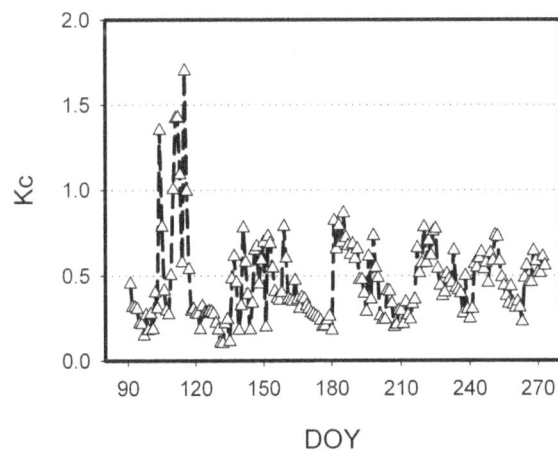

Figure 5. Seasonal variation of crop coefficient (Kc) over the semi-arid farmland ecosystem in the Loess Plateau.

The response ways of Kc to Ws, RH, SWC and VPD were shown in Figure 6. For the sake of analysis expediently, Ws, RH, SWC and VPD were grouped into bins with the following criterion: 0.5 m s^{-1}, 10%, 2%, and 0.2 kPa, respectively. According to the regression analysis result, Kc decreased exponentially with Ws, but increased exponentially with RH and SWC, and decreased linearly with VPD (Figure 6, Table 1).

2.5 Development of Kc Model

Based on the correlation and regression analysis between Kc and its major environmental factors, an empirical daily Kc model for the semi-arid farmland in the Loess Plateau that driven by Ws, RH, SWC and VPD, could be given by the following equation:

$$Kc = a\exp(-bWs) \times c\exp(dRH) \\ \times e\exp(fSWC) \times (g - hVPD) \qquad (4)$$

After mathematical deduction, and then:

$$Kc = a\exp(-bWs + cRH + dSWC) \times (e - fVPD) \qquad (5)$$

where a, b, c, d, e, and f are fitting parameters, Ws is wind speed; RH is relative air humidity; SWC is soil water content; VPD is vapor pressure deficit; and Kc is crop coefficient. The data of clear days (n = 132) during 2010 growing season was divided into two groups according to the order of Kc magnitude alternately, one data group for fitness, and another data group for verification. Based on daily environmental data and Kc data for the growing season in 2010, an empirical Kc model was developed (n = 66, $R^2 = 0.70$):

$$Kc = 0.231\exp(-0.362Ws + 0.015RH + 0.033SWC) \\ \times (0.958 - 0.042VPD) \qquad (6)$$

Eq. 6 can be considered as the empirical model for farmland Kc estimating in the semi-arid area of the Loess Plateau. Then, another group of data during the growing season 2010 was used to validate the empirical Kc model (Eq. 6). Statistical indices of root-

Table 1. Regression analysis between daily crop coefficient (Kc) and mainly environmental factors over the semi-arid farmland ecosystem in the Loess Plateau.

Factors	Regression equation	n	R^2	F	P
Ws (m s^{-1})	Kc = 1.06exp(−0.52Ws)	8	0.97	182.0	<0.0001
RH (%)	Kc = 0.16exp(0.02RH)	7	0.86	132.9	<0.0001
SWC (%)	Kc = −0.22exp(0.05SWC)	6	0.67	8.0	<0.05
VPD (kPa)	Kc = 0.51−0.12VPD	9	0.85	38.6	<0.001

Ws, wind speed (m s^{-1}); RH: air relative humidity (%); SWC, soil water content (%); VPD, vapor pressure deficit (kPa).

mean-square error, index of agreement, coefficient of determination and regression coefficient through the origin were 0.09, 0.90, 0.67, and 1.00, respectively, indicating that this Kc model can be able to well describe the Kc dynamics (Figure 7). However, it can be seen from Figure 7 that some scatters discussive from the 1:1 line, which might be that the biotic factors were not took into account in the Kc model, such as LAI, biomass.

2.6 Magnitude of ET and Kc with other Agroecosystems

The peak spring wheat ET value (4.69 mm day^{-1}) of the semi-arid farmland ecosystem in the Loess Plateau was approached to

the peak ET value of 4.47 mm day^{-1} for the winter wheat on a loess loam soil in a semi-humid region of the northwest China reported by Kang et al. [33]. However, the peak ET in the Loess Plateau farmland ecosystem was lower than that of the irrigated winter wheat in the semi-humid region of the North China Plain, which between 6.6 and 7.8 mm day^{-1} found by Lei et al. [34], and as large as 6–9 mm day^{-1} reported by Liu et al. [35]. The peak spring wheat ET value was also lower than that of 7.0 mm day^{-1} for the cultivated wheat ecosystem in north-central Oklahoma USA [36]. In the semi-arid area of the Loess Plateau, the drought weather condition and few rainfalls, constrained the soil water

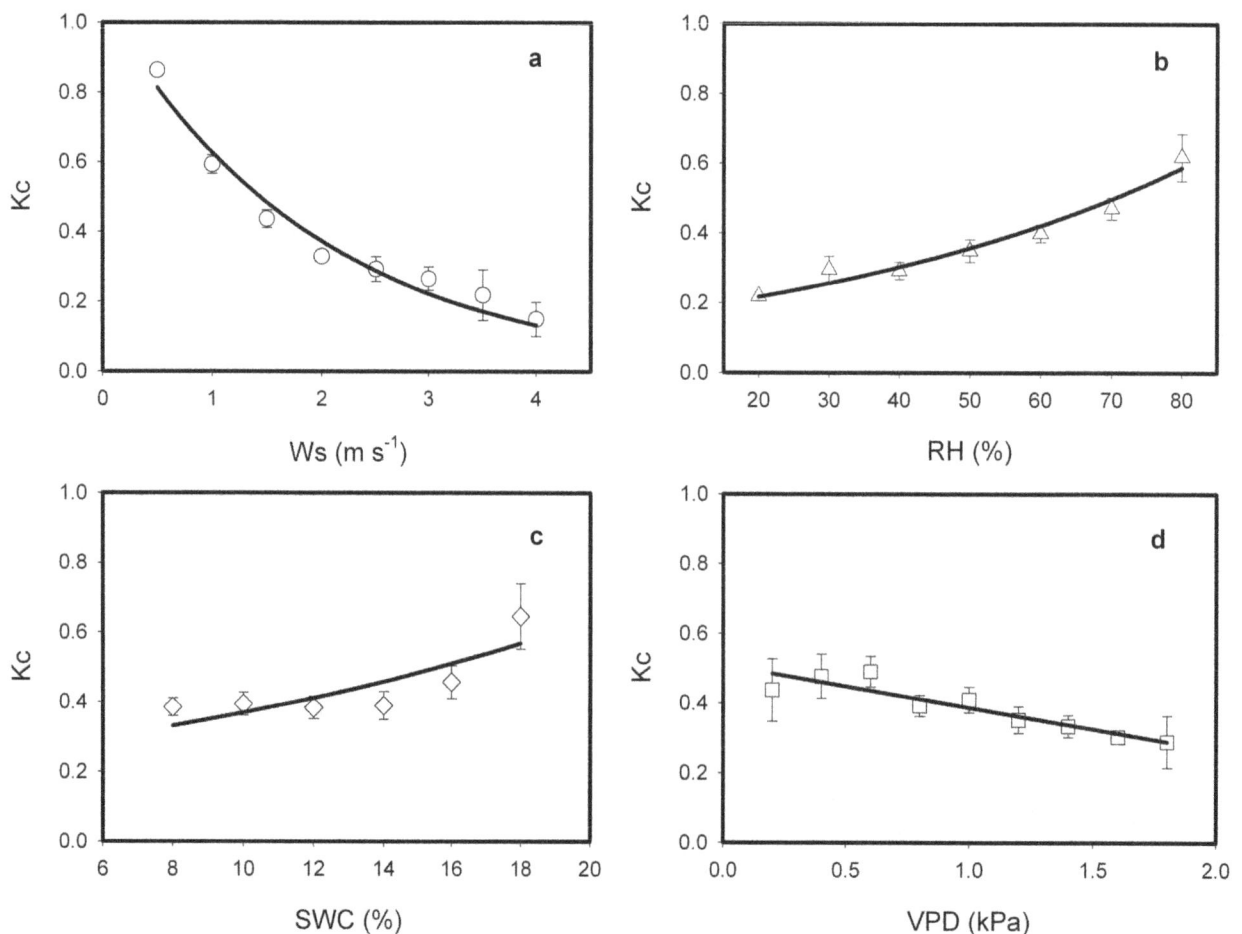

Figure 6. The response of crop coefficient (Kc) to wind speed (Ws), relative humidity (RH), soil water content (VPD), and vapor pressure deficit (VPD), respectively. Kc data were averaged with Ws bins of 0.5 m s^{-1} (a), RH bins of 10% (b), SWC bins of 2% (c) and VPD bins of 0.2 kPa, respectively, during the growing season over the semi-arid farmland ecosystem in the Loess Plateau. Error bars represent one standard error.

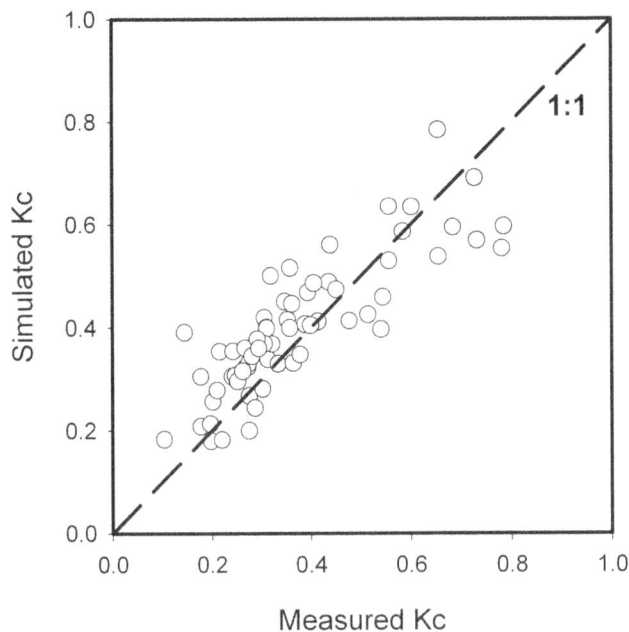

Figure 7. Comparison of simulated and measured crop coefficient (Kc) values.

availability for plant growth, may be the main environmental limiting factor for the farmland ecosystem ET. Moreover, the lack of irrigation management in the traditional farming system in the study region could also result in the relative low ET.

Li *et al.* [37] reported that the spring wheat Kc in an Inner Mongolia of China was 0.55 (initial stage), 1.03 (crop development stage), 1.19 (mid-season stage) and 0.65 (late-season stage), respectively, indicating the average Kc higher than that of 0.46 during the whole season in the Loess Plateau. The average Kc from this study was also significantly lower than 0.93 for winter wheat and 1.1 for corn in North China Plain found by Liu *et al.* [35]. The relatively low Kc value may be related with the lower precipitation and ET in the semi-arid area of the Loess Plateau. However, the averaged Kc was higher than Kc value reported for grassland field in the Liudaogou basin of the Loess Plateau (below 0.3) [38]. Based on the obvious seasonal variation of daily Kc value, it is necessary to be cautious about applying Kc as a constant for ET estimation in the Loess Plateau. The Kc

variability might explain possibly that the lower ET estimation derived from Penman-Monteith model with fixed Kc values than that observed through eddy covariance and lysimeter in the Loess Plateau reported by Zhang *et al.* [18].

Conclusions

The diurnal latent heat flux of the semi-arid farmland ecosystem in the Loess Plateau showed a single-peak shape. There was significant variance in daily ET rate with a maximum daily ET rate of 4.69 mm day^{-1}. The relative low daily ET could be related to the drought weather condition and lack of irrigation in study region. Radiation was the main driver of ET of the semi-arid farmland ecosystem in the Loess Plateau. Cumulative ET during the study period estimated directly by eddy covariance method was lower than precipitation received during the same period in this farmland ecosystem. The farmland Kc values showed the seasonal fluctuation, and significantly correlated with Ws, RH, SWC and VPD. Regression analysis results showed that Kc decreased exponentially with increasing Ws, increased exponentially with RH and SWC, and decreased linearly with increasing VPD. The four environmental factors can explain most of the day-to-day variation in Kc, and then an empirical daily Kc model for the semi-arid farmland ecosystem in the Loess Plateau was developed, which showed a good performance in consistency between the simulated and measured Kc.

Supporting Information

Table S1 Pearson product-moment correlation coefficients between daily crop coefficient (Kc) and daily average values for other variables: on a daily basis for clear days during growing season of 2010 over the semi-arid farmland ecosystem in the Loess Plateau.

Acknowledgments

We gratefully acknowledge the Drs. Heling Wang, Sheng Wang, Ping Yue, Kai Zhang and Hong Zhao for their contributions during the experiment and data processing assistance.

Author Contributions

Conceived and designed the experiments: QZ FY. Performed the experiments: FY RW. Analyzed the data: FY JZ. Contributed reagents/materials/analysis tools: FY JZ. Wrote the paper: FY JZ.

References

1. Jung M, Reichstein M, Ciais P, Seneviratne SI, Sheffield J, et al. (2010) Recent decline in the global land evapotranspiration trend due to limited moisture supply. Nature 467: 951–954.
2. Wang K, Dickinson RE (2012) A review of global terrestrial evapotranspiration: Observation, modeling, climatology, and climatic variability. Rev Geophys 50: RG2005, doi:10.1029/2011RG000373.
3. Kite G (2000) Using a basin-scale hydrological model to estimate crop transpiration and soil evaporation. J Hydrol 229: 59–69.
4. Suyker AE, Verma SB (2008) Interannual water vapor and energy exchange in an irrigated maize-based agroecosystem. Agric Forest Meteorol 148: 417–427.
5. Trezza R, Allen RG, Tasumi M (2013) Estimation of actual evapotranspiration along the Middle Rio Grande of New Mexico using MODIS and landsat imagery with the METRIC model. Remote Sens 5: 5397–5423.
6. Pauwels V, Samson R (2006) Comparison of different methods to measure and model actual evapotranspiration rates for a wet sloping grassland. Agric Water Manage 82: 1–24.
7. Qin Z, Yu Q, Xu SH, Hu BL, Sun XM, et al. (2004) Water, heat fluxes and water use efficiency measurement and modeling above a farmland in the North China Plain. Sci China Ser D 34: 183–192. (in Chinese).

8. Zhang YQ, Sheng YJ, Liu CM, Yu Q, Sun HY, et al. (2002) Measurement and analysis of water, heat and CO$_2$ flux from a farmland in the north China plain. Acta Geogr Sinica 57: 333–342. (in Chinese with English abstract).
9. Yang XG, Liu HL, Wang YL, Yu FN (2003) Research on evapotranspiration of field ecological system of summer maize in northern plain. Chinese J Eco-Agric 11: 66–68. (in Chinese with English abstract).
10. Li YJ, Xu ZZ, Wang YL, Zhou L, Zhou GS (2007) Latent and sensible heat fluxes and energy balance in a maize agroecosystem. J Plant Ecol 31: 1132–1144. (in Chinese with English abstract).
11. Ding R, Kang S, Li F, Zhang Y, Tong L (2013) Evapotranspiration measurement and estimation using modified Priestley-Taylor model in an irrigated maize field with mulching. Agric Forest Meteorol 168: 140–148.
12. Zhang B, Liu Y, Xu D, Zhao N, Lei B, et al. (2013) The dual crop coefficient approach to estimate and partitioning evapotranspiration of the winter wheat-summer maize crop sequence in North China Plain. Irrig Sci 31: 1303–1316.
13. Xin ZB, Xu JX, Zheng W (2007) Effects of climate changes and human activities on the variations of vegetation cover on the Chinese Loess Plateau. Sci China Ser D 37: 1504–1514. (in Chinese).
14. Wang YQ, Fan J, Shao MA (2010) Rules of soil evaporation and millet evapotranspiration in rain-fed region of Loess Plateau in Northern Shaanxi. T Chinese Soc Agric Eng 26: 6–10. (in Chinese with English abstract).

15. Wang G, Huang J, Guo W, Zuo J, Wang J, et al. (2010) Observation analysis of land-atmosphere interactions over the Loess Plateau of northwest China. J Geophys Res 115(D00K17): doi:10.1029/2009JD013372.

16. Li Z (2012) Applicability of simple estimating method for reference crop evapotranspiration in Loess Plateau. T Chinese Soc Agric Eng 28: 106–111. (in Chinese with English abstract).

17. Li Z (2012) Spatiotemporal variations in the reference crop evapotranspiration on the Loess Plateau during 1961–2009. Acta Ecol Sinica 32: 4139–4145. (in Chinese with English abstract).

18. Zhang Q, Zhang ZX, Wen XM, Wang S (2011) Comparison of observational methods of land surface evapotranspiraion and their influence factors. Adv Earth Sci 26: 538–547. (in Chinese with English abstract).

19. Allen RG, Pereira LS, Raes D, Smith M (1998) Crop evapotranspiration: guidelines for computing crop water requirements. FAO Irrigation and Drainage Paper. Rome: Food and Agriculture Organization. 17–28 p.

20. Yang FL, Zhou GS (2011) Characteristics and modeling of evapotranspiration over a temperate desert steppe in Inner Mongolia, China. J Hydrol 396: 139–147.

21. Lockwood JG (1999) Is potential evapotranspiration and its relationship with actual evapotranspiration sensitive to elevated atmospheric CO_2 levels?. Climatic Change 41: 193–212.

22. Yu GR, Wen XF, Sun XM, Tanner BD, Lee XH, et al. (2006) Overview of ChinaFLUX and evaluation of its eddy covariance measurement. Agric Forest Meteorol 137: 125–137.

23. Webb EK, Pearman GI, Leuning R (1980) Correction of flux measurements for density effects due to heat and water vapour transfer. Q J Roy Meteor Soc 106: 85–100.

24. Papale D, Reichstein M, Aubinet M, Canfora E, Bernhofer C, et al. (2006) Towards a standardized processing of Net Ecosystem Exchange measured with eddy covariance technique: algorithms and uncertainty estimation. Biogeosciences 3: 571–583.

25. Falge E, Baldocchi D, Olson R, Anthoni P, Aubinet M, et al. (2001) Gap filling strategies for long term energy flux data sets. Agric Forest Meteorol 107: 71–77.

26. Yang K, Wang JM (2008) A temperature prediction-correction method for estimating surface soil heat flux from soil temperature and moisture data. Sci China Ser D 51: 721–729.

27. Villalobos FJ, Testi L, Orgaz F, Omar GT, Alvaro LB, et al. (2013) Modelling canopy conductance and transpiration of fruit trees in Mediterranean areas: A simplified approach. Agric Forest Meteorol 171: 93–103.

28. Guo JX, Mei XR, Lin Q, Zhou QZ, Lu ZG (2006) Diurnal variation of water and heat flux under transient water stress in a winter wheat field. Acta Ecol Sinica 26: 130–137. (in Chinese).

29. Zhang Q, Sun ZX, Wang S (2011) Analysis of variation regularity of land-surface physical quantities over Dingxi Region of Loess Plateau. Chinese J Geophys 54: 1727–1737. (in Chinese with English abstract).

30. Baldocchi DD, Meyers TP (1998) On using eco-physiological, micrometeorological and biogeochemical theory to evaluate carbon dioxide, water vapor and trace gas fluxes over vegetation: a perspective. Agric Forest Meteorol 90: 1–25.

31. Li SG, Lai CT, Lee G, Shimoda S, Yokoyama T, et al. (2005) Evapotranspiration from a wet temperate grassland and its sensitivity to microenvironmental variables. Hydrol Process 19: 517–532.

32. Yu WY, Zhou GS, Chi DC, Zhou L, He QJ (2008) Evapotranspiration of Phragmites communis community in Panjin Wetland and its controlling factors. Acta Ecol Sinica 28: 4894–4601. (in Chinese with English abstract).

33. Kang S, Gu B, Du T, Zhang J (2003) Crop coefficient and ratio of transpiration to evapotranspiration of winter wheat and maize in a semi-humid region. Agric Water Manage 59: 239–254.

34. Lei H, Yang D (2010) Interannual and seasonal variability in evapotranspiration and energy partitioning over an irrigated cropland in the North China Plain. Agric Forest Meteorol 150: 581–589.

35. Liu C, Zhang X, Zhang Y (2002) Determination of daily evaporation and evapotranspiration of winter wheat and maize by large-scale weighing lysimeter and micro-lysimeter. Agric Forest Meteorol 111: 109–120.

36. Burba GG, Verma SB (2005) Seasonal and interannual variability in evapotranspiration of native tallgrass prairie and cultivated wheat ecosystems. Agric Forest Meteorol 135: 190–201.

37. Li YL, Cui JY, Zhang TH, Zhao HL (2003) Measurement of evapotranspiration of irrigated spring wheat and maize in a semi-arid region of north China. Agric Water Manage 61: 1–12.

38. Kimura R, Fan J, Zhang X, Takayama N, Kamichika M, et al. (2006) Evapotranspiration over the grassland field in the Liudaogou Basin of the Loess Plateau, China. Acta Oecol 29: 45–53.

7

Intercropping Competition between Apple Trees and Crops in Agroforestry Systems on the Loess Plateau of China

Lubo Gao[1], Huasen Xu[1], Huaxing Bi[1,2]*, Weimin Xi[3], Biao Bao[1], Xiaoyan Wang[1], Chao Bi[1], Yifang Chang[1]

1 College of Water and Soil Conservation, Beijing Forestry University, Beijing, P.R. China, **2** Key Laboratory of Soil and Water Conservation, Ministry of Education, Beijing, P.R. China, **3** Department of Biological and Health Sciences, Texas A&M University-Kingsville, Kingsville, Texas, United States of America

Abstract

Agroforestry has been widely practiced in the Loess Plateau region of China because of its prominent effects in reducing soil and water losses, improving land-use efficiency and increasing economic returns. However, the agroforestry practices may lead to competition between crops and trees for underground soil moisture and nutrients, and the trees on the canopy layer may also lead to shortage of light for crops. In order to minimize interspecific competition and maximize the benefits of tree-based intercropping systems, we studied photosynthesis, growth and yield of soybean (*Glycine max* L. Merr.) and peanut (*Arachis hypogaea* L.) by measuring photosynthetically active radiation, net photosynthetic rate, soil moisture and soil nutrients in a plantation of apple (*Malus pumila* M.) at a spacing of 4 m × 5 m on the Loess Plateau of China. The results showed that for both intercropping systems in the study region, soil moisture was the primary factor affecting the crop yields followed by light. Deficiency of the soil nutrients also had a significant impact on crop yields. Compared with soybean, peanut was more suitable for intercropping with apple trees to obtain economic benefits in the region. We concluded that apple-soybean and apple-peanut intercropping systems can be practical and beneficial in the region. However, the distance between crops and tree rows should be adjusted to minimize interspecies competition. Agronomic measures such as regular canopy pruning, root barriers, additional irrigation and fertilization also should be applied in the intercropping systems.

Editor: Randall P. Niedz, United States Department of Agriculture, United States of America

Funding: This paper was supported by National Scientific and Technology Program of China (No. 2011BAD38B02) and CFERN & GENE Award Funds on Ecological paper. The funders had no role in study design, data collection and analysis, decision to publish, or preparation of the manuscript.

Competing Interests: The authors have declared that no competing interests exist.

* E-mail: bhx@bjfu.edu.cn

Introduction

The Loess Plateau is the birthplace of China's primitive agriculture. However, because of unsound land use and destruction of forests, the Loess Plateau has suffered serious soil erosion. At the same time, rapid population growth has also brought greater pressure to the environment in the region. The ensuing ecological and environmental problems have slowed down the economic development and living standards of local people. These problems lead to further deterioration of ecological environment, forming a vicious cycle. The local government is facing dual pressures from both economy and ecology.

Agroforestry systems have been considered as an effective practice to alleviate the conflicts between the rapidly growing population and the limited arable land resources [1,2]. In recent years, agroforestry management has been widely applied in the Loess Plateau region for reducing soil erosion and water loss, restoring ecological balance, raising land utilization rate and increasing economic benefits [3,4]. However, in most agroforestry systems, competition for light, moisture and nutrients exists at the interface between trees and crops which can cause a reduction of crop yield [5]. It is a major constraint that has affected stability of the structure and the function of the agricultural ecosystems. The competition between woody tree species and understory crop

species not only exists aboveground (competition for light) but also comes from belowground (competition for soil moisture and nutrients), leading to lower crop yield. According to Friday and Fownes, the competition between trees and crops is overwhelmingly for light which is the main reason for the reduction of maize in alley cropping system in Hawii, USA [6]. Similar results were reported by Peng et al. in loess area of Weibei in Shaanxi Province, China [7]. Elsewhere in southern Australia, studies showed that reduced crop yields are associated with the competition for water in windbreak and alley systems [8,9]. Kowalchuk and Jong found that, especially in drought years, competition for water is the principal factor affecting the yield of spring wheat intercropped with shelterbelts in Western Saskatchewan [10]. In some related studies, the results indicated that competition for nutrients does not exist in intercropping systems [11–13]. However, others reported that as one of the main reasons leading to the reduction of crop yield, the competition for soil nutrients does exist in the interface of trees and crops and has a negative impact [14,15]. It is very important to explore the competitive mechanism in intercropping systems, in order to provide optimum management strategies and technologies for managing intercropping system with high-yield, high-efficiency and stabilization.

Table 1. Characteristics of apple trees intercropped with soybean and peanut in the experimental sites in July 2011.

Measurement	Intercropped with soybean	Intercropped with peanut
Tree height (m)	2.4	2.5
DBH (cm)	4.1	4.2
Depth of live crown (m)	1.7	1.8
Mean radius of crown (m)	1.3	1.2

Apple-crop intercropping system is one of the most commonly applied agroforestry systems in the Loess Plateau region owing to its good ecological, social and economic benefits. However, only few studies focused on this intercropping system in the area. In order to explore the biological reasons of the competition in typical intercropping systems and to provide effective management techniques, we report on a study of two apple-crop intercropping systems (apple-soybean, apple-peanut) on the Loess Plateau region in the western portion of Shanxi Province. The objectives of our research were (1) to analyze the interspecies competition relationship between trees and crops; (2) to find the limiting factors in the development of intercropping systems in this area; (3) to offer possible solutions to minimize the interspecies competitions and maximize resource utilization; (4) to enrich the related study and to improve the management of the intercropping systems in this region.

Materials and Methods

Study site

The study site was located in the Baidong Village, Jixian County, Shanxi Province, China (36°06' N, 110°35' E, 1025 m a.s.l.). The area is a typical hill and gully region of the Loess Plateau. The annual mean rainfall is about 575 mm, and the mean annual temperature is 10°C (1991–2010). The precipitation is unevenly distributed seasonally, with an average rainfall of 463 mm from June to August (1991–2010), which contributed about 80% of annual precipitation. The parent material of the soil is loess, and the soil properties are uniform. The bulk density, pH, total porosity, $CaCO_3$ content, cation exchange capacity, organic C, total N and available P of the top soil layer (100 cm) were 1.32 $Mg•m^{-3}$, 8.24, 50.16%, 18.35%, 18.43 $cmol•kg^{-1}$, 6.27 $g•kg^{-1}$, 0.39 $g•kg^{-1}$ and 4.39 $mg•kg^{-1}$, respectively. The main intercropping tree species are Apple (*Malus pumila* M.), Apricot (*Prunus armeniaca* L.), Pear (*Pyrus bretschneideri* R.), Chinese arborvitae (*Platycladus orientalis* (L.) and Franco) and Black locust (*Robinia pseudoacacia* L.).

Ethics Statement

No specific permits were required for the described field studies. The sampling locations were not privately-owned or protected in any way and the field studies did not involve endangered or protected species.

Treatments and Crop Cultivation

Two typical intercropping systems of apple-soybean and apple-peanut were chosen for this study during the crop growing season of 2011 and 2012. The apple trees were planted in an East-West orientation in 2007. The characteristics of the apple trees intercropped with soybeans and peanuts in July 2011 are listed in Table 1. There were four treatments in this study: apple-soybean intercropping treatment (AS), soybean monoculture

served as control (CS), apple-peanut intercropping treatment (AP) and peanut monoculture served as control (CP). Each treatment had three replicates. Each replicate of intercropping treatment (AS and AP) was an 8 × 10 m plot that included 12 trees planted in three rows with 4 m between trees and 5 m between rows. Each replicate of control treatment (CS and CP) was the same size of 8 × 10 m. For all treatments, the crops were planted at a spacing of 0.4 m with in rows and 0.5 m between rows and received the same agricultural management practices. Soybean and peanut were grown 0.3 m from an adjacent tree row in the intercropping systems. All plots received 147 kg N ha^{-1}, 30 kg P ha^{-1} and 30 kg K ha^{-1} as basal fertilizer and no additional fertilizer or irrigation in the rest of the year.

Measurements of Plant Photosynthesis, Soil Moisture and Nutrients

For the sampling of plant photosynthesis, soil moisture and soil nutrients, six sampling locations at distances of 0.5 m, 1.5 m and 2.5 m, respectively, from both side of tree row were identified as sampling points in each intercropping plot (Figure 1). The sampling points were further divided into three equal groups and denoted as F0.5, F1.5 and F2.5 based on the distance (0.5 m, 1.5 m and 2.5 m) from the tree row. Measurement parameters of F0.5, F1.5 and F2.5 were used to represent the major locations of 0.5 m, 1.5 m and 2.5 m away from apple tree row. For each control plot, five selected points were established with an S-shaped sampling method.

Photosynthetically active radiation (PAR) and net photosynthetic rate (NPR) of crops were performed by two portable Li-6400 photosynthesis systems which had a 6 cm^2 clamp-on leaf chamber connected to the main engine (Li-6400, Li-Cor Inc., Lincoln, NE, USA) under ambient humidity, temperature and irradiance. One fully expanded leaf from the upper part of the crop canopy in each sampling point was selected and measured five times with 2 h intervals during daytime (0900–1700 h). During each measurement period, all sampling points of intercropping treatment and control treatment were visited. These treatments were measured in mid-August 2011 and again in late August 2012, the typical phenological phases of peanut and soybean. For all measurements, the flow velocity was set at 500 $μmol•s^{-1}$ and the airstream entering the chambers was kept at the growth CO_2 concentration (370 $μmol•mol^{-1}$) by a computer-controlled CO_2 injector system supplied with Li-6400. PAR and CO_2/H_2O exchanged by the leaf were measured concurrently with the quantum sensor and the infrared gas analyzer on LI-6400. The data were recorded and calculated automatically with the software in the photosynthesis system.

For soil moisture, the samples were taken at different phenological phases of soybean and peanut: 8 July, 23 August, and 23 September in 2011; and 4 July, 11 August, and 22 September in 2012. A drill was used to remove the soil from 0–100 cm in 20 cm intervals in soil profile. The soil moisture content

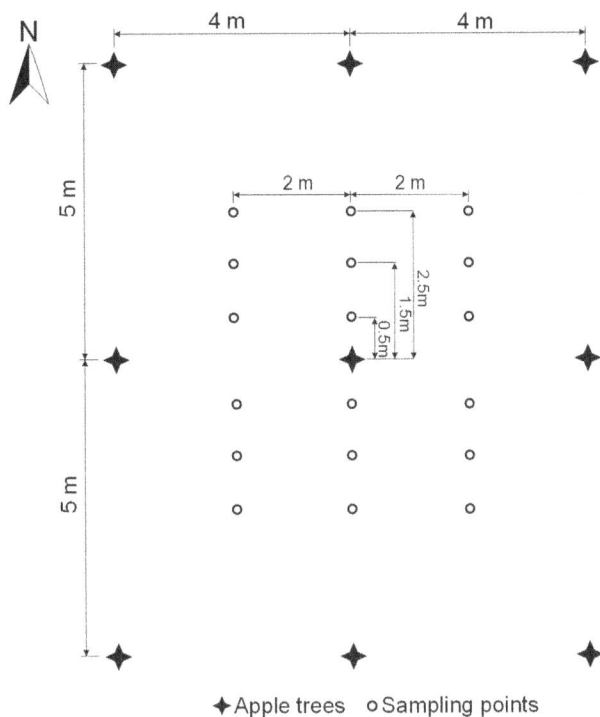

Figure 1. Sampling points of plant photosynthesis, soil moisture and nutrients in the intercropping study sites.

was determined gravimetrically in each layer. The mean soil moisture content of the five layers in all sampling time (2011 and 2012) was calculated and used as the final value of the sampling point.

For the sampling of soil nutrients, the soil samples were taken on 23 August, 2011 and 11 August, 2012, the typical phenological phases of peanut and soybean. The soil samples were collected from a depth of 0–100 cm in soil profile with a drill. Organic matter content was determined by H_2SO_4-$K_2Cr_2O_7$ pyrogenation. Total N was determined using the Kjeldahl method, with a KDY-9830 N Analyzer. Available P was determined by Olsen sodium-bicarbonate extraction. Available K was determined by flame photometer.

Measurements of Crop Growth and Yields

For the sampling of crop growth, we used the same sample locations as for soil moisture. A single crop plant was sampled at each sampling point on 24 August 2011 and 12 August 2012. A total of 69 soybean plants and 69 peanut plants were harvested during each measurement period. In the lab, plant height, hundred leaf dry weight and total above-ground biomass of all plants were measured and recorded.

At the end of the growing season, in each intercropping plot, peanuts and soybeans were harvested from both sides of the tree row in two rectangular areas. The rectangular area was 4.0 m long and 2.7 m wide. As the convenience of the study, the two rectangular areas were divided into three groups: (1) the area of 0.3–1.0 m away from tree row; (2) the area of 1.0–2.0 m away from tree row; and (3) the area of 2.0–3.0 m away from tree row. The yields of the three groups were used to represent the crop yield of F0.5, F1.5 and F2.5, respectively. In the control plots, 2 m × 2 m quadrates of soybean and peanut were harvested to get the grain production. The peanuts and soybeans were dried at 70 °C

and then weighed to obtain an average dry weight. Yield values were reported on a per hectare basis.

Data analysis

All parameters (PAR, NPR, soil moisture, soil nutrients content, crop growth and yields) measured for control treatments and three major locations (F0.5, F1.5 and F2.5) of intercropping treatments were described in terms of mean values followed by respective standard deviations. Simple regression analysis was used to examine the relationships between the data of PAR, NPR, soil moisture and the distance from the tree row. Differences among groups for each crop (soybean or peanut) were determined by one-way ANOVA, and the results of the multiple comparisons were performed with least significant difference (LSD) test at $P<0.05$. NPR, total above-ground biomass and yield values of soybean and peanut had a correlated analysis with environmental parameters to decide the effect of apple trees competition on crop growth and productivity via bivariate correlation (Pearson) analysis at $P<0.05$ and $P<0.01$. All the analyses were performed by using the software IBM SPSS Statistics 20.0 for Windows.

Results

Light Interception and Plant Photosynthesis

For both crops, diurnal variation of photosynthetically active radiation (PAR) in the intercropping systems and the monoculture configurations (control treatments) showed a single peak curve with time (Figure 2). The peak of PAR appeared at 13:00 pm and the minimum value appeared in 17:00 pm. Because of reflectance, absorbance and transmittance by the apple tree canopy, the PAR of crops in the intercropping systems were lower than that in the monoculture configurations during the same period. On the horizontal distribution, the general trend was that the closer the crops to the tree rows, the lower the PAR received. The same tendency was found in diurnal variation of net photosynthetic rate (Figure 3).

The daily mean values of PAR showed a clear linear relationship with distance from the apple tree row in both intercropping treatments (Figure 4A). The trend lines of PAR (Y, $\mu mol \cdot s^{-1} \cdot m^{-2}$) and distance from trees rows (X, m) were $Y=78.5 \times +865.3$ ($R^2=0.999$) in apple-soybean intercropping treatment (AS) and $Y=82.5 \times +881.9$ ($R^2=0.873$) in apple-peanut intercropping treatment (AP). The slopes of both regression lines suggested that the PAR in AP treatment had a higher growth than that in AS treatment as the distance from the tree increased. As shown in Figure 4A, PAR reaching the upper parts of the crop canopy in AP treatment also had higher values than that in AS treatment at the same distance away from the tree row. It indicated that peanut canopy could obtain more solar radiation in AP treatment than soybean canopy in AS treatment. At confidence level of 95%, the control treatment PAR mean fell within the confidence intervals of F2.5 in the corresponding intercropping systems. Compared with the corresponding control treatment, PAR at F0.5 and F1.5 showed a reduction of 17.9% and 10.4% in AS treatment, respectively, 17.8% and 5.4% in AP treatment. Similar linear relationships were also obtained through regression analysis of the relationship between NPR and distance from the apple tree row (Figure 4B). The trend lines of NPR (Y, $\mu mol \cdot s^{-1} \cdot m^{-2}$) and distance from trees rows (X, m) were $Y=1.025 \times +12.003$ ($R^2=0.902$) in AS treatment, and $Y=0.940 \times +10.983$ ($R^2=0.951$) in AP treatment. The NPR in AS treatment had higher values and growth than that in AP treatment as the distance from the tree increased which was different from the measurement of PAR. The control treatments

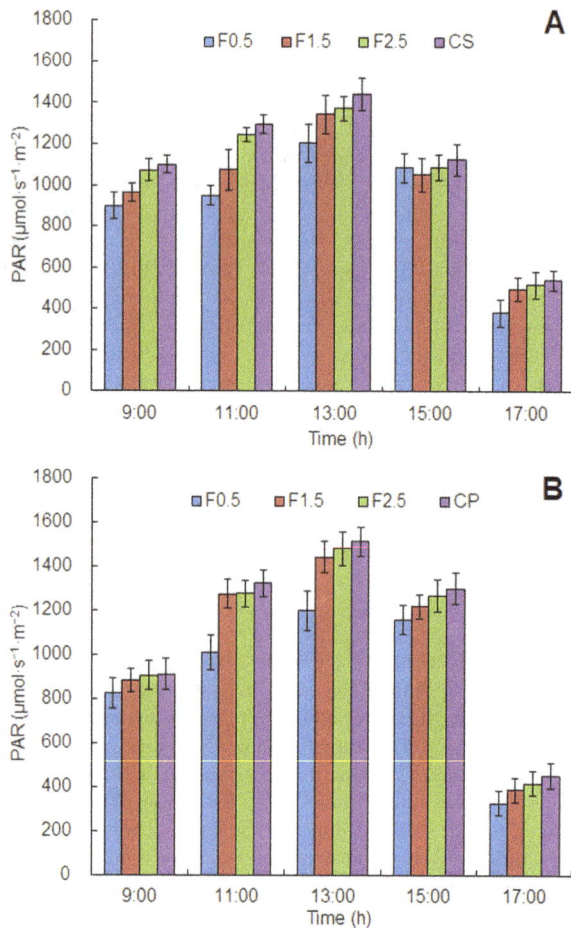

Figure 2. Diurnal variation of photosynthetically active radiation (PAR) for the intercropping systems and its control (A. apple–soybean and B. apple–peanut). F0.5, F1.5 and F2.5 were used to represent the sampling points which had different distance (0.5 m, 1.5 m and 2.5 m) from the tree row. Error bars indicate standard deviation.

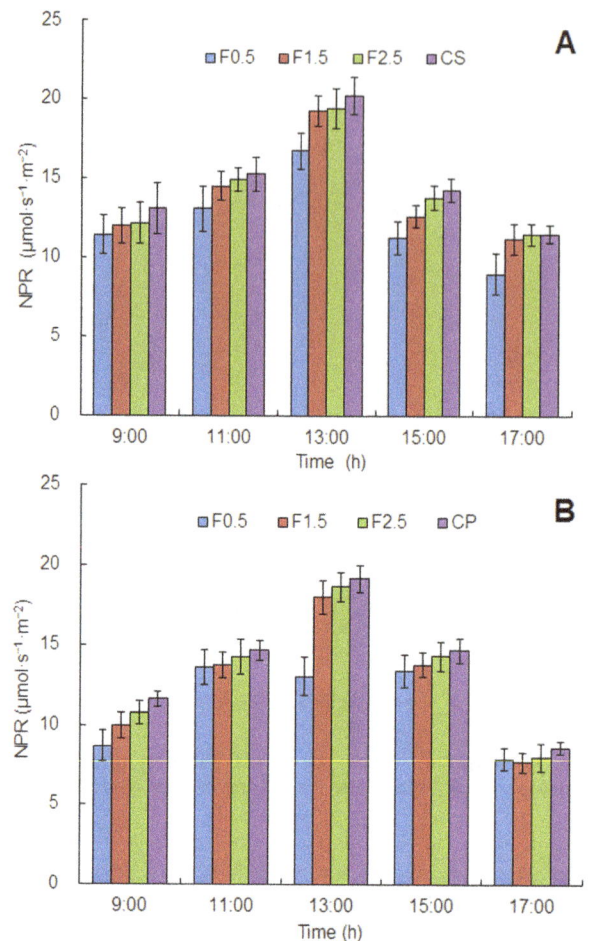

Figure 3. Diurnal variation of net photosynthetic rate (NPR) for the intercropping systems and its control (A. apple–soybean and B. apple–peanut). F0.5, F1.5 and F2.5 were used to represent the sampling points which had different distance (0.5 m, 1.5 m and 2.5 m) from the tree row. Error bars indicate standard deviation.

mean fell within the confidence intervals of F2.5 in the corresponding intercropping systems at confidence level of 95%.

Spatial Distribution of Soil Moisture

Although the soil moisture content in the whole soil profile (0 to 100 cm in depth) in AS was different from AP, the trend of spatial distributions of soil moisture was similar (Figure 5). Soil moisture content in AS was related to distance from the apple tree row and showed a clear linear relationship ($Y = 0.465 \times +11.602$, $R^2 = 0.999$), and AP showed the same trend ($Y = 0.590 \times +11.002$, $R^2 = 0.900$). Compared with AP, AS had higher values at the same distance away from the tree row. However, with increasing distance from the tree row, soil moisture in AP had a higher growth than that in AS. The lowest soil moisture content was 11.83% in AS and 11.41% in AP, showed a decrease of 10.31% and 11.14% when compared with the corresponding control treatments. The soil moisture at F2.5 in both intercropping systems also had slightly lower values than that in monoculture configurations, however no difference was observed at significance level of 5%, since the control treatments mean fell within the confidence intervals of F2.5 in the corresponding intercropping

systems (confidence level 95%). Otherwise, the average soil moisture content in AP was lower than that in AS.

Spatial Distribution of Soil Nutrients

The soil nutrients content in the 0 to 100 cm interval was calculated (Table 2). It represented that organic matter, total N, available P and available K in AS had different degrees of reduction when compared with CS, and showed significant differences ($P<0.05$). Similar results were found between AP and CP, except that no significant difference was observed for total N and available K at the location of F2.5. The average content of organic matter, total N, available P and available K in AS decreased by 30.77%, 63.24%, 56.08% and 27.83% when compared with CS–the monoculture configuration. For AP and CP, the decreased percentages were 18.32%, 21.05%, 36.27% and 7.49% respectively. In addition, except available K, soil nutrients content in AP was higher than that in AS at the same spatial location. With the increasing distance from tree row, the distribution trend of soil nutrients was different from that of PAR, NPR or soil moisture in the same intercropping condition. The lowest content of organic matter, total N, available P and available K in AS were present at the location of F1.5. The similar

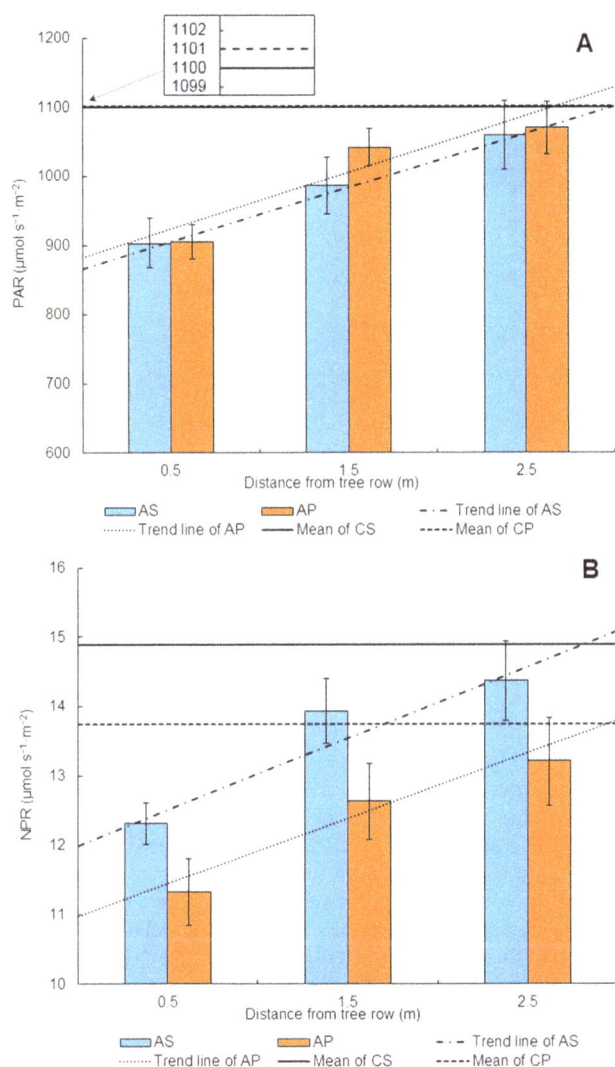

Figure 4. Daily mean of photosynthetically active radiation (PAR) and net photosynthetic rate (NPR) for the intercropping systems and its control (A. PAR and B. NPR). Vertical lines indicate confidence interval at 95% level.

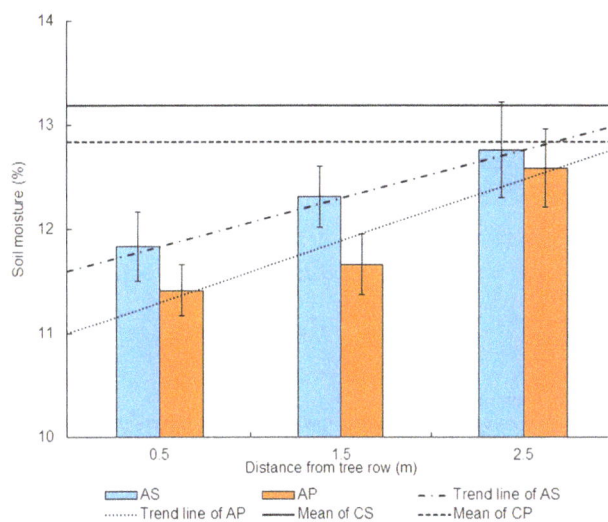

Figure 5. Soil moisture of 0 to 100 cm depth for the intercropping systems and its control. Vertical lines indicate confidence interval at 95% level.

CP. No differences were observed between the locations of F2.5 in both intercropping systems and the corresponding monoculture configuration ($P<0.05$).

Within plot differences in these parameters were significantly correlated (Table 4). NPR was highly correlated with PAR, and soil moisture. The total above-ground biomass of soybean was highly correlated with PAR, soil moisture and available P. The total above-ground biomass of peanut was correlated with PAR, soil moisture, total N and available P. Yield of soybean was highly correlated with PAR, soil moisture, available P and total N, with a trend of soil moisture >PAR>available P>total N. For peanut, the trend was soil moisture > PAR > total N > available P. It showed that, for both of the intercropping system in the study region, the primary factor affecting the yield is soil moisture, and the secondary factor is photosynthetically active radiation, and soil nutrient also have an impact on crop yield in some depth.

Discussion

Agroforestry system has been studied for a long time and has been widely used in the agricultural production practices in China [3,16,17]. However, there has been little research done on the agroforestry system in the Loess Plateau region. The main intercropping models which have been studied are always walnut-wheat and apple-wheat [18–20]. The types of fruit trees intercropping with economic crops such as soybean and peanut has not been well studied. In fact, compared with wheat, soybean and peanut could bring more economic income to farmers. At the same time, these two crops could be rotated with wheat in order to improving land-use efficiency, and re-establishing the economic viability of the Loess Plateau.

Our study observed a clearly positive linear relationship between distance from the apple tree rows and the daily mean values of PAR and NPR in the intercropping systems. For both apple-crop intercropping systems, the shading of the 4–5 years old apple trees had a significant negative effect on the crops in the range of 1.5 m away from the tree rows and further caused the reduction of crop yield. In other researches of temperate agroforestry systems, the similar results were reported by Reynolds et al. [21] about maize and soybean intercropped with poplar and

result was found in AP, except that the lowest value of available K was the location of F0.5.

Crop Growth and Yields

Plant height, hundred leaf dry weight and total above-ground biomass in both intercropping systems had lower values when compared with the monoculture configuration ($P<0.05$; Table 3). The locations of F0.5 and F1.5 in all these parameters showed significant differences with corresponding monoculture configuration ($P<0.05$); however, there were no difference observed in the location of F2.5.

The yield of soybean in AS was significantly related to the distance from the row of apple trees ($Y=0.180\times+1.400$, $R^2=0.991$), and the yield of peanut in AP showed the same trend ($Y=0.095\times+1.554$, $R^2=0.900$) which showed that yield of soybean had greater impacted by distance from the tree row. The yields at F0.5 and F1.5 in AS were lower than that in CS ($P<0.05$), with a reduction of 22.45% and 11.95%, and in AP the yields of reduction were 13.31% and 11.03% when compared with

Table 2. Soil nutrients for the intercropping systems and control configurations.

Measurement	AS			CS	AP			CP
	F0.5	F1.5	F2.5		F0.5	F1.5	F2.5	
Organic matter (g·kg^{-1})	4.63±0.27a	3.40±0.29b	4.93±0.27a	6.24±0.31c	6.26±0.36a	5.21±0.26b	6.37±0.36a	7.28±0.27c
Total N (g·kg^{-1})	0.28±0.05a	0.22±0.04b	0.25±0.04ab	0.68±0.03c	0.31±0.06a	0.24±0.05b	0.35±0.07ac	0.38±0.04c
Available P (mg·kg^{-1})	2.82±0.56a	2.50±0.54a	3.93±0.31b	7.02±0.22c	5.06±0.61a	3.44±0.44b	4.75±0.42a	6.93±0.33c
Available K (mg·kg^{-1})	97.33±1.77a	84.02±2.09b	95.43±2.97a	127.84±2.75c	87.45±2.81a	90.73±2.45a	98.11±2.42b	99.55±2.19b

Data were given as the means ± SD.
Different lowercase letters within a row of each crop indicate significant differences (LSD, $P<0.05$).

silver maple in Canada and Peng et al. [22] about mungbean and pepper intercropped with walnut and plum in Weibei area, China. For total above-ground biomass and yield of both crops, PAR of soybean had higher correlations than that of peanut, which indicated that soybean is more adversely impacted by tree shading. Within tree-based intercropping systems, many factors such as tree species, tree height, crown shape, tree row orientation and distance between tree rows can influence tree shading of adjoining agricultural crops. Light reduction would depend on the extent and duration of the shade of trees [21]. Regular pruning of fruit trees could reduce light competition within the intercropping system, improving crop yields.

In semiarid and arid regions, it is still a focus of studies whether intercropping system has an overall negative or positive effect on soil moisture [23]. In some related studies, it was considered that the trees can improve soil moisture holistic conditions in intercropping systems [17,24]. In other studies, the opposite results were reported [11,12,25,26]. However, little research has been carried out in this aspect on the Loess Plateau. Our research confirmed that the competition of water between trees and crops do exist, and showed adverse effects in the study site. A clear linear relationship was observed between the distance from the tree row and soil moisture in both of the intercropping systems. The closer to the tree row, the more intense the competition. The lowest soil moisture content in apple-soybean intercropping system and apple-peanut intercropping system showed a reduction of 10.31% and 11.14%, respectively. Only considering competition of water, the mainly affected region of the apple trees was 1.5 m away from the tree rows under the current tree age.

Another key factor of crop growth is soil nutrients in the intercropping systems. Elsewhere, Thomas et al. [27] and Thevathasan et al. [13] have reported that competition for nutrients in intercropping systems does not exist. In our study, it identified that there were competition for soil nutrients between trees and crops in the intercropping systems. The average content of organic matter, total N, available P and available K showed different degrees of reduction in both of the apple-crops intercropping systems than that of the corresponding control treatments. In particular, total N and available P had higher reduction rate than organic matter and available K, and had significantly correlation with yield of crops. As leguminous plants, soybean and peanut could fix nitrogen from the air via a symbiotic relationship with rhizobium bacteria and increase the mineral soil nitrogen content [28,29]. However, the nitrogen coming from biologically fixed N_2 of symbiosis could not meet all the demand of crops growth, and any gaps between N supply by N_2 fixation and crop N demand must be met by N uptake from soil [30]. The deficiency of light and water in the intercropping systems reduced the physiological activity of the crop, and then affected the N fixation capacity, resulting more intense competition for nitrogen between trees and crops. Compared with soybean and peanut, the growth of other non-nitrogen-fixing crop species (i.e. wheat, maize and millet) would be more severely affected because of nitrogen deficiency in the intercropping systems. Different from understory light distribution and soil moisture, soil nutrients had a different variation pattern in both intercropping systems. The main reasons for this phenomenon might be: (1) the crops close to tree row were seriously affected by tree shading, soil moisture stress and human activities, resulting in low physiological activity and low absorption

Table 3. Crop growth, biomass and yield for soybean and peanut intercropped with apple trees and its control.

Measurement	AS			CS	AP			CP
	F0.5	F1.5	F2.5		F0.5	F1.5	F2.5	
Crop height (cm)	43.7±2.8a	44.4±2.3a	47.7±3.2b	50.2±2.6b	19.1±0.7a	21.9±1.0b	22.6±0.9bc	23.5±0.8c
Hundred leaf dry weight (g)	14.12±0.81a	14.14±0.51a	14.95±0.38b	15.37±0.57b	4.39±0.23a	5.79±0.22b	5.90±0.31bc	6.14±0.28c
Total above-ground biomass (g)	54.55±4.66a	66.06±3.81b	76.96±4.30c	79.76±4.22c	50.10±2.62a	51.57±2.45a	79.16±3.28b	79.65±2.46b
Yield (t/ha)	1.48±0.06a	1.69±0.04b	1.84±0.06c	1.91±0.04c	1.62±0.04a	1.66±0.04a	1.81±0.05b	1.86±0.04b

Data were given as the means ± SD.
Different lowercase letters within a row of each crop indicate significant differences (LSD, $P<0.05$).

Table 4. Correlations of soybean and peanut net photosynthetic rate, biomass, and yield with environmental or physiological parameters measured in Jixian, China.

Independent variable	NPR ($\mu mol \cdot s^{-1} \cdot m^{-2}$)	Total above-ground biomass (g)	Yield (kg/ha)
Soybean			
PAR ($\mu mol \cdot s^{-1} \cdot m^{-2}$)	0.973**	0.996**	0.952**
Soil moisture (%)	0.953**	0.977**	0.957**
Organic matter (g·kg^{-1})	0.441	0.575	0.566
Total N (g·kg^{-1})	0.537	0.555	0.601*
Available P (mg·kg^{-1})	0.697*	0.750**	0.763**
Available K (mg·kg^{-1})	0.469	0.538	0.565
Peanut			
PAR ($\mu mol \cdot s^{-1} \cdot m^{-2}$)	0.986**	0.773**	0.816**
Soil moisture (%)	0.926**	0.965**	0.959**
Organic matter (g·kg^{-1})	0.450	0.562	0.583
Total N (g·kg^{-1})	0.513	0.843**	0.770**
Available P (mg·kg^{-1})	0.424	0.628*	0.646*
Available K (mg·kg^{-1})	0.479	0.531	0.446

*Significant at 5% level.
**Significant at 1% level.

of soil nutrients; (2) the decomposition of tree litter leaded to high nutrients content in the area near the tree row; (3) the overlapping of apple tree roots and crop roots resulting in lower nutrient content at F1.5; (4) the tree roots reduced with the increase of the distance from the tree, therefore, the soil nutrients had relatively high content F2.5. Therefore, in the area of 1.5 m away from tree row, strengthen the application of fertilizer (especially nitrogen and phosphorus) would be helpful to alleviate interspecific competition for soil nutrients.

In the apple-crop intercropping systems, the competition of light, water and nutrients resulted in a greater negative impact on crop growth and yields. For the two apple-crop intercropping systems in our study, the primary factor affecting the crop yield was soil moisture, and the secondary factor was light, and deficiency of the soil nutrient also had a negative impact on crop yields. In the same study area, Yun et al. reported a similar research with a different conclusion: the light is the primary limiting factor leading to reduction of crops, followed by soil moisture [31]. In their research, the apple trees had greater crown width, canopy density and root depth due to elder age (9-year-old) and smaller tree spacing (3 m×4 m). Affected by the impact of canopy structure, the obvious microclimate effect inhibited evapotranspiration of soil moisture to some extent [32], in the same time, the low transmittance led to more intense light stress to crops. Furthermore, the effect of hydraulic lift by tree roots also alleviated the interspecific competition for soil water [24]. Combined with these reasons, different results were found. For different intercropping patterns and tree ages, the intensity of competition for resources would be different in the intercropping system. Therefore, a long-term observation should be carried out in this region to obtain more details about the mechanism of interspecific competition in the intercropping systems. In our study, under the current tree age and growth conditions, the influence scope of the apple trees was 1.5 m away from the tree rows. Compared with the corresponding monoculture configuration, the yield of peanut in the intercropping system had a lower reduction than that of soybean. With comprehensive consideration, peanut is more suitable for intercropping with apple trees in this region.

As we have demonstrated in this study, soil moisture, light and soil nutrients were the limiting factors of crop yield. In order to obtain more production, appropriate management measures were needed to minimize competition between trees and crops. Namirembe [33] and Friday [6] have suggested that the competition for light between trees and crops could be alleviated by pruning of trees crown and increasing the intercropping distance. In general, the aboveground competition could be intuitively observed and managed. However, the competition belowground is invisible and easily ignored by farmers or managers. To avoid these yield losses, root barrier in the intercropping interface is considered to be a useful agricultural management practice according to some related studies [12,34,35]. Combined with their research achievements and our experiment result, we offered several specific recommendations to reduce the competition exist in apple-crop intercropping systems: (1) the selection of crop varieties which is more suitable for apple-crop intercropping systems; (2) appropriate distance increase between the crops and apple tree rows; (3) regular pruning of fruit trees, in order to increase canopy light transmittance rate; (4) additional fertilization and irrigation in the key phenological phase of the crops; (5) differences of irrigation and fertilization based on the distance from the apple trees. Management measures such as plastic film and straw mulching have been widely used in agricultural production. Whether these measures have overall positive effects on intercropping system would be one of the focus of our future research work in this region.

Conclusions

As an effective method to increase the efficiency of land use and economic returns, tree-based intercropping systems are particularly important on Loess Plateau. We concluded that the competitions exist both above-ground and below-ground between apple trees and crops. The competition for soil moisture is the primary limiting factor for the crop productivity in this region. Furthermore, the tree shading and the competition for soil nutrients in the interface of trees and crops also have a negative impact on the understory crops. However, it could be minimized by better agricultural technology and management measures.

In summary, our study suggests that there is great potential for intercropping systems in the Loess Plateau. Therefore, in order to relieve the shortage of arable land and promote the sustainable development of natural resources, the intercropping systems would continue to be the hot spot for future research. Canopy structure, roots distribution of trees, the application of different agronomic measures and the role they play in the competition process in the intercropping systems will be the focus of our future research.

References

1. Burel F (1996) Hedgerows and their role in agricultural landscapes. Critical Reviews in Plant Sciences 15: 169–190.
2. Gene Garrett HE, Buck L (1997) Agroforestry practice and policy in the United States of America. Forest Ecology and Management 91: 5–15.
3. Li W, Lai S (1994) Agroforestry in China. Beijing: Chinese Science Press. pp. 14–18.
4. Zhu Q, Zhu J (2003) Sustainable management technology for conversion of cropland to forest in loess area. Beijing:Chinese Forestry Press. pp. 160–165.
5. Ong CK, Huxley P (1996) Tree-crop interactions: a physiological approach. Wallingford:CAB International. pp. 386.
6. Friday JB, Fownes JH (2002) Competition for light between hedgerows and maize in an alley cropping system in Hawaii, USA. Agroforestry Systems 55: 125–137.
7. Peng X, Zhang Y, Cai J, Jiang Z, Zhang S (2009) Photosynthesis, growth and yield of soybean and maize in a tree-based agroforestry intercropping system on the Loess Plateau. Agroforestry Systems 76: 569–577.
8. Hall DJM, Sudmeyer RA, McLernon CK, Short RJ (2002) Characterisation of a windbreak system on the south coast of Western Australia. 3. Soil water and hydrology. Australian Journal of Experimental Agriculture 42: 729–738.
9. Unkovich M, Blott K, Knight A, Mock I, Rab A, et al. (2003) Water use, competition, and crop production in low rainfall, alley farming systems of south-eastern Australia. Australian Journal of Agricultural Research 54: 751–762.
10. Kowalchuk TE, Jong E (1995) Shelterbelts and their effect on crop yield. Canadian Journal of Soil Science 75: 543–550.
11. Jose S, Gillespie AR, Seifert JR, Biehle DJ (2000) Defining competition vectors in a temperate alley cropping system in the midwestern USA: 2. Competition for water. Agroforestry Systems 48: 41–59.
12. Miller AW, Pallardy SG (2001) Resource competition across the crop-tree interface in a maize-silver maple temperate alley cropping stand in Missouri. Agroforestry Systems 53: 247–259.
13. Thevathasan NV, Gordon AM, Simpson JA, Reynolds PE, Price G, et al. (2004) Biophysical and ecological interactions in a temperate tree-based intercropping system. Journal of Crop Improvement 12: 339–363.
14. Newman SM, Bennett K, Wu Y (1997) Performance of maize, beans and ginger as intercrops in Paulownia plantations in China. Agroforestry Systems 39: 23–30.
15. Yun L, Bi H, Gao L, Zhu Q, Ma W, et al. (2012) Soil moisture and soil nutrient content in walnut-crop intercropping systems in the Loess Plateau of China. Arid Land Research and Management 26: 285–296.
16. Meng P, Zhang J, Fan W (2003) Research on agroforestry in china. Beijing:Chinese Forestry Press. pp.235.
17. Meng P, Zhang J (2004) Effects of pear-wheat inter-cropping on water and land utilization efficiency. Forest Research 17: 167–171.
18. Zhang J, Meng P, Yin C (2002) Spatial distribution characteristics of apple tree roots in the apple-wheat intercropping. Scientia Silvae Sinicae 38: 30–33.
19. Zhang J, Meng P (2004) Model on wheat potential evapotranspiration in apple-wheat intercropping. Forest Research 17: 284–290.
20. Yun L, Bi H, Ren Y, Ma W, Tian X (2009) Soil moisture distribution at fruit-crop intercropping boundary in the Loess region of Western Shanxi. Journal of Northeast Forestry University 37: 70–78.
21. Reynolds PE, Simpson JA, Thevathasan NV, Gordon AM (2007) Effects of tree competition on corn and soybean photosynthesis, growth, and yield in a temperate tree-based agroforestry intercropping system in southern Ontario, Canada. Ecological Engineering 29: 362–371.
22. Peng X, Cai J, Jiang Z, Zhang Y, Zhang S (2008) Light competition and productivity of agroforestry system in loess area of Weibei in Shaanxi. Chinese Journal of Applied Ecology 19: 2414–2419.
23. Zhang J, Meng P, Yin C, Cui G (2003) Summary on the water ecological characteristics of agroforestry system. World Forestry Research 16: 10–14.
24. Hirota I, Sakuratani T, Sato T, Higuchi H, Nawata E (2004) A split-root apparatus for examining the effects of hydraulic lift by trees on the water status of neighbouring crops. Agroforestry Systems 60: 181–187.
25. Lott JE, Howard SB, Ong CK, Black CR (2000) Long-term productivity of a Grevillea robusta-based overstorey agroforestry system in semi-arid Kenya: II. Crop growth and system performance. Forest Ecology and Management 139: 187–201.
26. Lehmann J, Peter I, Steglich C, Gebauer G, Huwe B, et al. (1998) Below-ground interactions in dryland agroforestry. Forest Ecology and Management 111:157–169
27. Thomas J, Kumar BM, Wahid PA, Kamalam NV, Fisher RF (1998) Root competition for phosphorus between ginger and Ailanthus triphysa in Kerala, India. Agroforestry Systems 41: 293–305.
28. Cheng D (1994) Resource microbiology. Harbin:Northeast Forestry University Press. pp. 32.
29. Wani SP, Rupela OP, Lee KK (1995) Sustainable agriculture in the semi-arid tropics through biological nitrogen fixation in grain legumes. Plant Soil 174: 29–49.
30. Salvagiotti F, Cassman KG, Specht J E, Walters DT, Weiss A, et al. (2008). Nitrogen uptake, fixation and response to fertilizer N in soybeans: A review. Field Crops Research 108:1–13.
31. Yun L, Bi H, Tian X, Cui Z, Zhou H, et al. (2011) Main interspecific competition and land productivity of fruit-crop intercropping in Loess Region of West Shanxi. Chinese Journal of Applied Ecology 22:1225–1232
32. Zhang J, Meng P, Song Z, Gao J (2004) An overview on micro-climatic effects of agro-forestry systems in plain agricultural areas in China. Agricultural Meteorology 25:52–55
33. Namirembe S (1999) Tree management and resource utilization in agroforestry systems with Senna spectabilis in the drylands of Kenya. Bangor:University of Wales. pp. 206.
34. Singh RP, Saharan N, Ong CK (1989) Above and below ground interactions in alley-cropping in semi-arid India. Agroforestry Systems 9: 259–274.
35. Hou Q, Brandle J, Hubbard K, Schoeneberger M, Nieto C, et al. (2003) Alteration of soil water content consequent to root-pruning at a windbreak/crop interface in Nebraska, USA. Agroforestry Systems 57: 137–147.

Acknowledgments

We are grateful for the support from the Shanxi Jixian Forest Ecosystem Research Station. We also would like to thank the three anonymous reviewers and the editors for their helpful comments.

Author Contributions

Conceived and designed the experiments: LG HX HB BB. Performed the experiments: LG HX HB BB. Analyzed the data: LG HX WX BB XW CB YC. Contributed reagents/materials/analysis tools: XW CB YC. Wrote the paper: LG HB WX.

8

Soil Organic Carbon Redistribution by Water Erosion – The Role of CO_2 Emissions for the Carbon Budget

Xiang Wang*, Erik L. H. Cammeraat, Paul Romeijn, Karsten Kalbitz

Earth Surface Science, Institute for Biodiversity and Ecosystem Dynamics, University of Amsterdam, Amsterdam, The Netherlands

Abstract

A better process understanding of how water erosion influences the redistribution of soil organic carbon (SOC) is sorely needed to unravel the role of soil erosion for the carbon (C) budget from local to global scales. The main objective of this study was to determine SOC redistribution and the complete C budget of a loess soil affected by water erosion. We measured fluxes of SOC, dissolved organic C (DOC) and CO_2 in a pseudo-replicated rainfall-simulation experiment. We characterized different C fractions in soils and redistributed sediments using density fractionation and determined C enrichment ratios (CER) in the transported sediments. Erosion, transport and subsequent deposition resulted in significantly higher CER of the sediments exported ranging between 1.3 and 4.0. In the exported sediments, C contents (mg per g soil) of particulate organic C (POC, C not bound to soil minerals) and mineral-associated organic C (MOC) were both significantly higher than those of non-eroded soils indicating that water erosion resulted in losses of C-enriched material both in forms of POC and MOC. The averaged SOC fluxes as particles (4.7 g C m^{-2} yr^{-1}) were 18 times larger than DOC fluxes. Cumulative emission of soil CO_2 slightly decreased at the erosion zone while increased by 56% and 27% at the transport and depositional zone, respectively, in comparison to non-eroded soil. Overall, CO_2 emission is the predominant form of C loss contributing to about 90.5% of total erosion-induced C losses in our 4-month experiment, which were equal to 18 g C m^{-2}. Nevertheless, only 1.5% of the total redistributed C was mineralized to CO_2 indicating a large stabilization after deposition. Our study also underlines the importance of C losses by particles and as DOC for understanding the effects of water erosion on the C balance at the interface of terrestrial and aquatic ecosystems.

Editor: Ben Bond-Lamberty, DOE Pacific Northwest National Laboratory, United States of America

Funding: This work was supported by China Scholarship Council (CSC) and University of Amsterdam (UvA). The funders had no role in study design, data collection and analysis, decision to publish, or preparation of the manuscript.

Competing Interests: The authors have declared that no competing interests exist.

* E-mail: X.Wang@uva.nl

Introduction

Climate change will likely modify current precipitation regimes influencing the global carbon (C) cycle in relation to erosion processes [1,2]. The length and intensity of droughts and the intensity of more sporadic rainfall events are predicted to increase for Western Europe [3], which will accelerate soil erosion. Soil erosion has significant impacts on the redistribution and transformation of soil organic carbon (SOC) within a landscape [4,5]. Even now, there is no consensus whether soil erosion is acting as a net C sink [5,6] or source [7] of atmospheric CO_2. Therefore, quantitative assessments of soil organic C redistribution along geomorphic gradients and the processes involved become increasingly important in a changing climate to resolve this controversy [8]. It is crucial that such studies comprise the different processes associated with the redistribution of C along the slope including CO_2 emissions as a result of changes in C mineralization upon erosion, transport and subsequent deposition. Based on such studies, complete C budgets of soils affected by erosion processes can be determined.

Soil erosion seems to preferentially remove fresh and more labile materials from C rich topsoils in upslope eroding positions, i.e. SOC with low density (e.g. free light fraction) and dissolved organic C (DOC) [7–10]. However, the fate of this organic C has rarely been studied. It is well known that most of the eroded sediments are re-deposited close to the source areas and in the catchment (e.g. [4,11]). Deposition of C enriched sediments lead to accumulation of SOC in the downslope positions. The eroded and deposited C can be stabilized by interaction with minerals thereby decreasing mineralization of deposited C in soil profiles [12]. In addition, soil erosion could affect dissolved organic carbon (DOC) dynamics in soils. Wang et al. [12] found higher DOC concentration at eroding sites in comparison to depositional sites.

Soil erosion drastically influences not only lateral SOC distribution within a landscape but also vertical CO_2 fluxes into the atmosphere [7,10]. Van Oost et al. [5] summarized at least three key mechanisms controlling the net flux of C between the soil and atmosphere: 1) dynamic replacement of SOC at the eroding sites [6]; 2) deep burial of SOC rich topsoils at depositional sites [4,13]; 3) enhanced decomposition of SOC because of the chemical or physical breakdown of soil during detachment and transport [7]. Particularly, the second and the third mechanisms should be susceptible to changes in the precipitation regime.

A key uncertainty of erosion-induced C loss is C mineralization resulting from the breakdown of soil aggregates as a direct response to extreme precipitation [7,14,15]. During a given erosion event, rainfall leads to breakdown of aggregates and releases the encapsulated C due to flow shear and raindrop impact [14]. Some studies suggest that aggregates breakdown by raindrop impact and wetting is mainly caused by initial fast slaking [16] or welding [17]. However, the extent of additional CO_2 fluxes from breakdown of aggregates due to erosion is still largely unknown. Franzluebbers [18] estimated a 10–60% increase in CO_2 evolution from various soils after breakdown of aggregates during 0–3 days. Polyakov and Lal [14] suggested that mainly the breakup of initial soil aggregates by erosive forces is responsible for increased CO_2 emission. However, conducting a set of rainfall simulation experiments, Bremenfeld et al. [19] recently suggested that interill erosion and associated soil aggregates breakdown have no prominent effect on soil respiration *in situ*. Therefore, effects of erosion-induced breakdown of aggregates on CO_2 evolution need to be further assessed.

Estimates of soil and SOC redistribution and associated CO_2 emissions show a large spatial and temporal variability. As field SOC and CO_2 fluxes of soils under erosion strongly depend on temporal variability of environmental conditions (e.g. location, soil management, initial soil moisture, and rainfall event characteristics) rainfall simulations under controlled laboratory conditions may help to shed light on C flux processes. Several rainfall simulation experiments have attempted to investigate soil erosion and associated SOC dynamics [20–24]. Jacinthe et al. [24] determined mineralization of SOC in runoff under no-till, chisel till and moldboard plow conditions with rainfall simulation approach. Van Hemelryck et al. [23] experimentally simulated three typical agriculture erosion events to quantify CO_2 emission. So far, however, there is no direct process assessment on combining effects of erosion, transport and subsequent deposition on C redistribution including vertical CO_2 fluxes. Changes in SOC pools indicative for important mechanisms of SOC redistribution and differing in their stability against microbial decay are not well known.

To get a better process understanding of soil erosion, transport and deposition on the redistribution and mineralization of SOC, the main objective of the present study was to determine SOC redistribution and a complete C budget of a loess soil affected by water erosion using a pseudo-replicated rainfall simulation experiment under standardized conditions. The following processes were studied and considered in our C budget:

(i) We determined SOC mineralization by measuring CO_2 emissions at different slope positions.

(ii) We analyzed soil and C redistribution along the slope including potential export into aquatic ecosystems. We measured C enrichment in the redistributed sediment. In order to test the hypothesis that POC is preferentially eroded and exported into aquatic ecosystem we fractionated SOC by density into particulate organic C (free POC, C not bound to minerals) and mineral associated organic C (MOC).

(iii) Finally, we analyzed concentrations of DOC in soil solutions at different positions of the slope and in runoff and determined above and belowground lateral DOC fluxes.

Materials and Methods

Ethics Statement

The experimental station 'Proefboerderij Wijnandsrade' (The Netherlands) permitted access to their land and allowed for taking soil sample material from their cereal fields for the research carried out.

Site Description and Sampling

The loess soil was collected from an agricultural field with winter wheat in South Limburg (50°53′58. 42″N, 5°53′16. 23″E), The Netherlands in May 2011. South Limburg is part of the European loess belt and has a temperate maritime climate. This region has a mean annual precipitation of 825–850 mm [25] and a mean annual temperature of 10.2°C. The sampled soil has a silty loam texture, and is classified as a Haplic Luvisol [26]. In the present study, the top 10 cm of the *Ap* horizon was collected and sieved over an 8 mm mesh to homogenize the soils and to keep aggregates intact as much as possible. Agricultural management at the sampled site is characterized by a potato-winter wheat-beet-winter wheat rotation. Soils are plowed 30 cm by a cultivator in spring and conventional tillage was applied in winter (including 30 cm plowing). The basic physical and chemical properties of the used soil are shown in Table 1.

Soil Analysis

Field bulk density was estimated from undisturbed 100 cm^{-3} cores that were oven-dried at 105°C for 24 hours [27]. Grain size distribution of soils was obtained using a particle size analyser (Micromeritics, SediGraph 5100, Norcross, USA). Soil pH (1:2.5 in H_2O) was measured with a multi-parameter analyser (CONSORT C832, Abcoude, The Netherlands). Soil water content was continually determined by a multi-channel Metallic TDR cable tester system [28]. Carbon and nitrogen (N) contents in bulk soils, sediments and density fractions were determined using a C and N analyser (Elementar VarioEL, Hanau, Germany).

Experimental Design

The erosion experiment was carried out using a 1.25 m×3.75 m experimental stainless steel flume (Figure 1). The upper 1.75 m had a slope of 15° (upslope position) and the lower 2 m had a slope of 2° (downslope position). To assess the effects of erosion, transport and subsequent deposition on redistribution of soils and C along the erosion slope, the experimental flume was divided into three zones according to the positions of the slope and observed results of sediments redistribution (Figure 1): 1) the eroding zone, at the upper half of the upslope position; 2) the transport zone, at the lower half of the upslope position and the upper half of the downslope position; 3) the depositional zone, at the lower half of the downslope position of the flume. We used a static definition of the different zones as dynamic measurement locations would have disrupted the soil surface. We recognize that these zones can change during the event and between events and that during events in every zone also local deposition and re-entrainment will occur.

The entire flume was subdivided into three parallel replicates of 40±2 cm wide. The soil was laid on top of a 2 cm thick layer of inert quartz sand to allow water to drain away. On top there was a 20 cm layer of soil on the upper (erosion) section where soil was supposed to erode and a thinner (5 cm) soil layer on the lower deposition section to allow for material deposition. On the transport section there was a gradual transition from 20 to 5 cm soil layer. While placing the air-dried soil it was compacted for every 2 cm, using a hammer and wooden piece of board (30×30 cm) to distribute the applied force. The compaction was such that it approached bulk density under field conditions (1.28 g cm^{-3}). In addition, there were three controls. Three control buckets (diameter 34 cm) were filled with a 20 cm loess soil layer on top of a 2 cm quartz sand layer, similar to the main flume. These control buckets were also compacted to the same bulk density. The buckets were placed next to the flume so that they

Table 1. Basic properties of the loess soil used in the experiment. Results are shown as mean and standard error of three replicates.

Depth (cm)	Bulk density (g cm^{-3})	pH	SOC[a] (%)	TN[b] (%)	C/N	Soil texture (%)		
						Sand	Silt	Clay
0–10	1.28 (0.05)	6.5 (0.06)	1.07 (0.06)	0.11 (0.01)	10.3 (0.7)	8.6	82.2	9.2

[a]: Soil organic carbon.
[b]: Total nitrogen.

received the same rainfall as well, but no lateral displacement of soil material took place.

The soil layer was pre-wetted to an initial standard moisture contents (Table 2) in 10–15 min to initiate runoff generation prior to commencing the real rainfall experiment. Four 18-minutes rainfall events were carried out at a monthly time interval. Measurements were carried out every 2 minutes during rainfall simulation. Rainfall was simulated with two nozzles (Lechler 460 788) applying at 1600 hPa demineralized water using an average rainfall intensity of 41.8 ± 1.9 mm h^{-1}. A rainfall event with this intensity and duration of 18 minutes has a return period of about 2 years [29]. Mean drop size of the applied rainfall was 2.0 mm ($D_{50} = 2.0$). With an average falling height of 1.8 m, the kinetic energy applied on the soil surface was 12.5 J m^{-2} mm^{-1}. Demineralized water was used instead of tap water to prevent flocculation problems with dispersible soil material [30,31]. As the total load of ions in rainwater is very low (the annual average electrical conductivity EC_{25} was below 20 μS cm^{-1} at the official Dutch sampling site Beek [32], about 10 km from the soil sampling site) the physico-chemical impact of demineralized water on soil particles is considered to be the same as for rain water. The temperature was kept as constant as possible ($18.1 \pm 0.9°$C).

Sampling during Erosion Experiments

Sediment traps were installed in the middle of the eroding, transport and depositional zones respectively with entrance of the traps at the upslope side and at the same level as the soil surface to capture mobilized sediment in overland flow (Figure 1). The sediment traps were modified 12 ml Polypropylene screw cap tubes (Greiner Bio-One GmbH, Frickenhausen, Germany). The traps had a small diameter to minimize disturbance to overland flow and resulting erosion patterns. An opening in the side was made to collect mobilized sediments. The sediment traps were sampled every two minutes and the collected materials were transferred to containers, oven-dried at 35°C, weighed and later analysed for C and N contents.

Runoff and sediments were collected from weirs at the end of the flumes at 2-min intervals once continuous runoff had developed. Total runoff was collected using a polystyrene gutter that was installed at the lowest part of the experimental flume. The contents of the flume were then pumped into V-notched bottles to measure flow rates using a simple siphon pump made of Tygon R-3603 tubes (Saint-Gobin, Courbevoie, France). The lower end was constrained to 4 mm diameter to provide a constant flow velocity, without risking clogging by larger soil particles and keeping effects on the aggregation of the sediments limited. The V-notched bottles overflowed into sampling boxes which were replaced every two minutes or when the sampling box was full.

At the lowest end of the flume, three holes per replicate flume were present at the level of the sand drainage layer to collect through flow. Through flow was defined as the lateral underground flow in contrast to the overland flow. The holes were covered from the inside by a 63-μm stainless steel mesh allowing water to pass through, but to prevent clogging up. On the outside of the walls attached tubes drained into bottles, similar to the runoff setup.

Sampling after Erosion Experiments

Density Fractionation of Bulk Soils and Exported Sediments. After four rainfall events the 0–2 mm topsoils at the eroding, transport and depositional zones and sediments exported during the first and fourth events were fractionated into three fractions by a sodium polytungstate (NaPT) solution with a density of 1.6 g cm^{-3}: the free light fraction (fLF) which consisted

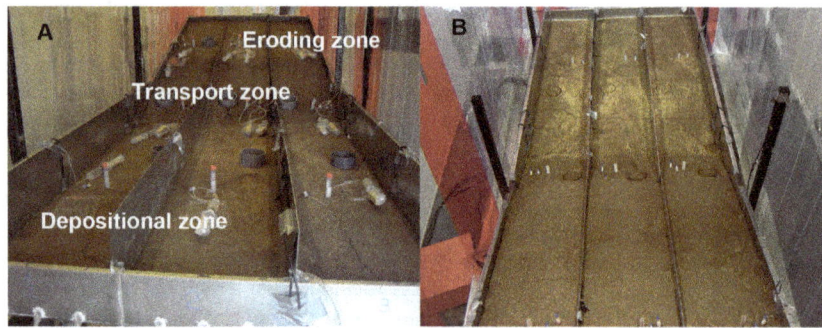

Figure 1. Photographs of the experimental setup and sampling locations along the experimental flume. It included the eroding, transport and depositional zones of the flume. A shows the lateral view; B shows the vertical view.

of large, undecomposed or partly decomposed root and plant fragments, the light fraction occluded in aggregates (oLF) and the heavy fraction (HF), which was associated with minerals [33,34]. Soil organic C in fLF, oLF and HF are defined as fPOC, oPOC and MOC, respectively. Particulate organic C (POC) is the C not bound to soil minerals including both fPOC and oPOC. The oPOC represents C sequestered in aggregates. Methods and procedures were followed as described in Cerli et al. [34]. All fractions were freeze-dried, homogenized and later analysed for C and N contents. Density fractionation was done in triplicate.

Dissolved Organic Carbon (DOC)

To investigate dynamics of dissolved organic C in different soil depths and positions as affected by soil redistribution, soil moisture samplers (MACRO RHIZON 19.21.35, 9 cm porous, 4.5 mm OD, 0.2 μm, Wageningen, The Netherlands) were inserted in the eroding, transport and depositional zones of the flume. Each sampler was connected to a syringe (50 mL) to collect the soil solution. At the eroding and transport zones of the flume, soil solutions were collected at 4 cm and 9 cm depths. At the depositional zones soil solutions were sampled at 4 cm only because of the thinner soil layer on the lower deposition section. Soil solutions were sampled twice per week during the first week immediately after one rainfall event because of higher soil water moisture. As the soil dried, soil solutions were collected once per week. Concentrations of DOC were determined by a TOC analyser (TOC-V CPH, Shimadzu, Kyoto, Japan).

Soil CO_2 Efflux Measurements

Soil respiration was measured using a Portable Gas Exchange and Fluorescence System (LI- 6400XT; LICOR Biosciences, Lincoln, NE USA). In order to enhance the comparability of data, most CO_2 efflux measurements were conducted in the afternoon between 17:00 and 19:00 at local time in PVC collars (10.2 cm in diameter and 7 cm in height). Soil CO_2 efflux was determined before and after each rainfall simulation event. As the 7 cm high collars, necessary for the CO_2 efflux measurements, would strongly affect the overland flow and erosion patterns during the rainfall event, the 7 cm high collars were replaced by smaller collars (same diameter but 1.5 cm tall). These were inserted at exactly the same place, to temporary fill the imprint of the high collar in the soil surface. The top of the collar was placed exactly equal to the soil surface, to minimize the disturbance of the sampling location by the CO_2 measurements but still enabling to measure the CO_2 efflux exactly at the same position later on. Overland flow was possible and erosion, transport and deposition processes at the surface of the area used for measuring CO_2 were hardly affected

by this strategy. Two to three measurements per site (i.e. per collar) were carried out each time. The number of replicated measurements per collar depended on the variation after the first two analysis with an additional measurement if the relative deviation of the second one was larger than 10%. Additionally, pre-experiments were carried out using the same loess soil to test impacts of soil depth on soil CO_2 efflux. In these experiments the CO_2 efflux was measured in columns with increasing soil thickness under constant soil moisture and temperature conditions. Results showed that soil depth did not have significant effect on soil respiration per soil weight up to a depth of 30 cm (data not published). Based on these results, all data measured in different experimental zones, and control soils, having different soil depths, were corrected to 20 cm soil layers in order to directly compare effects of erosion, transport and deposition on CO_2 effluxes.

Erosion-induced Carbon Budget

Fluxes of SOC and DOC were calculated by multiplying concentrations of SOC and DOC with the volume of the overland flow. Other parameters were calculated as follows:

$$Carbon\ enrichment\ ratio\ (CER) = \frac{C_{sediment}}{C_{control\ soils}}$$

$$Total\ C\ losses = Lateral\ C\ exported$$
$$+ Vertical\ CO_2\ emission$$

$$Lateral\ C\ exported = SOC\ exported\ in\ overland\ flow$$
$$+ DOC\ exported\ in\ overland\ flow$$
$$+ DOC\ exported\ in\ through\ flow$$

$$Net\ additional\ CO_2\ emission$$
$$= CO_2\ emission\ from\ soil\ in\ flume$$
$$- CO_2\ emission\ from\ soil\ in\ control\ treatment$$

Based on the 4-month data we calculated annual C fluxes by linear extrapolation making comparisons with the literature easier. However, the shortcomings of such budgets based on short-term laboratory experiments only are obvious.

The definition of C source and sink areas for calculating the C budget was based on two experimental observations. After the fourth rainfall event, soil layers with relocated materials were clearly visible in the flume, particularly in the downslope part of

Table 2. Initial soil water contents (m³/m³) before and after pre-wetting before starting the rainfall simulation.

Zones	Event 1		Event 2		Event 3		Event 4	
	Before	After	Before	After	Before	After	Before	After
Eroding	0.25	0.32	0.27	0.33	0.26	0.31	0.26	0.31
Transport	0.39	0.44	0.36	0.42	0.33	0.39	0.30	0.37
Depositional	0.36	0.48	0.33	0.46	0.30	0.42	0.23	0.44

the depositional zone (Figure 1B). In addition, we found that SOC was significantly depleted in the transport zone comparing with controls soils (cf. section results). Based on these two observations, the eroding and transport zones were defined as the C source area and the depositional zone and the runoff leaving the flume (exported into aquatic system) were defined as the C sink area. We calculated an erosion-induced SOC budget for the four rainfall events over the entire period using a mass balance approach (i.e. source = sink area). Changes in C distribution between the density fractions were appropriately considered by using the data of the original soil for the source area. This approach enabled us to include any changes in C redistribution between density fractions induced by erosion.

Statistical Analyses

Differences in C enrichment ratios, amounts of sediment exported and DOC concentrations in overland flow were tested with one-way ANOVA and the Post-hoc Duncan test to differentiate between individual differences. The difference of CO_2 effluxes measured in the 4-month period at eroding, transport and depositional zones of the gutter was tested by repeated measurement ANOVA. Averaged CO_2 efflux in different experimental zones was compared using a one-way ANOVA. For all tests, a significance level of $P = 0.05$ was set using the Post-hoc Duncan test, unless otherwise indicated. The relationship between cumulative CO_2 emission and DOC concentration was tested by two-tailed Pearson test. All statistical tests were performed using SAS software (Version 8.1) and SPSS (IBM Statistics 20).

Results

Loss of Sediment and Carbon Enrichment Ratios in Overland Flow

Total sediment losses in the overland flow increased during the course of the experiment from 9.5 g m^{-2} in the first event to 31.0 g m^{-2} in the fourth event (Figure 2). During the first rainfall event the average sediment concentration was 1.1 ± 0.2 g L^{-1} and doubled to 2.3 ± 0.8 g L^{-1} in the fourth event.

Carbon enrichment ratios (CER) of sediment loads of overland flow trapped at the eroding, transport and depositional zones of the flume ranged from 0.8 to 2.9. The CER was significantly higher at the depositional zone compared to those of the eroding and transport zones (Table 3). Carbon enrichment was even stronger in the sediments of the runoff with CER between 1.3 and 4.0. Carbon enrichment ratios decreased with increasing concentrations of suspended solids in the overland flow (Figure 3). Concentrations of suspended solids were smaller at the beginning of each rainfall event, resulting in larger C enrichment but also in larger variation of the data.

Preferential Erosion and Deposition of Organic Carbon at the Soil Surface

After four rainfall events, a thin sedimentation layer was present in the depositional zone (approximately 2 mm thick) without any layering. However the depositional zone clearly showed patterns of deposition of finer grained materials along the flow lines of overland flow and the whole lower part of the gutter. Soil organic C concentration (mg^{-1} g soil) of the surface soil (2 mm) decreased by 6.0% at the eroding zone and increased by 3.9% at the depositional zone if compared to control soils (Table 3). Nevertheless, soil organic C concentration did not differ significantly between the control, the eroding, transport and depositional zones of the gutter. Also the relative distribution of C in density fractions of the soil was not affected by soil erosion. Most of the C (86% to

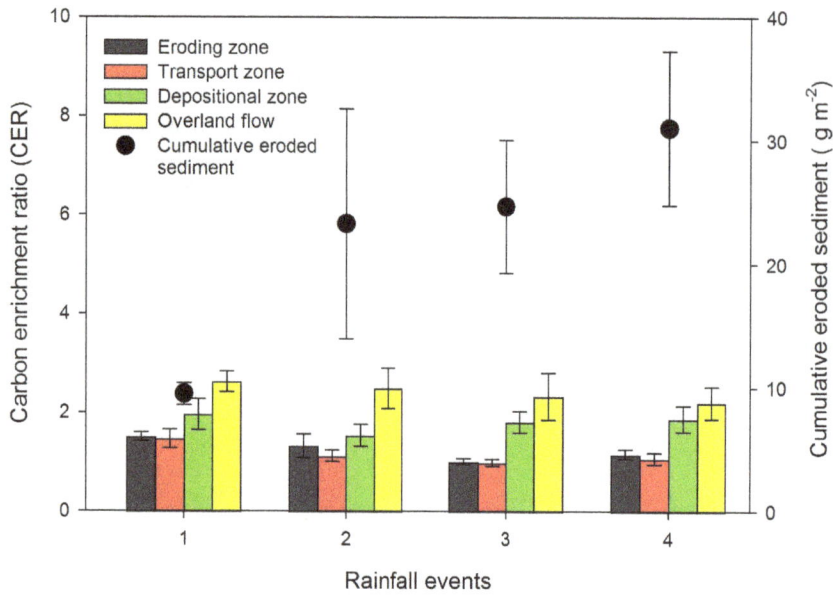

Figure 2. Average total eroded sediment per rainfall event exported by overland flow and carbon enrichment ratios (CER) during four rainfall events.

91%) was found in the heavy fraction, i.e. mineral associated organic C (MOC; Table 3). The rest was almost equally distributed between the free light fraction (particulate organic C in free light fraction = fPOC) and the fraction occluded within aggregates (oPOC). The free light fraction was significantly enriched in C at the transport and the depositional zone whereas the occluded light fraction (oLF) was depleted in C at the eroding zone (Table 3). The heavy fractions of surface soils in the flume did not significantly change in C contents.

In the sediments, all fractions were strongly enriched in C with the largest enrichment in the free light fraction. This C enrichment was smaller in the occluded and smallest in the heavy fraction and also decreased from the first to the last event (Table 3). However, the C content of the heavy fraction of the sediments (first event) was more than double the C content of the heavy fraction of the control soil (Table 3).

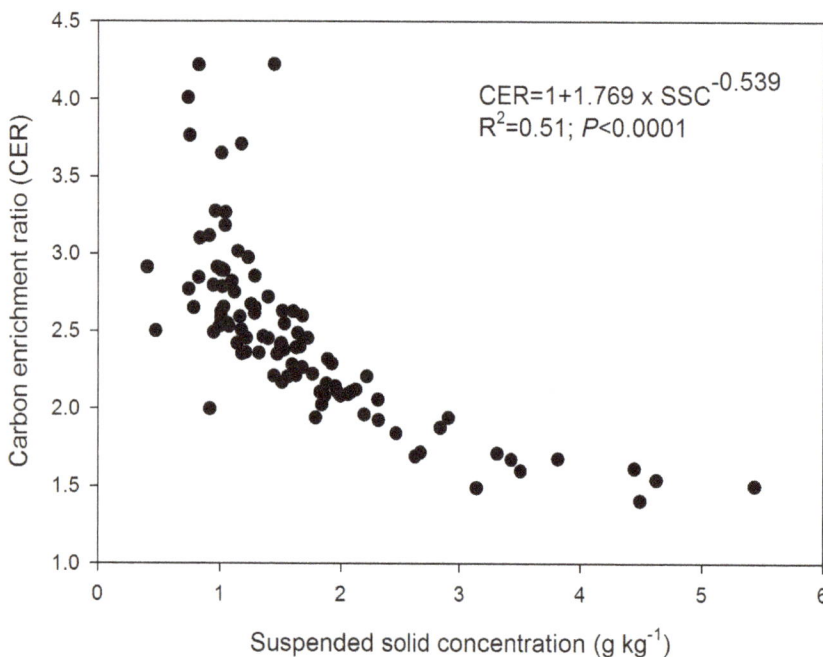

Figure 3. Relationship between carbon enrichment ratio (CER) and suspended solid concentration (SSC) in the overland flow.

Table 3. Carbon concentrations and specific carbon fractions of soils and sediments for different zones and events.

Zones	C concentration[a] (mg g⁻¹ soil)	C concentration[b] (mg C g⁻¹ specific density fraction)			C enrichment ratio (CER)[c] (-)				Relative proportion of MOC[d] (% SOC)
	Bulk soils	fPOC	oPOC	MOC	Bulk soils	fPOC	oPOC	MOC	MOC
Control	10.0 (0.5)	134.3 (28.9)	162.3 (24.4)	8.0 (0.1)					91
Eroding	9.4 (0.2)	189.8 (26.3)	175.3 (20.4)	8.0 (0.4)	0.94	1.1	0.8	0.9	91
Transport	9.7 (0.2)	220.3 (60.3)	143.1 (31.5)	7.7 (0.2)	0.97	1.9	1.0	0.9	87
Depositional	10.4 (0.5)	205.0 (61.8)	175.5 (25.1)	7.9 (0.3)	1.04	1.6	0.9	1.0	90
Overland flow 1	22.9 (0.9)	151.2 (42.2)	345.5 (20.1)	17.3 (0.6)	2.30	3.9	3.2	2.2	86
Overland flow 4	16.6 (1.6)	219.1 (54.3)	296.4 (39.3)	13.9 (0.9)	1.67	2.2	2.3	1.6	88

C in free light fraction = free particulate organic C, fPOC; C in occluded light fraction = occluded particulate organic C, oPOC; C in heavy fraction = mineral-associated organic C, MOC. Results are shown as mean and standard error of three replicates.
a. C concentration of bulk soils (mg C g⁻¹ soil).
b. Carbon concentration of the three density fractions fPOC, oPOC and MOC in relation to the total weight of that specific soil fraction (mineral + C parts) (mg C g⁻¹ soil fraction).
c. Carbon enrichment ratios, calculated on the basis of mg C soil fraction g⁻¹ soil organic C.
d. Relative proportion of MOC (%SOC) in bulk soils, density fractions and sediments of overland flow for the first (Overland flow 1) and fourth rainfall event (Overland flow 4).

Soil CO_2 Efflux

All measured CO_2 efflux rates for the whole experiment ranged from 0.12 to 4.34 g C m^{-2} day^{-1} (Figure 4). During the entire experimental period, rates of CO_2 emissions exhibited a similar behaviour in the eroding, transport and depositional zones and the non-eroded control with a sharp initial increase immediately after each rainfall event, followed by continuously decreasing rates thereafter. Rates of CO_2 efflux significantly decreased with time during the four events ($P = 0.001$). The spatial and temporal variability of CO_2 efflux rates was larger in the first rainfall event than during the other events.

The largest mean CO_2 efflux was observed in the transport zone during the first three rainfall events (Figure 5). In the fourth event, however, the depositional zone had the largest mean CO_2 efflux. The relative differences of the mean CO_2 efflux between the depositional and the eroding zones increased during the course of the whole experiment and became significant in the fourth event.

Cumulative CO_2 fluxes in the eroding, transport and depositional zones ranged from 80 to 180, 116 to 317, and 146 to 204 g C m^{-2} yr^{-1}, respectively. The largest mean cumulative CO_2 fluxes (221 g C m^{-2} yr^{-1}) were observed in the transport zone. Mean cumulative CO_2 fluxes in the depositional zone (181 g C m^{-2} yr^{-1}) were significantly larger than those in the control soils ($P = 0.02$) while CO_2 fluxes in the eroding zone were similar in comparison to the control. The total losses of C as CO_2 emission during the entire experiment accumulated to 1.8 to 2.9% of total soil organic C stocks.

Dissolved Organic Carbon (DOC)

Concentrations of DOC in soil solutions at eroding, transport and depositional zones ranged from 7.1 to 25.9 mg L^{-1} during four rainfall events (Figure 6). In the shallow soil (4 cm depth), mean concentration of DOC decreased in the following order: transport zone (15.1 mg L^{-1}) > control soils (14.3 mg L^{-1}) > depositional zone (12.3 mg L^{-1}) > eroding zone (11.8 mg L^{-1}). However, only DOC concentrations in the depositional and eroding zones were significantly lower than those in the transport zone and the control. Mean concentrations of DOC in the deeper soil (10 cm) were almost equal as in the shallow soil and decreased in the following order (not statistically significant): control soils (16.8 mg L^{-1}) > transport zone (15.2 mg L^{-1}) > eroding zones (12.3 mg L^{-1}). Concentrations of DOC in soil solutions of both depths showed distinct temporal patterns in all zones of the gutter. They increased at the beginning of each rainfall event, then decreased and increased again with time. This trend was less obvious during the first rainfall.

Concentration of DOC in overland flow remained constant during each single event, ranging from 0.3 to 8.3 mg L^{-1} and significantly decreased from the first to the third rainfall event (means of the four rainfall events: 7.2 ± 0.4, 2.6 ± 0.4, 0.9 ± 0.7, 0.7 ± 0.4 mg L^{-1}, no further data shown). Cumulated DOC fluxes transported by overland flow were on average 0.23 g C m^{-2} yr^{-1} (Figure 7). The amount of C exported as DOC by overland flow was small, accounting for 0.014% of the total SOC stocks in the flume. Fluxes of DOC in through flow (i.e. 0.002% of total SOC stocks) were significantly smaller than overland flow.

Discussion

Preferential Transport and Deposition of Organic Carbon

As expected from the literature [8,12], the soil of the eroding zone was depleted in C whereas the soil of the depositional zone and the sediments of the overland flow were enriched in C after the four rainfall events (Table 3). The results of the density

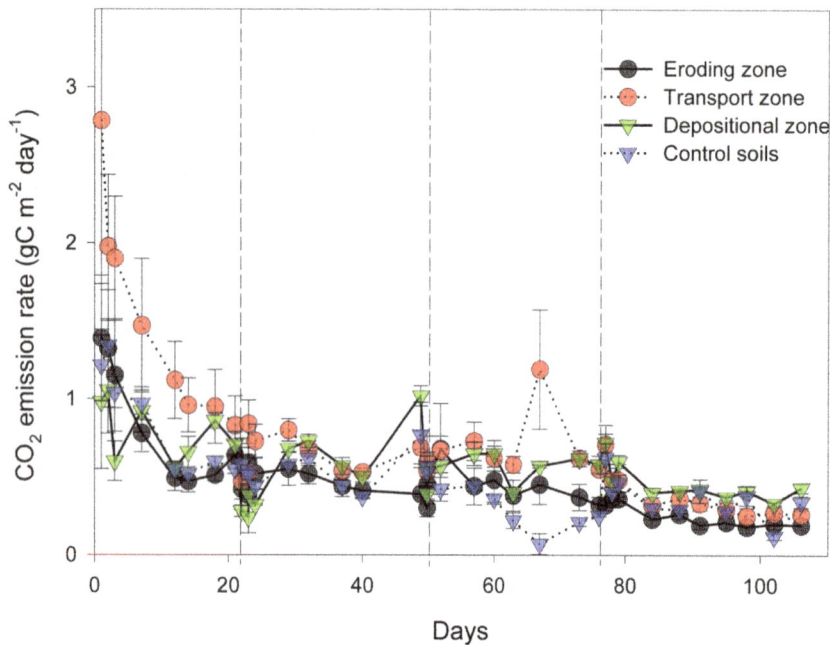

Figure 4. CO_2 efflux at different zones of the gutter and control soil during four rainfall events. Solid line + circle represents the eroding zone; dotted line + circle represents the transport zone; solid line + triangle represents the depositional zones; and dotted line + triangle represents control soils. Values are mean± standard error of three replicates.

fractionation clearly showed a large loss of C occluded in aggregates in the eroding zone, which was accompanied by an enrichment of C in the fPOC fraction in the other zones of the flume and in overland flow (Table 3). We assume that the disruption of macro-aggregates by raindrop peeling [35] and aggregate welding and development of a structural crust [17] resulted in the liberation of fPOC, which was preferentially

transported [8]. The disruption of macro-aggregates will result in the release of micro-aggregates (smaller than 250 µm) from the macro-aggregates too. The C content of micro-aggregates within macro-aggregates is usually larger than that of macro-aggregates [36–38]. The release of such small aggregates and selective transport of small aggregates with low density [39] could be the reasons for the observed significant C enrichment of oPOC in

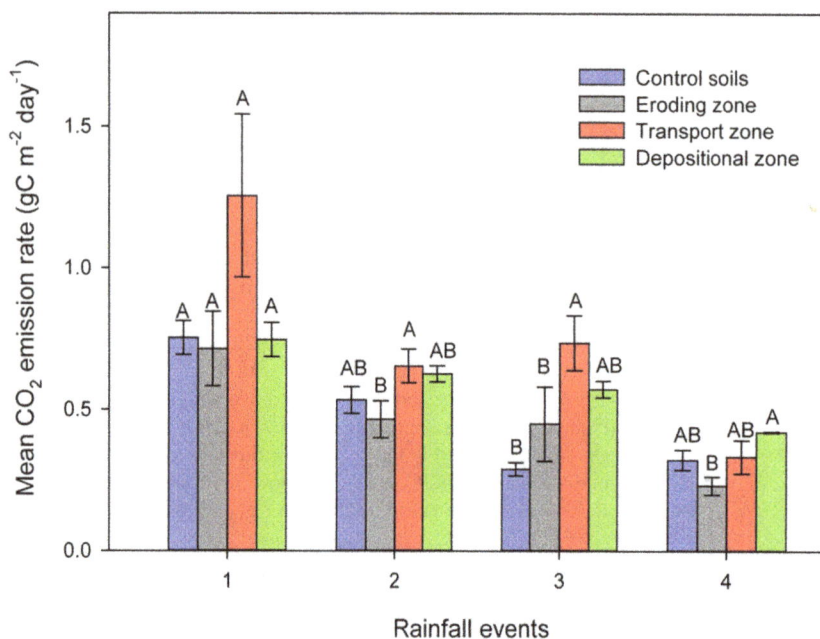

Figure 5. Mean cumulative CO_2 emission at the eroding, transport and depositional zones and control soil. Different capital letters mean significant difference at a single rainfall event between the different zones. Values are mean± standard error of three replicates.

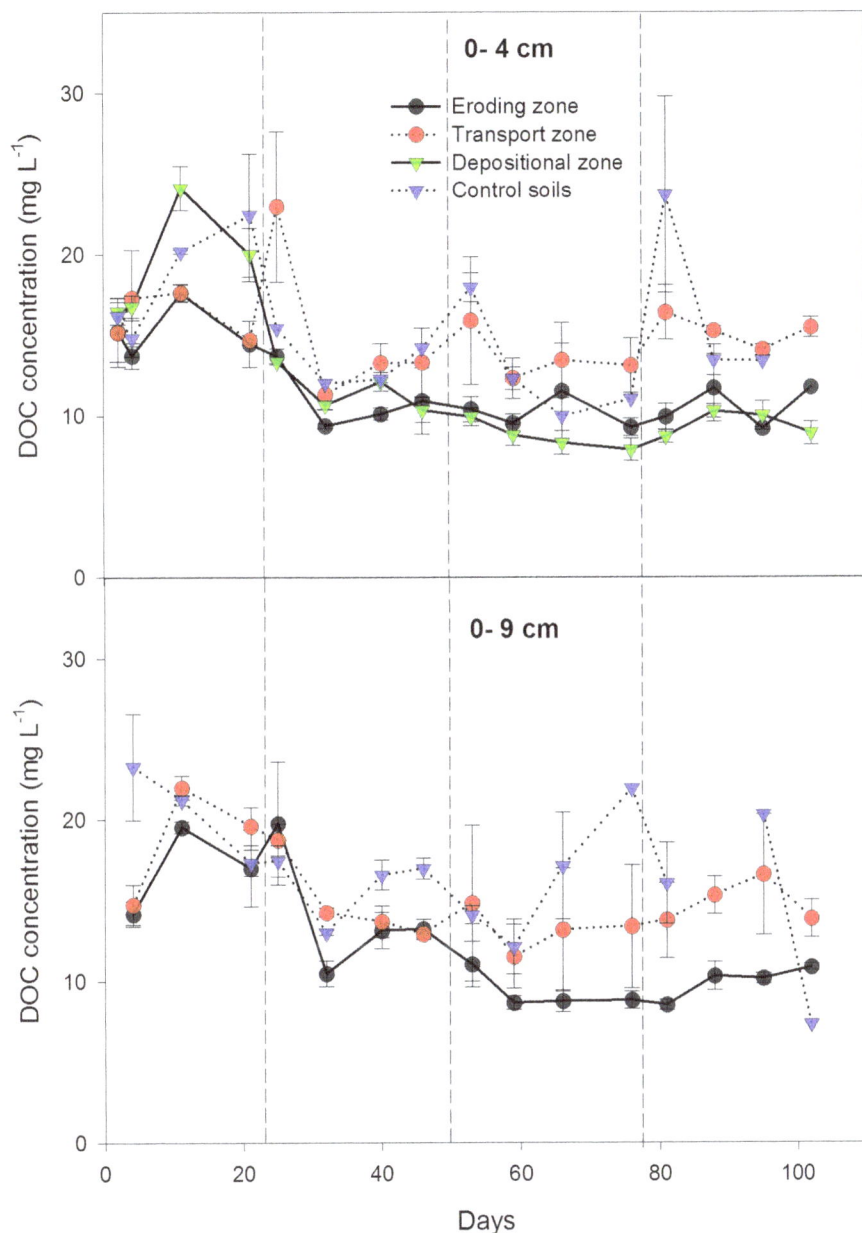

Figure 6. Dissolved organic carbon (DOC) concentrations. DOC solutions were collected at 0–4 cm and 0–9 cm depths of the eroding, transport and depositional zones of the flume during four rainfall events. Solid line + circle represents the eroding zone; dotted line + circle represents the transport zone; solid line + triangle represents the depositional zones; and dotted line + triangle represents control soils. Values are mean± standard error.

sediments ranging from 2.3 to 3.2 (Table 3). However, we did not study aggregate stability and the detailed processes resulting in breakdown of the aggregates neither the related preferential erosion, transport and deposition of different sizes of aggregates and particles. That should be done in follow-up experiments.

The calculated mass balance of the experiment illustrates the disruption of aggregates in the eroding zone and the redistribution of C from aggregates to fPOC with an erosion-induced accumulation of fPOC in the sink area of 0.24 g C (Table 4). This accumulation is equal to an increase in fPOC by 48% comparing the source and the sink area. One logical source of this additional POC would be C occluded within aggregates in the eroding zone at the beginning of the experiment. This large

accumulation of fPOC in the sink area contributed to the observed relative increase by 6% (Table 3) in the SOC content of the first two mm layer of the depositional zone.

Particulate organic C already present in the soils and formed by disruption of aggregates (cf. above) was preferentially eroded and transported by overland flow as indicated by the largest CER ratio of the density fractions in any of the sampled soils and sediments. Per definition, fPOC is the lightest fraction, not associated with minerals and therefore easier to be translocated by water than soil particles with a higher density [23,40,41]. The high C enrichment of mineral-associated organic C (MOC) in the sediments of the overland flow suggested that water erosion separated the whole soil particles according to their density (Table 3). This fraction-

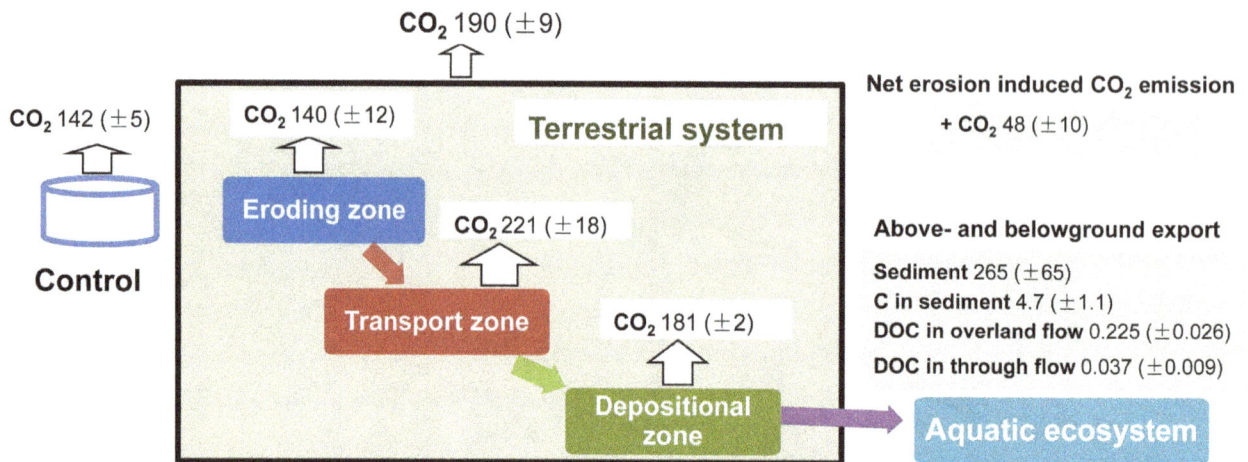

Figure 7. Conceptual diagram illustrating the total carbon budget as affected by soil erosion, transport and deposition in the four months rainfall simulation experiment. Fluxes were calculated on an annual base (interpolated from the 4-months experiments). The values were expressed as mean values and standard error of three replicates.

ation occurred between the different density fractions. Increasing C concentrations ($MOC<oPOC<fPOC$; Table 3) resulted in increasing CER of the sediments in the same order.

A significant portion of the eroded and transported C enriched sediment was not retained in the downslope parts of the depositional zone and was exported by overland flow and left the flume (Table 4). Particularly the weakly decomposed C of the fPOC should be a readily available C and nutrient source for aquatic organisms [16,42] contributing to CO_2 emission from aquatic ecosystems. This process linking terrestrial and aquatic systems cannot be neglected for modeling the C cycle and has to be studied in more detail in future.

Relationship between Erosion Rate and Carbon Enrichment

The inverse, non-linear relationship between the erosion rate and C enrichment of the sediments we found (Figure 3) is in agreement with previous studies [21,40,41,43]. This inverse relationship is the result of increasing sediment concentration in the overland flow during each single event and from the first to the fourth rainfall event. One of the most important reasons for this relationship should be the breakdown of macro-aggregates by the raindrops as already discussed to be the main reason for the preferential erosion of fPOC [8,10,35]. This process should be particularly important at the beginning of each rainfall event because rewetting of dry soils results in the disruption of aggregates and the release of organic matter [44]. It is also reasonable to assume that the importance of this process will decrease with increasing number of rainfall events. Heavy rainfall causes compaction, welding and crust formation resulting in reduced infiltration and increased erosion and suspended solid concentration with time [15]. The preferential removal of C enriched soil will result in C enriched sediments particularly at the beginning of the experiment where the erosion rate was still small. After removal of this soil enriched in C, the erosion rate increases because of decreasing infiltration and generation of more overland flow. That will result in even increasing erosion rates because soils are less protected by organic matter and aggregation. In field situations, the relationship between erosion rate and C enrichment might be weaker because of continuous above and belowground C input and its positive effect on aggregation [45].

Decreasing C enrichment with large erosion rates, i.e. increasing sediment concentration, indicated that an increasing erosion rate does not result in proportionally increasing C losses. However, this does not mean that more severe erosion events lead to less impact on soil C. Very strong erosion events will translocate large amounts of C. However, this C might be better protected against further mineralization after deposition because C is mostly deposited as mineral associated C (Table 4). The C loading of mineral surfaces should be low as well, resulting in a more efficient stabilization against microbial decay [46,47]. In addition, long-term erosion-induced C sequestration or depletion might depend on the precipitation frequency and intensity.

Soil CO_2 Effluxes

This study provides new data and insight on C decomposition under controlled conditions in an artificial landscape setting at eroding, transport and depositional positions allowing for a better process understanding. Although we did not scale up our results to the landscape level, it is important to know whether the fluxes measured do compare with observed field measurements and make any sense, also in comparison with previous indirect measurements on eroded sediments and soil profile investigations [5,12]. In the present study, CO_2 efflux rates measured (0.12 to 4.34 g C m^{-2} day^{-1}) were in the range of soil respiration rates from agricultural loess soils [5,19]. Initial increases of CO_2 emissions immediately after each of our rainfall events might be explained by the increase in microbiological activity after re-wetting the dry soil and/or increased bioavailability due to aggregate breakdown [44]. Aggregate breakdown and subsequent exposure of previously encapsulated SOC provide substrates for microbial decomposition [8]. The re-wetting effect was particularly important after the first event and decreased during the course of the experiment. This is in line with a decreasing capacity of soil to release C from aggregates over time [48,49].

Transport of topsoil and associated C influenced SOC decomposition rates at the different positions of the artificial slope. The small cumulative CO_2 emission from the eroding zone should be the result of the observed preferential removal of C enriched materials (i.e. higher CER of the sediments at the depositional zone in comparison to the eroding zone, Table 3), which was either preferentially deposited or left the gutter. This

Table 4. Soil organic C redistribution in three density fractions due to erosion (mass balance approach); C in free light fraction, fPOC; C in occluded light fraction, oPOC; C in heavy fraction, MOC).

Fraction	Source area (g C)[a]	Sink area (g C)			Relative value (% of SOC redistributed)			Erosion-induced fPOC[b] ΔC (g)	Aggregate Breakdown oPOC[c] ΔC (g)
		Depositional zone	Overland flow	CO$_2$ emission	Depositional zone	Overland flow	CO$_2$ emission		
fPOC	0.5 (0.0)	0.6 (0.4)	0.14 (0.05)		4.5	1.0		+0.24	
oPOC	0.8 (0.0)	0.5 (0.1)	0.19 (0.04)		3.8	1.4			−0.11
MOC	12.3 (0.4)	9.8 (0.3)	2.13 (0.47)		72.1	15.7			
Total SOC/CO$_2$	Σ 13.6	Σ 10.9 (0.6)	Σ 2.46 (0.60)	0.2 (0.0)	Σ 80.4	Σ 18.1	1.5		
SOC redistributed	Σ 13.6	Σ 13.6							

Results are given as mean and standard error of three replicates.
[a]. Original soil data were used to exclude any effect of soil erosion.
[b]. Erosion induced formation of fPOC (disruption of aggregates) ΔC (g) = fPOC in depositional zone + fPOC in overland flow- fPOC in source area.
[c]. Erosion-induced breakdown of aggregates (decline in oPOC) ΔC (g) = oPOC in depositional zone+oPOC in overland flow- oPOC in source area.

preferential removal of more labile C (POC: fPOC and oPOC) left behind less C, which was relatively more stable (Table 3). In turn, accumulation of labile C fractions in the depositional zone contributed to an increasing difference in cumulative CO$_2$ emission between the eroding and depositional zone over time (Figure 5).

The cumulative CO$_2$ emission was significantly and positively related to fPOC ($R^2 = 0.94$; $P = 0.03$) illustrating the more labile character of this SOC fraction. The large accumulation of fPOC probably explained that the transport zone had the largest cumulative CO$_2$ emissions (Table 3 and Figure 7). Larger CO$_2$ emissions from the depositional zone were anticipated because the deposited labile C (fPOC, oPOC) could be used as substrate and a source of energy for microbial respiration [50,51]. Although CO$_2$ emissions were large at the depositional zone, the SOC content increased by 6% in comparison to the control soils after four erosion events. Obviously, parts of the eroded and deposited SOC were preserved (Figure 5).

Considering all positions of the slope, mean DOC concentration in near-surface layers was positively correlated to median soil CO$_2$ efflux rate ($P = 0.02$). The largest CO$_2$ efflux was accompanied by largest DOC concentration in the transport zone – a second parameter (first fPOC) explaining the large CO$_2$ efflux in this zone. Creed et al. [52] found that substrates (i.e. DOC) in the near-surface soil were strongly related to median soil CO$_2$ efflux. Considering each position of the slope separately, median soil CO$_2$ efflux rates were not significantly related to mean DOC concentrations at the eroding ($P = 0.18$), transport ($P = 0.49$) and depositional zones ($P = 0.22$). However, DOC was significantly correlated to the median soil CO$_2$ efflux rate in the control soil ($P = 0.05$), which indicated DOC could be mineralized during the experimental period. Thus, DOC dynamics could not explain the observed additional C decomposition at the depositional zone. This might suggest a fast turnover of DOC or/and a direct use of POC by the microbial community.

Total Carbon Budget

We estimated an erosion-induced C loss of 53 g C m^{-2} yr^{-1} calculated as the sum of net erosion-induced CO$_2$ emission, C losses by overland flow (C in sediments, DOC) and through flow (cf. materials and methods, figure 7). During the entire experimental period, the averaged SOC fluxes leaving the flume with overland flow were 18 times larger than DOC fluxes including lateral fluxes by the through flow (Figure 7). Fluxes of DOC (0.26 g C m^{-2} yr^{-1}) were rather low particularly due to decreasing DOC concentration during the experiment, i.e. with increasing number of events. Fluxes of sediment associated C were equivalent to 8.9% of the erosion-induced C loss while DOC fluxes were equivalent to 0.5% of those C losses. Therefore, sediment associated C played a much larger role than DOC in the erosion-induced linking of terrestrial and aquatic ecosystems. However, erosion-induced DOC flux should not be neglected because DOC might be particularly important for aquatic food webs [42,53].

Erosion-induced CO$_2$ emission was the dominant form of C loss, representing 90.5% of erosion-induced C loss. Based on the assumption made (cf. material and methods), 1.5% of total C redistributed (deposited C at the depositional zone plus C exported to aquatic ecosystems) was mineralized to CO$_2$ (Table 4 and Figure 7). Previous estimates of decomposition of eroded SOC showed large variations, ranging from 0 to 100% (e.g. [7,14,15,23]). In modelling studies the assumption is often used that at least 20% of the eroded SOC is decomposed as a consequence of soil erosion [7,54]. Our measured values were much smaller than this conventional view of erosion effects on the C cycle. Palyakov and Lal [14] estimated 8% of SOC displaced by erosion had a potential to be mineralized and van Hemelryck et al. [23] estimated mineralization of 2% to 12% of the eroded SOC in a loess soils using laboratory rainfall simulation experiments. We propose three main reasons for the large difference. Firstly, the effects of the disruption of aggregates on extra CO_2 efflux were relatively short-lived [48]. Secondly, C stabilization as affected by soil erosion and deposition might be underestimated in the previous assumption [5,8]. Thirdly, the artificial slope was relatively short in our experimental setting in comparison to the field, which may result in an underestimation of transport effects on C mineralization.

Conclusions

Rainfall simulation experiments are a useful approach to determine the role of soil erosion for the C cycle. The data of our 4-months experiment were comparable to field situations, despite of well-known shortcomings of laboratory approaches. The erosion rate was estimated to be 2.1 mm yr^{-1} (26 t ha^{-1} yr^{-1}), which was comparable with estimations in this region based on field data [55]. Also C enrichment of exported sediments and soil CO_2 efflux were in the range of field measurements.

Erosion-induced CO_2 emission was the dominant form of C loss, representing about 90.5% of the erosion-induced C loss. In addition, a considerable amount of C rich sediments (265 g m^{-2} yr^{-1}) was laterally exported by overland flow. Carbon associated with sediments was the main form of erosion-induced lateral C loss and not DOC. This exported C plays an important role in the connection of terrestrial and aquatic ecosystems.

In our experiment, this redistribution of C rich materials resulted in a net additional CO_2 emission during transport and deposition. However, this enhanced CO_2 emission is much smaller than previously thought. Most of the redistributed C by overland flow was bound to soil minerals (heavy fraction), which might be one reason for the unexpected small mineralization. As a consequence, the induced C sink by deposition could be larger than assumed.

Our study clearly demonstrated a fractionation of SOC upon erosion, transport and deposition controlling C mineralization. Disruption of macro-aggregates was identified as the main process responsible for the observed preferential redistribution of labile particulate organic C. Future studies should determine the conditions and processes resulting in breakdown of the aggregates and related preferential erosion, transport and deposition of different sizes of aggregates and particles. Furthermore, the replacement of carbon at eroding zones has to be included in future studies determining the role of soil erosion as a potential C source or sink.

Acknowledgments

We gratefully acknowledge Leen de Lange, Dr. Chiara Cerli, Dr. Gillian Kopittke, Dr. Sebastiaan de Vet, Caridad Díaz López, and Bianca Pricope for their help during the rainfall simulation experiments. We also thank Leo Hoitinga and Bert de Leeuw for their lab support. The experimental station 'Proefboerderij Wijnandsrade' (The Netherlands) is acknowledged for providing soil material. Dr. John Parsons is also acknowledged for checking the language. Three anonymous reviewers are acknowledged for their useful and constructive suggestions.

Author Contributions

Conceived and designed the experiments: XW ELHC PR KK. Performed the experiments: XW PR. Analyzed the data: XW ELHC PR KK. Wrote the paper: XW ELHC KK.

References

1. Huxman TE, Snyder KA, Tissue D, Leffler AJ, Ogle K, et al. (2004) Precipitation pulses and carbon fluxes in semiarid and arid ecosystems. Oecologia 141: 254–268.
2. Chapin FS, McFarland J, McGuire AD, Euskirchen ES, Ruess RW, et al. (2009) The changing global carbon cycle: linking plant-soil carbon dynamics to global consequences. Journal of Ecology 97: 840–850.
3. IPCC (2007) Climate Change Fourth Assessment Report, Cambridge University Press, Cambridge, UK.
4. Stallard RF (1998) Terrestrial sedimentation and the carbon cycle: Coupling weathering and erosion to carbon burial. Global Biogeochemical Cycles 12: 231–257.
5. Van Oost K, Quine TA, Govers G, De Gryze S, Six J, et al. (2007) The impact of agricultural soil erosion on the global carbon cycle. Science 318: 626–629.
6. Harden JW, Sharpe JM, Parton WJ, Ojima DS, Fries TL, et al. (1999) Dynamic replacement and loss of soil carbon on eroding cropland. Global Biogeochemical Cycles 13: 885–901.
7. Lal R (2003) Soil erosion and the global carbon budget. Environment International 29: 437–450.
8. Berhe AA, Harden JW, Torn MS, Kleber M, Burton SD, et al. (2012) Persistence of soil organic matter in eroding versus depositional landform positions. Journal of Geophysical Research-Biogeosciences 117.
9. Zhang JH, Quine TA, Ni SJ, Ge FL (2006) Stocks and dynamics of SOC in relation to soil redistribution by water and tillage erosion. Global Change Biology 12: 1834–1841.
10. Gregorich EG, Greer KJ, Anderson DW, Liang BC (1998) Carbon distribution and losses: erosion and deposition effects. Soil & Tillage Research 47: 291–302.
11. Smith SV, Sleezer RO, Renwick WH, Buddemeier R (2005) Fates of eroded soil organic carbon: Mississippi basin case study. Ecological Applications 15: 1929–1940.
12. Wang X, Cammeraat LH, Wang Z, Zhou J, Govers G, et al. (2013) Stability of organic matter in soils of the Belgian Loess Belt upon erosion and deposition. European Journal of Soil Science 64: 219–228.
13. Smith SV, Renwick WH, Buddemeier RW, Crossland CJ (2001) Budgets of soil erosion and deposition for sediments and sedimentary organic carbon across the conterminous United States. Global Biogeochemical Cycles 15: 697–707.
14. Polyakov VO, Lal R (2008) Soil organic matter and CO_2 emission as affected by water erosion on field runoff plots. Geoderma 143: 216–222.
15. Jacinthe PA, Lal R, Kimble JM (2002) Carbon dioxide evolution in runoff from simulated rainfall on long-term no-till and plowed soils in southwestern Ohio. Soil & Tillage Research 66: 23–33.
16. Wan Y, El-Swaify SA (1998) Sediment enrichment mechanisms of organic carbon and phosphorus in a well-aggregated Oxisol. Journal of Environmental Quality 27: 132–138.
17. Kwaad FJPM, Mucher HJ (1994) Degradation of soil-structure by welding - a micromorphological study. Catena 23: 253–268.
18. Franzluebbers AJ (1999) Potential C and N mineralization and microbial biomass from intact and increasingly disturbed soils of varying texture. Soil Biology & Biochemistry 31: 1083–1090.
19. Bremenfeld S, Fiener P, Govers G (2013) Effects of interrill erosion, soil crusting and soil aggregate breakdown on in situ CO_2 effluxes. Catena 104: 14–20.
20. Strickland TC, Truman CC, Frauenfeld B (2005) Variable rainfall intensity effects on carbon characteristics of eroded sediments from two coastal plain ultisots in Georgia. Journal of Soil and Water Conservation 60: 142–148.
21. Truman CC, Strickland TC, Potter TL, Franklin DH, Bosch DD, et al. (2007) Variable rainfall intensity and tillage effects on runoff, sediment, and carbon losses from a loamy sand under simulated rainfall. Journal of Environmental Quality 36: 1495–1502.
22. Palis RG, Ghandiri H, Rose CW, Saffigna PG (1997) Soil erosion and nutrient loss. 3. Changes in the enrichment ratio of total nitrogen and organic carbon under rainfall detachment and entrainment. Australian Journal of Soil Research 35: 891–905.
23. Van Hemelryck H, Fiener P, Van Oost K, Govers G, Merckx R (2010) The effect of soil redistribution on soil organic carbon: an experimental study. Biogeosciences 7: 3971–3986.
24. Jacinthe PA, Lal R (2001) A mass balance approach to assess carbon dioxide evolution during erosional events. Land Degradation & Development 12: 329–339.
25. Koninklijk Nederlands Meteorologisch Instituut (KNMI) website. Available: http://www.klimaatatlas.nl/klimaatatlas.php?wel = neerslag. Assessed 2014 April 10.
26. WRB (2006) World reference base for soil resources. FAO, ISRIC, ISSS, Rome.
27. Blake GR HK (1986) Bulk density. In: Klute A (ed) Methods of soil analysis: Part 1 physical and mineralogical methods. American Society of Agronomy and Soil Science Society of America, Madison

28. Heimovaara TJ, Bouten W (1990) A computer-controlled 36-channel time domain reflectometry system for monitoring soil-water contents. Water Resources Research 26: 2311–2316.

29. Buishand TA, Velds CA (1980) Extreme neerslaghoeveelheden, Neerslag en Verdamping. KNMI, de Bild: pp. 104–118.

30. Borselli L, Torri D, Poesen J, Sanchis PS (2001) Effects of water quality on infiltration, runoff and interrill erosion processes during simulated rainfall. Earth Surface Processes and Landforms 26: 329–342.

31. Kuhn NJ (2007) Erodibility of soil and organic matter: independence of organic matter resistance to interrill erosion. Earth Surface Processes and Landforms 32: 794–802.

32. RIVM Landelijk meetnet luchtkwaliteit website (National measurement network air quality (includes rainwater quality). Available: http://www.lml.rivm.nl/data/gevalideerd/. Accessed 2014 April 10.

33. Golchin A, Oades JM, Skjemstad JO, Clarke P (1994) Study of free and occluded particulate organic matter in soils by solid state C13 Cp/MAS NMR spectroscopy and scanning electron microscopy. Australian Journal of Soil Research 32: 285–309.

34. Cerli C, Celi L, Kalbitz K, Guggenberger G, Kaiser K (2012) Separation of light and heavy organic matter fractions in soil - Testing for proper density cut-off and dispersion level. Geoderma 170: 403–416.

35. Ghadiri H, Rose CW (1993) Water eosion processes and the enrichment of sorbed pesticides. 2. Enrichment under rainfall dominated erosion process. Journal of Environmental Management 37: 37–50.

36. Chen FS, Zeng DH, Fahey TJ, Liao PF (2010) Organic carbon in soil physical fractions under different-aged plantations of Mongolian pine in semi-arid region of Northeast China. Applied Soil Ecology 44: 42–48.

37. Allison SD, Jastrow JD (2006) Activities of extracellular enzymes in physically isolated fractions of restored grassland soils. Soil Biology & Biochemistry 38: 3245–3256.

38. Denef K, Six J, Paustian K, Merckx R (2001) Importance of macroaggregate dynamics in controlling soil carbon stabilization: short-term effects of physical disturbance induced by dry-wet cycles. Soil Biology & Biochemistry 33: 2145–2153.

39. Nadeu E, Berhe AA, de Vente J, Boix-Fayos C (2012) Erosion, deposition and replacement of soil organic carbon in Mediterranean catchments: a geomorphological, isotopic and land use change approach. Biogeosciences 9: 1099–1111.

40. Wang ZG, Govers G, Steegen A, Clymans W, Van den Putte A, et al. (2010) Catchment-scale carbon redistribution and delivery by water erosion in an intensively cultivated area. Geomorphology 124: 65–74.

41. Schiettecatte W, Gabriels D, Cornelis WM, Hofman G (2008) Enrichment of organic carbon in sediment transport by interrill and rill erosion processes. Soil Science Society of America Journal 72: 50–55.

42. Cole JJ, Prairie YT, Caraco NF, McDowell WH, Tranvik LJ, et al. (2007) Plumbing the global carbon cycle: Integrating inland waters into the terrestrial carbon budget. Ecosystems 10: 171–184.

43. Ghadiri H, Rose CW (1991) Sorbed chemical-transport in overland-flow. 2. Enrichment ratio variation with erosion processes. Journal of Environmental Quality 20: 634–641.

44. Denef K, Six J, Bossuyt H, Frey SD, Elliott ET, et al. (2001) Influence of dry-wet cycles on the interrelationship between aggregate, particulate organic matter, and microbial community dynamics. Soil Biology & Biochemistry 33: 1599–1611.

45. Six J, Elliott ET, Paustian K, Doran JW (1998) Aggregation and soil organic matter accumulation in cultivated and native grassland soils. Soil Science Society of America Journal 62: 1367–1377.

46. Feng W, Plante AF, Aufdenkampe AK, Six J (2014) Soil organic matter stability in organo-mineral complexes as a function of increasing C loading. Soil Biology & Biochemistry 69: 398–405.

47. Kaiser K, Guggenberger G (2003) Mineral surfaces and soil organic matter. European Journal of Soil Science 54: 219–236.

48. Van Hemelryck H, Govers G, Van Oost K, Merckx R (2011) Evaluating the impact of soil redistribution on the in situ mineralization of soil organic carbon. Earth Surface Processes and Landforms 36: 427–438.

49. Casals P, Gimeno C, Carrara A, Lopez-Sangil L, Sanz MJ (2009) Soil CO_2 efflux and extractable organic carbon fractions under simulated precipitation events in a Mediterranean Dehesa. Soil Biology & Biochemistry 41: 1915–1922.

50. Doetterl S, Six J, Van Wesemael B, Van Oost K (2012) Carbon cycling in eroding landscapes: geomorphic controls on soil organic C pool composition and C stabilization. Global Change Biology 18: 2218–2232.

51. Fontaine S, Barot S, Barre P, Bdioui N, Mary B, et al. (2007) Stability of organic carbon in deep soil layers controlled by fresh carbon supply. Nature 450: 277–U210.

52. Creed IF, Webster KL, Braun GL, Bourbonniere RA, Beall FD (2013) Topographically regulated traps of dissolved organic carbon create hotspots of soil carbon dioxide efflux in forests. Biogeochemistry 112: 149–164.

53. Bianchi TS (2011) The role of terrestrially derived organic carbon in the coastal ocean: A changing paradigm and the priming effect. Proceedings of the National Academy of Sciences of the United States of America 108: 19473–19481.

54. Lal R (2004) Soil carbon sequestration impacts on global climate change and food security. Science 304: 1623–1627.

55. Kwaad FJPM, de Roo APJ, Jetten VG (2006) The Netherlands. In: Boardman J and Poesen J (eds) Soil Erosion in Europe. Wiley and Sons, Chichester.

9

Seed Dormancy, Seedling Establishment and Dynamics of the Soil Seed Bank of *Stipa bungeana* (Poaceae) on the Loess Plateau of Northwestern China

Xiao Wen Hu[1]*, Yan Pei Wu[1], Xing Yu Ding[1], Rui Zhang[1], Yan Rong Wang[1]*, Jerry M. Baskin[2], Carol C. Baskin[2,3]

1 State Key Laboratory of Grassland Agro-ecosystems, College of Pastoral Agriculture Science and Technology, Lanzhou University, Lanzhou, 730020, China, 2 Department of Biology, University of Kentucky, Lexington, Kentucky 40506-0225, United States of America, 3 Department of Plant and Soil Sciences, University of Kentucky, Lexington, Kentucky 40546-0312, United States of America

Abstract

Studying seed dormancy and its consequent effect can provide important information for vegetation restoration and management. The present study investigated seed dormancy, seedling emergence and seed survival in the soil seed bank of *Stipa bungeana*, a grass species used in restoration of degraded land on the Loess Plateau in northwest China. Dormancy of fresh seeds was determined by incubation of seeds over a range of temperatures in both light and dark. Seed germination was evaluated after mechanical removal of palea and lemma (hulls), chemical scarification and dry storage. Fresh and one-year-stored seeds were sown in the field, and seedling emergence was monitored weekly for 8 weeks. Furthermore, seeds were buried at different soil depths, and then retrieved every 1 or 2 months to determine seed dormancy and seed viability in the laboratory. Fresh seeds (caryopses enclosed by palea and lemma) had non-deep physiological dormancy. Removal of palea and lemma, chemical scarification, dry storage (afterripening), gibberellin (GA_3) and potassium nitrate (KNO_3) significantly improved germination. Dormancy was completely released by removal of the hulls, but seeds on which hulls were put back to their original position germinated to only 46%. Pretreatment of seeds with a 30% NaOH solution for 60 min increased germination from 25% to 82%. Speed of seedling emergence from fresh seeds was significantly lower than that of seeds stored for 1 year. However, final percentage of seedling emergence did not differ significantly for seeds sown at depths of 0 and 1 cm. Most fresh seeds of *S. bungeana* buried in the field in early July either had germinated or lost viability by September. All seeds buried at a depth of 5 cm had lost viability after 5 months, whereas 12% and 4% seeds of those sown on the soil surface were viable after 5 and 12 months, respectively.

Editor: Jian Liu, Shandong University, China

Funding: This study was supported by Program for Changjiang Scholars and Innovative Research Team in University (IRT13019), the National Key Technology Research and Development Program (2011BAD17B02) and the Gansu Provincial Key Grant Project (1203FKDA035). The funders had no role in study design, data collection and analysis, decision to publish, or preparation of the manuscript.

Competing Interests: The authors have declared that no competing interests exist.

* Email: huxw@lzu.edu.cn (XWH); yrwang@lzu.edu.cn (YRW)

Introduction

Soil erosion is the main cause of land degradation in arid and semiarid regions, and it is a widespread problem on the Chinese Loess Plateau [1]. One way to restore degraded soils and reduce soil erosion is by revegetation [2,3]. The first step in any program of rehabilitation of soils degraded by erosion is to select the most suitable species to use for revegetation, based on the capacity of the seeds to germinate and the seedlings to become established [4,5].

Stipa bungeana Trin. (Poaceae) is a perennial grass that mainly occurs in semi-arid areas of the temperate steppe zone in Eurasia. The species is widely distributed on the Loess Plateau and other areas of western China. It is the main wild forage species in natural grasslands of northwestern China and also plays important roles in protecting the soil from erosion and reducing water loss by runoff. Due to its environmental benefits and economic value, Cheng *et*

al. [6] and Hu *et al.* [7] suggested that *S. bungeana* is a potential key species for revegetation of degraded land on the Loess Plateau.

Seed dormancy is the failure of viable seeds to germinate in a specified period of time under conditions suitable for their germination after they become nondormant [8,9]. Dormancy could prevent or delay germination even under favorable conditions, thus enabling seeds to accumulate in the soil seed bank and preventing plants from expending their entire reproductive outputs at a given time [10]. As such, then, seed dormancy is expected to be important in optimizing the timing of germination to maximize seedling establishment. Although *S. bungeana* produces up to 1430 seeds m^{-2} in typical *S. bungeana*-dominated rangeland, only a low portion of them germinate in the field; thus, few seedlings became established following seed dispersal [7]. Two reasons may contribute to low seedling establishment: 1) fresh seeds of *S. bungeana* exhibit primary dormancy, which prevents seed germination immediately after dispersal; and 2) environmental conditions during the dispersal season prevent seeds from germinating. In the first case, no seeds, or only a small portion of them, germinate even under otherwise favorable conditions until dormancy release. Thus, we expected that primary dormancy plays a role in regulating the time of seed germination and seedling recruitment in the field. Hu *et al.* [7] showed that germination of *S. bungeana* seeds was inhibited by light and sensitive to water stress, implying that most seeds would germinate slowly or not at all on the soil surface. Thus, seed burial in the soil may play a key role in determining whether they can germinate after dispersal.

The effects of temperature [11], light [11], water stress [7] and burial depth [7] on germination and of fungicide pretreatment on seed survival in the field have been determined for seeds of *S. bungeana* stored (afterripened) in the lab for 1 year. However, no studies have been done to test for seed dormancy in *S. bungeana* and its underlying mechanism, and consequently its effect on seedling emergence and seed survival in the soil. Moreover, seeds stored in the lab for 1 year were used in previous studies. As reported by Baskin and Baskin [12], results from studies initiated after seeds have been stored dry may have little ecological relevance. That is, the germination responses of seeds may have changed through time, and thus interpretation of results obtained using stored seeds may differ from fresh seeds dispersed in the natural environment.

Thus, the aims of this study were to determine: 1) whether fresh seeds are dormant and if so why; 2) the effect of storage condition on seed dormancy of fresh seeds; 3) the effect of seed dormancy on seedling emergence in relation to sowing depth; and 4) seed dormancy and survival of fresh seeds in relation to burial depth and duration in the field.

Materials and Methods

Seed collection

Stipa bungeana flowers in the early May, and seeds mature and are dispersed in late June in the study area. The dispersal unit of *S. bungeana* is a caryopsis tightly enclosed by the palea and lemma. It is 5–6 mm in length and 0.7–1.0 mm in width. Hereafter, the dispersal unit of *S. bungeana* will be referred to as a seed.

S. bungeana seeds were collected on 23 June 2012 and 28 June 2013 from a field on the Yuzhong Campus of Lanzhou University, Gansu Province (35°57′N, 104°10′E). The mean annual temperatures is 6.7°C and mean annual rainfall 350 mm, most of which falls from July to September. The soils consist of silt (66.9%), clay (20.8%) and sand (12.3%), and natural vegetation is dominated by *S. bungeana*. Other species growing at this site include *Achnatherum inebrians*, *Artemisia* spp., *Glycyrrhiza* spp. and *Lespedeza davurica*. Infructescences with ripe seeds were collected from several hundred plants and taken to the laboratory, where the seeds were separated from them, cleaned and dried at room temperature for one week (RH, 20–35%, 18–25°C) and stored at 4°C until used in experiments. The experiments were conducted within two weeks after seed collection except for the one on seedling emergence.

Seed viability

Viability of fresh seeds collected in 2012 and 2013 was determined. Seeds were soaked in distilled water for 12 h, after which hulls and half of endosperm were removed. Then the remaining part of the seed containing the embryo was soaked in 1% tetrazolium phosphate-buffer solution for 6–8 hours at 30°C in the dark. Seeds with embryos that stained red were considered to be viable and those with unstained embryos nonviable. For fresh seeds collected in 2012 and 2013, four replicates of 50 seeds were tested.

Effect of light and temperature on germination

The aim of this experiment was to determine whether fresh seeds are dormant. Fresh seeds were tested at four constant (10°C, 15°C, 20°C and 25°C) and three alternating (10/20°C, 15/25°C and 20/30°C) temperature regimes (12 h/12 h). At each temperature, seeds were incubated at a 12 h/12 h daily photoperiod (hereafter light) or in continuous darkness. For treatments in light, seeds were exposed to light produced by white fluorescent tubes with a photon irradiance of 60 μmol·m^{-2}·s^{-1} (400–700 nm). For continuous darkness, Petri dishes were covered with two layers of aluminum foil, and seeds were monitored for germination (root emergence) daily under a LED green safe light (520 nm±10 nm, Sanpai, Shanghai, China). Photon irradiance at Petri dishes level was 10 μmol·m^{-2}·s^{-1}, as determined by use of a quantum sensor (LI-190SA) connected to a LI-6400 portable photosynthesis system (LI-COR, USA). For each treatment, four replicates of 50 seeds each were placed in 11-cm-diameter Petri dishes on two sheets of filter paper (Shuangquan, Hangzhou) moistened with 8 mL of distilled water. Seeds incubated in both light and dark were examined for germination daily for 14 days, and any seedlings present were removed from the Petri dishes.

Effect of hulls, half-endosperm removal and scarification with NaOH on germination

The aim of this experiment was to determine the role of hulls and endosperm in controlling seed dormancy and effect of NaOH on breaking dormancy. The mechanical removal experiment consisted of a control (caryopsis with intact lemma and palea) and three treatments: 1) hulls (lemma and palea) removed; 2) hulls removed and put back in their original position enclosing the caryopsis; and 3) half of endosperm (and enclosing portions of hulls) removed from the endosperm-end of the seed without damaging the embryo. Before performing the removal treatments, seeds were immersed in distilled water for 12 h, and then the hulls were removed from the caryopses using forceps. For treatment 2, the hulls were removed from the seed and then re-attached loosely to their original position starting at the embryo end. For treatment 3, half of the endosperm (along with enclosing portions of hulls) was removed from the endosperm-end of the seed with a scalpel.

For scarification with NaOH, seeds (with hulls) were soaked in 50 mL of a 30% NaOH solution at 20°C for 20, 40 or 60 min. Then the seeds were rinsed thoroughly with tap water five times and allowed to dry on filter paper for 48 h on the laboratory bench. For each treatment, four replicates of 50 seeds were used for testing germination at 20°C in dark. Seeds without pretreatment were used as a control. Germination was monitored daily for 14 days as described above.

Effect of dry storage on seed dormancy

To determine the effect of storage duration and temperature on seed dormancy, fresh seeds were placed in a paper bag and stored in darkness at 5°C and at 20°C for 1, 3 and 6 months. After each treatment, germination was tested in light and in darkness at 20°C. There were four replicates of 50 seeds per treatment. Germination was monitored daily for 14 days as described above.

Effect of fluridone, GA$_3$ and KNO$_3$ on seed germination

To determine the role of plant growth regulator and potassium nitrate (KNO$_3$) in controlling seed dormancy of *S. bungeana*, seeds were treated with gibberellic acid (GA$_3$, Sigma, China);

fluridone (FLU, Sigma, China), an inhibitor of abscisic acid (ABA) biosynthesis, or KNO$_3$. Fluridone and GA$_3$ each were dissolved in 2 mL ethanol prior to dilution in water, and the final concentration of FLU and GA$_3$ was 200 μM. A preliminary experiment showed that a low concentration of ethanol did not affect germination of S. bungeana seeds. The potassium nitrate solution was 1 mM. Seeds incubated in distilled water were used as control. For each treatment, four replicates of 50 seeds were incubated at 20°C in dark. Germination was monitored daily for 14 days as described above.

Seedling emergence in field

To determine the effect of seed dormancy on seedling emergence, the seedling emergence experiment was conducted at the Yuzhong Campus from 16 July 2013 to 12 September 2013. Seeds that had been stored dry at 20°C for 1 year and fresh seeds collected on 28 June 2013 were used in this experiment. There were two seed lots (fresh, afterripened) × three burial depths (0, 1, 5 cm) in a completely randomized design. Seeds were sown on 16 July directly in PVC pots (15 cm in diameter, 11 cm in height) that were buried in the field with the rim 5 cm above the soil surface. Soil level in the pot was even with the soil surface. To avoid seed contamination, soil was passed through a 0.5 mm-mesh wire sieve to remove any S. bungeana seeds present in it. Then, the soil was placed in the pots to the desired depth, and seeds were placed at this depth and covered with soil. The pots were covered with nylon mesh to prevent animal predation and contamination by extraneous seeds of S. bungeana. Ten replicates of 50 seeds each were used for each treatment. Seedling emergence was monitored weekly for 8 weeks, and any seedlings present were counted and removed. The vegetation surrounding the pots was removed by hand every week. The speed of seedling emergence was calculated using the emergence index (EI):

$$EI = \sum \left(\frac{Et}{Tt}\right)$$

where Et is the number of seeds emerged on tth week, and Tt is the weeks of seedling emergence from sowing [13].

Seed burial experiment

To determine the effect of burial depth and burial duration on seed survival and seed dormancy of S. bungeana, seeds collected on 23 June 2012 were put into 96 15 cm×10 cm nylon mesh bags and buried at the Yuzhong Campus of Lanzhou University on 5 July 2012. The permeable nylon fabrics allowed movement of water, air and microbes between inside and outside of bags. The burial site is about 300 m from the seed collection site and has a sparse vegetation cover. The vegetation within and surrounding burial site was removed before burial. Forty-eight bags with 50 seeds each were placed on the soil surface (0 cm) and 48 at a soil depth of 5 cm. They were arranged in a randomized complete block design. For 0 cm burial depth, the bags were fixed to the ground by iron nails so that each bag was in contact with the soil and won't move to the other soil profile. Physical removal of weeds within burial site was applied every week during experimental period. Six bags each were retrieved from 0 cm and 5 cm burial depths on 28 July, 30 August, 30 September, 31 October and 30 December 2012 and on 2 March, 9 May and 5 July 2013. For each of these dates, the seeds in each of 12 bags were put into one 11-cm-diameter Petri dish with two layers of filter paper moistened with 8 mL of distilled water and incubated in darkness at 20°C. Thus, there were six replications each for the 0 cm and 5 cm

burial depth treatments. The number of germinated seeds was counted after 14 days. Germination of 6 replicates of 50 seeds was tested before burial as described above. For all germination tests, seeds failed to germinate were tested for viability.

Temporal changes in soil seed bank size

To determine soil seed bank size in the field, soil samples were taken on 28 July and 28 August, 2012 and 2 March and on 20 May, 2013 on the Yuzhong Campus of Lanzhou University. The sampling site is 2000 m^2 and dominant plants were S. bungeana and Medicago sativa. Other species at the site included Achnatherum inebrians and Artemisia spp. Sixteen 1 m×1 m quadrats were haphazardly established at the study site for each sampling time. In each quadrat, five soil subsamples 0–5 cm deep were collected using a 10 cm×10 cm×5 cm soil sampler, mixed together, air dried and sieved through a 0.5 mm sieve. Seeds of S. bungeana were separated from the litter and incubated in 11-cm-diameter Petri dishes with two layers of filter paper moistened with 8 mL of distilled water in darkness at 20°C. Germination percentages were determined after 14 days of incubation, and seeds that failed to germinate were tested for viability.

Statistical analysis

A two way ANOVA at a significance level of $P<0.05$ was used to analyze the effect of light, temperature and their interaction on seed germination and of burial time, burial depth and their interaction on seed viability and seed dormancy. Duncan's multiple range tests was used to compare means of germination percentage between treatments when significant differences were found. Independent t-test was used to compare the mean viability of seeds collected in 2012 and 2013. Germination percentage data were arcsine transformed to increase homogeneity of variance prior to analysis, but nontransformed data are shown in all figures and in tables. All analyses were conducted in SPSS 15.0 software.

Results

Seed viability

Fresh seeds collected in 2012 and 2013 showed no significant difference in terms of seed viability which was 85±4.6% and 87±3.7%, respectively.

Effect of light and temperature on germination

Light, temperature and their interaction had significant effects on germination (Fig. 1, Table 1). Light inhibited germination at the three temperatures at which germination occurred. The highest germination was 25%, in darkness at 20°C, and only 7% of the seed germinated in light. Seeds germinated to significantly lower percentage at 15/25°C and 20/30°C in both darkness and light. No seeds germinated at 10, 15, 25 or 10/20°C in darkness or in light.

Effect of hulls, half-endosperm removal and scarification with NaOH on seed germination

Hulls, endosperm and scarification with NaOH had a significant effect on release of seed dormancy (P<0.05, Fig. 2). Twenty-five percent of fresh seeds (with hulls) without pretreatment germinated, and dormancy was completely released when the hulls were removed. However, seeds in which the hulls were put back to their original position germinated to a significantly lower percentage (46) than those hulls removed (94). Removal of half of the endosperm increased germination from 25% to 45%

Table 1. Two way ANOVA of the effects of temperature, light and their interaction on germination of fresh seeds of *Stipa bungeana* (n = 4).

Source	Sum of Squares	df	F	P-value
Temperature(T)	1790	6	39.2	.000
Light(L)	330	1	43.4	.000
T * L	623	6	13.6	.000

(P<0.05). Seeds treated with NaOH for 20, 40 and 60 minutes germinated to 45, 63 and 82%, respectively (Fig. 2).

Effect of dry storage on dormancy break

Storage at 5°C and 20°C significantly increased germination percentages in light and in dark. However, there was little difference in afterripening at the two storage temperatures, although seeds afterripened of 20°C for 3 months germinated to significantly higher percentage in light and dark than those afterripened at 5°C, and seeds stored at 20°C for 6 months germinated to a significantly higher percentage in light than those afterripened at 5°C (Fig. 3).

Effect of fluridone, GA₃ and KNO₃ on germination

GA$_3$ and KNO$_3$ significantly increased germination from 25% to 36% and 44%, respectively (P<0.05), but FLU had no effect. Germination percentage was significantly higher for seeds treated with KNO$_3$ than for those treated with GA$_3$ (Fig. 4).

Effect of dry storage and burial depth on seedling emergence

Percentage and speed of seedling emergence in the field varied with burial depth and seed lot (fresh vs. one-year-stored seeds) (Fig. 5, Table 2). At 0, 1 and 5 cm burial depths, seeds stored one year had a significantly higher emergence speed (higher emergence index) than fresh seeds. Most seeds stored for 1 year germinated within 3 weeks after sowing. However, fresh seeds had just begun to germinate by the third week, and they continued to do so for another 4–5 weeks. On the other hand, there was no significant difference in final seedling emergence between stored and fresh seeds sown at 0 cm and 1 cm (Table 2). In contrast, seedling emergence percentage at 5 cm was significantly higher for stored seeds than for fresh seeds, but emergence percentages were <10% for both burial depths. The highest final seedling emergence percentage was for 1 cm burial depth, with 33% for one year stored seeds and 27% for fresh seeds.

Effect of burial on seed viability and dormancy

Burial depth, burial duration and their interaction had a significant effect on seed viability (germinated seeds in the lab +

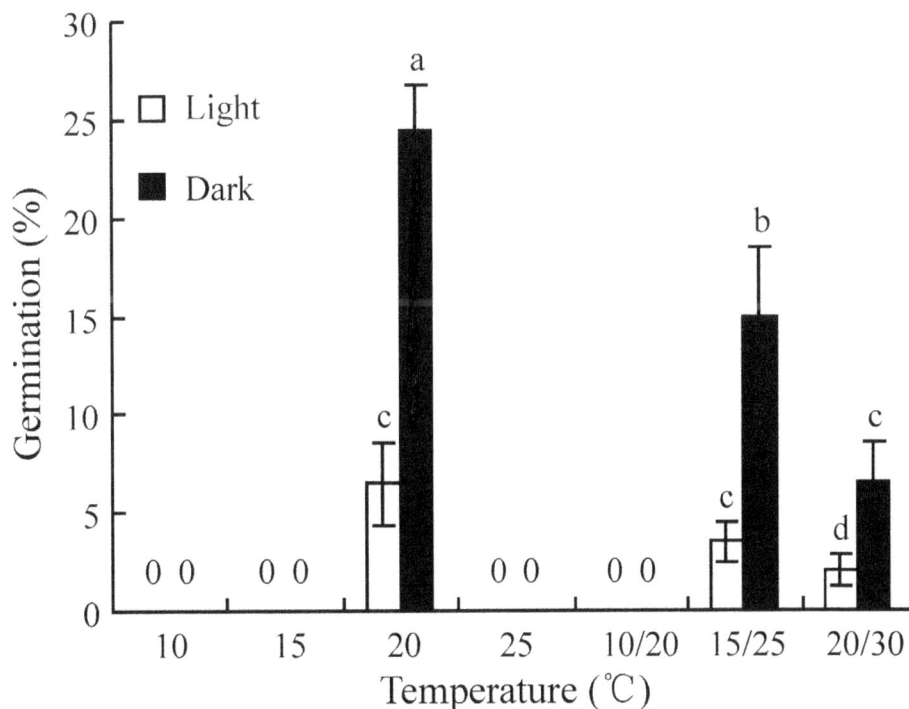

Figure 1. Effect of temperature on germination of fresh seeds of *Stipa bungeana* in a 12 h/12 h photoperiod and in dark. Different letters indicate significant difference (P<0.05) among all treatments (n = 4).

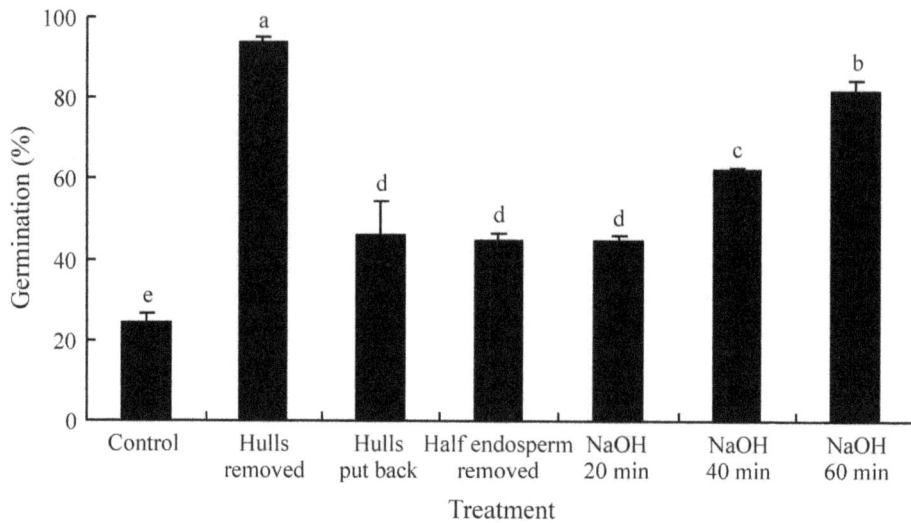

Figure 2. Effect of hulls, half-endosperm removal and NaOH scarification on germination of fresh seeds of *Stipa bungeana* at 20°C in dark. Different letters indicate significant difference (P<0.05) among all treatments (n = 4).

dormant seeds) and on seed dormancy (Fig. 6, Table 3). After five months of burial at 5 cm, all seeds had lost viability, whereas 12% and 4% of those on the soil surface were viable after 5 and 12 months, respectively. Further, seeds buried at 5 cm lost dormancy more quickly than those sown on the soil surface. For example, almost 99% of seeds buried at 5 cm depth had lost dormancy after 3 months, while 87% seeds of those on the soil surface had done so.

Temporal changes in soil seed bank size

The size of the soil seed bank of *S. bungeana* declined with time and significantly so between August and March. The highest density of seeds in the seed bank was 869 m^{-2}, in July, and the lowest density was 31 m^{-2}, in May (Fig. 7).

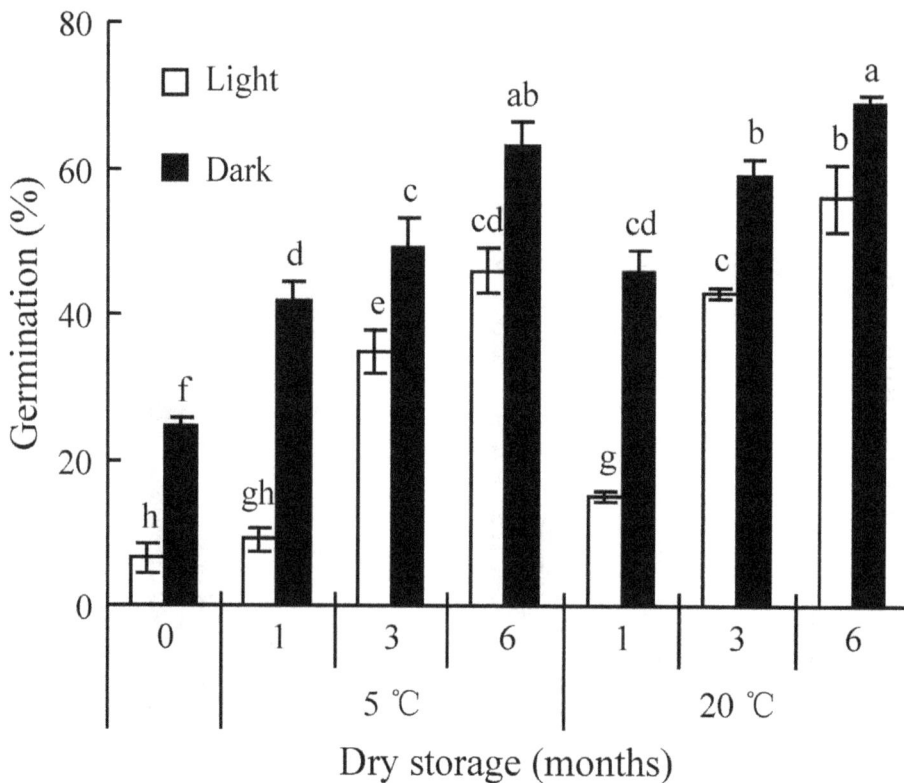

Figure 3. Effect of dry storage (afterripening) on germination of seeds of *Stipa bungeana* in a 12 h/12 h photoperiod and in dark at 20°C. Different letters indicate significant difference (P<0.05) among all treatments (n = 4).

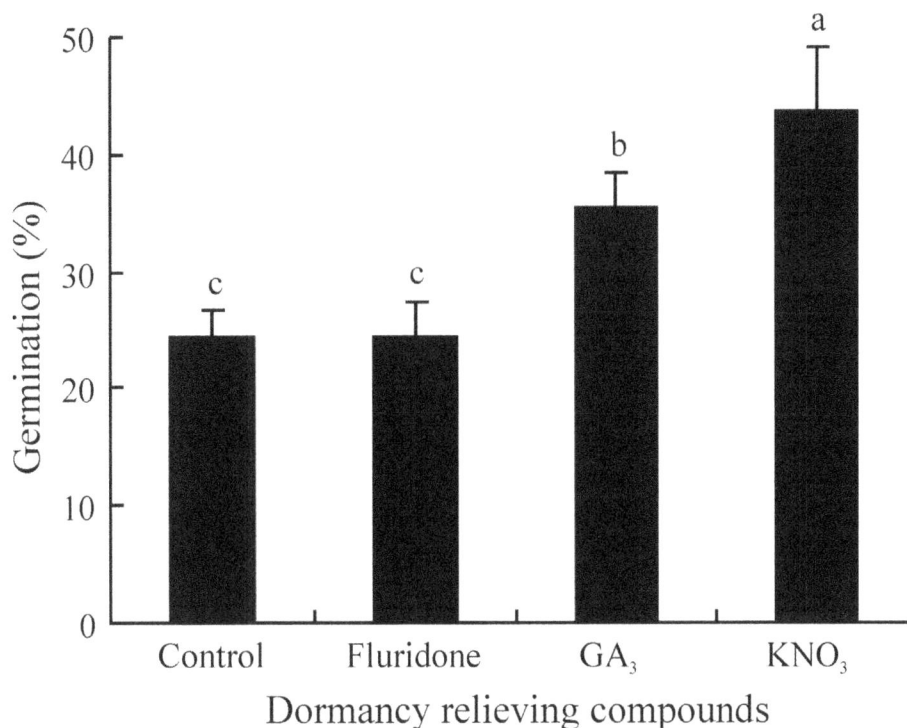

Figure 4. Effect of fluridone, GA₃ and KNO₃ on germination of fresh seeds of *Stipa bungeana* **in dark at 20°C.** Different letters indicate significant difference (P<0.05) among all treatments (n=4).

Discussion

Seed dormancy and its underlying mechanism

Seed dormancy prevents or delays germination even under favorable conditions and spreads the risk of recruitment over time. It generally is accepted that seed dormancy plays a significant role in ensuring seed germination at the right time and in the proper sites to maximize the probability of successful seedling establishment [9,14]. It is clear from the present study that *S. bungeana* seeds exhibit primary dormancy, since germination percentages were low at various combinations of light and temperature, and they were increased significantly by several dormancy breaking treatments. According to the seed dormancy classification system of Baskin & Baskin [8,9], physiological dormancy can be caused by the tissues surrounding the embryo, by low growth potential of the embryo or by a combination of the two. The palea and lemma clearly play an important role in regulating germination of fresh *S.*

bungeana seeds since their removal released dormancy completely. This is consistent with other studies that reported hulls imposed dormancy in other grass species, for example, *Stipa viridula* [15], *S. tenacissima* [16], *Leymus secalinus* [17], *L. chinensis* [18–20], *Hordeum spontaneum* [21,22] and *H. vulgare* [23].

When hulls were removed from the dispersal unit of *S. bungeana* and then loosely reattached in their original position, seeds germinated to significantly lower percentage than those with hulls removed and not reattached. This suggests the possibility of the presence of germination inhibitors in the hulls, since the loosely-attached hulls should not have inhibited germination via mechanical restriction. Chemical germination inhibitors have been isolated from the hulls of grass seeds [24,25]. ABA in hulls of *Leymus chinensis* inhibited germination of this species in vitro [26,27]. Soluble germination inhibitors in the hulls of *Aegilops geniculata* [28,29] may regulate germination response to amount of rain [30]. However, seed dormancy of *Stipa viridula* was not

Table 2. Seedling emergence percentages and index for fresh one-year-stored seeds of *Stipa bungeana* at sowing depths of 0, 1 and 5 cm.

Seed lots	Sowing depth (cm)	Emergence percentage	Emergence index
Fresh seed	0	18±2.9c	4.1±1.2d
	1	27±2.6ab	6.2±1.4c
	5	1±0.8e	0.2±0.2e
Stored seed	0	22±2.6bc	9.5±1.5b
	1	33±2.9a	14.2±1.2a
	5	8±1.5d	2.8±0.5d

Different letters within a column indicate significant differences (P<0.05, n=10).

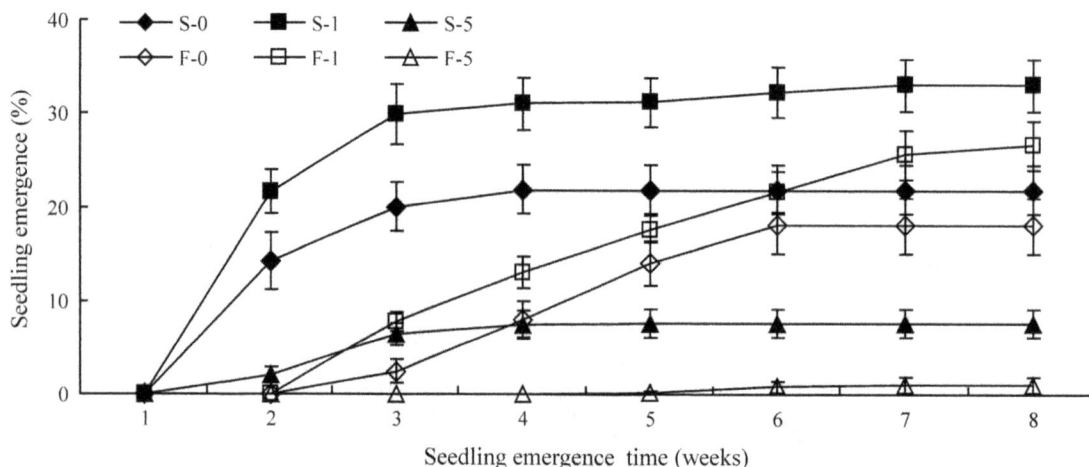

Figure 5. Seedling emergence from fresh (F) and one-year-stored (S) seeds of *Stipa bungeana* **at 0, 1 and 5 cm sowing depths.** F-0, F-1 and F-5 indicate fresh seeds sown at a depths of 0, 1 and 5 cm, respectively (n = 10). S-0, S-1 and S-5 indicate seeds stored for one year and then sown at depths of 0, 1 and 5 cm, respectively (n = 10).

caused by chemical inhibitors in the palea and lemma since seeds with palea and lemma clipped on both ends without damage to the enclosed caryopsis and seeds with the lemma and palea removed germinated equally well [15]. The presence vs. absence of chemical inhibition of hulls on germination may vary among and within species [16,24]. Huang *et al.* [31], Ma *et al.* [18], and He [32] suggested that the endosperm was responsible for seed dormancy in *Leymus racemosus* and *L. chinensis*, respectively. However, this obviously was not the case in *S. bungeana* since almost all fresh intact caryopses (with hulls removed) germinated.

Scarification with acids or alkalis has been shown to be effective in breaking seed dormancy in grasses [24,25], and scarification with NaOH significantly increased germination of *S. bungeana* seeds. This may be due to damage to the hulls, thus decreasing their mechanical resistance to germination. Also, an increase in permeability of embryo covering tissues after treatment with NaOH may favor leaching of germination inhibitors from them. A combination of NaOH soaking and exogenous GA₃ completely broke seed dormancy in *Leymus chinensis* [32].

ABA plays a role in the induction and maintenance of seed dormancy, whereas gibberellins (GAs) are associated with dormancy breaking and germination [33–35]. GA₃ has been reported to stimulate germination of seeds of many grass species [36], and it does so by increasing the growth potential of the

embryo [25,33]. The small but significant increase in germination percentage of *S. bungeana* seeds by GA₃ indicates that this plant growth regulator increased the growth potential of embryos in only some of the seeds to the point where the embryo overcame the mechanical restriction of its covering layers. The failure of fluridone, to promote germination suggests that ABA biosynthesis during imbibition may not be the primary cause of dormancy in *S. bungeana* seeds. Gianinetti & Vernieri [37] concluded that ABA is not the primary mediator of dormancy in imbibed rice seeds. However, a correlation between embryonic ABA level, sensitivity to ABA and hull imposed dormancy was found in barley [23].

Potassium nitrate is used extensively to break grass seed dormancy under laboratory conditions [38]. It alleviated the light inhibition of germination of *S. bungeana* seeds [27] and significantly increased germination of fresh seeds of this species (this study). Nitrate is one of the most ubiquitous inorganic ions in soils [39], and it can release seed dormancy and stimulate germination [9,25,36]. The much better regeneration of *Plantago lanceolata* in gaps than in closed vegetation may be attributed to higher nitrate level in gaps (0.2–1.1 mM) than that in closed vegetation (0.1 mM) [40]. Peaks in seedling emergence of *Capsella bursa-pastoris* in summer were correlated with increases in nitrate level in the soil [41]. Sensitivity of *Arabidopsis thaliana* seed to nitrate level at 20°C was highest in summer-early autumn when

Table 3. Two way ANOVA of the effects of burial time, burial depth and their interaction on seed viability and seed dormancy of *Stipa bungeana* (n = 6).

Source	Sum of Squares	df	F	P-value
Seed viability				
Burial time (BT)	24348	7	131.4	.000
Burial depth (BD)	1226	1	46.3	.000
BT * BD	638	7	3.4	.003
Seed dormancy				
Burial time (BT)	800	7	28.8	.000
Burial depth (BD)	224	1	56.3	.000
BT * BD	234	7	8.4	.001

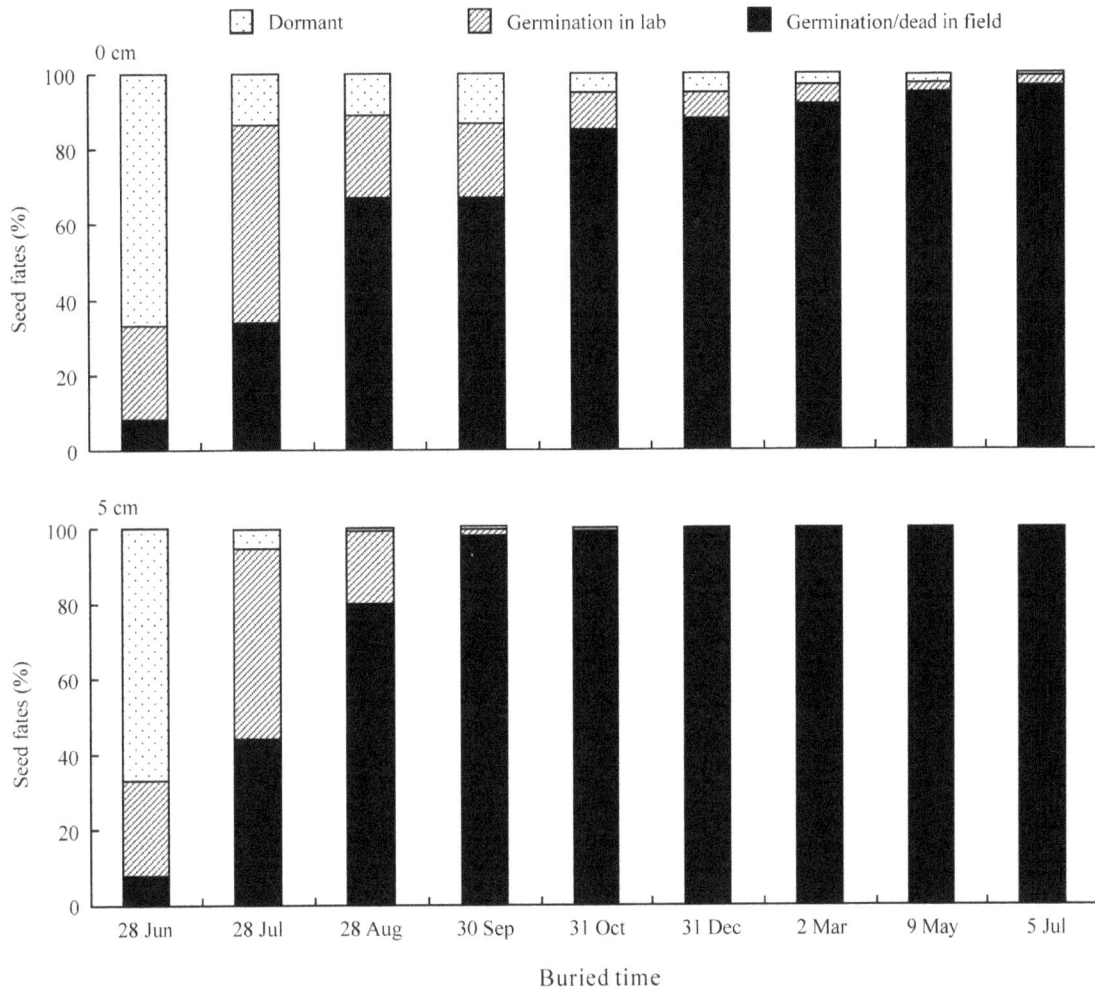

Figure 6. Fates of seeds in the field at 0 (surface) and 5 cm soil depths (n = 6).

dormancy level in the seeds was low and expression of nitrate transporter 1.1 (*NRT1.1*) and nitrate reductase 1 (*NR1*) genes was highest [42].

Seed afterripening can be characterized by dormancy release and an increase of germination speed during dry storage [9,43,44]. Seeds of many grass species come out of dormancy during dry storage [16,25,36]. Dry storage for 4 months significantly increased germination percentage of *Stipa tenacissima* seeds [16], and the germination percentage of *S. bungeana* seeds increased significantly during dry storage for 1, 3 and 6 months at 5°C and 20°C. Effect of storage temperature on germination varies with the species [45]. A positive relationship between dormancy release rate and afterripening temperature has been found in seeds of some grass species [46–48]. Rate of afterripening of *Bromus tectorum* seeds was approximately the same at 20°C and 30°C [49]. Overall, however, storage temperature had only a small effect on germination of *S. bungeana* seeds, which germinated to 63% and 69% at 20°C in dark after dry storage for 6 months at 5°C and 20°C, respectively.

In sum, fresh seeds of *S. bungeana* have hull-imposed dormancy, and it can be released completely by removal of the hulls and in part by GA₃, KNO₃, NaOH scarification and dry storage. These results suggest that seeds of *S. bungeana* have non-deep physiological dormancy [8], as have been reported for seeds of many other species of grasses [9].

Effect of seed dormancy on seedling emergence

On the Chinese Loess Plateau, caryopses of *S. bungeana* normally mature and are dispersed by the end of June (before rainy season). It is expected that seeds will germinate and seedlings become established in summer and autumn due to suitable temperature and precipitation for them to do so from July to September. Indeed, seedlings recruit mainly from July to September [7,50]. Although they were dormant at the time of burial, and most seeds of *S. bungeana* that germinated from July to September (Fig. 6), and most of them were depleted from the soil during August and September (Fig. 7). There was no significant difference in the final percentage of seedling emergence in the field between stored and fresh seeds sown at 0 cm and those buried 1 cm (Table 2).

Although in final emergence percentage of stored and fresh seeds sown in the field did not differ, fresh seeds germinated much slower than stored seeds, indicating that primary dormancy delayed germination. The timing of germination has pronounced effects on subsequent survival and phenology of seedling, and, in turn, affect the reproductive output of adult plants [9,14]. Water stress is one of the most important factors limiting seedling establishment in arid and semi-arid area. The long term (1957–2009) data showed that the humid index of study area ranked as September > August > July. Thus, seeds germinated in August or September will have an advantage for seedling growth and

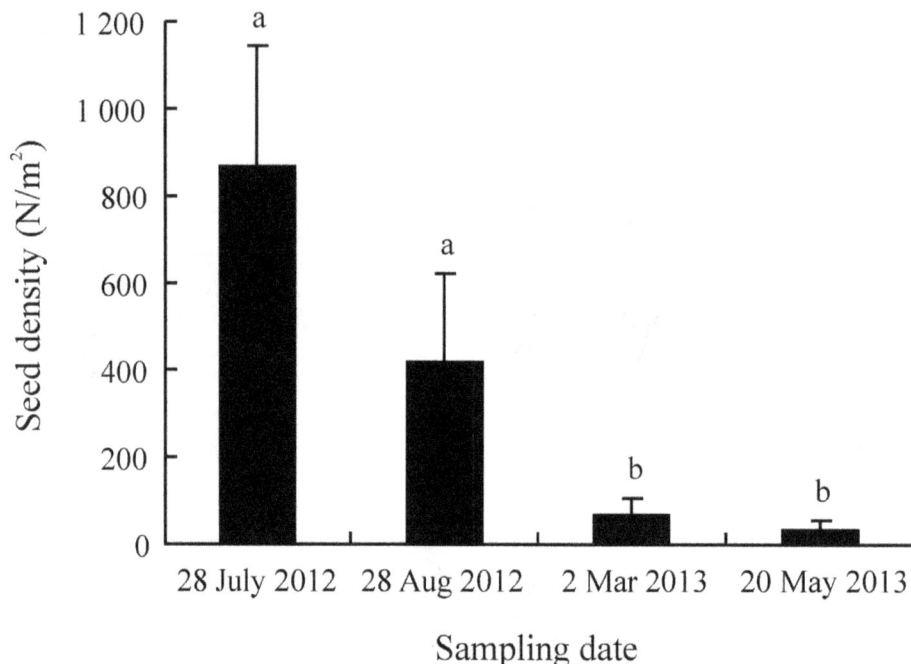

Figure 7. Seed density on different sampling dates in a *S. bungeana*-dominated grassland. Different letters indicate significant difference (P<0.05) among sampling dates (n = 16).

development with less water stress. However, severe winter conditions can over-ride this advantage through increasing seedling mortality. In present study, fresh seeds sown at the soil surface mostly emerged during 4–6 weeks after sown which corresponding the late July to mid-August after seed dispersal. However, stored seeds mostly emerged during 2–3 weeks after sown which corresponding the mid-July after seed dispersal. Thus, the delay in germination by seed dormancy of *S. bungeana* seems to be a compromise between water stress avoidance and seedling overwinter. This may provide an adaptive advantage of *S. bungeana* in arid environment.

Timing of seed germination largely depends on dormancy release which regulated by various environmental factors [9]. Dormancy of fresh seeds of *S. bungeana* buried in the soil was released more quickly than those on the soil surface (Fig. 6). There are at least three possible reasons for this difference. One is that nitrate in the soil may promote dormancy release (see above). Second, seeds were exposed to a higher level of hydration than those on the soil surface, which via microbial decay activity might have decreased mechanical resistance of the hull to embryo growth and thus germination. Third, seeds had high moisture content, which favored seed afterripening. Increasing the moisture content of rice seeds from 8% to 11% resulted in a 2.5-fold reduction in the storage period required for a given level of germination [51]. Seed dormancy release speed of *Lolium rigidum* increased as seed water content increased from 6% to 18% [48]. Increased dormancy release by burial may have ecological significance in that the seeds moisture conditions beneath the soil surface are more suitable for germination and subsequent seedling establishment than they are on the surface [7,52].

Effect of seed burial depth on seedling emergence and soil seed bank

The vertical distribution of a seed in the soil plays an important role in determining whether it remains dormant, germinates or

dies [14]. Emergence percentage and speed (Table 2) were significantly higher for both fresh and stored seeds of *S. bungeana* buried 1 cm deep than they were for seeds sown on the soil surface (Table 2), probably due to the negative effect of light and water stress on germination [7]. Further, 16% seeds on the soil surface were dormant after one month of burial, but in only 5% of those buried at 5 cm (Fig. 6). Thus, buried seeds germinated to a higher percentage and speed than those on the soil surface.

All seeds buried 5 cm in soil and 88% of those on the soil surface on 28 June had either germinated or died by 31 December (Fig. 6). Further, the number of seeds in the soil seed bank decreased from 869 m^{-2} to 31 seeds m^{-2} between July and May (Fig. 7). Thus, our study indicates that only a small number of seeds produced by *S. bungeana* may have the potential to form a persistent seed bank, i.e. remain viable for ≥1 year in/on soil. Hou [50] reported that *S. bungeana* had a transient seed bank when seeds were hand-buried in nylon bags. However, *S. bungeana* formed a natural persistent seed bank in a typical prairie [53] and on eroded slopes [54], respectively, on the Loess Plateau.

Practical implications

Fresh *S. bungeana* seeds exhibit primary dormancy, and thus seed pretreatment to release dormancy is important for quick establishment of *S. bungeana* plants. Although most seeds will germinate in 4–6 weeks after dispersal if soil moisture is suitable for them to do so, we recommended sowing one-year-stored (after-ripened) seeds 1 cm deep to attain a uniform stand. An option is to pretreat fresh seeds with a 30% NaOH solution for 60 min before sowing them. Trampling by livestock during a short-term grazing period in early July immediately following seed dispersal may promote seed burial in the soil, which promotes germination. However, disturbance, such as by grazing, during the stand establishment period should be avoided in order to prevent seedlings from being injured or destroyed.

Author Contributions

Conceived and designed the experiments: XWH YRW. Performed the experiments: YPW XYD RZ. Analyzed the data: YPW XWH JMB CCB.

Contributed reagents/materials/analysis tools: YRW XWH. Contributed to the writing of the manuscript: XWH JMB CCB.

References

1. Zheng FL (2006) Effect of vegetation changes on soil erosion on the Loess Plateau. Pedosphere 16: 420–427.
2. Mensching HG (1986) Desertification in Europe? A critical comment with examples from Mediterranean Europe. In: Fantechi R, NS Margaris NS, editors. Desertification in Europe. Dordrecht, Holland: Reidel Publishing Company. 3–8 p.
3. Francis CF, Thornes JB (1990) Matorral: erosion and reclamation. In: Albaladejo J, Stocking MA, Díaz E (editors) Soil degradation and rehabilitation in Mediterranean environmental conditions. Murcia: Consejo Superior de Investigaciones Científicas. 87–116 p.
4. Morgan RPC, Rickson RJ, Wright E (1990) Regeneration of degraded soils. In: Albaladejo J, Stocking MA, Díaz E, editors. Soil degradation and rehabilitation in Mediterranean environmental conditions. Murcia: Consejo Superior de Investigaciones Científicas. 69–85 p.
5. Albaladejo J, Castillo V, Roldán A (1996) Rehabilitation of degraded soils by water erosion in semiarid environments. In: Rubio JL, Calvo A, editor. Soil degradation and desertification in Mediterranean environments. Logrono, Spain: Geoforma Ediciones. 265–278 p.
6. Cheng J, Hu TM, Cheng JM, Wu GL (2010) Distribution of biomass and diversity of Stipa bungeana community to climatic factors in the Loess Plateau of northwestern China. Afr J Biotechnol 9: 6733–6737.
7. Hu XW, Zhou ZQ, Li TS, Wu YP, Wang YR (2013) Environmental factors controlling seed germination and seedling recruitment of Stipa bungeana on the Loess Plateau of northwestern China. Ecol Res 28: 801–809.
8. Baskin JM, Baskin CC (2004) A classification system for seed dormancy. Seed Sci Res 14: 1–16.
9. Baskin CC, Baskin JM (2014) Seeds: ecology, biogeography and evolution of dormancy and germination. Second edition. San Diego: Elsevier/Academic Press.
10. Koller D (1969) The physiology of dormancy and survival of plants in desert environments. In: Woolhouse HW, editor. Dormancy and Survival. Symposium of the Society for Experimental Biology 23: 449–469.
11. Zhou ZQ, Li TS, Wu YP, Hu XW (2013) A study of optimum germination condition of Stipa bungeana seeds. Pratacultural Science, 30: 218–222.
12. Baskin CC, Thompson K, Baskin JM (2006) Mistakes in germination ecology and how to avoid them. Seed Sci Res 16: 165–168.
13. Wang YR, Zhang JQ, Liu HX, Hu XW (2004). Physiological and ecological responses of lucerne and milkvetch seed to PEG priming. Acta Ecologia Sinica 24: 402–408.
14. Fenner M, Thompson K (2005) The ecology of seeds. Cambridge, UK: Cambridge University Press.
15. Fulbright TE, Redente EF, Wilson AM (1983) Germination requirements of green needle-grass (Stipa viridula Trin.). J Range Manage 36: 390–394.
16. Gasque M, Garcia-Fayos P (2003) Seed dormancy and longevity in Stipa tenacissima (L.) Poaceae. Plant Ecol 168: 279–290.
17. Zhu YJ, Dong M, Huang ZY (2007) Caryopsis germination and seedling emergence in an inland dune dominant grass Leymus secalinus. Flora 202: 249–257.
18. Ma HY, Liang ZW, Wang ZC (2008) Lemmas and endosperms significantly inhibited germination of Leymus chinensis (Trin.) Tzvel. (Poaceae). J Arid Environ 72: 573–578.
19. Ma HY, Liang ZW, Wu HT, Huang LH, Wang ZC (2010) Role of endogenous hormones, glumes, endosperm and temperature on germination of Leymus chinensis (Poaceae) seeds during development. J Plant Ecol 3: 269–277.
20. He XQ, Hu XW, Wang YR (2010) Study on seed dormancy mechanism and breaking technique of Leymus chinensis. Acta Botanica Boreali -Occidentalia Sinica 30: 120–125.
21. Gutterman Y, Corbineau F, Côme D (1996) Dormancy of Hordeum spontaneum caryopses from a population on the Negev Desert Highlands. J Arid Environ 3: 337–345.
22. Zhang FC, Gutterman Y (2003) The trade-off between breaking of dormancy of caryopses and revival ability of young seedlings of wild barley (Hordeum spontaneum). Can J Bot 81: 375–382.
23. Benech-Arnold RL, Giallorenzi MC, Frank J, Rodriguez V (1999) Termination of hull-imposed dormancy in developing barley grains is correlated with changes in embryonic ABA levels and sensitivity. Seed Sci Res 9: 39–47.
24. Simpson GM (1990) Seed dormancy in grass. Cambridge: Cambridge University Press.
25. Adkins SW, Bellairs SM, Loch DS (2002) Seed dormancy mechanisms in warm season grass species. Euphytica 126: 13–20.
26. Yi J, Li QF, Tian RH (1997) Seed dormancy and hormone control of germination in Leymus. Acta Agrestia Sinica 5: 93–100.
27. Hu XW, Huang XW, Wang YR (2012) Hormonal and temperature regulation of seed dormancy and germination in Leymus chinensis. Plant Growth Regul 67: 199–207.
28. Lavie D, Levy EC, Cohen A, Evenari M, Gutterman Y (1974) New germination inhibitor from Aegilops ovata L. Nature 249: 388.
29. Gutterman Y, Evenari M, Cooper R, Levy EC, Lavie D (1980) Germination inhibition activity of a naturally occurring lignin from Aegilops ovata L. in green and infrared light. Experientia 36: 662–663.
30. Gutterman Y (2002) Survival strategies of annual desert plants. Berlin: Springer-Verlag.
31. Huang ZY, Dong M, Gutterman Y (2004) Caryopses dormancy, germination and seedling emergence in sand, of Leymus racemosus (Poaceae), a perennial sand dune grass inhabiting the Junggar Basin of Xinjiang, China. Aust J Bot 52: 519–528.
32. He XQ (2011) Seed coat permeability and location of semipermeable layer in seeds of several grass species. Ph.D. Thesis, Lanzhou University, Lanzhou.
33. Kucera B, Cohn MA, Leubner-Metzger G (2005) Plant hormone interactions during seed dormancy release and germination. Seed Sci Res 15: 281–307.
34. Graeber K, Nakabayashi K, Miatton E, Leubner-Metzger G, Soppe WJ J (2012) Molecular mechanisms of seed dormancy. Plant Cell Environ 35: 1769–1786.
35. Linkies A, Leubner-Metzger G (2012) Beyond gibberellins and abscisic acid: how ethylene and jasmonates control seed germination. Plant Cell Rep 31: 253–270.
36. Baskin CC, Baskin JM (1998) Ecology of seed dormancy and germination in grasses. In: Cheplick GP, editor. Population biology of grasses. Cambridge, UK: Cambridge University Press. 30–83 p.
37. Gianinetti A, Vernieri P (2007) On the role of abscisic acid in seed dormancy of red rice. J Exp Bot 58: 3449–3462.
38. International Seed Testing Association (2012) International rules for seed testing. Seed Sci Technol 27 (Supplement).
39. Brady NC, Weil RR (2008) The nature and properties of soils. 14th Edition. Upper Saddle River: Prentice Hall.
40. Pons TL (1989) Breaking of seed dormancy by nitrate as a gap detection mechanism. Ann Bot 63: 139–143.
41. Popay AI, Roberts EH (1970) Ecology of Capsella bursa-pastoris (L.) Medik and Senecio vulgaris L. in relation to germination behavior. J Ecol 58: 123–139.
42. Footitt S, Douterelo-Soler I, Clay H, Finch-Savage WE (2011) Dormancy cycling in Arabidopsis seeds is controlled by seasonally distinct hormone-signaling pathways. PNAS 108: 20236–20241.
43. Finch-Savage Leubner-Metzger (2006) Seed dormancy and the control of germination. New Phytol 171: 501–23.
44. Finkelstein R, Reeves W, Ariizumi T, Steber C (2008) Molecular aspects of seed dormancy. Ann Rev Plant Biol 59: 387–415.
45. Liu K, Baskin JM, Baskin CC, Bu HY, Liu MX, et al (2010) Effect of storage conditions on germination of seeds of 489 species from high elevation grasslands of the eastern Tibet Plateau and some implications for climate change. Am J Bot 98: 12–19.
46. Roberts EH (1962) Dormancy in rice seed. III. The influence of temperature, moisture and gaseous environment. J Exp Bot 13: 75–94. editor.
47. Probert RJ (2000) The role of temperature in the regulation of seed dormancy and germination. In: Fenner M, editor. Seeds: the ecology of regeneration in plant communities. Wallingford: CAB International. 261–292 p.
48. Steadman KJ, Crawford AD, Gallagher RS (2003) Dormancy release in Lolium rigidum seeds is a function of thermal after-ripening time and seed water content. Funct Plant Biol 30: 345–352.
49. Bair NB, Meyer SE, Allen PS (2006) A hydrothermal after-ripening time model for seed dormancy loss in Bromus tectorum L. Seed Sci Res 16: 17–28.
50. Hou JW (2009) Effects of fungicide seed treatment on soil seed bank under various conditions. Ph.D. Thesis, Lanzhou University, Lanzhou.
51. Ellis RH, Hong TD, Roberts EH (1983) Procedure for the safe removal of dormancy from rice seed. Seed Sci Technol 11: 77–112.
52. Thanos CA, Georghiou K, Douma DJ, Marangaki CJ (1991) Photoinhibition of seed germination in Mediterranean maritime plants. Ann Bot 68: 469–475.
53. Zhao LH, Cheng JM, Wan HE (2008) Dynamic analysis of the soil seed bank in typical prairie on the Loess Plateau. Bulletin of Soil and Water Conservation 28: 60–66.
54. Wang N, Jiao JY, Jia YF, Wang DL (2011) Seed persistence in the soil on eroded slopes in the hilly-gullied Loess Plateau region, China. Seed Sci Res 21: 295–304.

10

Changes in Soil Carbon and Nitrogen following Land Abandonment of Farmland on the Loess Plateau, China

Lei Deng[1], Zhou-Ping Shangguan[1]*, Sandra Sweeney[2]

1 State Key Laboratory of Soil Erosion and Dryland Farming on the Loess Plateau, Northwest A&F University, Yangling, Shaanxi, China, **2** Institute of Environmental Sciences, University of the Bosphorus, Istanbul, Turkey

Abstract

The revegetation of abandoned farmland significantly influences soil organic C (SOC) and total N (TN). However, the dynamics of both soil OC and N storage following the abandonment of farmland are not well understood. To learn more about soil C and N storages dynamics 30 years after the conversion of farmland to grassland, we measured SOC and TN content in paired grassland and farmland sites in the Zhifanggou watershed on the Loess Plateau, China. The grassland sites were established on farmland abandoned for 1, 7, 13, 20, and 30 years. Top soil OC and TN were higher in older grassland, especially in the 0–5 cm soil depths; deeper soil OC and TN was lower in younger grasslands (<20 yr), and higher in older grasslands (30 yr). Soil OC and N storage (0–100 cm) was significantly lower in the younger grasslands (<20 yr), had increased in the older grasslands (30 yr), and at 30 years SOC had increased to pre-abandonment levels. For a thirty year period following abandonment the soil C/N value remained at 10. Our results indicate that soil C and TN were significantly and positively correlated, indicating that studies on the storage of soil OC and TN needs to focus on deeper soil and not be restricted to the uppermost (0–30 cm) soil levels.

Editor: Ben Bond-Lamberty, DOE Pacific Northwest National Laboratory, United States of America

Funding: The study was funded by the Strategic Priority Research Program of the Chinese Academy of Sciences (XDA05060403) and the Scholarship Award for Excellent Doctoral Student granted by Ministry of Education. The funders had no role in study design, data collection and analysis, decision to publish, or preparation of the manuscript.

Competing Interests: The authors have declared that no competing interests exist.

* E-mail: shangguan@ms.iswc.ac.cn

Introduction

Changes in land use have important effects on regional ecological processes and global climate change [1–2]. During the past two decades, many studies have focused on the effects of land-use change on soil organic carbon (SOC) and total nitrogen (TN) in terrestrial ecosystems [3–5]. The large differences in climatic conditions [6], soil properties [7], and type of land use change [8] are three factors whose effects are not yet well understood.

The revegetation of degraded land is one of the principal strategies for the control of both soil erosion and ecosystem recovery in fragile regions where either anthropogenic activities or severe environmental conditions can lead to disturbance [9]. Revegetation also greatly influences soil quality, C and N cycling, land management, as well as regional socioeconomic development [10–11]. The development of managed grassland and forest accelerates ecosystem restoration [12], affecting C and N cycles and C and N pools stored in soils [10,13]. Altered C and N cycles and C and N pools influence the production of biomass and ecosystem function [14]. Furthermore, C–N interactions are very important in determining whether the C sink in land ecosystems can be sustained over the long term [15–16]. Luo et al. [15] have proposed that N dynamics are a key factor in the regulation of long-term terrestrial C sequestration. N may become progressively more limiting as C accumulates in ecosystems under elevated CO_2 if the total N content in an ecosystem does not change [15]. Therefore, studying the dynamics of organic carbon (OC) and N in soils along a restoration succession gradient and analyzing the

relationships between C and N storage dynamics following restoration may be of importance in improving our knowledge of the sustainable management of land resources and predictions of future global C and N cycling.

Large-scale monocropping and over-grazing [17] have affected the semi-arid northern Loess Plateau in China. In addition, over the last century, the expanding human population, combined with a changing lifestyle, has accelerated ecosystem fragmentation and degradation [9]. To stabilize the fragile natural ecosystems characteristic of the Loess Plateau and to alleviate the degradation of land, the Chinese government has launched a series of nationwide conservation projects. One such project converts degraded farmland to either grassland or forest [18–19] to control soil erosion, increase storage of SOC and N, and prevent the occurrence of soil desiccation on the Loess Plateau [11]. Restoration succession may affect SOC and N decomposition. With regard to these ecosystem functions, the retention of C and N in soil is crucial [20]. This is of particular concern at sites with substantial N saturation, which are becoming increasingly widespread due to elevated atmospheric N deposition.

Previous studies report that the revegetation of degraded land can increase OC and N storages in soil [9,13]. Wang et al. [21] reported that both SOC and TN increase as a linear function of years abandoned. However, many previous studies have focused primarily on the topsoil (0–30 cm) [21–24]. Little is known about long-term changes to SOC and N in the deeper soil layers of the restoration succession on the Loess Plateau. This information can be useful in estimating the temporal distribution of storage of SOC and N and for evaluating OC and N dynamics throughout the conversion from managed to natural communities in semi-arid regions.

The objectives of this work were to investigate changes in SOC and TN concentration, soil OC and N storage and the relationship between SOC and TN with time since abandonment of farmland with depth in the soil profile.

Materials and Methods

Site description

This study was conducted in the Zhifanggou watershed in Ansai County, Shaanxi Province, NW China (36°46′28″–36°46′42″N, 109°13′03″–109°16′46″E; 1,010–1,431 m a.s.l., 8.27 km²) (Figure 1). The study area is characterized by a semi-arid climate and a deeply incised hilly-gully Loess landscape. Slopes vary between 0°–65°. The mean annual temperature range is 9.1±0.1°C from 1970 to 2010, and the annual mean temperature has increased over time (Figure 2a), in summer, the highest temperature is 35.3±2.3°C and in winter, the lowest temperature is −20.3±3.7°C from 1970 to 2010; the average frost-free period is 157 days. Mean annual precipitation is 503±15 mm (from 1970 to 2010) (Figure 2b), of which 70% falls between July and September. The loess-derived soils are fertile but extremely susceptible to erosion. The sand, silt and clay contents are 65%, 24% and 11%, respectively [21]. The main grassland species are *Stipa bungeana, Bothriochloa ischaemum, Artemisia sacrorum, Potentilla acaulis, Stipa grandis, Androsace erecta, Heteropappus altaicus, Lespedeza bicolor, Artemisia capillaries*, and *Artemisia frigid*, of which, *S. bungeana* is the most widely distributed. In addition, shrubland species such as *Rosa xanthina, Spiraea pubescens*, and *Hippophae rhamnoides* can be found in gullies. The primary planted trees in the study area are *Robinia pseudoacacia, Populus simonii, Caragana microphylla*, and *Platycladus orientalis*.

Experimental design and sampling

A common method used to study vegetation restoration is to monitor plants and soils under similar climatic conditions following the sequence of vegetation development [25]. This chronosequence method is widely adopted in applied ecosystem research [26–27] and is considered a "retrospective" research method because it compares existing conditions with original conditions and treatments [28]. The substitution of "space" for "time" is an effective way of studying change over time [28–29]. It makes the critical assumption that each site in the sequence differs only in age and that each site has traced the same history in both its abiotic and biotic components [27]. Before revegetation soil conditions are largely driven by geomorphological processes.

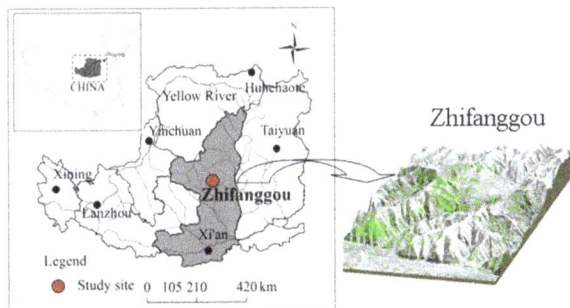

Figure 1. Location and DEM model of the Zhifanggou watershed (adapted from Wang et al., 2011).

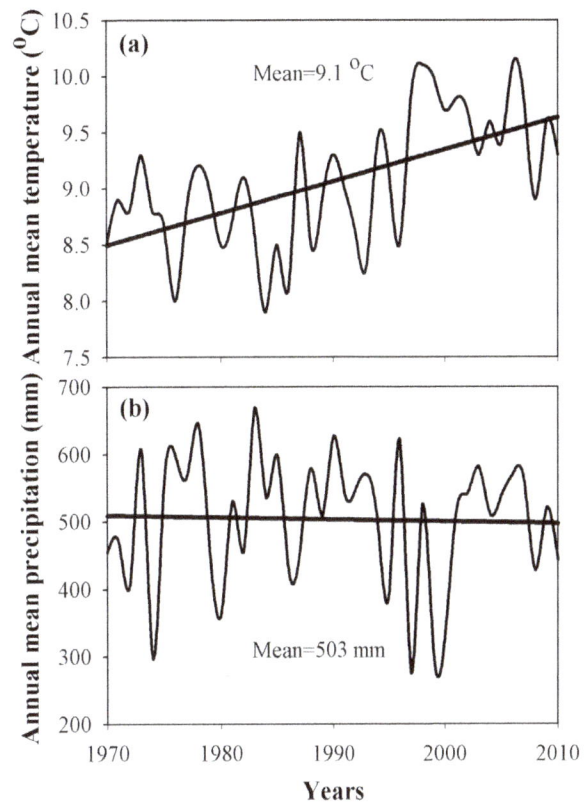

Figure 2. Annual mean temperature (a) and Annual mean precipitation (b) at the study site 1970–2010. Note: the solid lines in the figures a, b are the dynamic fitting curve for annual mean temperature, annual mean precipitation changes with time.

Thus, sites stabilized through revegetation for different periods of time offer an ideal opportunity to understand vegetation succession processes in extreme environments.

An abandoned farmland chronosequence in the watershed was selected for study after the history of the sites was determined through interviews with local farmers (Mr. Haibin Zhang, Soil and Water Conservation Experiment Station, Northwest A&F University, Ansai County, Shaanxi, NW China). Five age classes, 1, 7, 13, 20, and 30 years were selected. In August 2011, when the grassland community biomass peaked, five sites were established in each of the age classes, and the sites were separated by 0.5–1.5 km apart. At each site we set up a plot of 20 m×20 m. In each plot, five quadrats (1 m×1 m) were separately chosen in each of the four corners and center of the plot. In total, we surveyed five plots with twenty-five quadrats in each age class, and twenty-five plots with one hundred and twenty-five quadrats for our study. In each quadrat, the coverage and height, above- and belowground biomass, litter biomass, and soil samples in 0–100 cm soil cores were observed. The morphological traits of the herbage in each age group are listed in Table 1. The plots were all located near the top of loess mounds and there was little difference among the sites in altitude (1185–1341 m), aspect (half south-facing and half north-facing), gradient (18°–37°), or previous farming practices. In the study area, the soils were loess-derived. In addition, five sites on maize farmland (CK) were selected for comparison. Before the farmland was abandoned maize (*Zea mays*) had been widely seeded. The average amount of fertilizer applied was 225–300 kg ha⁻¹ of sheep manure as the base fertilizer in April and 300–450 kg ha⁻¹ of urea which was applied in June as topdressing. We state clearly

Table 1. Dominant species and their biomass, total cover, height at different number of years abandoned.

Years abandoned (yr)	Above- ground biomass (g m^{-2})	Below- ground biomass (g m^{-2})	Accumulative litter biomass (g m^{-2})	Coverage (%)	Height (cm)	Dominant species
1	169.34±49.67a	78.92±13.57c	28.90±4.26c	8.8±0.58c	66±6.96ab	*Artemisia scoparia*
7	73.73±7.96a	138.73±6.72c	118.51±14.93bc	19.4±2.86bc	41.6±7.88b	*Lespedeza bicolor+ Setaira viridis*
13	100.26±19.01a	218.71±36.29c	130.78±33.00bc	24.4±5.61abc	68.2±6.01a	*Agropyron cristatum+ Heteropappus altaicus*
20	210.42±56.73a	575.21±129.16b	209.79±35.99ab	41±10.77ab	69±5.79a	*Artemisia sacrorum+ Bothriochloa ischaemum*
30	159.08±12.03a	1080.34±111.18a	281.86±25.77a	48±6.04a	52±2.07ab	*Bothriochloa ischaemum+ Stipa bungeana*

Different letters indicate significant differences at $P<0.05$ among years abandoned.
Values are mean ±SE of 5 sites.

that no specific permissions were required for the location. We confirm that the location is not privately-owned or protected in any way. We confirm that the field studies do not involve endangered or protected species.

In each quadrat, all the aboveground parts of the green plants were cut, collected from the ground and put into envelops and tagged, as was all litter. Because the biomass samples were large, they were weighed fresh and then a part of each sample was dried and weighed. The aboveground biomass of the samples was calculated by multiplying the ratio of the dry weight/fresh weight ratio by the fresh weight.

Soil samples were taken at five points in the quadrats of each plot. These were the four corners and center of the biomass sampling sites as described above. Litter horizons were removed before soil sampling. Soil sampling, using a soil drilling sampler (9 cm inner diameter), was done in seven soil layers, 0–5, 5–10, 10–20, 20–30, 30–50, 50–70, and 70–100 cm. We then mixed the

same layers together to make one sample. All samples were sieved through a 2 mm screen, and roots and other debris were removed. Each sample was air-dried and stored at room temperature for the determination of soil physical and chemical properties. The soil bulk density (g cm^{-3}) of the different soil layers was measured using a soil bulk sampler with a 5 cm diameter and 5 cm high stainless steel cutting ring (5 replicates) at points adjacent to the soil sampling quadrats. The original volume of each soil core and its dry mass after oven-drying at 105°C were measured.

To measure belowground biomass, soil sampling was done three times in seven soil layers, 0–5, 5–10, 10–20, 20–30, 30–50, 50–70, and 70–100 cm at a depth of 0–100 cm in each quadrat using a 9 cm diameter root auger. The majority of the roots were found in the soil samples thus obtained and then isolated using a 2 mm sieve. The remaining fine roots taken from the soil samples were isolated by spreading the samples in shallow trays, overfilling the trays with water and allowing the outflow from the trays to pass

Table 2. One-way analysis of variance (ANOVA) for soil properties in the seven soil layers among abandoned farmland in different years.

Soil layer (cm)	Df	SOC		TN	
		F	sig. (P)	F	sig. (P)
0–5	5	10.367	0.0001**	14.557	0.0001**
5–10	5	2.604	0.0668	3.651	0.0217*
10–20	5	0.88	0.4936	0.397	0.8081
20–30	5	5.322	0.0044**	0.132	0.9688
30–50	5	6.892	0.0012**	3.348	0.0298*
50–70	5	5.787	0.0029**	6.963	0.0011**
70–100	5	2.567	0.0697	1.825	0.1636

Soil depth (cm)	Df	C storages		N storages	
		F	sig. (P)	F	sig. (P)
0–5	5	5.71	0.0031**	5.809	0.0029**
0–10	5	4.053	0.0145*	4.527	0.0091**
0–20	5	0.783	0.5496	0.732	0.5811
0–30	5	1.404	0.2688	0.139	0.966
0–50	5	4.483	0.0095**	1.014	0.4237
0–70	5	7.042	0.0010**	3.089	0.0392*
0–100	5	7.905	0.0005**	3.783	0.0190*

*indicates significant at $P<0.05$ and **indicates significant at $P<0.01$.

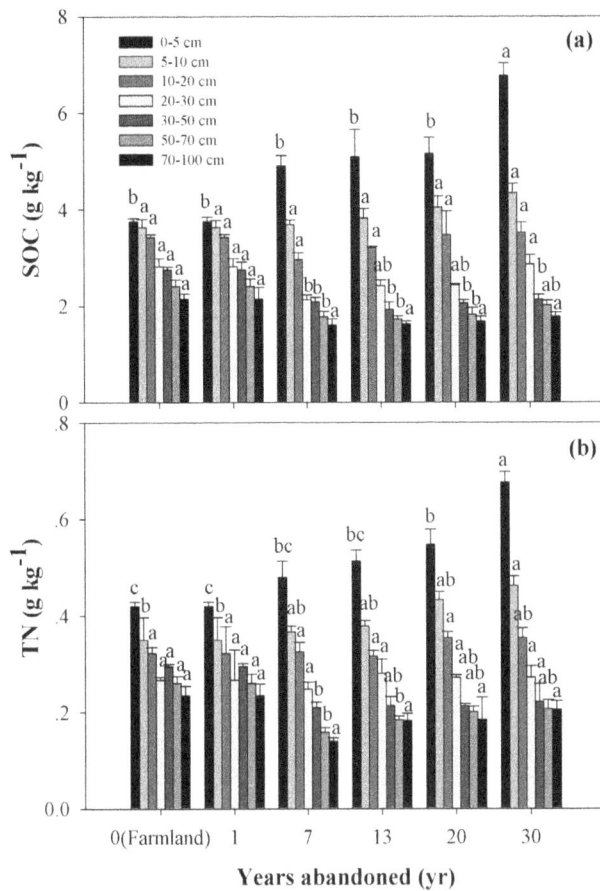

Figure 3. SOC and TN dynamics for seven soil layers of 0–100 cm following the conversion of farmland to grassland. Values are in the form of Mean ± SE and the sample size n = 5. Different lower-case letters above the bars mean significant differences in the same soil layers among abandoned land in different years (P<0.05).

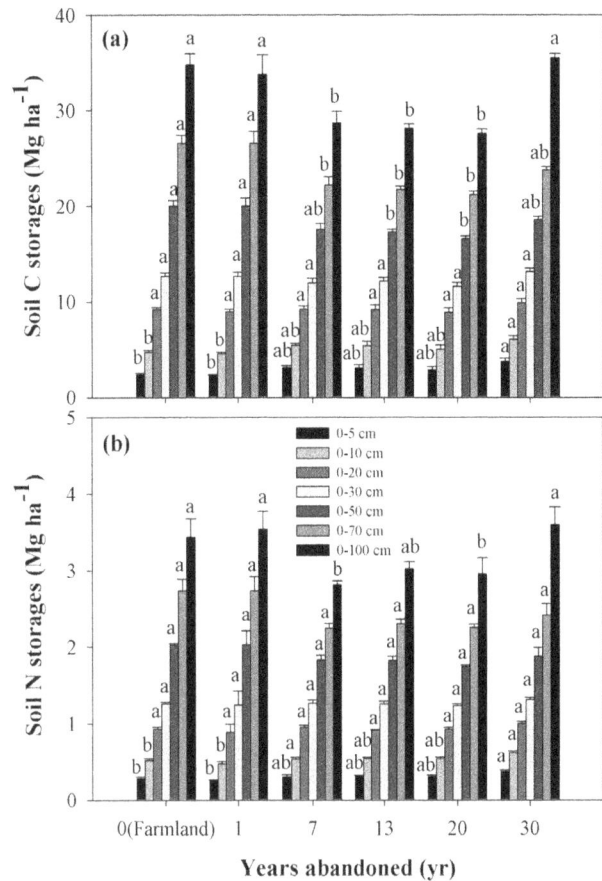

Figure 4. Dynamics of soil OC and N storages in 0–100 cm soil depth following the conversion of farmland to grassland. Values are in the form of Mean ± SE and the sample size n = 5. Different lower-case letters above the bars mean significant differences in the same soil layers among abandoned land in different years (P<0.05).

through a 0.5 mm mesh sieve. No attempts were made to distinguish between living and dead roots. All the roots thus isolated were oven-dried at 65°C and weighed to within 0.01 g.

Physical and chemical analysis

Soil bulk density (BD) was calculated depending on the inner diameter of the core sampler, sampling depth and the oven dried weight of the composite soil samples [30]. Soil OC content was assayed by dichromate oxidation [31] and soil TN content was assayed using the Kjeldahl method [32].

Calculation of soil C and N storages

Our sample soils did not have any coarse fraction (>2 mm). Therefore, the study used the following equation to calculate soil organic carbon storage (Cs) [22]:

$$Cs = BD \times SOC \times D/10 \qquad (1)$$

in which, Cs is SOC storages (Mg ha^{-1}); BD is soil bulk density (g cm^{-3}); SOC is soil organic carbon concentration (g kg^{-1}); and D is soil thickness (cm).

The following equation was used to calculate soil N storage (Ns) [33]:

$$Ns = BD \times TN \times D/10 \qquad (2)$$

in which, Ns is soil N storage (Mg ha^{-1}); BD is soil bulk density (g cm^{-3}); TN is soil TN concentration (g kg^{-1}); and D is soil thickness (cm).

Statistical analysis

One-way ANOVA was used to analyze the means of the same soil layers among the different abandoned years. Differences were evaluated at the 0.05 significance level. When significance was observed at the P<0.05 level, Tukey's post hoc test was used to carry out the multiple comparisons. All statistical analyses were performed using the software program SPSS, ver. 17.0 (SPSS Inc., Chicago, IL, USA).

Results

Dynamics of SOC and TN

SOC in 0–5 cm soil was higher in older grassland following abandonment (Table 2, Figure 3a). In the first 20 years, 0–5 cm SOC showed no significant changes; after 20 years, it had significantly increased (Figure 3a). The 5–20 cm SOC had not significantly increased (Table 2). The 20–70 cm SOC was significantly lower in the younger grasslands (<20 yr) and higher

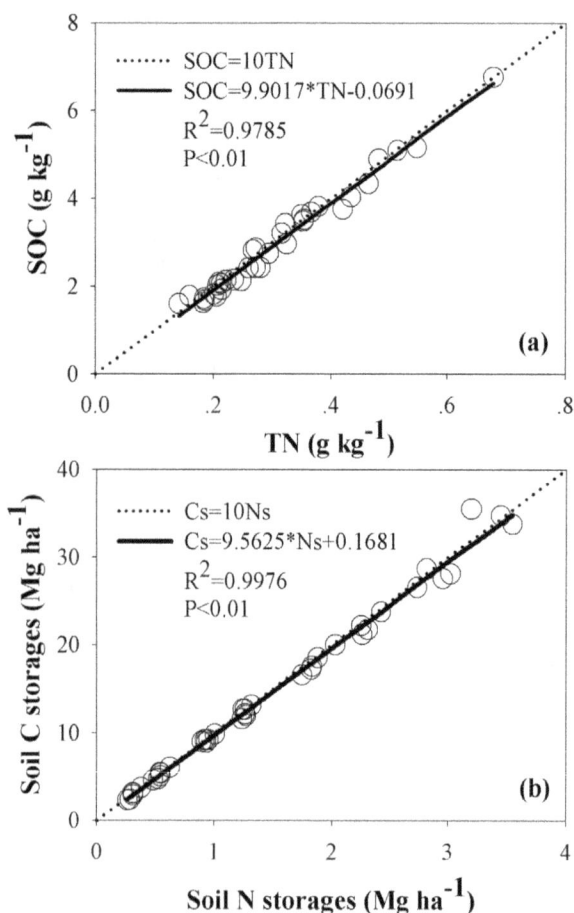

Figure 5. Relationship between SOC and TN, soil OC storages and soil N storages, and accumulative soil C storages and soil N storage.

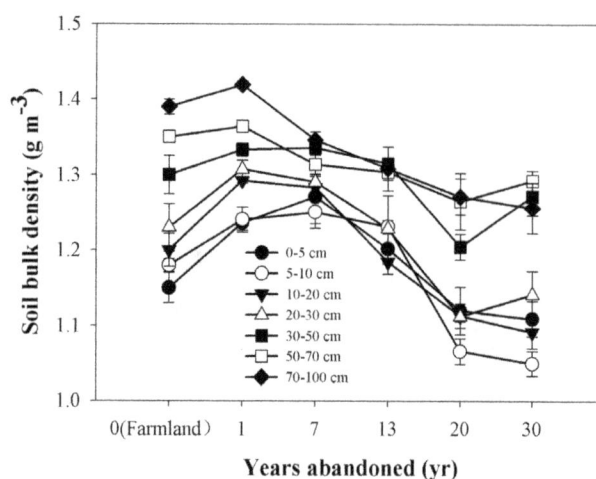

Figure 6. Dynamics of soil bulk density in 0–100 cm soil depth following the conversion of farmland to grassland. Values are in the form of Mean ± SE and the sample size n = 5.

in the older grasslands (30 yr) (Figure 3a). When abandoned for 30 years, 20–70 cm SOC had increased to the level prior to abandonment (maize) (Figure 3a). The 70–100 cm SOC showed no significant changes since having been abandoned, but also showed a tendency to be low in younger grasslands and higher in the older grasslands (Table 2, Figure 3a).

Soil TN in 0–5 cm soil was also higher in older grassland following abandonment (Table 2, Figure 3b). In the first 13 years after having been abandoned, TN in the 0–5 cm soil layer showed no significant changes; after 13 years, it had significantly increased (Figure 3b). The TN in the 5–10 cm soil layer showed no significant changes in the first 13 years, whereas after 13 years, it had significantly increased (Figure 3b). The 10–30 cm soil TN did not increase significantly (Table 2, Figure 3b). The 30–70 cm soil TN first decreased significantly and then increased to the level prior to abandonment (maize) (Figure 3b). Similar to the SOC, 70–100 cm soil TN had showed no significant changes since abandonment, but also showed a tendency to be lower in younger grasslands and higher in older grasslands (Table 2, Figure 3b).

C and N storages dynamics

Soil OC storage in the 0–10 cm soil was also higher in older grassland since abandonment (Table 2, Figure 4a). After having been abandoned for 30 years, the 0–10 cm soil OC storage had increased significantly. 0–30 cm soil OC storage had not increased

significantly (Table 2, Figure 4a). 0–100 cm soil OC storage was significantly lower in the younger grasslands (<20 yr) and higher in the older grasslands (30 yr); at 30 years it had increased to pre-abandonment levels (Figure 4a). Soil C storage changed mainly in the top 10 cm of soil following the conversion of farmland to grassland for a period of 30 years.

Similar to the accumulative soil OC storage, soil N storage had changed mainly in the top 10 cm soil layer 30 years after having been abandoned. The conversion of farmland into grassland significantly increased the soil N storage in the 0–10 cm soil (Table 2, Figure 4b). After having been abandoned for 30 years, 0–10 cm soil N storage had significantly increased. 0–50 cm soil N storage had not significantly increased (Table 2, Figure 4b). 0–100 cm soil N storage was also significantly lower in younger grasslands (<20 yr) and higher in older grasslands (30 yr), until after thirty years it had increased to pre-abandonment levels (maize) (Figure 4b).

Relationship between Soil C and N

Soil C and N showed significant positive correlations (Figure 5). The relationship between SOC and TN, soil C storage and soil N storage were significant ($P<0.01$). In the process of the revegetation, soil OC and TN, soil C storage and N storage approximately represents SOC = 10TN (Figure 5a) and Cs = 10Ns (Figure 5b).

Discussion

The soil C and N results supported the hypothesis that soil organic C and N conditions in both the top soil and in the deeper soil layer are significantly affected by land use change on the northern Loess Plateau. In our study, top soil OC and TN were higher in older post-abandonment grassland, especially in the 0–5 cm soil depth (figure 3), indicating the accumulation of soil OC and TN by revegetation [23–24]. These results agree with those of Wang et al. [21], who studied changes to the physico-chemical properties of top soil during natural succession on abandoned farmland in the Zhifanggou watershed. The evident increase may be partly attributed to a lower fraction of non-soluble materials in more readily decomposed plant residues. In the farmland, cultivation breaks up soil aggregates, decreases total soil porosity, and accelerates composition and mineralization of soil organic

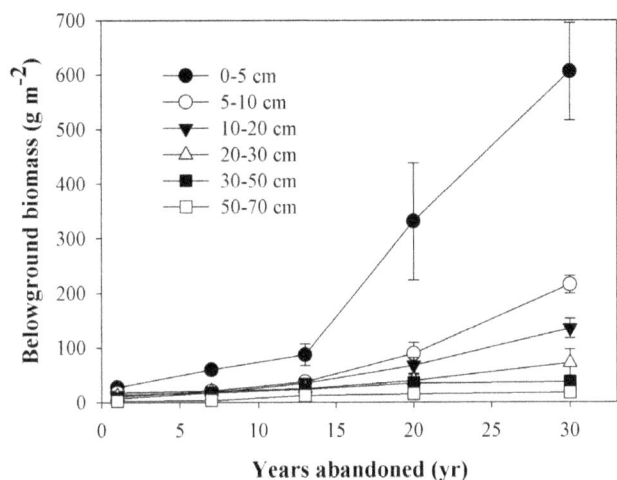

Figure 7. Dynamics of belowground biomass in 0–100 cm soil depth following the conversion of farmland to grassland. Values are in the form of Mean ± SE and the sample size n = 5. It didn't have roots in the 70–100 cm soil layers.

matter (SOM) due to exposure of previously accessible SOM to microbial attack [34]. This results in a reduction in the amounts of intra-aggregate light fraction organic carbon (LFOC) and some organomineral SOC [35]. In addition, the reduction of crop residue return to soil may also be a factor as farmers take away straw with grain harvesting each year. This speculation can be supported by the results of Wu et al. [36] in this region where plant residue in the top soil layer was reduced considerably after native grasslands were cultivated, contributing to the decrease in LFOC and SOC. Conversely, the conversion of farmland into grassland increases SOC and its fractions [37], and increases total soil porosity, thus resulting in a reduction in soil BD (Figure 6, [21]).

Aboveground vegetation plays an important role in regulating the biogeochemistry of ecosystems by fixing C and nutrients and preventing the loss of nutrients under disturbance, such as plant decline, acid-rain and climate change [38]. It is also clear from these results that revegetated grassland has a great impact on the storage of soil OC and N. In our study, both soil OC and N storage in the 0–10 cm soil depth was higher in older grassland that had been abandoned for thirty years (Figure 4). This indicates that during the 30 year period soil OC and N storage mainly had changed in the top 10 cm of soil since abandoned. There may be a range of potential mechanisms through which soil OC and N in the top soil increased with revegetation. A prime candidate is the return of both C and N from increased aboveground biomass and litter. As soil OC and N input are mainly derived from the decomposition of litter [21], primary productivity is the main driver of soil carbon sequestration [39], which resulted primarily in soil OC and N storage increasing in the top soil. Secondly, belowground biomass (dead roots, mycorrhizae, and exudates) is an important element of soil carbon sequestration [40]. Belowground biomass increased in the time after the farmland had been abandoned (Table 1, Figure 7). Thirdly, changes in vegetation composition and the dominant plant functional group could affect the sequestration of C and N in the soil [39,41]. Plant functional composition strongly influences the chemical and physical composition of litter inputs, and thereby their decomposability, carbon loss through soil respiration and leaching, and carbon immobilization in hummified plant residues [39], and the increase of C3 plants can increase soil C and N accumulation in meadow

soils [41]. Revegetation had a direct effect on the dominant vegetation species, vegetation cover, height, above- and below-ground biomass (Table 1).

SOC and TN at deeper soil levels (>30 cm) were lower in younger grasslands and higher in older grasslands (Figure 3), this is probably due to long-term natural organic fertilizer and inorganic fertilizer inputted into the soil resulting in higher SOC and N in the farmland stage. So, this initial loss of soil OC and N following the conversion of farmland has been commonly attributed to the net effect of decreased organic matter inputs and losses through decomposition [42]. During the course of grasslands development, we observed an increase in the soil SOC and N storage, after 30 years of revegetation, where deeper soil C and N had returned to pre-abandonment levels (Figure 3); the return of belowground biomass to the deeper soil layer is another reason (Figure 7). Previous studies of soil C dynamics have emphasized the role of physical protection from different particle fractions (sand, silt and clay) [43], microbes and enzymes within aggregates [44], microaggregates and macroaggregates [45], and bacterial and fungal [46], therefore, the mechanism in the soil C and N dynamics in the deeper soil during the process of conversion from farmland into grassland probably relate to those factors. However, our lack of a full understanding of this process calls for more attention to be paid to soil C and N dynamics in the deeper soil layers. Wang et al. [21] found that SOC and TN had a negative relationship with soil bulk density, and Singh et al. [23–24] reported that SOC and TN were significantly greater while BD was significantly lower during the restoration of degraded lands, our results agreed with them (Figures 3 and 6). So, we can infer the trend of SOC and TN according to the trend of soil bulk density.

We observed that soil C and N were significantly and positively correlated (Figure 5). In the restoration process, soil OC and TN, soil C storage and N storage approximately represents SOC = 10TN (Figure 5a) and Cs = 10Ns (Figure 5b). So, we can conclude that soil C/N value was 10 in the process of 30 years of conversion from farmland to grassland in the Loess Plateau, a value greater that reported by Liu et al. [35] (C/N = 7.62). Ammonium-N and the sum of NO_3^--N and NO_2^--N are the readily available forms of soil N for root uptake. Liu et al. [35] reported that ammonium-N and the sum of NO_3^--N and NO_2^--N were lower in grassland than in cropland with the addition of chemical fertilizer N, because the conversion of farmland into grassland increased the C/N ratio and reduced soil mineral N by enhancing soil N immobilization.

Grasslands play an important role in the global C and N cycles [3–4,10]. Grassland in good condition should be in balance in terms of C and N input and output or in the state where C and N input is greater than their output [41]. At least one study has shown that carbon input is greater than carbon output in enclosed grassland [47]. In our study, 0–100 cm soil OC and N storage was significantly lower in the younger grasslands (<20 yr) and higher in the older grasslands (30 yr), at 30 years they had increased to pre-abandonment levels, and that the values in farmland were higher compared to grasslands abandoned for 20 years. Our results indicate that the study of the storage of soil OC and N need to include the deeper soil layer and not focus solely on the top soil. Li et al. [48] also found that SOC density in the deeper soil layer was significantly higher than that of the farmland in the Zhifanggou watershed, Loess Plateau. Li et al. [49] found that soil C and N storage in the deeper layers (mineral layer) show significant difference in top soil layers (organic layer) with time after afforestation at the global scale. Therefore, estimating soil OC and N input or output requires that research consider not only soil depth but also time since abandonment.

Conclusions

The results of this study indicate that plant succession after land has been abandoned resulted in a significant improvement in the physico-chemical properties of soil. Thirty years following abandonment, soil OC and N storage had increased primarily in the top 10 cm of the soil depth. After 30 years of restoration, deeper soil C and N storage had increased to pre-abandonment levels. This finding indicates that deeper soil has a higher potential to fix both C and N in the future (>30 yr). Thus, in the semi-arid environment of the Loess Plateau, vegetation recovery following abandonment is slow and the improvement of soil properties is likely to require a considerably long period of time (>30 yr). Therefore, the findings are important for assessing the resilience of

these degraded ecosystems and developing a more effective strategy of vegetation restoration for the management of degraded grassland from a long-term perspective. More research, for example, on soil physico-chemical properties, soil enzyme activities, soil microbial, animal and plant function and composition, is required to better understand the mechanism behind how the soil fixes C and N in the deeper soil profile of sub-arid regions, for example, the Loess Plateau, China.

Author Contributions

Conceived and designed the experiments: LD ZS. Analyzed the data: LD. Contributed reagents/materials/analysis tools: LD. Wrote the paper: LD ZS SS.

References

1. Kalnay E, Cai M (2003) Impact of urbanization and land-use change on climate. Nature 423: 528–531.
2. Ficetola GF, Maiorano L, Falcucci A, Dendoncker N, Boitani L, et al. (2010) Knowing the past to predict the future: land-use change and the distribution of invasive bullfrogs. Global Change Biol 16: 528–537.
3. Russell AE, Laird DA, Parkin TB, Mallarino AP (2005) Impact of nitrogen fertilization and cropping system on carbon sequestration in Midwestern Mollisols. Soil Sci Soc Am J 69: 413–422.
4. Brown J, Angerer J, Salley SW, Blaisdell R, Stuth JW (2010) Improving estimates of rangeland carbon sequestration potential in the US Southwest. Rangeland Ecol Manag 63: 147–154.
5. Qiu SJ, Ju XT, Ingwersen J, Qin ZC, Li L, et al. (2010) Changes in soil carbon and nitrogen pools after shifting from conventional cereal to greenhouse vegetable production. Soil Till Res 107: 80–87.
6. Jobbágy EG, Jackson RB (2000) The Vertical Distribution of Soil Organic Carbon and Its Relation to Climate and Vegetation. Ecol Appl 10: 423–436.
7. Piao SL, Fang JY, Ciais P, Peylin P, Huang Y, et al. (2009) The Carbon balance of terrestrial ecosystems in China. Nature 458: 1009–1013.
8. Arora VK, Boer GJ (2010) Uncertainties in the 20th century carbon budget associated with land use change. Global Change Biol 16: 3327–3348.
9. Jia XX, Wei XR, Shao MA, Li XZ (2012) Distribution of soil carbon and nitrogen along a revegetational succession on the Loess Plateau of China. Catena 95: 160–168.
10. Eaton JM, McGoff NM, Byme KA, Leahy P, Kiely G (2008) Land cover change and soil organic carbon stocks in the Republic of Ireland 1851–2000. Climate Change 91: 317–334.
11. Fu XL, Shao MA, Wei XR, Horton R (2010) Soil organic carbon and total nitrogen as affected by vegetation types in Northern Loess Plateau of China. Geoderma 155: 31–35.
12. She DL, Shao MA, Timm LC, Reichardt K (2009) Temporal changes of an alfalfa succession and related soil physical properties on the Loess Plateau, China. Pesquisa Agropecuária Brasileira 44: 189–196.
13. Wei XR, Shao MA, Fu XL, Horton R (2010) Changes in soil organic carbon and total nitrogen after 28 years of grassland afforestation: effects of tree species, slope position, and soil type. Plant Soil 331: 165–179.
14. Foster D, Swanson F, Aber J, Burke I, Brokaw N (2003) The importance of land-use legacies to ecology and conservation. BioScience 53: 77–88.
15. Luo YQ, Field CB, Jackson RB (2006) Does nitrogen constrain carbon cycling, or does carbon input stimulate nitrogen cycling? Ecology 87: 3–4.
16. Luo YQ, Su B, Currie WS, Dukes JS, Finzi A, et al. (2004) Progressive nitrogen limitation of ecosystem responses to rising atmospheric carbon dioxide. BioScience 54: 731–739.
17. Fu BJ, Chen LD, Ma KM, Zhou HF, Wang J (2000) The relationships between land use and soil conditions in the hilly area of the Loess Plateau in northern Shaanxi, China. Catena 39: 69–78.
18. Deng L, Shangguan ZP (2011) Food security and farmers' income: impacts of the Grain for Green Programme on rural households in China. J Food Agric Environ 9: 826–831.
19. Deng L, Shangguan ZP, Li R (2012) Effects of the grain-for-green program on soil erosion in China. Int J Sediment Res 27: 120–127.
20. Prietzel J, Bachmaan S (2012) Changes in soil organic C and N stocks after forest transformation from Norway spruce and Scots pine into Douglas fir, Douglas fir/spruce, or European beech stands at different sites in Southern Germany. Forest Ecol Manag 269: 134–148.
21. Wang B, Liu GB, Xue S, Zhu BB (2011) Changes in soil physico-chemical and microbiological properties during natural succession on abandoned farmland in the Loess Plateau. Environ Earth Sci 62: 915–925.
22. Guo LB, Gifford RM (2002) Soil carbon storage and land use change: a meta analysis. Global Change Biol 8: 345–360.
23. Singh K, Pandey VC, Singh B, Singh RR (2012) Ecological restoration of degraded sodic lands through afforestation and cropping. Ecol Eng 43: 70–80.
24. Singh K, Singh B, Singh RR (2012) Changes in physico-chemical, microbial and enzymatic activities during restoration of degraded sodic land: Ecological suitability of mixed forest over monoculture plantation. Catena 96: 57–67.
25. Bhojvaid PP, Timmer VR (1998) Soil dynamics in an age sequence of Prosopis Juliflora planted for sodic soil restoration in India. Forest Ecol Manag 106: 81–193.
26. Fang W, Peng SL (1997) Development of species diversity in the restoration process of establishing a tropical man-made forest ecosystem in China. Forest Ecol Manag 99: 185–196.
27. Johnson EA, Miyanishi K (2008) Testing the assumptions of chronosequences in succession. Ecol Let 11: 419–431.
28. Li XR, Kong DS, Tan HJ, Wang XP (2007) Changes in soil and vegetation following stabilization of dunes in the southeastern fringe of the Tengger Desert, China. Plant Soil 300: 221–231.
29. Sparling GP, Schipper LA, Bettjeman W, Hill R (2003) Soil quality monitoring in New Zealand: practical lessons from a 6-year trial. Agr Ecosyst Environ 104: 523–534.
30. Jia GM, Cao J, Wang CY, Wang G (2005) Microbial biomass and nutrients in soil at the different stages of secondary forest succession in Ziwulin, northwest China. Forest Ecol Manag 217: 117–125.
31. Kalembasa SJ, Jenkinson DS (1973) A comparative study of titrimetric and gravimetric methods for the determination of organic carbon in soil. J Sci Food Agr 24: 1085–1090.
32. Bremner JM (1996) Nitrogen-total. In: Sparks, D.L. (Ed.), Methods of Soil Analysis, Part 3. America Society of Agronomy, Madison, pp. 1085–1121 (SSSA Book Series: 5).
33. Rytter RM (2012) Stone and gravel contents of arable soils influence estimates of C and N stocks. Catena 95: 153–159.
34. Shepherd TG, Saggar S, Newman RH, Ross CW, Dando JL (2001) Tillage-induced changes to soil structure and organic carbon fractions in New Zealand soils. Aust J Soil Res 39: 465–489.
35. Liu X, Li FM, Liu DQ, Sun GJ (2010) Soil Organic Carbon, Carbon Fractions and Nutrients as affected by Land Use in Semi-Arid Region of Loess Plateau of China. Pedosphere 20: 146–152.
36. Wu T, Schoenau JJ, Li F, Qian P, Malhi SS, et al. (2004) Influence of cultivation and fertilization on total organic carbon and carbon fractions in soils from the Loess Plateau of China. Soil Till Res 77: 59–68.
37. Zeng ZX, Liu XL, Jia Y, Li FM (2007) The effect of conversion of cropland to forage legumes on soil quality in a semiarid agroecosystem. J Sustain Agr 32: 335–353.
38. Bormann BT, Sidle RC (1990) Changes in productivity and distribution of nutrients in a chronosequence at Glacier Bay National Park, Alaska. Journal of Ecology 78: 561–578.
39. De Deyn GB, Cornelissen JHC, Bardgett RD (2008) Plant functional traits and soil carbon sequestration in contrasting biomes. Ecol Lett 11: 516–531.
40. Langley JA, Hungate BA (2003) Mycorrhizal controls on belowground litter quality. Ecology 84: 2302–2312.
41. Wu GL, Liu ZH, Zhang L, Chen JM, Hu TM (2010) Long-term fencing improved soil properties and soil organic carbon storage in an alpine swamp meadow of western China. Plant Soil 332: 331–337.
42. Zhao WZ, Xiao HL, Liu ZM, Li J (2005) Soil degradation and restoration as affected by land use change in the semiarid Bashang area, Northern China. Catena 59: 173–186.
43. He NP, Wu L, Wang YS, Han XG (2009) Changes in carbon and nitrogen in soil particle-size fractions along a grassland restoration chronosequence in northern China. Geoderma 150: 302–308.
44. Udawatta RP, Kremer RJ, Adamson BW, Anderson SH (2008) Variation in soil aggregate stability and enzyme activities in a temperate agroforestry practice. Appl Soil Ecol 39: 153–160.
45. Chen FS, Zeng DH, Fahey TJ, Liao PF (2010) Organic carbon in soil physical fractions under different-aged plantations of Mongolian pine in semi-arid region of Northeast China. Appl Soil Ecol 44: 42–48.

46. Six J, Frey SD, Thiet RK, Batten KM (2006) Bacterial and fungal contributions to carbon sequestration in agroecosystems. Soil Sci Soc Am J 70: 555–569.

47. Li YQ, Zhao HL, Zhao XY, Zhang TH, Chen YP (2006) Soil respiration, carbon balance and carbon storage of sandy grassland under post-grazing natural restoration. Acta Prataculturae Sinica 15: 25–31. (in Chinese with English abstract).

48. Li MM, Zhang XC, Pang GW, Han FP (2013) The estimation of soil organic carbon distribution and storage in a small catchment area of the Loess Plateau. Catena 101: 11–16.

49. Li DJ, Niu SL, Luo YQ (2012) Global patterns of the dynamics of soil carbon and nitrogen stocks following afforestation: a meta-analysis. New Phytol 195: 172–181.

11

Stratification of Carbon Fractions and Carbon Management Index in Deep Soil Affected by the Grain-to-Green Program in China

Fazhu Zhao, Gaihe Yang*, Xinhui Han*, Yongzhong Feng, Guangxin Ren

College of Agronomy, Northwest A&F University, Yangling, Shaanxi, China; and The Research Center of Recycle Agricultural Engineering and Technology of Shaanxi Province, Yangling, Shaanxi, China

Abstract

Conversion of slope cropland to perennial vegetation has a significant impact on soil organic carbon (SOC) stock in A horizon. However, the impact on SOC and its fraction stratification is still poorly understood in deep soil in Loess Hilly Region (LHR) of China. Samples were collected from three typical conversion lands, *Robinia psendoacacia* (RP), *Caragana Korshinskii Kom* (CK), and abandoned land (AB), which have been converted from slope croplands (SC) for 30 years in LHR. Contents of SOC, total nitrogen (TN), particulate organic carbon (POC), and labile organic carbon (LOC), and their stratification ratios (SR) and carbon management indexes (CMI) were determined on soil profiles from 0 to 200 cm. Results showed that the SOC, TN, POC and LOC stocks of RP were significantly higher than that of SC in soil layers of 0–10, 10–40, 40–100 and 100–200 cm ($P<0.05$). Soil layer of 100–200 cm accounted for 27.38–36.62%, 25.10–32.91%, 21.59–31.69% and 21.08–26.83% to SOC, TN, POC and LOC stocks in lands of RP, CK and AB. SR values were >2.0 in most cases of RP, CK and AB. Moreover, CMI values of RP, CK, and AB increased by 11.61–61.53% in soil layer of 100–200 cm compared with SC. Significant positive correlations between SOC stocks and CMI or SR values of both surface soil and deep soil layers indicated that they were suitable indicators for soil quality and carbon changes evaluation. The Grain-to-Green Program (GTGP) had strong influence on improving quantity and activity of SOC pool through all soil layers of converted lands, and deep soil organic carbon should be considered in C cycle induced by GTGP. It was concluded that converting slope croplands to RP forestlands was the most efficient way for sequestering C in LHR soils.

Editor: Raffaella Balestrini, Institute for Plant Protection (IPP), CNR, Italy

Funding: This work was supported by Special Fund for forest-scientific Research in the Public Interest (201304312). The funders had no role in study design, data collection and analysis, decision to publish, or preparation of the manuscript.

Competing Interests: The authors have declared that no competing interests exist.

* E-mail: ygh@nwsuaf.edu.cn (GY); hanxinhui@nwsuaf.edu.cn (XH)

Introduction

Soil organic carbon (SOC) is a dynamic component of the terrestrial system, with internal changes in both vertical and horizontal directions and external exchanges between the atmosphere and the biosphere [1]. SOC storage is estimated at approximately 1500 Pg globally, which is about two and three times the size of carbon pools in the atmosphere and vegetation, respectively [2]. Since carbon uptake and storage is tightly linked to the nitrogen (N) cycle, it is equally important to understand how N pools and fluxes are affected by land use change [3]. Moreover, more than 50% of the total SOC is stored in the subsoil [4]. The proportion of soil organic matter (SOM) stored in the first meter of the world soils below 30 cm depth ranges 46%~63%, except for Podzoluvisols, where 30% of SOC is stored below the depth of 30 cm [4]. A recent study also suggests that in the northern circumpolar permafrost region, at least 61% of the total soil C is stored below 30 cm [5]. Therefore, subsoil C may be even more important in terms of source or sink for CO_2 than topsoil C [6]. Considering the potential role of SOC in atmospheric CO_2 sink, it is important to understand what leads to sequestration of large amounts of SOC in the subsoil or even in deep soil. However, the

SOC contents in deep soil layers are not fully understood in LHR of China to date.

As an indicator of soil quality, SOM stratification, which is related to the rate and amount of SOC sequestration [7], is common in many natural ecosystems [8] and managed grasslands and forests [9–10]. Stratification ratio (SR) is defined as the ratio of a soil property at the surface layer to that at a deeper layer. In general, high SR values indicate good soil quality and are usually used to assess agricultural practices [7]. For instance, SR values for SOC at depths of 0–5 cm and 20–40 cm range from 1.1 to 1.5 under traditional tillage (TT) while from 1.6 to 2.6 under conservation tillage (CT) [11]. Little information is available on natural ecosystems and managed shrubs or forests land. Additionally, under semiarid climate, SOC in active fractions, such as particulate organic carbon (POC) and labile organic carbon (LOC) was more sensitive to soil management practices than total SOC [12]. Previous researches have indicated that changing rate of POC and LOC was faster than SOC in whole soil [11], and they could be an early indicator for SOC change in soil [13]. Meanwhile, the carbon management index (CMI), which is derived from the total soil organic C pool and C lability, had been extensively used as a sensitive indicator of SOC variation rate in response to soil management changes [14–15]. Therefore, under semi-arid climate, using SR of total SOC and of different SOC fractions may be useful to reveal how soil management affects soil quality and helpful to understand the mechanism of SOC transformation and cycling in subsoil as well as in deep soil. In LHR of China, soil erosion and desertification are causing a loss of net primary productivity that was estimated as high as 12 kg C ha^{-1}y^{-1} [16]. To counteract soil erosion and other environmental problems, an environmental protection policy was implemented by Chinese central government, which was known as the Grain to Green Program (GTGP). The purpose of GTGP was to convert up to 26.87 million ha low-yield sloped croplands (>25°) into forests, shrubs or grasslands by the end of 2008 [17]. It is the first and the most ambitious "payment–for–ecosystem–services" program in China to date [18]. Although the initial goal of GTGP was to control soil erosion in China, it also plays a significant role in circulation of SOC and total nitrogen (TN). In recent years, a few studies estimated the effects of GTGP on vegetation structure, economic benefits, soil physiochemical properties, and niche characteristics [19–21]. However, SR values of SOC and/or TN and CMI value among different land use types are rarely reported. Especially, information on dynamics of C in deep soil is largely ignored in this region.

This study aimed to: 1) analyze the contents of SOC, TN, POC and LOC and their vertical distributions at the depths of 0–200 cm; 2) assess the stocks of SOC and TN at different soil depths of three land use types; and 3) evaluate the soil quality of different land use types using SR and CMI values as the main assessment parameters.

Materials and Methods

All sites in the watershed we were selected for study was determined through interviews with local farmers (Mr. Yibin Zhang, Soil and Water Conservation Experiment Station, Northwest A&F University, Ansai County, Shaanxi, NW China).We state clearly that no specific permissions were required for the location. We confirm that the location is not privately-owned or protected in any way. We confirm that the field studies do not involve endangered or protected species.

Research area

The study was conducted in the Zhifanggou catchment (36°46′42″–36°46′28″N, 109°13′46″–109°16′03″E), which is located in Ansai county, central LHR (see Fig. 1). Ansai is a typical county characterized by semi-arid climate and hilly loess landscape in the Loess Plateau with an annual average temperature of 8.8°C, and an average annual precipitation of 505 mm. 60% of precipitation occurs between July and September (~300 mm in dry years while >700 mm in wet years). Accumulated temperatures above 0°C and 10°C are 3733°C and 3283°C, respectively. On average, there are about 157 frost-free days and 2415 h sunny time each year. Arable farming mostly occurs on

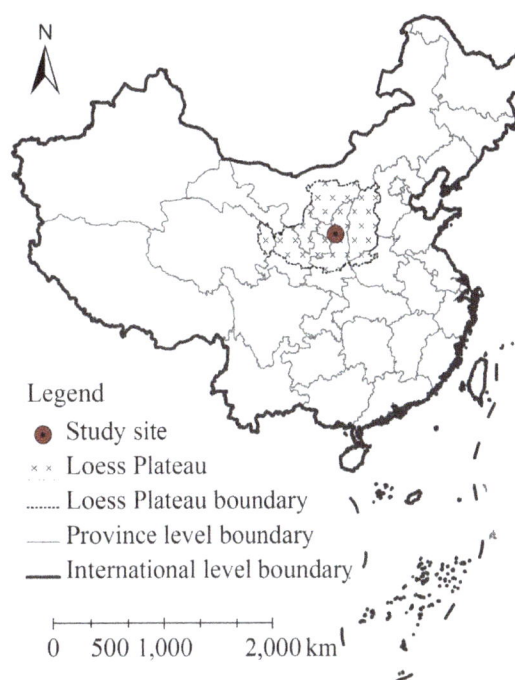

Figure 1. Location of the Loess Plateau and the study site.

sloping lands without irrigation. The loess parent material at the site has an average thickness of approximately 50–80 m and the soil in this region is classified as Calciustepts soil [22]. Sand (2–0.05 mm) and silt (0.05–0.002 mm) account for approximately 29.22% and 63.56% in soil depth of 0–20 cm, respectively. The soil is highly erodible, with an erosion modulus of 10,000–12,000 Mg·km^{-2}·yr^{-1} before the start of restoration efforts [23]. After 30 years vegetation restoration, the area of forest lands significantly increased from 5% to 40% [24].

The Zhifanggou catchment has been an experimental site of the Institute of Soil and Water Conservation, Chinese Academy of Science (CAS) since 1973 [25]. The major agricultural land use type in LHR is slope cropland. Agricultural management in this region, including the major crop types grown, has not been changed significantly since the 1970s. The main crops grown in these sites were millet (*Setaria italica*) and soybean (*Glycine max*) rotation, and no irrigation was provided in grown season (depend on rainfall). One crop was grown each year, and fertilizer was applied (mainly manure). After more than 30 years of comprehensive management, the ecological environment of the catchment has been significantly improved [26]. Since late 1970s, slope cropland is replanted with shrubs and woods, mainly *Robinia pseudoacacia L.* (RP) and *Caragana Korshinskii Kom (CK)*, to control soil erosion (see Table 1). Abandoned cropland was also generated during this period due to its extremely low productivity and long

Table 1. Characteristics of different vegetation types.

Vegetation types	Age	Canopy closure (%)	Litter accumulation (t.ha^{-2})	Undergrowth Vegetation[a]	Species diversity indices
Robinia pseudoacacia L.	30	58	20.5	*Lespedeza dahurica - Stipa bungeana*	6.5
Caragana Korshinskii Kom	30	50	13.3	*Achillea capillaries, Stipa bungeana*	3.9

[a]means the main vegetation in forest/shrub land.

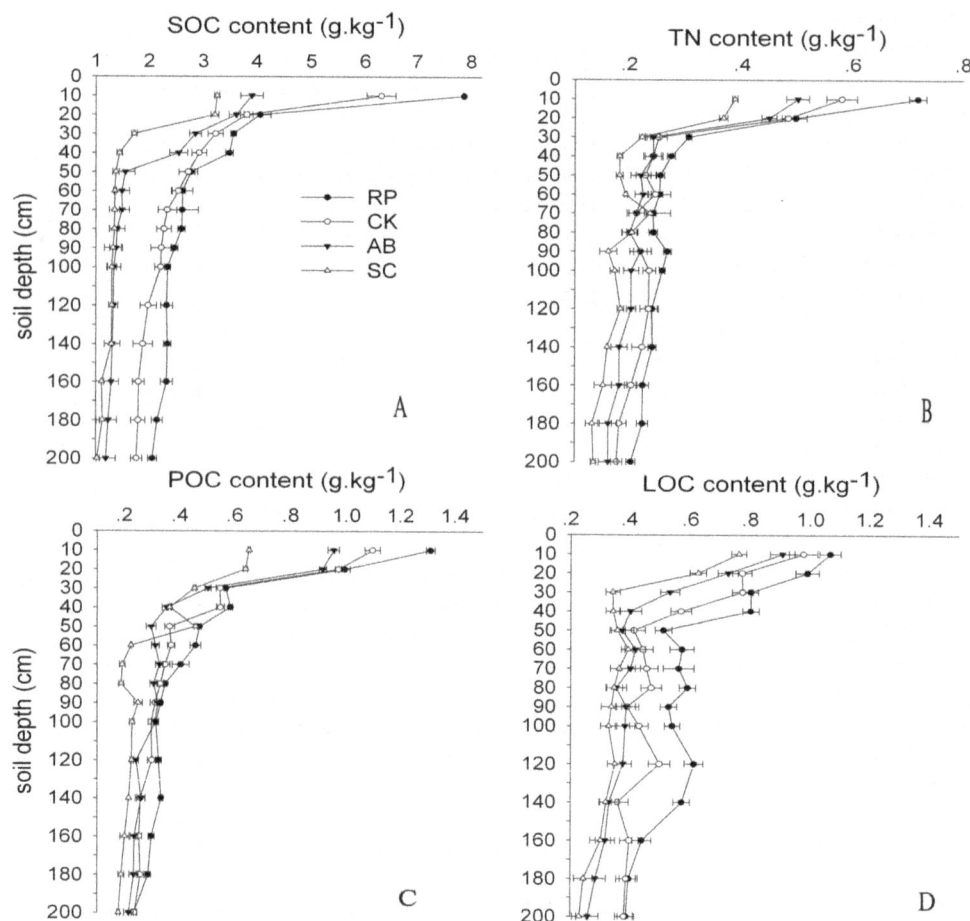

Figure 2. Distribution of soil organic carbon (SOC, A), total nitrogen (TN, B), particulate organic carbon (POC, C), and labile organic carbon (LOC, D) contents of different land used types in soil depth of 0–200 cm. The error bars are the standard errors.

distance from farmers' residences [27–28]. Despite wild grasslands and shrub lands were usually found on steep slopes, these sites were used for firewood collection as well. So the wild vegetation was of limited coverage or even barren for long periods. In 1999, most slope lands were closed for vegetation restoration under the GTGP [29].

Soil sampling

In September 2012, based on land use history, 30 year old *Robinia psendoacacia* (RP), *Caragana Korshinskii Kom* (CK), abandoned land (AB) and slope cropland (SC) in the Zhifanggou catchment were selected. Three 30 m×20 m plots were established for each land use type. All sites were located on the same physiographical units with same slope aspects, same elevation of 1250 m and a spatial distance of 1200 m.

Soil samples were taken at several soil depths using a soil auger (diameter 5 cm) from 10 points within "S" shape at each plot (0–10 cm, 10–20 cm, 20–30 cm, 30–40 cm, 40–50 cm, 50–60 cm, 60–70 cm, 70–80 cm, 80–90 cm, 90–100 cm, 100–120 cm, 120–140 cm, 140–160 cm, 160–180 cm, and 180–200 cm). Then after removing the litter layer, ten soil samples at each depth of each plot were mixed to make one sample. Samples were collected at least 80 cm away from the trees. All samples were sieved through a 2 mm screen, and roots and other debris were removed. Soil samples were air-dried and stored at room temperature for the

determination of soil chemical properties. A ring tube was used to determine the bulk density in each soil depth.

Laboratory analysis

SOC content (g.kg^{-1}) and TN content (g.kg^{-1}) were determined using $K_2Cr_2O_7$ oxidation method and Kjeldhal method, respectively [30].

To determine POC content, 25 g soil was dispersed with 100 mL of 5 g L^{-1} sodium hexametaphosphate before being. Then, the mixed soil solution was shaken for 1 h at high speed on an end-to-end shaker and screened by a 0.053 mm sieve with several deionized water rinses. The soil remained on the sieve was backwashed into a pre-weighed aluminum box and dried at 60°C for 24 h, then it was grounded for analysis of C [31].

Soil labile organic carbon (LOC) was measured following the method described in Graeme et al. [32]. A 2–6 g air dried soil sample was put into a 50 mL centrifuge tube, and 25 mL of 333 mmolL^{-1} KMnO$_4$ solution was added before being shaken with a rate of 120 rpm for 1 h, and centrifuged for 5 min with a rate of 5,000×g. The upper clear solution was transferred, and diluted by 250 times, and then the absorbance at 565 nm wavelength was determined. The absorbances at values 565 nm with different KMnO$_4$ concentrations were also determined for preparation of standard curve, which was used for the determination of the KMnO$_4$ concentrations. Difference between the

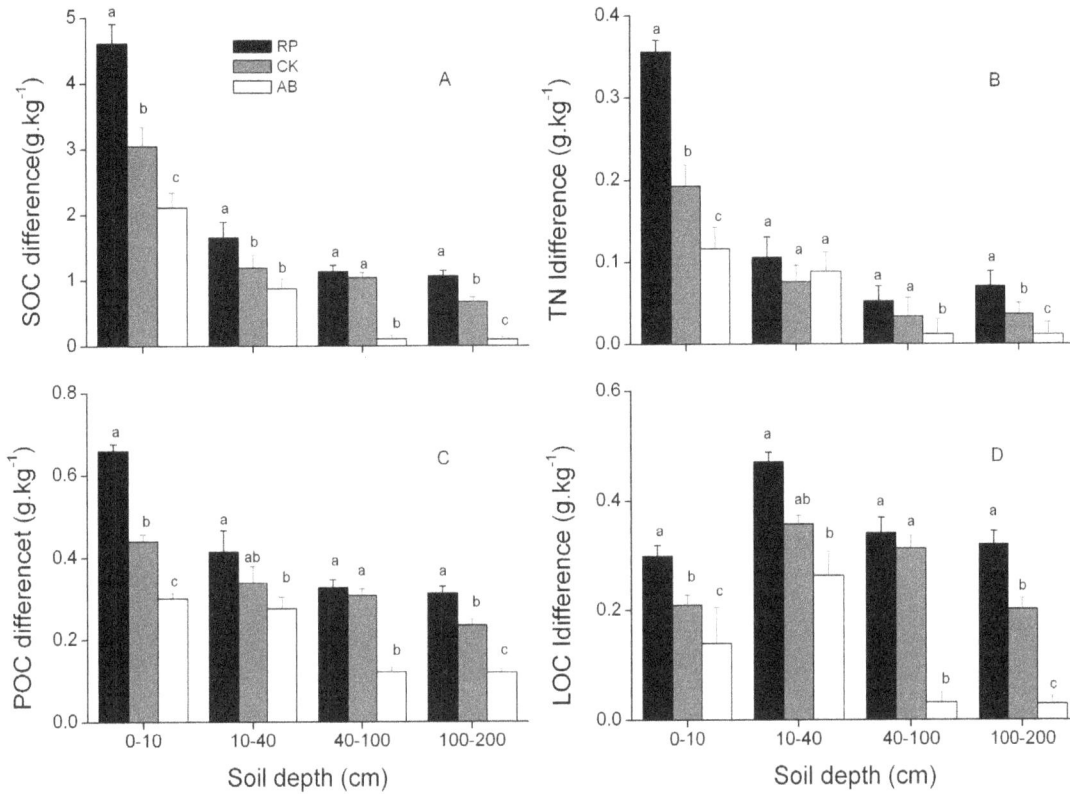

Figure 3. Differences in soil organic carbon (SOC, A), total nitrogen (TN, B), particulate organic carbon (POC, C), labile organic carbon (LOC, D) contents between SC and RP, CK or AB (RP/CK/AB - SC). Error bars are the standard errors. Different lowercase letters indicate significant difference among different land use types within same soil layer (P<0.05). The same for Fig 4

amounts of $KMnO_4$ added and remained was used to calculate labile C concentration in the soil sample.

Calculation of SOC (TN) stocks, SR of SOC (TN, POC, and LOC) and CMI

SOC density (SOCD, TND) represents the total SOC (TN) storage of overall certain sampling depth. SOCD(TND) of different sampling depths were calculated:

$$SOCD(TND) = C_{SOC,TN} \times \rho \times H \times (1 - \delta/100) \times 10^{-1} \quad (1)$$

where SOCD (TND) is the density ($Mg \cdot ha^{-1}$) of SOC (TN) and $C_{SOC,TN}$ is the content ($g \cdot kg^{-1}$) of SOC (TN). ρ is the bulk density ($g \cdot cm^{-3}$), H is the soil horizon thickness (cm), and δ is the fraction (%) of gravels >2 mm in size in soil. Because the soil gravel size of loess in China is mostly below 2 mm, this fraction was assumed to be 0 [33].

SR values (0–10 cm: 10–40 cm, 0–10 cm: 40–100 cm and 0–10 cm: 100–200 cm) were calculated from the contents of SOC, TN, POC and LOC following the method in Franzluebbers (2002).

CMI values were calculated using following procedures:

Firstly, a C pool index (CPI) was calculated:

$$CPI = \frac{sample\ total\ organic\ C\ (g/kg)}{reference\ sample\ total\ C\ (g/kg)} \quad (2)$$

where reference sample is SC soil. Then, a lability index (LI) was calculated:

$$LI = \frac{L\ in\ each\ sampled\ soil}{L\ in\ the\ reference\ soil} \quad (3)$$

where reference soil is SC soil, and L was calculated from the C lability:

$$L = \frac{content\ of\ labile\ C}{content\ of\ non-labile\ C} \quad (4)$$

At last, CMI was calculated:

$$CMI = CPI \times LI \times 100 \quad (5)$$

Statistical analyses

All statistical analyses were carried out with SPSS 17.0. Analysis of variance (ANOVA) and Duncan's Multiple Range Test (DMRT) at 5% level of significance were used to compare the difference in contents and/or stocks of SOC, TN, POC, LOC, CMI, and SR among different land use types or soil depths. A sample linear-regression analysis was used to estimate the relationships between carbon stocks with CMI or SR values.

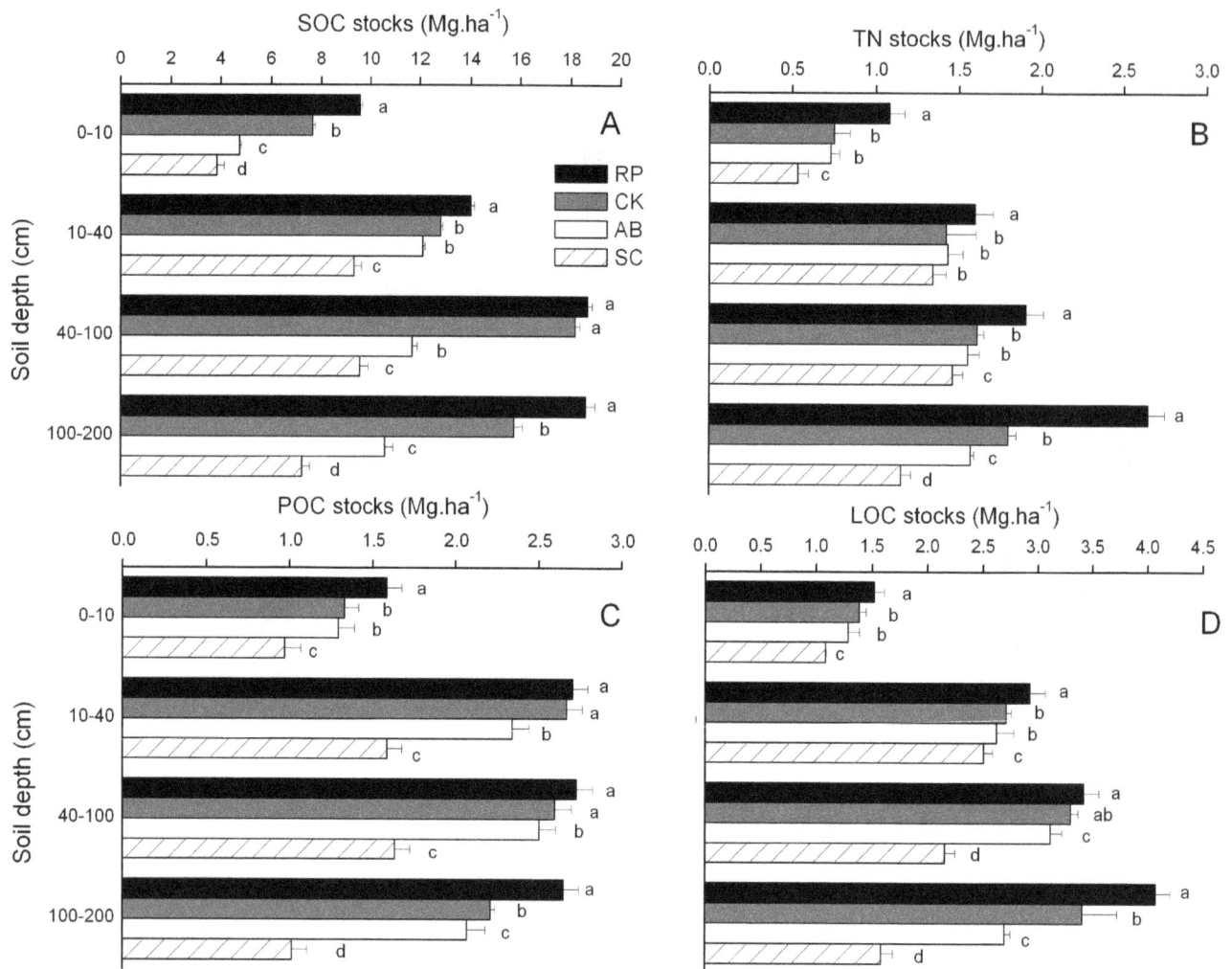

Figure 4. Stocks of soil organic carbon (SOC, A), total nitrogen (TN, B), particulate organic carbon (POC, C), labile organic carbon (LOC, D) of different land use types. The error bars are the standard errors.

Results

Changes in contents of SOC, TN, POC and LOC

The contents of SOC, TN, POC and LOC responded differently as the change of soil depth (Fig. 2). In all land use types, contents of SOC, TN, POC and LOC in top soil (0–10 cm) were 3.26–7.86 g.kg^{-1}, 0.39–0.72 g.kg^{-1}, 0.65–1.31 g.kg^{-1} and 0.76–1.07 g.kg^{-1}, respectively, which were significantly higher than other soil layers (P<0.05). The contents of SOC, TN, POC and LOC decreased significantly in soil depth of 10–40 cm while the decreases trended to be flatter in subsoil (40–100 cm). Additionally, the differences in contents of SOC, TN, POC and LOC in deep subsoil (100–200 cm) were negligible (P<0.05).

The differences in contents of SOC, TN, POC and LOC between three forest/shrub types (RP, CK and AB) and SC are shown in Fig. 3. The differences in SOC, TN, POC and LOC of RP and SC in soil depths of 0–10 cm and 100–200 cm were significantly higher than that between other land use types and SC (P<0.05). The differences in SOC and TN of RP were 33.78% and 45.97% larger than that of CK, and 54.13% and 67.28% larger than that of AB in soil depth of 0–10 cm (P<0.05), while the differences in POC and LOC were 32.8%, 54.0% higher than that of CK, and 23.3% and 45.0% higher than that of AB (P<0.05).

Moreover, the differences in SOC, TN, POC and LOC of RP were 25.05–85.29% higher than that of CK, and 61.78–90.70% higher than that of AB in soil depth of 100–200 cm. Additionally, significant differences in SOC, TN, POC and LOC contents were observed between RP and CK in soil depths of 10–40 cm and 40–100 cm, but there was no difference between CK and AB (P<0.05).

Changes and distribution of SOC, TN, POC and LOC stocks

SOC, TN, POC and LOC stocks of RP, CK and AB were higher than SC in all soil profiles (Fig 4). The SOC, TN, POC and LOC stocks of RP were significantly increased (P<0.05), which were 0.43–5.8 Mg.ha^{-1}, 0.25–4.70 Mg.ha^{-1}, 0.44–9.14 Mg.ha^{-1} and 1.49–11.38 Mg.ha^{-1} higher than that of SC in soil layers of 0–10, 10–40, 40–100 and 100–200 cm, respectively. Moreover, the stocks of SOC, TN, POC and LOC in soil layer of 100–200 cm of RP were higher than that of CK and AB by 15.4–32.1% and 21.8–43.1%, respectively (P<0.05).

The SOC, TN, POC and LOC stocks responded differently as the change of soil depth (Fig. 5). Although the distribution of SOC, TN, POC and LOC stocks in soil depths of 0–10 cm and 10–

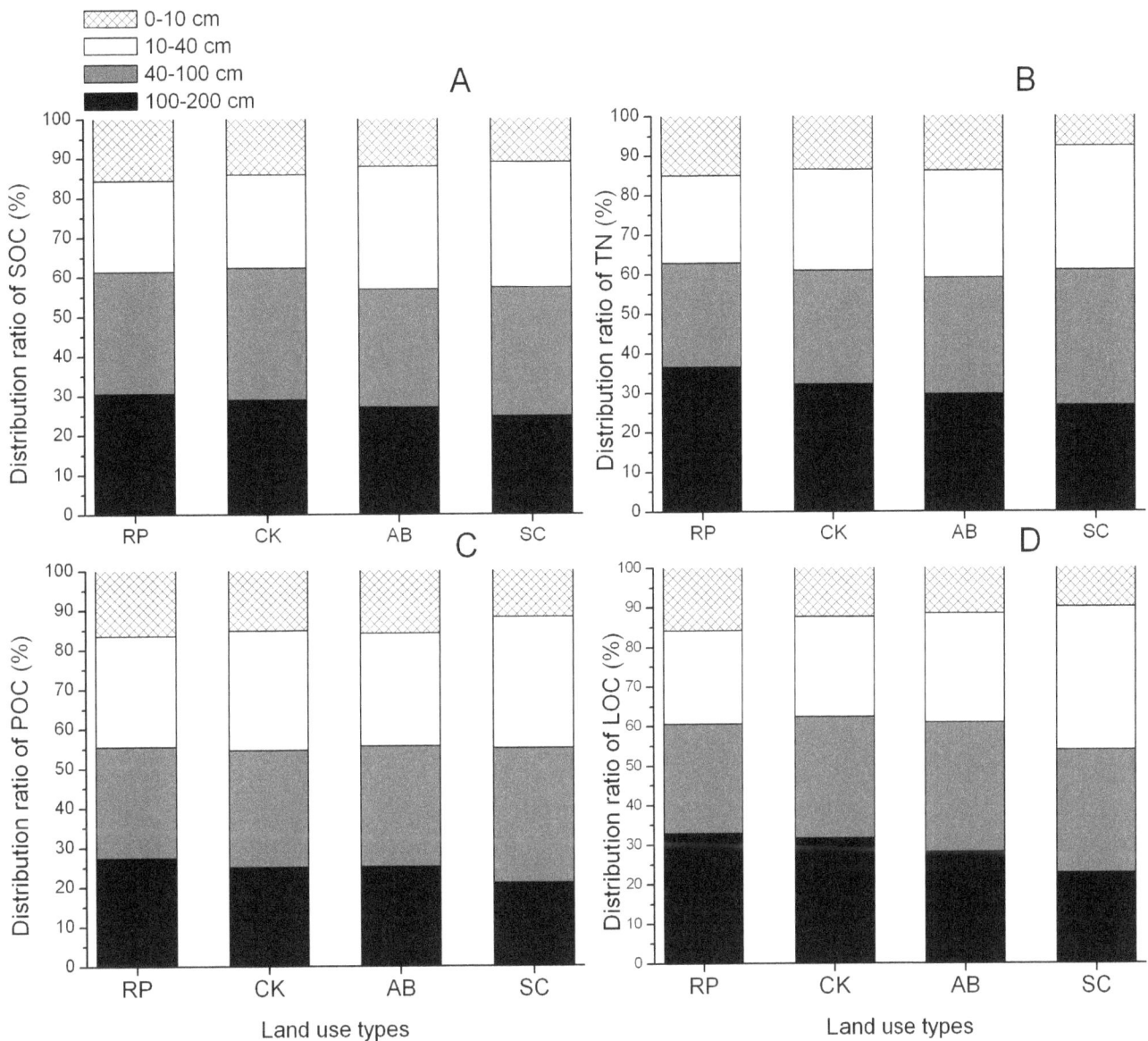

Figure 5. Distribution ratios of soil organic carbon (SOC, A), total nitrogen (TN, B), particulate organic carbon (POC, C), labile organic carbon (LOC, D) in soil depth of 0–200 cm under different land use types.

40 cm accounted for the majority, 26.36–34.06% and 21.08–36.62% were distributed in soil layers of 40–100 cm and 100–200 cm, respectively. Among four land use types, the highest proportion of SOC, TN, POC and LOC stocks were found in RP, while the lowest were in soil depths of SC in 0–10 cm and 100–200 cm of SC. The proportion of SOC, TN, POC and LOC stocks under RP were higher than SC by 4.68%, 7.32%, 4.65% and 5.96% respectively in soil depth of 0–10 cm soil depth, whereas by 5.90%, 9.78%, 6.30% and 10.06% was higher in soil depth of 100–200 cm soil depth respectively.

Change in SR and CMI values

Responses of SR in different land use types to change of soil depth were different (Fig 6). The SR values of SOC, TN and LOC differed significantly among different soil depths (P<0.05), while the SR values of LOC differed only between 0–10:10–40 cm, 0–10:40–100 cm and 0–10:100–200 cm. Among four land use types, the SR values of SOC, TN, POC and LOC of RP were the

highest, but that of SC were the lowest in each soil depth(P<0.05). The SR values of SOC, TN, POC and LOC were in a decreasing order of CK>AB>SC. The SR values differed significantly between CK or AB with SC (P<0.05), while there was no significant difference between CK and AB. Additionally, the ratios of SR values of SOC, TN, POC and LOC in the surface layer (0–10 cm) to that in layer of 10–40 cm were >2.0 in most cases.

The CMI values were significantly affected by land use types. In our study, the CMI values were in a decreasing order of RP>CK>AB>SC in four soil profiles and CMI values were significantly enhanced by RP compared with SC (Fig 7). Averaged CMI values of RP, CK, and AB were 40.60%, 50.54%, 37.81%, and 14.1% higher than that of SC in soil layers of 0–10 cm, 10–40 cm, 40–100 cm and 100–200 cm.

Regression equations to assess CMI/SR values of TN, POC, and LOC (Y) were showed in Table 2. There was a significant

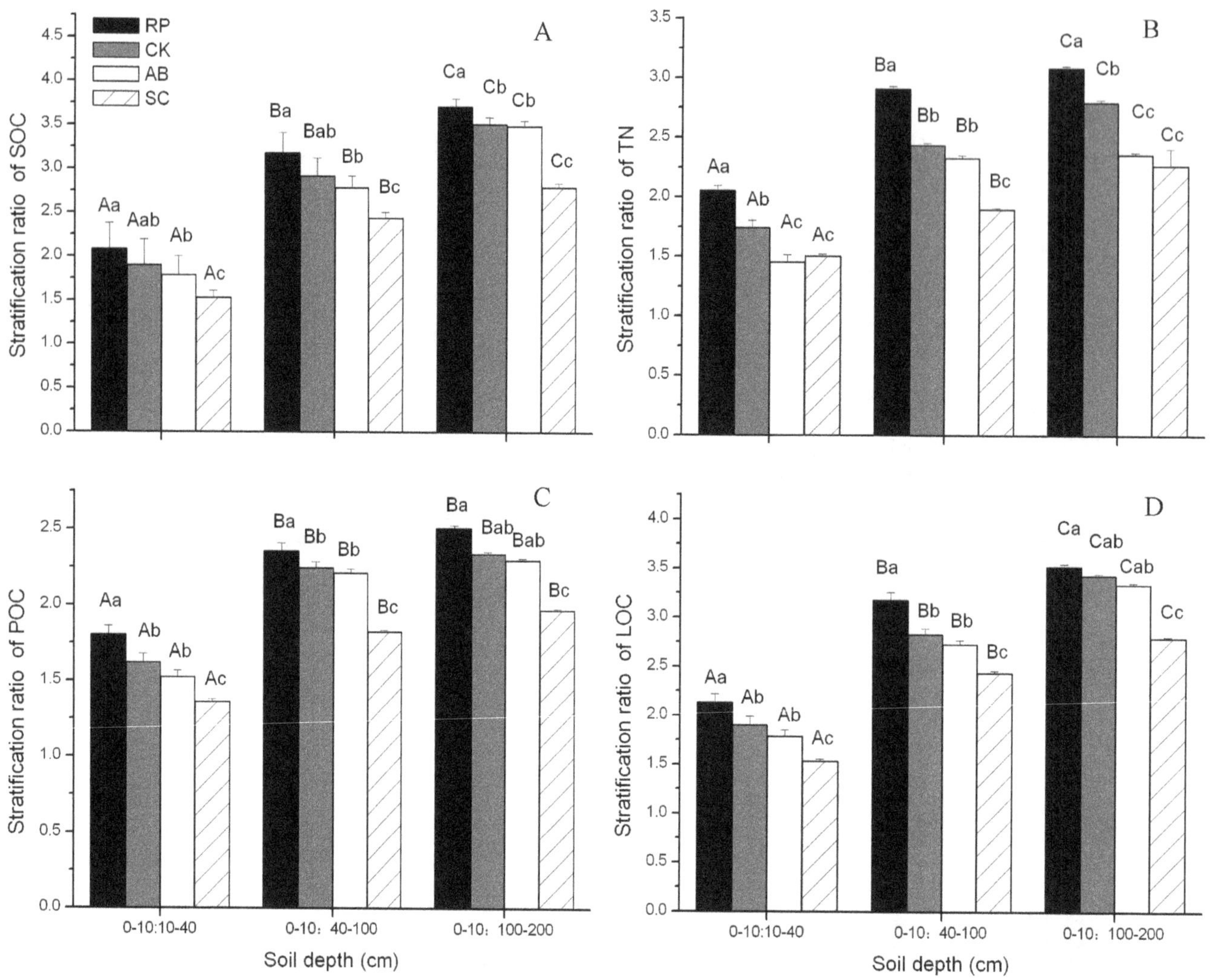

Figure 6. Comparison of stratification ratio of soil organic carbon (SOC, A), total nitrogen (TN, B), particulate organic carbon (POC, C), labile organic carbon (LOC, D) under different land use types. Different uppercase letters indicate significant difference among different soil depths within same land use type while the different lowercase letters indicate significant difference among different land use types within same soil depth. The error bars are the standard errors.

positive correlation between CMI/SR values of TN, POC, and LOC with SOC stocks in surface soil and deep soil.

Discussion

SOC, TN, POC and LOC contents and SOC and TN stocks

Vegetation can greatly influence soil quality, C and N cycling, and regional socioeconomic development [34–35]. It is also reported that converting cropland into land with perennial vegetation would increase the SOC content [36]. Our results showed that land use type and soil depth significantly affected the contents of SOC and TN (Fig. 2). The conclusion that both land use type and soil depth are important factors influencing the soil carbon and nitrogen distribution was consistent with previous studies [35,37]. We also observed that the lowest SOC, TN, POC and LOC contents were found in slope cropland (Fig 4), which essentially agree with a previous study [38], indicating that the conversion of slope cropland to vegetation improves the C and N contents. A possible reason is that the lower residue input into the soil in slope cropland leads to lower SOC and TN contents.

Additionally, our results showed that SOC, TN, POC and LOC contents of RP were greater than that of CK and AB (Fig 4). It infers that the effects of RP on soil C and N play a significant role in land use and ecosystem management. The conclusion was consistent with Qiu et al [39], who reported that RP has potential to improve SOC content in the loessial gully region of the Loess Plateau and the improvements are greater in long-term than middle-term.

Recently it was reported that the depth of sampling is an important factor for the measurement of change in SOC stocks [40], and land use could influence subsoil C pools [41]. We found that SOC, TN, POC and LOC stocks of RP, CK, and AB were higher than SC for different soil profiles, especially in depths of 40–100 cm and 100–200 cm (Fig. 4 and 5). It is demonstrated that converting slope cropland into woodland and shrubland not only affects SOC and TN stocks in surface soil, but also largely influences that in deep soil. The result was consistent with Wang et al [42], who reported that deep layer (50–200 cm) SOC stocks were equivalent to approximately 25% of that in the shallow layer (0–50) in Hilly Loess Plateau. This is mainly due to the fact that

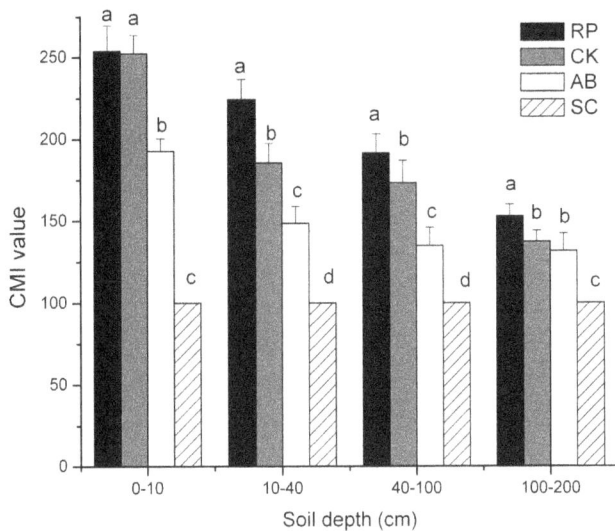

Figure 7. Carbon management index (CMI) values of different land use types at different soil depths. The error bars are the standard errors. Different lowercase letters indicate significant difference among different land use types within same soil depth.

SOC input into subsoil is largely affected by plant roots and root exudates, dissolved organic matter and bioturbation. In addition, most important factors leading to protection of SOC in subsoil include the spatial separation of SOM, microorganisms and extracellular enzyme activity related to the heterogeneity of C input [43]. As a result, stabilized SOC in subsoil is horizontally stratified.

Stratification ratios of SOC, TN, POC, and LOC

According to Franzluebbers [7], SOC SR values >2 in degraded conditions is uncommon, and the SR values of SOC are generally low and seldom reach 2.0. And SR values of soil organic C and N pools with value of >2 would be an indicator that soil quality might be improved [7]. In our study, the most of SR in SOC, TN, POC and LOC was more than 2 after convert slope cropland to forest or shrub land (Fig. 6). This means soil quality was improved in these afforested soils without disturbance. Greater C stratification ratios could be related to the fact that, during soil recovery by re-vegetation or land abandonment, soil was undisturbed thus reducing oxidation and favoring soil C [44]. The result was consistent with Sá et al [45]. Similar results were also reported by Moreno et al [46] and Franzluebbers [7], who reported that stratification of SOC occurs over time when soil tillage and disturbance is stopped and it is usually greater in undisturbed soils than in disturbed soils. In addition, the stratification may increase with time, and SOC, TN, POC and LOC contents are still aggrading but have not reached soil C saturation yet. That is the reason why SR values of SOC, TN, POC and LOC under CK and AB were higher but no significant differences were observed compared with SC (Fig. 6). Sá et al [45] concluded that the SOC pool stabilization may be attained in about 40 years after long-term no-tillage adoption.

Carbon management index

CMI value was calculated to obtain indications of the C dynamics of the system and provide an integrated measure for quantity and quality of SOC [15]. Soils with higher CMI values are considered as better managed [47]. We found that CMI values were significantly enhanced by RP forest compared with CK, AB and SC in both surface soil and subsoil and deep soil (Fig. 7). Soil management under RP plot was more appropriate to improve the SOC status than other land use types. Similar result was reported by Qiu et al [39], who illustrated that RP forest has significantly increased SOC, total nitrogen, ratio of carbon to nitrogen and ratio of carbon to phosphorus compared to other vegetation types. Our result showed that there were significant positive correlations between SOC stocks and CMI/SR in both surface soil and deep soil (Table 2). These findings showed that SR values of SOC, TN,

Table 2. Regression equations among SOC stocks and CMI/SR for different soil layers.

Axis		Soil depth (cm)	Equations	R^2	Significant level
X, CMI/SR[a]	Y, SOC stocks[b]				
CMI	SOC	0–10	Y = 53.5+23.20X	0.91	P = 0.048
		10–40	Y = −146+25.81X	0.97	P = 0.027
		40–100	Y = 24.2+8.60X	0.93	P = 0.023
		100–200	Y = 77.5+4.05X	0.93	P = 0.046
TN	SOC	0–10:10–40	Y = 1.12+0.08X	0.93	P = 0.047
		0–10:40–100	Y = 1.26+0.07X	0.96	P = 0.032
		0–10:100–200	Y = 1.67+0.07X	0.98	P = 0.017
POC	SOC	0–10:10–40	Y = 1.18+0.06X	0.97	P = 0.023
		0–10:40–100	Y = 1.55+0.04X	0.93	P = 0.041
		0–10:100–200	Y = 1.74+0.04X	0.92	P = 0.047
LOC	SOC	0–10:10–40	Y = 1.30+0.08X	0.96	P = 0.031
		0–10:40–100	Y = 1.95+0.05X	0.97	P = 0.025
		0–10:100–200	Y = 2.53+0.05X	0.98	P = 0.013

[a]CMI = carbon management index, SR = stratification ration, TN = SR of total nitrogen, POC = SR of particulate organic carbon, LOC = SR of labile organic carbon.
[b]For the Y-axis, the SOC stocks (0–10 cm, 40–100 cm, and 100–200 cm) were used to analyze correlations between SOC stocks and SR values of TN, POC, and LOC (0–10:10–40, 0–10:40–100, 0–10:100–200).

POM, and LOC, and CMI are suitable indicators for evaluating soil quality and C changes induced by GTGP in surface soil and deep soil.

Conclusion

In this study, the SOC, TN, POC and LOC contents of RP, CK and AB in soil layer of 100–200 cm were higher than SC, especially for RP plot. Although the SOC, TN, POC and LOC stocks in soil layer of 100–200 cm were lower, there was more than 27.38–36.62%, 25.10–32.91%, 21.59–31.69% and 21.08–26.83% of SOC, TN, POC and LOC stocks were distributed in 100–200 cm soil depth under RP, CK and AB. Meanwhile, the SR of SOC, TN, POC and LOC in the surface to lower depth ratio (i.e., 0–10:10–40 cm) was >2.0 in most of case. And SR and as well CMI values were significantly enhanced by RP compared with SC in deep soil (100–200 cm) ($P<0.05$). Indicating that soil quality

was improved after converting slope land into perennial vegetation, especially under RP plot from surface soil to deep soil. Moreover, there were significant and positive correlations between SOC stocks and CMI or SR of TN, POC, LOC both surface soil and deep soil indicated that the SR and CMI value are suitable indicators for evaluating soil quality and C changes in surface soil as well as in deep soil. We, therefore, propose deep soil organic carbon should be considered in C cycle induced by Grain-to-Green Program (GTGP) and under RP forest is more appropriate strategy to improve the SOC status than other land use types in surface soils and deep soil.

Author Contributions

Conceived and designed the experiments: GY XH. Analyzed the data: FZ. Contributed reagents/materials/analysis tools: YF GR. Wrote the paper: FZ.

References

1. Zhang CS, McGrath D (2004) Geostatistical and GIS analyses on soil organic carbon concentrations in grassland of southeastern Ireland from two different periods. Geoderma 119: 261–275.
2. Jobbágy EG, Jackson RB (2000) The vertical distribution of soil organic carbon and its relation to climate and vegetation. Ecological Applications 10: 423–436.
3. Cole CV, Duxbury J, Freney J, Heinemeyer O, Minami K, et al. (1997) Global estimates of potential mitigation of greenhouse gas emissions by agriculture. Nutrient Cycling in Agroecosystems 49: 221–228.
4. Amundson R (2001) The soil carbon budget in soils. Annual Reviews of Earth and Planetary Sciences 29: 535–562.
5. Guo L, Gifford RM (2002) Soil carbon stock s and land use change: a meta analysis. Global Change Biology 8: 345–360.
6. IPCC (2007) Climate change 2007: the physical Science basis. In: Solomon, S., Qin, D., Manning, M., Chen, Z., et al. (Eds.), Contribution of Working Group I to the Fourth Assessment Report of the Intergovernmental Panel on Climate Change. Cambridge University Press, Cambridge.
7. Franzluebbers AJ (2002) Soil organic matter stratification ratio as an indicator of soil quality. Soil Tillage Research 66: 95–106.
8. Prescott CE, Weetman GF, DeMontigny LE, Preston CM, Keenan RJ (1995) Carbon chemistry and nutrient supply in cedar–hemlock and hemlock –amabilis fir forest floors. In: McFee, W.W., Kelley, J.M. (Eds.), Carbon Forms and Functions in Forest Soils. Soil Sci. Soc. Am., Madison, WI, pp. 377–396.
9. Van Lear DH, Kapeluck PR, Parker MM (1995) Distribution of carbon in a Piedmont soil as affected by loblolly pine management. In: McFee, W.W., Kelley, J.M. (Eds.), Carbon Forms and Functions in Forest Soils. Soil Sci. Soc. Am., Madison, WI, pp. 489–501.
10. Schnabel RR, Franzluebbers AJ, Stout WL, Sanderson MA, Stuedemann,JA. (2001) The effects of pasture management practices. In: Follett, R.F., Kimble, J.M., Lal, R. (Eds.), The Potential of US Grazing Lands to Sequester Carbon and Mitigate the Greenhouse Effect. Lewis Publishers, Boca Raton, FL, pp. 291–322.
11. Sa JCM, Lal R (2009) Stratification ratio of soil organic matter pools as an indicator of carbon sequestration in a tillage chronosequence on a Brazilian Oxisol. Soil & Tillage Research 103: 46–56.
12. Haynes RJ (2005) Labile organic matter fractions as central components of the quality of agricultural soils: an overview. Adv. Agron 85: 221–268.
13. Franzluebbers AJ, Arshad MA (1992) Particulate organic carbon content and potential mineralisation as affected by tillage and texture. Soil Science Society of America Journal 61: 1382–1386.
14. Sparling GP (1997) Soil microbial biomass activity and nutrient cycling: an indicator of soil health. In: Pankhurst, C.E., Doube, B.M., Gupta, V.V.S.R. (Eds.), Biological Indicators of Soil Health. CAB International, Wallingford, UK, pp. 97–119.
15. Blair GJ, Lefroy RDB, Lisle L (1995) Soil carbon fractions based on their degree of oxidation and the development of a carbon management index for agricultural systems. Australian Journal of Agricultural Research 46: 1459–1466.
16. Bai ZG, Dent D (2009) Recent land degradation and improvement in China. Ambio 38: 150–156.
17. Jia ZB (2009) Investigation Report on Forestry major problem in 2008. Forestry Press in China. 267–273 (In Chinese)
18. LüY H, Fu BJ, Feng XM, Zeng Y, Liu Y, et al. (2012) A Policy-Driven Large Scale Ecological Restoration: Quantifying Ecosystem Services Changes in the Loess Plateau of China. PLoS ONE 7, e31782. doi:10.1371/journal.pone.0031782.
19. Zhao YT (2010) Analysis on the Necessity and Feasibility of Implementing the Project for Conversion of Cropland to Forest. Ecological Economy 7: 81–83.
20. Wei J, Cheng J, Li W, Liu W (2012) Comparing the Effect of Naturally Restored Forest and Grassland on Carbon Sequestration and Its Vertical Distribution in the Chinese Loess Plateau. PLoS ONE 7(7): e40123. doi:10.1371/journal.pone.0040123

21. Wei XR, Qiu LP, Shao MA, Zhang XC, Gale WJ (2012) The Accumulation of Organic Carbon in Mineral Soils by Afforestation of Abandoned Farmland. PLoS ONE 7(3): e32054. doi:10.1371/journal.pone.0032054
22. Gong ZT, Lei WJ, Chen ZC, Gao YX, Zeng SG, et al. (1999) Chinese Soil Taxonomy. Science Press, Beijing 36–38.
23. Liu G (1999) Soil conservation and sustainable agriculture on the Loess Plateau: challenges and prospective. Ambio 28: 663–668.
24. Xue S, Liu GB, Pan YP, Dai QH, Zhang C, et al. (2009) Evolution of Soil Labile Organic Matter and Carbon Management Index in the Artificial Robinia of Loess Hilly Area. Scientia Agricultura Sinica 4: 1458–1464
25. Jiao JY, Zhang ZG, Bai WJ, Jia YF, Wang N (2012) Assessing the Ecological Success of Restoration by Afforestation on the Chinese Loess Plateau. Restoration Ecology 20: 240–249.
26. Zhang F, Zhang SL, Cheng ZJ, Zhao HY (2007) Time structure and dynamics of the insect communities in bush vegetation restoration areas of Zhifanggou watershed in Loess hilly region. Acta Ecologica Sinica 27: 4555–4562. (in Chinese with English abstract)
27. Chen QB, Wang KQ, Qi S, Sun LD (2003) Soil and water erosion in its relation to slope field productivity in hilly gully areas of the Loess Plateau. Aata Ecologica Sinica 23: 1463–1469.
28. Li FM, Song QH, Jjemba PK, Shi YC (2004) Dynamics of soil microbial biomass C and soil fertility in cropland mulched with plastic film in a semiarid agro-ecosystem. Soil Biology and Biochemistry 36: 1893–1902.
29. Wang Z, Liu GB, Xu MX, Zhang J, Wang Y, et al. (2012) Temporal and spatial variations in soil organic carbon sequestration following revegetation in the hilly Loess Plateau, China. Catena 99: 26–33.
30. Bao SD (2000) Soil and Agricultural Chemistry Analysis. China Agriculture Press, Beijing, China (in Chinese).
31. Cambardella CA, Elliot ET (1992) Particulate soil organic matter changes a grassland cultivation sequence. Soil Science Society of America Journal 56: 777–783.
32. Graeme JB, Rod DBL, Leanne L (1995) Soil carbon fractions based on their degree of oxidation, and the development of a carbon management index for agricultural systems. Australian Journal of Agricultural Research 46:1459–1466
33. Wang YF, Fu BJ, Lu YH, Song CJ, Luan Y (2010) Local-scale spatial variability of soil organic carbon and its stock in the hilly area of the Loess Plateau, China. Qua-ternary Research 73: 70–76.
34. Eaton JM, McGoff NM, Byme KA, Leahy P, Kiely G (2008) Land cover change and soil organic carbon stocks in the Republic of Ireland 1851–2000. Climate Change 91: 317–334.
35. Fu XL, Shao MA, Wei XR, Robertm H (2010) Soil organic carbon and total nitrogen as affected by vegetation types in Northern Loess Plateau of China, Geoderma 155: 31–35.
36. Groenendijk FM, Condron LM, Rijkse WC (2002) Effect of afforestation on organic carbon, nitrogen, and sulfur concentration in New Zealand hill country soils. Geoderma 108: 91–100.
37. Davis M, Nordmeyer A, Henley D, Watt M (2007) Ecosystem carbon accretion 10 years after afforestation of depleted subhumid grassland planted with three densities of Pinus nigra. Global Change Biology 13: 1414–1422
38. Chen LD, Gong J, Fu BJ, Huang ZL, Huang YL, et al. (2007) Effect of land use conversion on soil organic carbon sequestration in the loess hilly area, loess plateau of China. Ecology. Research 22: 641–648.
39. Qiu LP, Zhang XC, Cheng JM, Yin XQ (2010) Effects of black locust (Robinia pseudoacacia) on soil properties in the loessial gully region of the Loess Plateau, China. Plant Soil 332: 207–217.
40. VandenBygaart AJ, Bremer E, McConkey BG, Ellert BH, Janzen HH, et al. (2010) Impact of Sampling Depth on Differences in Soil Carbon Stocks in Long-Term Agroecosystem Experiments. Soil Science Society of America Journal 75: 226–234

41. Strahm BD, Harpison RB, TeRPy TA, Harpington TB, Adams AB, et al. (2009) Changes in dissolved organic matter with depth suggest the potential for postharvest organic matter retention to increase subsurface soil carbon pools. Forest Ecology Management 258: 2347–2352

42. Wang Z, Liu GB, Xu MM (2010) Effect of revegetation on soil organic carbon concentration in deep soil layers in the hilly Loess Plateau of China. Acta Ecologica Sinica 14: 3947–3952 (in Chinese with English abstract)

43. Rumpel C, Kögel-Knabner I (2011) Deep soil organic matter-a key but poorly understood component of terrestrial C cycle. Plant Soil 338: 143–158

44. Fayez R (2012) Soil properties and C dynamics in abandoned and cultivated farmlands in a semi-arid ecosystem. Plant Soil 351: 161–175.

45. Sá JCM, Cerpi CC, Dick WA, Lal R, Vesnke-Filho SP, et al. (2001) Organic matter dynamics and carbon sequestration rates for a tillage chronosequence in a Brazilian Oxisol. Soil Science Society of America Journal 5: 1486–1499.

46. Moreno F, Murillo JM, Pelegrín F, Girón IF (2006) Long-term impact of conservation tillage on stratification ratio of soil organic carbon and loss of total and active CaCO3. Soil Tillage Research 85:86–93

47. Diekow J, Mielniczuk J, Knicker H, Bayer C, Dick DP, et al. (2005) Carbon and nitrogen stocks in physical fractions of a subtropical Acrisol as influenced by long-term no-till cropping systems and N fertilization. Plant Soil 268: 319–328.

12

Sediment Delivery Ratio of Single Flood Events and the Influencing Factors in a Headwater Basin of the Chinese Loess Plateau

Mingguo Zheng[1], Yishan Liao[2], Jijun He[3]*

1 Key Laboratory of Water Cycle and Related Land Surface Processes, Institute of Geographic Sciences & Natural Resources Research, Chinese Academy of Sciences, Beijing, China, **2** Guangdong Institute of Eco-environment and Soil Sciences, Guangzhou, China, **3** Base of the State Laboratory of Urban Environmental Processes and Digital Modelling, Capital Normal University, Beijing, China

Abstract

Little is known about the sediment delivery of single flood events although it has been well known that the sediment delivery ratio at the inter-annual time scale is close to 1 in the Chinese Loess Plateau. This study examined the sediment delivery of single flood events and the influencing factors in a headwater basin of the Loess Plateau, where hyperconcentrated flows are dominant. Data observed from plot to subwatershed over the period from 1959 to 1969 were presented. Sediment delivery ratio of a single event (SDR_e) was calculated as the ratio of sediment output from the subwatershed to sediment input into the channel. It was found that SDR_e varies greatly for small events (runoff depth < 5 mm or rainfall depth <30 mm) and remains fairly constant (approximately between 1.1 and 1.3) for large events (runoff depth >5 mm or rainfall depth >30 mm). We examined 11 factors of rainfall (rainfall amount, rainfall intensity, rainfall kinetic energy, rainfall erosivity and rainfall duration), flood (area-specific sediment yield, runoff depth, peak flow discharge, peak sediment concentration and flood duration) and antecedent land surface (antecedent precipitation) in relation to SDR_e. Only the peak sediment concentration significantly correlates with SDR_e. Contrary to popular belief, channel scour tends to occur in cases of higher peak sediment concentrations. Because small events also have chances to attain a high sediment concentration, many small events (rainfall depth <20 mm) are characterized by channel scour with an SDR_e larger than 1. Such observations can be related to hyperconcentrated flows, which behave quite differently from normal stream flows. Our finding that large events have a nearly constant SDR_e is useful for sediment yield predictions in the Loess Plateau and other regions where hyperconcentrated flows are well developed.

Editor: Vanesa Magar, Centro de Investigacion Cientifica y Educacion Superior de Ensenada, Mexico

Funding: Financial support for this research was provided by Non-profit Industry Financial Program of MWR (201201083; www.cws.net.cn) and National Natural Science Foundation of China (41401302 and 41271306; www.nsfc.gov.cn). The funders had no role in study design, data collection and analysis, decision to publish, or preparation of the manuscript.

Competing Interests: The authors have declared that no competing interests exist.

* Email: hejiun_200018@163.com

Introduction

Sediment yield represents the total quantity of sediment observed at a certain point in a landscape or a river system, such as the watershed outlet, in a specified time interval. The sediment yield prediction is of key interest in stream and watershed management due to increasing concerns on water quality, aquatic habitat, biodiversity and life of man-made structures (dams, bridges, harbors, and water supply systems) [1,2]. The concept of sediment delivery ratio (SDR), commonly defined as the ratio of sediment yield to gross erosion [3,4], provides a convenient way to estimate sediment yield to a point of interest. In equation form SDR is expressed as

$$SDR = SY/SE, \qquad (1)$$

where SY (mass per unit time) represents sediment yield at a point of interest, and SE (mass per unit time) represents gross erosion rate of the area upstream of that point. It is believed that if the relationship between SDR and its influential factors is well established, equation (1) would be greatly helpful for estimating sediment yields in ungauged locations. Because of the simplicity in concept and the ability to link on-site erosion with downstream sediment yield, studies of SDR have received much attention [4–6]. SDR is affected by fluvial processes operating at a variety of spatial scales from slopes to channels. Factors affecting SDR almost includes all variables representing hydrological regime (e.g. flood and rainfall) and watershed prosperities (e.g. topography, vegetation and land use). Due to the multitude of the influencing factors and their interactions, it is difficult to identify the dominant controls on SDR [7–9]. As a result, the established relationships between SDR and the influencing factors are largely empirical and can hardly extrapolate beyond the data range with confidence [7–10].

The Chinese Loess Plateau is famous for its high-intensity soil erosion, which frequently exceeds 10, 000 t km^{-2} a^{-1}. Gong and Xiong [11] proposed that SDR is as high as 1 in the Loess Plateau. Mou and Meng [12] subsequently found that almost all sediments (>95%) are moved as wash load as a result of the fine texture of the loess in combination with the strong sediment transport capacity [13] of hyperconcentrated flows, which are well developed in the Loess Plateau [14] and behave quite differently from normal sediment-laden streamflow [15–17]; this mechanism physically enables a SDR as high as 1. Nowadays, it has been widely accepted that SDR in the Loess Plateau is close to 1 over a wide range of basin sizes at inter-annual time scale [5,7,14,18–20]. Nevertheless, knowledge of the sediment delivery and the influencing factors is currently lacking at the time scale of the flood event in the loess Plateau. Among more than 100 rainstorm events over a single year, only 2–7 rainstorm events are erosive in the Loess Plateau [21]. Research efforts are, therefore, needed to investigate sediment delivery processes at the event time scale.

One of great concerns in determining SDR is the enormous uncertainty introduced by estimating gross erosion [1,3], i.e. the denominator in Equation (1). To guarantee a reasonable estimation of the gross erosion and in turn, the SDR, we limited our study to the Tuanshangou subwatershed (Fig. 1), a headwater basin of the first-order channel in the Loess Plateau. The subwatersheds, where eroded sediments are primarily sourced, are the endmember unit for soil conservation practices in the Loess Plateau. The object of this study is to examine sediment delivery processes of single events in the Tuanshangou subwatershed, hoping to further the knowledge of fluvial processes under the control of hyperconcentrated flows. We firstly calculated the SDR for single events and then, examined a number of factors in relation to sediment delivery, including factors of rainfall, flood and antecedent land surface.

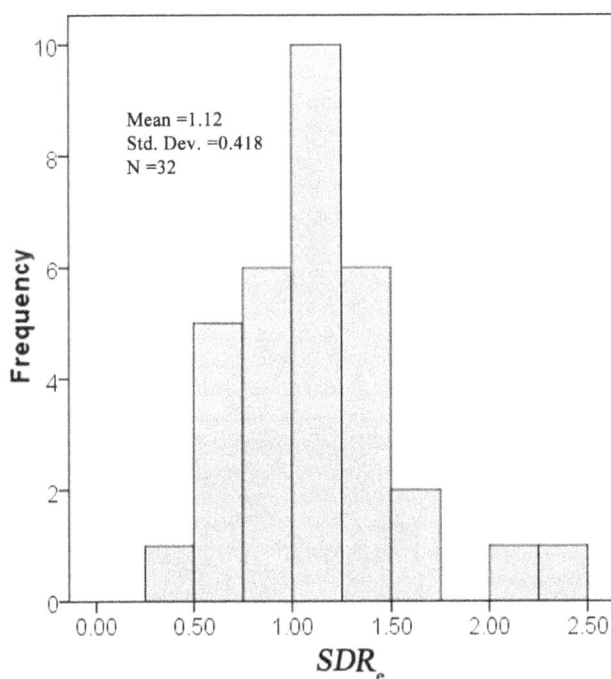

Figure 1. Histograms of *SDR*ₑ for 32 flood events observed at the Tuanshangou station.

Study Area and Data Source

The Tuanshangou Creek (latitude 37°41′N, longitude 109°58′E; See Fig. 1 in [22] for the location) drains an area of 0.18 km^2. Typical of the Loess Plateau, the local loess mantle is thicker than 100 m. As wind-borne dust in Quaternary times, loess is loosely compact and highly erodible. The climate is typically semi-arid. During the monitoring period (1959–1969), the mean average annual precipitation is approximately 450 mm and the maximum 30-min rainfall intensity is as high as 2.17 mm min^{-1}. The mean slope of the Tuanshangou subwatershed is as high as 26.8°. The valley side slope is particularly steep (>35°), allowing active mass wasting such as slumping, sliding and collapsing. Approximately 80% of the area was under arable without soil conservation practices. Other lands were abandoned due to the precipitous topography.

Observations at the subwatershed outlet (i.e. the Tuanshangou station) show that the annual sediment yield varied from 200 to 72 000 t km^{-2} a^{-1} with a mean of 19 700 t km^{-2} a^{-1} during the monitoring period. The instantaneous sediment concentrations of storm runoff frequently exceeded 1 000 kg m^{-3} and the mean sediment concentration of flood events was 742 kg m^{-3} (See Table 4 in [22]), much higher than the concentration threshold between the normal sediment-laden flow and the hyperconcentrated flow in the Loess Plateau (200 kg m^{-3} [23] or 300–400 kg m^{-3} [14]).

Data and Methods

Data

Unless stated otherwise, all data used were obtained from the Yellow River Water Conservancy Commission (YRWCC). The YRWCC stream-gauging crews conducted all measurements following national standard procedures of China [24], which have been described in [22] and [25].

This study primarily used data observed at three experimental sites: the Tuanshangou station and two runoff plots within the Tuanshangou subwatershed (i.e. Plots 7 and 9 in [22] and [25]). Both plots were under arable. Crops varied between years, including millet, potato, mung bean, clover, sorghum and wheat. The plots are composed of two parts: hill slope and valley side slope. Such slopes are conventionally termed as the entire slope in Chinese literatures. The plot lengths were 126 m and 164 m, respectively. Detailed information of the plots is available in Table 2 in [22]. Hyetograph data were obtained using a rainfall gauge near Plot 7.

Calculations of SDR

This study defined the gross erosion, i.e. the denominator in Equation (1), as sediment input into stream channels, as did in [18,26]. Such a definition in effect reflects the sediment transport efficiency of stream channels [27]. Plots 7 and 9 were large enough for gullies to develop, through which overland flows drain into the Tuanshangou Creek (See Fig. 2 in [25]). Plot 7 was in the lower part and Plot 9 was in the upper part of the Tuanshangou Creek. Hence, we can use the average of the collective discharges of sediment at Plots 7 and 9 to estimate the sediment input into the creek. We calculated SDR of a single flood event (*SDR*ₑ, dimensionless) as follows:

$$SDR_e = SSY/(0.5E_7 + 0.5E_9) \qquad (2)$$

where SSY (t km^{-2}) represents observed area-specific sediment yield at the Tuanshangou station, and E_7 and E_9 (t km^{-2})

represents erosion intensity at Plots 7 and 9 for a rainfall event, respectively. Comparisons between E_7 and E_9 on the same rainfall day produced a regression coefficient very close to 1 (0.94) and a R^2 of 0.93, suggesting that the erosion intensity may not vary greatly among entire slopes within the Tuanshangou subwatershed. We thus believe that the average of E_7 and E_9, i.e. the denominator of Equation (2), reasonably represents the sediment discharge into the Tuanshangou Creek per unit area.

The SDR_e values obtained from Equation (2) can be larger or smaller than 1. A SDR_e larger than 1 indicates channel degradation, while a SDR_e smaller than 1 indicates channel aggradation. When SDR_e is equal to 1, stream channels are in equilibrium.

Factors influencing sediment delivery

Factors influencing sediment delivery can be grouped into three categories: rainfall factors, flood factors and antecedent land surface factors. SDR is related to not only flow discharge but also the rheologic and fluid proprieties of flows, which largely depends on suspended load in flows. Flood factors we examined thus include five factors: SSY (t km^{-2}), h (runoff depth of a flood event, mm), q_{max} (peak flow discharge of a flood event, m^3 s^{-1}), C_{max} (maximum sediment concentration of a flood event, kg m^{-3}) and T_f (flood duration, min). All flood factors were measured at the Tuanshangou station.

Rainfall factors we examined also included 5 factors: P (rainfall depth, mm), I_{30} (the maximum 30-min rainfall intensity, mm min^{-1}), E (rainfall kinetic energy, J m^{-2}), EI_{30} (the product of E and I_{30}) and T_p (rainfall duration, min). EI_{30}, the rainfall erosivity index of the Universal Soil Loss Equation (USLE) [28], is the most common rainfall erosivity index. E was calculated using the following relationship:

$$E = \sum_{r=1}^{m} e_r p_r, \qquad (3)$$

where e_r is the rainfall kinetic energy per unit depth of rainfall per unit area (J m^{-2} mm^{-1}), and p_r is the depth of rainfall (mm) for the rth interval among m intervals of the storm hyetograph. e_r is calculated using an empirical equation for the Loess Plateau [29]:

$$e_r = 28.95 + 12.3 \lg i_r, \qquad (4)$$

where i_r (mm min^{-1}) represents the mean rainfall intensity for the rth interval. Equation (4), built on measurements of the drop size distribution of 195 storms, is almost identical to the rainfall intensity-energy equation of the USLE [28] after unit conversion.

Pre-event factors we examined only includes the antecedent precipitation index (P', mm), which is a surrogate for pre-event soil moisture and an important factor affecting runoff yield and soil erodibilities [30]. The vegetation cover rarely exceeded 25% in the Tuanshangou subwatershed. Hence, we do not take the vegetation cover into consideration. P' is defined as [31,32]:

$$P' = \sum_{i=1}^{n} k^i P_i, \qquad (5)$$

where n is the number of antecedent days, P_i (mm) is the daily precipitation for the ith day prior to the event, and k (dimensionless) is the decay constant representing the outflow of the regolith. In practices, k generally lie between 0.80 and 0.98 and n is typically 5, 7 or 14 days [31,33]. Here, k is set at 0.9

following [34]. The gauging crews made measurements of soil moisture near Plot 7. The correlation between P' and the observed soil moisture (top 30 cm) increases asymptotically with increasing n. Because the correlation varies little when n exceeds 11 days [35], n is taken as 11 days in this study. The resultant P' is well correlated with the observed soil moisture ($r = 0.85$, $p < 0.001$).

Results and Discussion

The calculation results of SDR

A total of 36 storm events were well monitored simultaneously at the three sites: the Tuanshangou station and Plots 7 and 9. We calculated SDR_e for all of the events. The bedrock is exposed at the channel bed of the Tuanshangou Creek. Because the bedrock is more prone to runoff production than loess slopes, runoff and sediment are primarily sourced from the channel bed in cases of small rainfall intensities. Among 11 small events with h smaller than 1 mm, four had a SDR_e (2.8, 5.7, 6.3 and 23.7 respectively) distinctly higher than others (the maximum is 1.2), implying that these events essentially conveyed pre-event sediment storage on the channel bed rather than soils eroded from uplands. The four events all had a runoff depth not greater than 0.1 mm at Plots 7 and 9, implying that the sediment discharge into the channel was small enough to be neglected. We removed the four events from the subsequent analyses and the subsequent calculation of SDR_e involves 32 events.

Histograms of SDR_e for the 32 events are presented in Fig. 1. SDR_e ranges from 0.46 to 2.39 with a mean of 1.12 and a median of 1.1. Twenty events have a SDR_e larger than 1. As shown in Fig. 2, SDR_e varies greatly for small events (approximately $h < 5$ mm or $P < 30$ mm) and remains fairly constant for large events ($h > 5$ mm or $P > 30$ mm). Only in cases of small events can significant channel degradation or aggradation occurs. For major sediment-producing events ($SSY > 5000$ t km^{-2}, $h > 5$ mm or $P > 30$ mm), SDR_e essentially lies between 1.1–1.3 (Fig. 2(a), (c) and (f)). Interesting to note is that many small events ($P < 20$ mm; Fig. 2(f)) have a SDR_e much larger than 1.

Flood factors in relation to sediment delivery

Among 11 factors we examined, only C_{max} are significantly correlated with SDR_e ($p = 0.004$; Fig. 2). Nevertheless, the considerable scatters, as shown in Fig. 2(b), prevent C_{max} from being a good predictor of SDR_e.

Contrary to popular belief, SDR_e increase with increasing C_{max} (Fig. 2(b)), a phenomenon also reported in [18]. When C_{max} is higher than 700 kg m^{-3}, most of the events (16 out of 19) have a SDR_e larger than 1, indicating channel scour. In contrast, most of the events (9 out of 13) correspond to a SDR_e smaller than 1 when C_{max} is smaller than 700 kg m^{-3}, indicating channel fill. This observation can be related with hyperconcentrated flows. Different from normal sediment-laden flows, the energy expenditure on suspended-load motion of hyperconcentrated flows decreases with increasing sediment load, as evidenced by laboratory experiments and field observations [14,17,36]. As a result, the sediment transport capacity of hyperconcentrated flows would increase with sediment concentrations and thus, the channel scour is more likely to occur at high rather than at small C_{max}, as has also been observed in the main stream of the Yellow River (See Fig. 3 in [14]).

Sediment delivery along a stream channel depends on not only sediment transport capacity of flows, but also sediment availability within channels. The time interval between large flood events, which occurs relatively infrequently, is generally longer than small ones. Numerous preceding runoff events can prepare a large

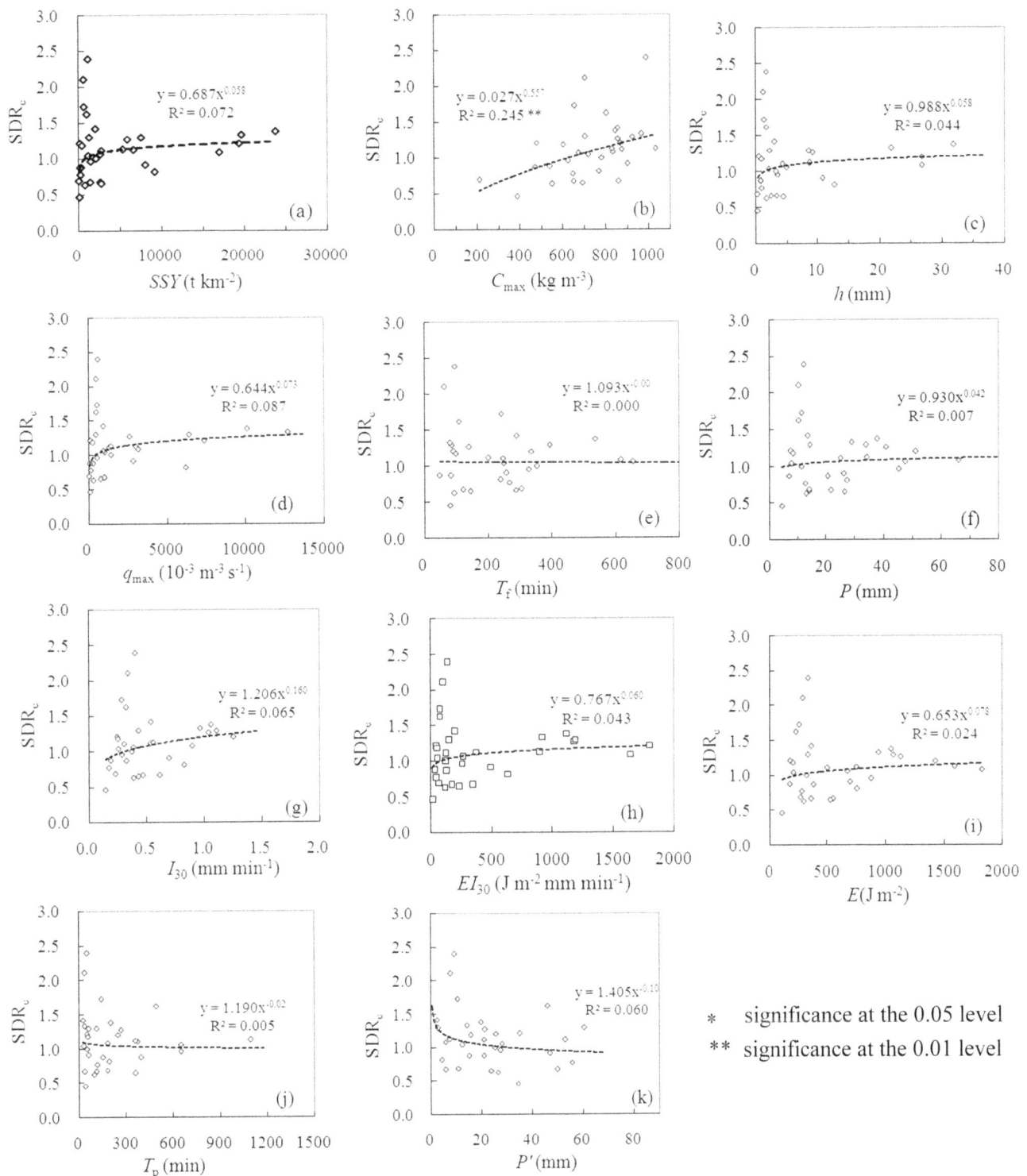

Figure 2. SDR_e **in relation to flood factors (a–e), rainfall factors (f–j) and the pre-event factor (k).**

amount of sediment storage within channels prior to a large flood event. Mass wasting also occurs more readily during large events. Hence, large flood events generally have a SDR_e larger than 1.

As indicated in Fig. 2(a), (c) and (d), SDR_e is larger than 1 not only for large events but also for many small events, as opposed to that observed in the Murray Darling Basin of Australia [7]. No direct relationship exists between sediment concentration and

water discharge for hyperconcentrated flows [13]. Small events also have chances to attain a high level of sediment concentration (Fig. 3) and thus, a high sediment transport capacity. In addition, antecedent sediment storage within channels may contribute a large part of the event sediment yield considering the minuscule sediment yield of a small event. In contrast, antecedent sediment storage within channels can hardly form the major sediment

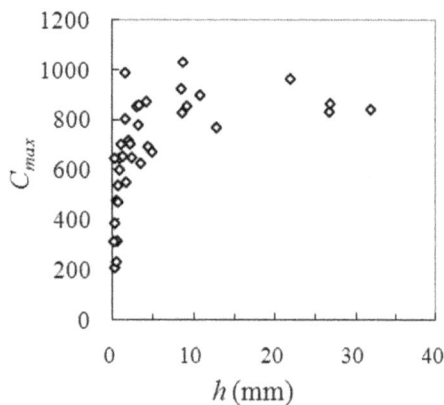

Figure 3. The relationship between C_{max} and h at the Tuanshangou station, showing that small runoff events also have chances to achieve a high level of sediment concentration.

source for large events due to their tremendous sediment yields. Only for small events, hence, can SDR_e be distinctively greater than 1. In contrast, SDR_e for large events falls within a narrow range between approximately 1.1 and 1.3.

Rainfall factors and antecedent precipitation in relation to sediment delivery

None of Rainfall properties show correlation with SDR_e (Fig. 2(f)–(j)). Raindrops splash soil particles and provide sediment for overland flows. Rainfall impact also increases the turbulence of sheet flow and thus, enhances its ability to detach soil and to transport sediment. In both ways, rainfall exerts direct effects on sediment delivery. However, rill erosion and gully erosion are strongly dominant over splash erosion in the Loess Plateau [37,38]. Meanwhile, raindrop impact has no effect on rill flows or other concentrated flows because the turbulent effect of raindrop impact is attenuated with increasing water depth [39–41]. Consequently, the both direct mechanisms that rainfall affects sediment delivery become ineffective. Though the large rainfall event corresponds to a high q_{max} resulting a high sediment transport capacity, the small event can achieve a high transport capacity by achieving a high sediment concentration. Consequently, the indirect mechanism also poses no impact on SDR_e.

For the same reason as above, it cannot be expected that a high antecedent soil moisture results in a high SDR_e by increasing q_{max}. Due to high vulnerability to water erosion at high antecedent soil moistures, sediment concentration and thus sediment transport capacity of hyperconcentrated flows may be enhanced. However, this enhancement should be obscured in a site where mass wasting events are active. Small-sized mass wasting events, such as bank failure and knickpoint retreat, even act as an important agent for rill development [42–44]. Indeed, there is no correlation between P' and C_{max} ($p = 0.37$). As a result, SDR_e are totally independent of P' (Fig. 2(k)). Similarly, due to intensive mass wastings in the Loess Plateau, vegetation and slope land measures for soil conservation, such as terraces and ridges, have no effect on sediment concentrations at the watershed outlet [45].

References

1. Rovira A, Batalla RJ (2006) Temporal distribution of suspended sediment transport in a Mediterranean basin: The Lower Tordera (NE SPAIN). Geomorphology 79(1): 58–71.

Conclusions

This study calculated the sediment delivery ratio of single flood events (SDR_e) and examined the factors of rainfall, flood and antecedent land surface in relation to SDR_e in a headwater basin of the Loess Plateau, where hyperconcentrated flows dominate the fluvial processes. SDR_e were calculated as the ratio of sediment output from the subwatershed to sediment input into the channel. Due to distinct behaviours of hyperconcentrated flows, the sediment delivery process of the examined subwatershed is quite different from that under the control of the normal stream flow:

1) SDR_e varies greatly for small events ($h < 5$ mm or $P < 30$ mm) and remains fairly constant for large events ($h > 5$ mm or $P > 30$ mm). Most of the examined events (20 out of 32) have a SDR_e higher than 1, implying channel degradation. Such high sediment transfer efficiency can be related to hyperconcentrated flows, which have very strong capacity to transport sediment.

2) Due to decreasing energy expenditure on suspended-load motion with increasing sediment load for hyperconcentrated flows, SDR_e show increasing trends with the increasing level of sediment concentration, as indexed by the maximum sediment concentration of a flood event (C_{max}). Channel degradation primarily occurs when C_{max} exceed 700 kg m^{-3}. Otherwise, channel aggradation occurs. Among 11 factors we examined, only C_{max} is correlated with SDR_e ($p < 0.01$).

3) Small events also have chances to attain a high C_{max}, thereby leading to a SDR_e higher than 1. Moreover, the extremely large SDR_e always corresponds to small events because pre-event sediment storage within channels can hardly form the major sediment source for large events. Because both large and small events are capable of achieve a high SDR_e, the peak flow discharge is poorly correlated with SDR_e.

4) Both rainfall factors (including rainfall amount, rainfall intensity, rainfall kinetic energy, rainfall erosivity and rainfall duration) and antecedent precipitation show no correlation with SDR_e. Due to poor correlations and considerable scatters, any factors we examined cannot be expected to be a good predictor of SDR_e. Nevertheless, our finding that large flood events ($h > 5$ mm or $P > 30$ mm) has similar values of SDR_e in a narrow range between approximately 1.1 and 1.3 should be a valuable aid to the sediment yield prediction in the Loess Plateau given the fact that large events contribute almost all sediments.

Acknowledgments

We appreciate the suggestions of the anonymous reviewer and the editor. Thanks the Data Sharing Infrastructure of Earth System Science-Data Sharing Infrastructure of Loess Plateau for providing our data.

Author Contributions

Conceived and designed the experiments: MGZ JJH. Performed the experiments: MGZ JJH YSL. Analyzed the data: MGZ JJH YSL. Contributed reagents/materials/analysis tools: MGZ YSL. Wrote the paper: MGZ JJH.

2. Shi ZH, Ai L, Fang NF, Zhu HD (2012) Modeling the impacts of integrated small watershed management on soil erosion and sediment delivery: a case study in the Three Gorges Area, China. J Hydrol 438: 156–167.

3. Lane LJ, Hernandez M, Nichols M (1997) Processes controlling sediment yield from watersheds as functions of spatial scale. Environ. Modell Softw 12(4): 355–369.

4. Alatorre LC, Beguería S, García-Ruiz JM (2010) Regional scale modeling of hillslope sediment delivery: A case study in the Barasona Reservoir watershed (Spain) using WATEM/SEDEM. J Hydrol 391: 109–123.

5. Walling DE (1983) The sediment delivery problem. J Hydrol 65(1): 209–237.

6. De Vente J, Poesen J, Arabkhedri M, Verstraeten G (2007) The sediment delivery problem revisited. Prog Phys Geog, 31(2): 155–178.

7. Lu H, Moran CJ, Prosser IP (2006) Modelling sediment delivery ratio over the Murray Darling Basin. Environ. Modell Softw 21(9): 1297–1308.

8. Shi ZH, Ai L, Li X, Huang XD, Wu GL, et al. (2013) Partial least-squares regression for linking land-cover patterns to soil erosion and sediment yield in watersheds. J Hydrol 498: 165–176.

9. Yan B, Fang NF, Zhang PC, Shi ZH (2013) Impacts of land use change on watershed streamflow and sediment yield: an assessment using hydrologic modelling and partial least squares regression. J Hydrol 484: 26–37.

10. Ferro V, Minacapilli M (1995) Sediment delivery processes at basin scale. Hydrol Sci J, 40:6, 703–717.

11. Gong SY, Xiong GS (1979) The origin and the regional distribution of sediment of the Yellow River. Yellow River (1): 7–11. (in Chinese).

12. Mou JZ, Meng QM (1982) Sediment delivery ratio as used in the computation of the watershed sediment yield. J Sediment Res (2): 223–230. (in Chinese).

13. Pierson TC (2005) Hyperconcentrated flow-transitional process between water flow and debris flow. In: Jakob M, Hungr O, editors. Debris-flow Hazards and Related Phenomena. Berlin Heidelberg: Springer. pp. 159–201.

14. Xu JX (1999) Erosion caused by hyperconcentrated flow on the Loess Plateau. Catena 36: 1–19.

15. Engelund F, Wan ZH (1984) Instability of hyperconcentrated flow. J Hydraul Eng 110(3): 219–233.

16. Pierson TC, Scott KM (1985) Downstream dilution of a lahar: Transition from debris flow to hyperconcentrated stream flow. Water Resour Res 21: 1511–1524.

17. Hessel R (2006) Consequences of hyperconcentrated flow for process-based soil erosion modelling on the Chinese Loess Plateau. Earth Surf. Processes Landforms 31: 1100–1114.

18. Cai QG, Wang GP, Chen YZ (1998) Processes of soil erosion and sediment yield and the related simulation for small catchments on the Loess Plateau. Beijing: Science Press. (in Chinese).

19. Jing K, Cheng YZ, Li FX (1993) Sediment and environment in the Huanghe River. Beijing: Science Press. (in Chinese).

20. Walling DE (1999) Linking land use, erosion and sediment yields in river basins. Hydrobiologia 410: 223–240.

21. Zhou PH, Wang ZL (1992) Study on Rainstorm causing erosion in the Loess Plateau. Journal of soil and water conservation 6(3): 1–5. (In Chinese).

22. Zheng MG, Yang JS, Qi DL, Sun LY, Cai QG (2012) Flow-sediment relationship as functions of spatial and temporal scales in hilly areas of the Chinese Loess Plateau. Catena 98: 29–40.

23. Wan ZH, Wang ZY (1994) Hyperconcentrated Flow, IAHR monograph series. Balkema: Rotterdam.

24. Ministry of Water Conservancy and Electric Power, PRC (1962) National Standards for Hydrological Survey of China. Beijing: Industry Press. (in Chinese).

25. Zheng MG, Qin F, Yang JS, Cai QG (2013) The spatio-temporal invariability of sediment concentration and the flow-sediment relationship for hilly areas of the Chinese Loess Plateau. Catena 109: 164–176.

26. Walling DE (1988) Erosion and sediment yield research-some recent perspectives. J Hydrol 100(1): 113–141.

27. Goudie A (2004) Encyclopedia of geomorphology (Vol. 2). Psychology Press. PP. 932–933.

28. Wischmeier WH, Smith DD (1978) Predicting rainfall erosion losses: a guide to conservation planning. USDA Handbook 537, Washington, DC.

29. Jiang ZS, Song WJ, Li XY (1983) Studies of the raindrop characteristics for Chinese loess area. Soil and Water Conservation in China (3): 32–36. (in Chinese).

30. Kinnell PIA (2010) Event soil loss, runoff and the Universal Soil Loss Equation family of models: A review. J Hydrol 385: 384–397.

31. Anctil F, Michel C, Perrin C, Andréassian V (2004) A soil moisture index as an auxiliary ANN input for stream flow forecasting. J Hydrol 286: 155–167.

32. Ma T, Li C, Lu Z,Wang B (2014) An effective antecedent precipitation model derived from the power-law relationship between landslide occurrence and rainfall level. Geomorphology 216: 187–192.

33. Heggen RJ (2001) Normalized antecedent precipitation index. J Hydrol Eng 6 (5): 377–381.

34. Li Q (1989) Variation of the decay constant of soil moisture and calculation of runoff yield in loess areas of China. Yellow River (3): 18–23. (in Chinese).

35. Cheng XA (2010) Study on soil erosion and erosion empirical model in hilly loess region on the Loess Plateau-as an example to Cheabagou. M.D. Dissertation. Wuhan: Huazhong Agricultural University, pp. 54–56. (in Chinese).

36. Chien N, Wan ZH (1999) Mechanics of sediment transport. Reston: ASCE Press.

37. Wang L, Shi ZH, Wang J, Fang NF, Wu GL, et al. (2014) Rainfall kinetic energy controlling erosion processes and sediment sorting on steep hillslopes. J Hydrol 512: 168–176.

38. Liu QJ, Shi ZH, Fang NF, Zhu HD, Ai L (2013) Modeling the daily suspended sediment concentration in a hyperconcentrated river on the Loess Plateau, China, using the Wavelet–ANN approach. Geomorphology 186: 181–190.

39. Foster GR, Lambaradi F, Moldenhauer WC (1982) Evaluation of rainfall-runoff erosivity factors for individual storms. Trans AM Soc Agric Eng 25:124–129.

40. Zhang KL (1999) Hydrodynamic characteristics of rill flow on loess slopes. J Sediment Res (1): 55–60. (in Chinese).

41. Schiettecatte W, Verbist K, Gabriels D (2008) Assessment of detachment and sediment transport capacity of runoff by field experiments on a silt loam soil. Earth Surf Processes Landforms 33, 1302 1314.

42. Chen YZ, Jing K, Cai QG (1988) Modern Erosion and Management in Loess Plateau. Beijing: Science Press. (in Chinese).

43. Han P, Ni JR, Wang XK (2003) Experimental study on gravitational erosion process. J Basic SCI Eng (1): 51–56. (in Chinese).

44. Wirtza S, Seegerb M, Riesa JB (2012) Field experiments for understanding and quantification of rill erosion processes. Catena 91: 21–34.

45. Zheng MG, Cai QG, Chen H (2007) Effect of vegetation on runoff-sediment yield relationship at different spatial scales in hilly areas of the Loess Plateau, North China. Acta Ecologica Sinica 27: 3572–3581.

13

Jack of All Trades, Master of All: A Positive Association between Habitat Niche Breadth and Foraging Performance in Pit-Building Antlion Larvae

Erez David Barkae[1]*, Inon Scharf[1,2], Zvika Abramsky[1], Ofer Ovadia[1]*

1 Department of Life Sciences, Ben-Gurion University of the Negev, Beer-Sheva, Israel, **2** Institute of Zoology, Johannes Gutenberg University of Mainz, Mainz, Germany

Abstract

Species utilizing a wide range of resources are intuitively expected to be less efficient in exploiting each resource type compared to species which have developed an optimal phenotype for utilizing only one or a few resources. We report here the results of an empirical study whose aim was to test for a negative association between habitat niche breadth and foraging performance. As a model system to address this question, we used two highly abundant species of pit-building antlions varying in their habitat niche breadth: the habitat generalist *Myrmeleon hyalinus*, which inhabits a variety of soil types but occurs mainly in sandy soils, and the habitat specialist *Cueta lineosa*, which is restricted to light soils such as loess. Both species were able to discriminate between the two soils, with each showing a distinct and higher preference to the soil type providing higher prey capture success and characterizing its primary habitat-of-origin. As expected, only small differences in the foraging performances of the habitat generalist were evident between the two soils, while the performance of the habitat specialist was markedly reduced in the alternative sandy soil. Remarkably, in both soil types, the habitat generalist constructed pits and responded to prey faster than the habitat specialist, at least under the temperature range of this study. Furthermore, prey capture success of the habitat generalist was higher than that of the habitat specialist irrespective of the soil type or prey ant species encountered, implying a positive association between habitat niche-breadth and foraging performance. Alternatively, *C. lineosa* specialization to light soils does not necessarily confer upon its superiority in utilizing such habitats. We thus suggest that habitat specialization in *C. lineosa* is either an evolutionary dead-end, or, more likely, that this species' superiority in light soils can only be evident when considering additional niche axes.

Editor: Zoltan Barta, University of Debrecen, Hungary

Funding: The authors have no support or funding to report.

Competing Interests: The authors have declared that no competing interests exist.

* E-mail: barkaeer@bgu.ac.il (EDB); oferovad@bgu.ac.il (OO)

Introduction

Habitat utilization spectrum is an important dimension of the ecological niche. A broadly accepted explanation for the variation in niche breadth among closely related species along such central niche axes, is the existence of a trade-off between the ability of a species to utilize a wide range of resources and its performance when exploiting only one or a few of them [1–6]. In other words, if adaptation to an additional habitat entails a fitness loss in the former, species having a narrow spectrum of habitat utilization (i.e., habitat specialists) should perform better than those utilizing a wider range of habitats (i.e., habitat generalists), but only within a narrower habitat spectrum. Habitat generalists, on the other hand, can inhabit more habitats, but they never achieve the performance of the habitat specialist on any one of them. Empirical studies, however, have not always been able to confirm this trade-off in performance associated with niche breadth (e.g. [7–10]), suggesting that this principle is less trivial and common than initially assumed.

Trap-building predators, such as web-building spiders or pit-building antlions, are opportunistic predators which depend heavily on their physical environments [11–13]. Pit-building antlion species can greatly differ in their habitat niche breadth and preferred habitats. Although antlions often prefer inhabiting shaded habitats,

they may also reside in open habitats exposed to direct sun [13–15]. In addition, antlions exhibit extensive variation in their preferences for soil/sand particle sizes ([16,17] see also [18] for a comparison between antlions and wormlions). Despite their preferences for different soil types, however, antlions will sometimes construct pits in less desirable habitats, but because such pits are usually smaller, they can cause reductions in prey capture rate [12,19].

We report here on the results of an empirical study whose aim was to test for a negative association between habitat niche breadth and foraging performance. As a model system we used two highly abundant species of pit-building antlions that vary in their habitat utilization spectrum: the habitat generalist *Myrmeleon hyalinus* inhabits a variety of soil types but occurs mainly in sandy soils, and the habitat specialist *Cueta lineosa* is restricted to light soils (finer textured soils) such as loess [14]. The two antlions are similarly sized and have comparable life cycles. We hypothesized that the habitat specialist, *C. lineosa*, would construct pits and respond to prey faster than the habitat generalist, *M. hyalinus*, in the loess soil, resulting in higher prey capture success. We also hypothesized that the habitat specialist's superiority in its preferred habitat of light soils would be significantly reduced in the sand compared to the habitat generalist, whose average performance should not vary between the soils.

Methods

Study species and habitats-of-origin

We collected *M. hyalinus* larvae under different tamarisk trees located in Nahal Secher (N 31°06′, E 34°49′), a sandy area 15 km south of the city of Beer-Sheva, Israel, and brought them to the laboratory. *M. hyalinus* is the most abundant pit-building antlion in Israel [14]. The larvae attain maximal lengths of about 10 mm and body masses of up to 0.06 g before pupating [20]. They inhabit a variety of soil types but occur mainly in sandy soils [14]. In addition, we collected *C. lineosa* larvae from the loessial plains near Beer-Sheva (N 31°16′, E 34°50′). Occurring mainly in the Israeli Negev desert, *C. lineosa* also exists in several small populations located in central and northern Israel, but is restricted to light soils, such as loess [14]. The two antlions are similarly sized and have comparable life cycles. Although they largely overlap in their geographical distribution, they rarely overlap in their microhabitat use. Specifically, *M. hyalinus* prefers shaded microhabitats [14,21], while *C. lineosa* is mainly found in open microhabitats exposed to direct sunlight [14]. Therefore, it is unlikely that interference competition exists between the two antlion species. However, it is possible that they indirectly compete for their arthropod prey (i.e., exploitation competition). All required permits and approvals for this work were obtained from Israel's Nature and National Parks Protection Authority, permit no. 2010/37830. In compliance with all the relevant laws and regulations prevailing in Israel, self-regulation and accountability of local programs by an Institutional Animal Care and Use Committee (IACUC) are not applicable for the use of invertebrates in research (Israel's Animal Welfare Act 1984).

Experimental design & statistical analysis

The study comprised of three complementary experiments: 1) Foraging behavior experiment, investigating the foraging behavior of both species in two different soil types, loess and sand, while using only one prey ant species. 2) To test if prey capture success is sensitive to prey species, we repeated the first experiment using three different prey ant species collected from the two field sites mentioned earlier. 3) Habitat selection experiment, testing whether the two species are capable of distinguishing between the two soils and choosing the soil type providing a higher prey capture success.

Prior to all experiments, antlions were fed with one flour beetle larva (mean larva mass ~1 mg), starved for 10 days in small plastic cups (diameter of 4.5 cm, filled with about 3 cm of sand or loess), and then weighed using an analytical scale (CP224S, Sartorius AG, Goettingen, Germany; accurate to 0.1 mg). Our previous experience with antlion larvae indicates that this procedure is useful for standardizing their hunger level and physiological state before they enter the experiment [22].

Foraging behavior experiment. Sixty individuals of each species were divided into two groups characterized by similar body size distributions (Kolmogorov-Smirnov two sample test, P = 0.20 and P = 0.52 for *C. lineosa* and *M. hyalinus*, respectively). Body size distribution also did not differ between species (Kolmogorov-Smirnov two sample test, P = 0.38). To avoid competition and potential cannibalism [23,24], we introduced single larvae into round plastic cups (diameter = 8.5 cm, depth = 6 cm) filled either with sand from Nahal Secher or with loess brought from the loessial plains near Beer-Sheva (i.e., soil type treatment, Fig. 1). All larvae were kept in the same room under an identical night/day photoperiod (12:12 h), temperature of 27.8°C and 70% r.h. Among the sand grains, 7.97% were larger than 0.25 mm, 78.65% were between 0.125–0.25 mm, 11.54% were between 0.062–

0.125 mm, and 1.84% were smaller than 0.062 mm [25]. The smaller loess particles comprised 2.8% grains larger than 0.2 mm, 32.4% grains between 0.05–0.2 mm, 48.8% grains between 0.05–0.002 mm, and 16% grains smaller than 0.002 mm [26].

Immediately after placing the larvae in the cups, we monitored the foraging behavior of the two species in both soils. Specifically, we documented the time to soil diving by measuring the time from placing the larvae on the soil surface to a complete disappearance of the larvae under the soil, time required to construct a pit measured as the time from initial movement of the larvae, until pit was completed and no further sand tossing was observed, and pit diameter and depth using a caliper (accuracy of 0.1 mm). Pit diameter was calculated as the average of two successive measurements of the diameter at the soil surface, while pit depth was measured from the soil surface to the bottom of the antlion pit (similarly to [22,27]) We also provided antlions with ants [*Messor aegyptiacus*; mean ant mass = 1.6 mg±0.2 mg (±1 S.D; N = 20)] and documented their response times to prey (similar to [28]) and their prey capture success. This ant species inhabits the loess plains of the Negev desert and the Arava valley, but does not occur in the sand dunes of the western Negev desert [29]. Each antlion received only one prey item, as the foraging behavior of antlions varies between fed and non-fed larvae [22] and between experienced and inexperienced larvae [28]. With the exception of prey capture success, all response variables (i.e., pit diameter, pit depth, time to soil diving, time to pit construction, and response time to prey) were analyzed using two-way ANCOVAs, with species and soil as the explanatory variables and body mass as the covariate. Our prey capture success analysis included only those individuals that responded to prey. Since the proportion of individual *C. lineosa* responding to prey was very low in the sand, we had to provide ants to a larger number of individuals (i.e., 90), to ensure that our analysis would be more balanced. Differences in prey capture success were tested using a logistic regression [30]. All

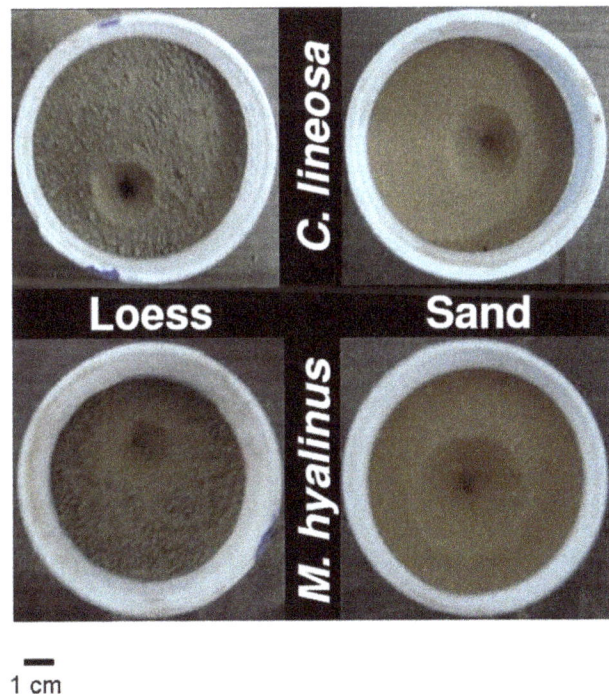

1 cm

Figure 1. Experimental design showing pit construction of both antlion species in the two soil types.

data were log transformed prior to the analysis. Finally, since individuals were randomly placed in plastic cups located in the same room (i.e., same conditions) and because observations on randomly selected individuals took place in the same day, there was no need to include block or time effects in the analysis.

Prey capture success experiment. There is a substantial variation in morphological and behavioral characteristics among prey ant species, such as thickness of cuticle [31], mandible properties [32], body size and running speed [33], behavioral defense mechanisms [34] and habitat use [35]. Such differences can be also reflected in the probability of being captured by antlion larvae. Thus, we carried out a second experiment whose aim was to test if the prey capture success of these antlions is sensitive to their prey ant species. Specifically, we collected 180 new larvae of each species from the field. Similarly to the foraging behavior experiment, larvae were individually stocked into plastic cups, identical to those described in the foraging behavior experiment, and were randomly assigned to one of the two soil type treatments (Fig. 1). In addition, we collected three different species of ants from the field: *M. aegyptiacus* [mean ant mass = 1.6 mg±0.2 mg (±1 S.D; N = 20)] mainly occurring in the loess plains of the Negev desert and the Arava valley, but absent from the sand dunes of the western Negev desert [29]; *Pheidole pallidula* [mean ant mass = 0.3 mg±0.1 mg (±1 S.D; N = 23)] and *Messor ebeninus* [mean ant mass = 4 mg±0.3 mg (±1 S.D; N = 22)]. These two latter ant species are distributed all over Israel while inhabiting both loess and sandy soil habitats [35]. Antlion larvae were divided into three groups, each provided with a different ant species as prey (i.e., prey species treatment). As in the foraging behavior experiment, differences in prey capture success were tested using a logistic regression [30].

Habitat selection experiment. To test if antlions are capable of distinguishing between the two soils, we collected 60 new larvae of each species from the field. We used 25×17 cm aluminum trays partitioned into two halves of equal sizes. Using cardboard as a barrier, we filled the trays with sand and loess at opposite halves, and then removed the cardboard. We placed a single antlion larva in the middle of the aluminum tray, and recorded the location of the antlion pit after 72 h (i.e., sand or loess), as a previous study indicated that a 2-day period is sufficient for pit construction [21], and that this pattern does not vary with time [36]. Trays were kept under identical conditions as in the previous experiments. We tested antlions habitat selection using a χ^2 test of independence. All statistical analyses were performed using SYSTAT v. 11 (SYSTAT Software, San Jose, CA, USA).

Results

Foraging behavior experiment

We could not detect significant differences in the proportion of pits constructed between species or soil types (*M. hyalinus* sand: 87%, *M. hyalinus* loess: 83%, *C. lineosa* loess: 87%, *C. lineosa* sand: 90%; $\chi^2 = 0.0001$, df = 1, P = 0.997).

There was an overall significant increase in pit diameter with body mass ($F_{1,96} = 52.682$, P<0.001), and this pattern was consistent between species (species×body mass interaction; $F_{1,96} = 1.783$, P = 0.209), but not among soil types (soil×body mass interaction; $F_{1,96} = 11.964$, P<0.001). This latter two-way interaction was caused by the faster increase in pit diameter, evident in both species, in the loess (Fig. 2A). The three-way soil×species×body mass interaction was not significant ($F_{1,96} = 1.677$, P = 0.198). In both species, pit diameter was larger in the sand than in the loess ($F_{1,96} = 7.991$, P = 0.006). Additionally, *M. hyalinus* pits were larger than those of *C. lineosa* irrespective

of soil type (saturated GLM: $F_{1,96} = 3.514$, P = 0.063; reduced GLM including all three main effects and the significant soil×body mass interaction: $F_{1,99} = 29.72$, P<0.001; Fig. 2A).

Pit depth increased significantly with body mass ($F_{1,96} = 20.59$, P<0.001) and differed between soil types (deeper in general in the loess; $F_{1,96} = 5.01$, P = 0.028; Fig. 2B). Notably, the three-way species×soil×body mass interaction was significant ($F_{1,96} = 3.86$, P = 0.052), indicating that the increase in pit depth with body mass was not consistent between species and soil types. *C. lineosa* pits in

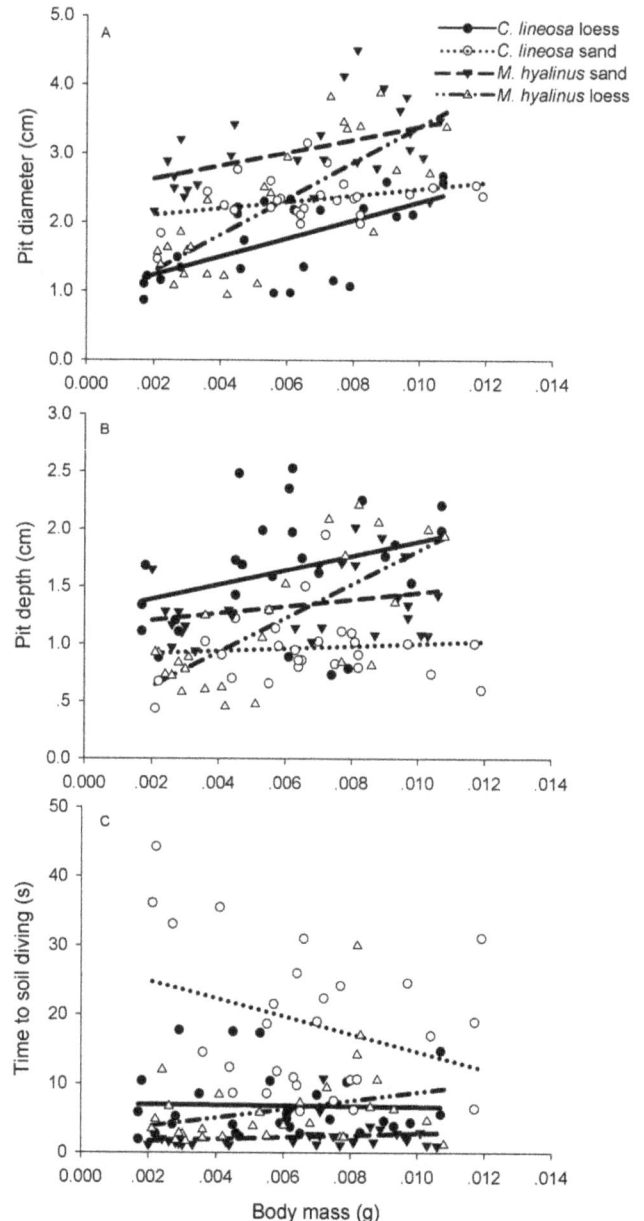

Figure 2. Pit diameter (A), pit depth (B), and time to soil diving (C) of both antlion species in the two soils as functions of their body masses. Both antlion species constructed pits with larger diameters in the sand than in the loess. *M. hyalinus* pits were larger than those of *C. lineosa*, irrespective of soil type ($R^2 = 0.625$). Loess pits of *C. lineosa* were deeper than those of *M. hyalinus*, but the sand pits of the latter were deeper than those of the former ($R^2 = 0.436$). *M. hyalinus* dives into the soil faster than *C. lineosa*, but this pattern is evident only in the sand ($R^2 = 0.612$).

loess were deeper than their pits in sand over the entire range of body masses examined. The depths of pits dug by *M. hyalinus* increased with body mass at a faster rate in the loess than in the sand (Fig. 2B). As a result, pits of larvae weighing <6.9 mg were deeper in the sand while those of larvae weighing >6.9 mg were deeper in the loess (Fig. 2B). *C. lineosa* dug significantly deeper loess pits than those of *M. hyalinus* ($F_{1,48} = 13.48$, P<0.001), but the sand pits of *M. hyalinus* were deeper than those of *C. lineosa* ($F_{1,50} = 26.49$, P<0.001; Fig. 2B).

The relationship between time to soil diving and body mass was not consistent between species and soil types (species×soil×body mass interaction: $F_{1,111} = 72.01$, P<0.001). Specifically, time to soil diving in *M. hyalinus* did not change significantly with body mass in either soil (r = 0.337, P = 0.068 and r = 0.227, P = 0.228 for sand and loess, respectively), but it was shorter in the sand than in the loess across the entire range of body masses examined (Fig. 2C). In contrast, *C. lineosa* larvae, again in a pattern that was consistent over the entire range of masses, dived faster in the loess than in the sand. Moreover, time to soil diving in this species decreased significantly with body mass in the sand but not in the loess (r = −0.375, P = 0.049 and r = 0.062, P = 0.749 for sand and loess, respectively; Fig. 2C). No significant differences in time to soil diving in the loess were evident between the two species ($F_{1,55} = 0.43$, P = 0.516); however, in the sand, *M. hyalinus* dived significantly faster than *C. lineosa* ($F_{1,55} = 198.58$, P<0.001; Fig. 2C).

Time to pit construction did not vary significantly with body mass ($F_{1,76} = 0.02$, P = 0.882) or between soil types ($F_{1,76} = 1.21$, P = 0.275). *M. hyalinus* constructed pits at a faster rate than *C. lineosa* ($F_{1,76} = 67.54$, P<0.001; Fig. 3), a pattern that was much stronger in sand (species×soil interaction, $F_{1,76} = 16.20$, P<0.001; Fig. 3) than in loess.

There was a significant negative correlation between response time to prey and body mass ($F_{1,125} = 4.3068$, P = 0.04) that was consistent between species (species×body mass interaction; $F_{1,125} = 0.0008$, P = 0.978). To control for the effect of body mass, we analyzed the residuals obtained by regressing response times against body masses. Between soil types, we could not detect significant differences in response time to prey ($F_{1,126} = 0.95$, P = 0.33). However, we found that the response of *M. hyalinus* to

prey was significantly faster than that of *C. lineosa* ($F_{1,126} = 18.79$, P<0.0001; Fig. 4), and this pattern was consistent between soil types (soil×species interaction; $F_{1,126} = 3.4$, P = 0.06; Fig. 4).

Using a logistic regression, we found that the effect of soil type on prey capture success was not consistent between the two antlions (species×soil type interaction; Table 1). Specifically, in *M. hyalinus* prey capture success was relatively high in both soil types (90% and 83% in the sand and loess, respectively; Fig. 5). In contrast, *C. lineosa* prey capture success was significantly lower in the sand than in the loess (23% and 70%, respectively; Fig. 5). Notably, the prey capture success of *M. hyalinus* was higher than that of *C. lineosa* in both soil types (Fig. 5).

Prey capture success experiment

Using a logistic regression, we found that prey capture success of the two antlions did not vary significantly among prey ant species (Table 2). Similarly to the results obtained in the foraging behavior experiment, we found that the effect of soil type on prey capture success was not consistent between the two antlions (species×soil type interaction; Table 2). Specifically, prey capture success of *M. hyalinus* did not vary significantly between soil types and was higher than that of *C. lineosa* irrespective of the prey ant species (Fig. 6). However, in *C. lineosa* prey capture success dropped by ~50% when switching from the loess to the sandy soil and this pattern was consistent among prey ant species (Fig. 6). Note that also when we examined only the loess data, prey capture success of *M. hyalinus* was significantly higher than that of *C. lineosa* (P = 0.011 after applying a Bonferroni correction for multiple testing).

Habitat selection choice experiment

Using a choice experiment we found that ~97% of *M. hyalinus* larvae preferred constructing pits in the sand ($\chi^2 = 52.27$, df = 1, P<0.0001). *C. lineosa* larvae, on the other hand, preferred to construct their pits in the loess (~80%; $\chi^2 = 21.60$, df = 1, P<0.0001; Fig. 7).

Discussion

We used two highly abundant species of pit-building antlions, varying in their habitat niche breadth, to test the classical

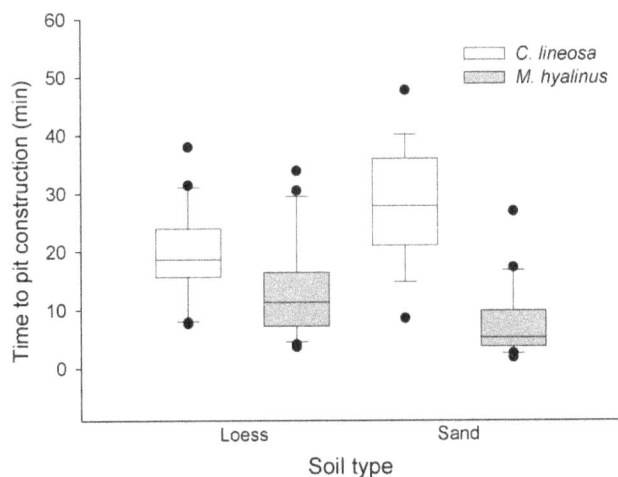

Figure 3. Time to pit construction of antlions. *M. hyalinus* larvae constructed pits faster than *C. lineosa* larvae, irrespective of the soil type. Key: median (horizontal lines in boxes), inter-quartile range (boxes), 95th and 5th percentiles (vertical bars), outliers (black dots).

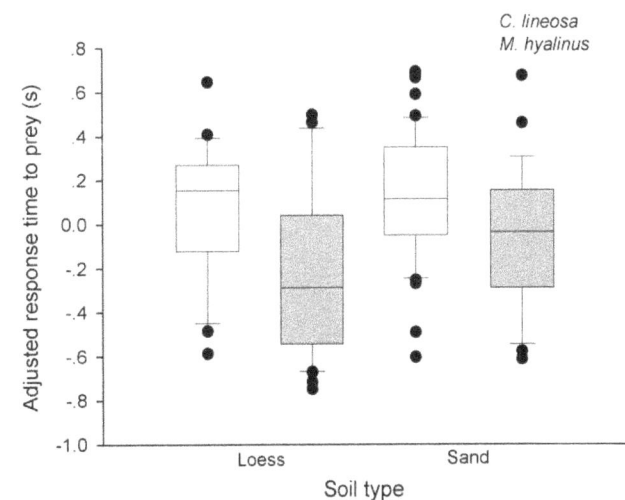

Figure 4. Response time of antlions to prey. *M. hyalinus* larvae responded to prey faster than *C. lineosa* larvae, irrespective of the soil type. Key: median (horizontal lines in boxes), inter-quartile range (boxes), 95th and 5th percentiles (vertical bars), outliers (black dots).

Table 1. Logistic regression analysis examining prey capture success in the foraging behavior experiment.

Parameter	D.F	Wald statistics	p-value
Species	1	23.46	<0.001
Soil type	1	4.632	0.031
Antlion mass	1	0.847	0.357
Species×Soil type	1	11.574	<0.001
Intercept	1	3.067	0.080

Prey capture success differed significantly between the two antlion species and soil types. However, the corresponding interaction term was also significant, implying that observed differences in prey capture success between antlion species were not consistent among soil types (see text for more detail).

Table 2. Logistic regression analysis testing for differences in prey capture success of antlions encountering different prey ant species.

Parameter	D.F	Wald statistics	p-value
Antlion species	1	60.231	<0.001
Soil type	1	15.215	<0.001
Prey species	2	1.703	0.427
Antlion mass	1	3.271	0.071
Antlion species×Soil type	1	19.312	<0.001
Prey species×Soil type	2	0.246	0.884
Antlion species×Prey species	2	1.427	0.490
Antlion species×Soil type×Prey species	2	0.592	0.744
Intercept	1	13.258	0.021

Antlion prey capture success was not affected by the prey ant species they encountered, but it differed significantly between antlion species and soil types. Again, there was a significant Antlion species×Soil type interaction, implying that differences in prey capture success between antlions were not consistent among soil types (see text for more detail).

assumption that adaptation to an additional habitat entails a fitness loss in the former one (e.g. [6,37]). We show that both antlions are capable of discriminating between the two soils, with each showing higher preference to the soil type providing higher prey capture success and characterizing its primary habitat-of-origin. As expected, only small differences in the foraging performances of the habitat generalist, *M. hyalinus*, were evident between the two soils, while the performance of the habitat specialist, *C. lineosa*, was markedly reduced in the alternative sandy soil (Fig. 6). In both species, pit diameter was larger in the sand than in the loess. *M. hyalinus* pits were larger than those of *C. lineosa* irrespective of soil type. Although loess pits of *C. lineosa* were deeper than those of *M. hyalinus*, the sand pits of the latter were deeper than those of the former. *M. hyalinus* dived into the soil faster than *C. lineosa*, but this pattern was evident only in the sand. Remarkably, in both soil types, the habitat generalist *M. hyalinus* constructed pits and responded to prey faster than the habitat specialist *C. lineosa*. As a result, the former enjoyed higher prey capture success, implying a positive association between habitat niche-breadth and foraging performance. Furthermore, this pattern was not sensitive to the prey ant species encountered.

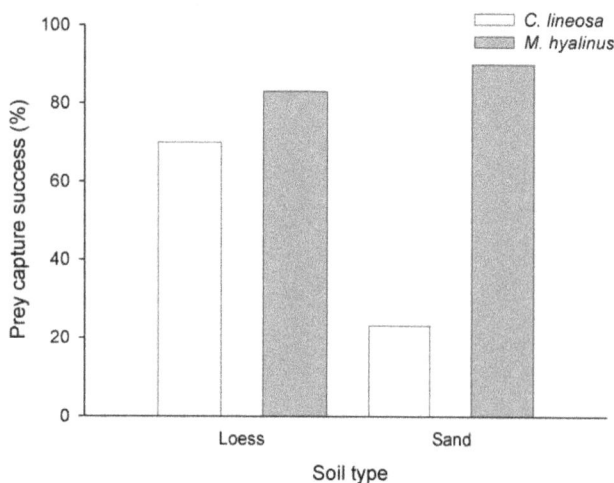

These findings clearly indicate that the habitat specialization of *C. lineosa* to light soils (e.g., loess) does not necessarily confer upon this species superiority in utilizing such habitats, at least under the temperature range of this study.

Remarkably, the widely accepted theoretical assumption, suggesting that adaptation to an additional habitat should confer inferiority in utilizing the former (see references in [38]), has been empirically demonstrated in some studies (e.g. [39–42]). For example, Laverty & Plowright [39], showed that a flower specialist bumblebee, *Bombus consobrinus*, is more effective in foraging on its specialized flower, specifically in handling time and ability to find nectar, compared with two closely related flower generalists,

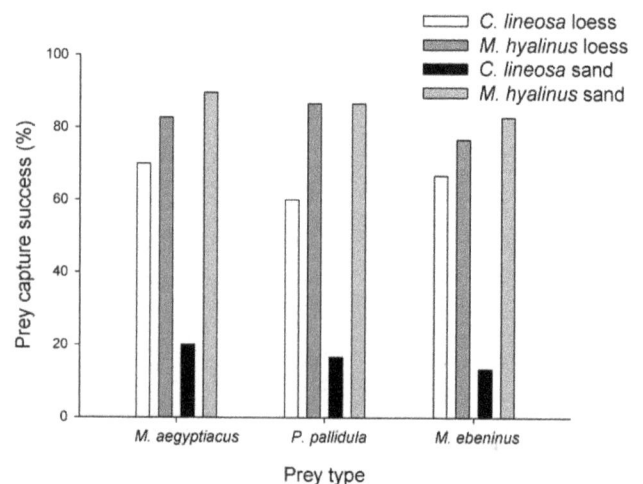

Figure 5. Prey capture success of antlions in the foraging behavior experiment. Prey capture success of *M. hyalinus* varied little between soils, but that of *C. lineosa* was markedly reduced when put in the sand. Notably, the prey capture success of the habitat generalist *M. hyalinus* was higher than that of the habitat specialist *C. lineosa* in both soil types.

Figure 6. Prey capture success of antlions encountering different prey ant species. Prey capture success of both species did not vary among the different ant prey species. Prey capture success of *M. hyalinus* varied little between soils, but that of *C. lineosa* was markedly reduced when put in the sand. Notably, the prey capture success of the habitat generalist *M. hyalinus* was higher than that of the habitat specialist *C. lineosa* in both soil types.

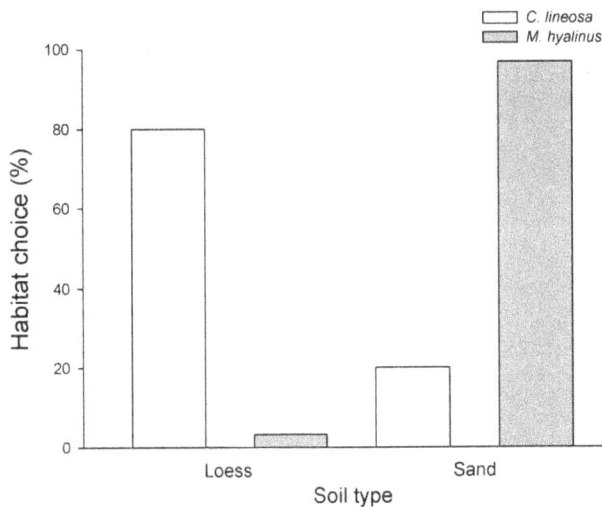

Figure 7. Habitat choice of antlions. Both antlion species discriminate between soils, choosing the soil type providing a higher prey capture success and characterizing their primary habitat-of-origin.

Bombus fervidus and *Bombus pennsylvanicus*. However, since such empirical support is limited and because several studies have even failed to detect it in different systems (e.g. [7–10,43,44]), this trade-off may be less trivial and common than initially assumed.

C. lineosa's inferiority in the loess soil environment, where it is supposedly a habitat specialist, can be clarified through several non-mutually exclusive explanations. First, although *M. hyalinus* can be found in a variety of soils, including in hyper-arid regions characterized by extremely high temperatures [20], it is restricted to shaded micro-habitats (i.e., under trees or bushes), minimizing its exposure to these high temperatures. *C. lineosa*, in contrast, is usually found in micro-habitats exposed to direct sun, and the soil surface in such places may reach extremely high temperatures during the summer. Therefore, it is possible that *C. lineosa* has adapted to function at extremely high temperatures in addition to being a light soil specialist. Second, *C. lineosa* may compensate for its poor performance by reducing its metabolic rate to better resist starvation periods. Such a trade-off between intense foraging activity and the loss of body mass during starvation has already been shown in antlions [45]. Third, the relatively small differences in foraging efficiency between the two species in the loess (e.g., capture success of *M. hyalinus* was ~16% higher than that of *C. lineosa*, irrespective of the prey ant species provided) may have little actual significance under stochastic natural conditions. Fourth, it is possible that *C. lineosa* inhabits light soils because its eggs or pupae better persist in these habitats. Alternatively, *C. lineosa* superiority may be evident only when considering other factors, such as growth rate and predation risk characterizing the different habitats. Clearly, these factors cannot be evident in short term behavioral experiments. Moreover, the role of predation in shaping the behavior of trap-building predators in general and pit building antlions in particular is still unclear (reviewed in [13,46]). Finally, it is possible that the deeper pits of *C. lineosa* in the loess enable it to capture specific prey items not tested for in this experiment, which is an unlikely explanation, as antlions usually feed on ants (~70% of their diet; [14]), and we have used different-sized ants from different locations, which are probably included in both species' diets.

Our prey capture success experiment clearly demonstrates that the success of both antlions is consistent among different prey ant species encountered, indicating that these opportunistic predators are diet generalists, while also suggesting that their spatial distribution should not be affected by the distribution of these ants. Obviously, prey capture success rates may change under natural conditions, as there are some ant species which help nest mates [47]. For example, Czechowski et. al. [48], observed that workers of *Formica sanguinea* caught by a larva of an antlion *Myrmeleon formicarius* can induce rescue behavior in their nest mates. Typical rescue behavior involves both the attempts to pull away the attacked ant by tugging at its limbs, and rapid, intense digging behavior. Such nest mate behavior, which can reduce the prey capture success of both antlion species, could not have been detected in our experiments, as each antlion received only one prey item. However, the fact that the response time to prey of *C. lineosa* is slower than that of *M. hyalinus*, strongly suggests that such rescue behavior may reduce the prey capture success of the former (i.e., habitat specialist) to a higher extent.

According to the theory of habitat selection, animals should select the habitat in which their fitness is higher [49]. Our habitat selection experiment indicates that when alone, each antlion species prefers the soil type providing higher prey capture success and characterizing its primary habitat-of-origin (Fig. 7). Surprisingly, although discrimination of the preferred habitat is critical, especially for *C. lineosa* due to its reduced ability to capture prey in the sandy soil, it appears, from our habitat selection experiment, that this species is significantly less selective compared to *M. hyalinus*. One possible explanation is that female *C. lineosa* oviposit their eggs in light soils far away from the alternative sandy soils, so that the larvae's probability of encountering a different soil type is relatively low. In other words, *C. lineosa* larvae less frequently exercise such discrimination between soils, and thus are more likely to make mistakes in choosing the correct habitat. Habitat selection practiced by the ovipositing females can greatly influence the future success of their progenies. Several studies have suggested that habitat selection in pit-building antlions is largely determined by the ovipositing female [50,51]. Nevertheless, this study demonstrates that larvae of both species are capable of correcting their mother's choice by relocating and selecting the habitat which maximizes their prey capture success. Active habitat selection of antlion larvae, although relatively limited in scale, has been shown among microhabitats of different substrates [12,17] and of different illumination levels [21,36].

Specialization for light soils such as loess is not trivial, especially in arid and semi-arid environments where the above-ground net productivity of this soil type is much lower than that of coarse-textured sandy soils (i.e., the inverse texture effect; [52,53]). Increased productivity is expected to correlate with increased potential prey biodiversity and abundance (e.g. [54]). We thus suggest that some mechanism compensates for this reduced insect abundance, such as low inter-specific competition. To summarize, the broad habitat niche breadth characterizing *M. hyalinus* may explain why its abundance, over large geographical scales, is higher than that of *C. lineosa*, which utilizes a narrower habitat range (i.e., being limited to light soil habitats). Since both antlions are opportunistic predators, their spatial distribution should be less affected by prey community structure. Finally, we suggest that habitat specialization in *C. lineosa* is either an evolutionary dead end [55], or, more likely, that this species' superiority in light soils may only be evident when considering additional niche axes such as starvation endurance and thermal conditions. In a broader context, we suggest that specialization should be examined while considering the multidimensional nature of the ecological niche.

Acknowledgments

We would like to thank Aziz Subach for his constructive comments and Ron Rotkopf for his assistance in the lab.

References

1. Lynch M, Gabriel W (1987) Environmental tolerance. Am Nat 129: 283–303.
2. Futuyma DJ, Moreno G (1988) The evolution of ecological specialization. Ann Rev Ecol Syst 19: 207–233.
3. Kawecki TJ (1994) Accumulation of deleterious mutations and the evolutionary cost of being a generalist. Am Nat 144: 833–838.
4. Van Tienderen PH (1997) Generalists, specialists, and the evolution of phenotypic plasticity in sympatric populations of distinct species. Evolution 51: 1372–1380.
5. Scheiner SM (1998) The genetics of phenotypic plasticity. VII. Evolution in a spatially-structured environment. J Evol Biol 11: 303–320.
6. Futuyma DJ (2001) Ecological specialization and generalization. In: Fox CW, Roff DA, Fairbairn DJ, eds. Evolutionary Ecology: Concepts and Case Studies Oxford University Press, Oxford, UK. pp 177–189.
7. Tollrian R (1995) Predator-indcued morphological defenses: Costs, life history shifts, and maternal effects in *Daphnia pulex*. Ecology 76: 1691–1705.
8. Fry JD (1996) The evolution of host specialization: Are trade-offs overrated? Am Nat 148: S84–S107.
9. Schlichting CD, Pigliucci M (1998) Phenotypic Evolution – A Reaction Norm Perspective. Sinauer Associates, Inc., Publishers, Sunderland, Massachusetts, USA. pp 273–277.
10. Garcia-Robledo C, Horvitz CC (2012) Jack of all trades master of novel host plants: positive genetic correlation in specialist and generalist insect herbivores expanding their diets to novel hosts. J Evol Biol 25: 38–53.
11. Gotelli NJ (1993) Ant lion zones: Causes of high density predator aggregation. Ecology 74: 226–237.
12. Farji-Brener AG (2003) Microhabitat selection by antlion larvae, *Myrmeleon crudelis*: Effect of soil particle size on pit trap design and prey capture. J Insect Behav 16: 783–793.
13. Scharf I, Ovadia O (2006) Factors influencing site abandonment and site selection in a sit-and-wait predator: A review of pit building. J Insect Behav 19: 197–218.
14. Simon D (1988) Ant-lions (Neuroptera: Myrmeleontidae) of the coastal plain: systematical, ecological, and zoogeographical aspects with the emphasis on the coexistence of a species guild of the unstable dune. PhD thesis, Tel-Aviv University, Israel.
15. Lucas JR (1989) Differences in habitat use between two pit-building antlion species: Causes and consequences. Am Midl Nat 121: 84–98.
16. Botz JT, Loudon C, Barger JB, Olafsen JS, Steeples DW (2003) Effects of slope and particle size on ant locomotion: Implications for choice of substrate by antlions. J Kans Entomol Soc 76: 426–435.
17. Devetak D, Spernjak A, Janzekovic F (2005) Substrate particle size affect pit building decision and pit size in the antlion larvae *Eroleon nostras* (Neuroptera: Myrmeleontidae). Physiol Entomol 30: 158–163.
18. Devetak D (2008) Substrate particle size preference of wormlion *Vermileo vermileo* (Diptera: Vermileonidae) larvae and their interaction with antlions. Eur J Entomol 105: 631–635.
19. Lucas JR (1982) The biophysics of pit construction by antlion larvae (Myrmeleon, Neuroptera). Anim Behav 30: 651–664.
20. Scharf I, Filin I, Golan M, Buchshtav M, Subach A, et al. (2008) comparison between desert and Mediterranean antlion population: differences in life history and morphology. J Evol Biol 21: 162–172.
21. Scharf I, Hollender Y, Subach A, Ovadia O (2008) Effect of spatial pattern and microhabitat on pit construction and relocation in Myrmeleon hyalinus (Neuroptera: Myrmeleontidae) larvae. Ecol Entomol 33: 337–345.
22. Scharf I, Golan B, Ovadia O (2009) The effect of sand depth, feeding regime, density, and body mass on the foraging behaviour of a pit-building antlion. Ecol Entomol 34: 26–33.
23. Griffiths D (1991) Intra-specific competition in larvae of the antlion *Morter*-Sp and inter-specific interactions with *Macroleon quinquemaculatus*. Ecol Entomol 16: 193–202.
24. Barkae ED, Scharf I, Subach A, Ovadia O (2010) The involvement of sand disturbance, cannibalism and intra-guild predation in competitive interactions among pit-building antlion larvae. Zoology 113: 308–315.
25. Danin A (1978) Plant species diversity and plant succession in a sandy area in the northern Negev. Flora 167: 409–422.
26. Singer A (2007) *The soils of Israel*. Springer - Verlag Berlin Heidelberg, New-York. pp 63–77.
27. Griffiths D (1980) Feeding biology of ant-lion larvae – prey capture, handling and utilization. J Anim Ecol 49: 99–125.
28. Scharf I, Barkae ED, Ovadia O (2010) Response of pit-building antlions to repeated unsuccessful encounters with prey. Anim Behav 79: 153–158.
29. Ofer J (2000) Let's go to the ant. A field guide to the ants of Israel. Yuval Ofer Publishing, Jerusalem, Israel [In Hebrew].
30. Christensen R (1991) *Log-linear models*. New York, NY, USA. Springer-Verlag.
31. Martin S, Drifhout F (2009) A review of ant cuticular hydrocarbons. J Chem Ecol 35: 1151–1161.
32. Schofield RMS, Nesson MH, Richardson KA (2002) Tooth hardness increase with zinc-content in mandibles of young adult leaf-cutter ants. Naturwissenschaften 89: 579–583.
33. Hurlbert AH, Ballantyne F, Powell S (2008) Shaking a leg and hot to trot: the effect of body size and temperature on running speed in ants. Ecol Entomol 33: 144–154.
34. Gibernau M, Dejean A (2001) Ant protection of heteropteran trophobiont against a parasitoid wasp. Oecologia 126: 53–57.
35. Segev U (2010) Regional patterns of ant-species richness in an arid region: The importance of climate and biogeography. J Arid Environ 74: 646–652.
36. Scharf I, Subach A, Ovadia O (2008) Foraging behaviour and habitat selection in pit-building antlion larvae in constant light or dark conditions. Anim Behav 76: 2049–2057.
37. Bernays EA, Funk DJ (1999) Specialists make faster decisions than generalists: experiments with aphids. Proc R Soc B 266: 151–156.
38. Egas M, Sabelis MW, Dieckmann U (2005) Evolution of specialization and ecological character displacement of herbivores along a gradient of plant quality. Evolution 59: 507–520.
39. Laverty TM, Plowright RC (1988) Flower handling by bumblebees: a comparison of specialists and generalists. Anim Behav 36: 733–740.
40. Larsson M (2005) Higher pollinator effectiveness by specialist than generalist flower-visitors of unspecialized *Knautia arvensis* (Dipsacaceae). Oecologia 146: 394–403.
41. Sorensen JS, Turnbull CA, Dearing MD (2004) A specialist herbivore (*Neotoma stephensi*) absorbs fewer plant toxins than does a generalist (*Neotoma albigula*). Physiol Biochem Zool 77: 139–148.
42. Caley MJ, Munday PL (2003) Growth trades off with habitat specialization. Proc R Soc B 270: S175–S177.
43. Strauss SY, Zangerl AR (2002) Plant-insect interactions in terrestrial ecosystems. In: Herrera CM, Pellmyr O, eds. Plant-Animal Interactions Blackwell Publishing, Oxford, UK. pp 77–106.
44. Berumen ML, Pratchett MS (2008) Trade-offs associated with dietary specialization in corallivorous butterfly fishes (Chaetodontidae: *Chaetodon*). Behav Ecol Sociobiol 62: 989–994.
45. Scharf I, Filin I, Ovadia O (2009) A trade-off between growth and starvation endurance in a pit-building antlion. Oecologia 160: 453–460.
46. Scharf I, Lubin Y, Ovadia O (2011) Foraging decisions and behavioral flexibility in trap-building predators: a review. Biol Rev 86: 626–639.
47. Nowbahari E, Scohier A, Durand JL, Hollis KL (2009) Ants, Cataglyphis cursor, Use Precisely Directed Rescue Behavior to Free Entrapped Relatives. PLoS ONE 4(8): e6573.
48. Czechowski W, Godzinska EJ, Kozlowski MW (2002) Rescue behaviour shown by workers of *Formica sanguinea* Latr., *F. fusca* L. and *F. cinerea* Mayr (Hymenoptera: Formicidae) in response to their nestmates caught by an ant lion larva. Annales Zoologici 52: 423–431.
49. Rosenzweig ML, Abramsky Z (1986) Centrifugal community organization. Oikos 46: 339–348.
50. Matsura T (1987) An experimental study on the foraging behavior of a pit-building antlion larva, *Myrmeleon bore*. Res Popul Ecol 29: 17–26.
51. Matsura T, Yoshitaka Y, Madoka I (2005) Substrate selection for pit making and ovipositing in an antlion, *Myrmeleon bore* (Tjeder), in terms of sand particle size. Entomol Sci 8: 347–353.
52. Noy-Meir I (1973) Desert Ecosystems: Environment and Producers. Annu Rev Ecol S 4: 25–51.
53. Lane DR, Coffin DP, Lauenroth WK (1998) Effects of soil texture and precipitation on above ground net productivity and vegetation structure across the central grassland region of the United States. J Veg Sci 9: 239–250.
54. Wenninger EJ, Inouye RS (2008) Insect community response to plant diversity and productivity in a sagebrush-steppe ecosystem. J Arid Environ 72: 24–33.
55. Kelley S, Farrel BD (1998) Is specialization a dead end? The phylogeny of host use in Dendroctonus bark beetles (Scolytidae). Evolution 52: 1731–1743.

Author Contributions

Conceived and designed the experiments: EDB IS ZA OO. Performed the experiments: EDB IS. Analyzed the data: EDB OO. Contributed reagents/materials/analysis tools: OO. Wrote the paper: EDB IS OO.

14

A Policy-Driven Large Scale Ecological Restoration: Quantifying Ecosystem Services Changes in the Loess Plateau of China

Yihe Lü[1], Bojie Fu[1]*, Xiaoming Feng[1], Yuan Zeng[2], Yu Liu[1], Ruiying Chang[1], Ge Sun[3], Bingfang Wu[2]

1 State Key Laboratory of Urban and Regional Ecology, Research Center for Eco-Environmental Sciences, Chinese Academy of Sciences, Beijing, China, 2 Institute of Remote Sensing Applications, Chinese Academy of Sciences, Beijing, China, 3 USDA-Forest Service, Southern Research Station, Raleigh, North Carolina, United States of America

Abstract

As one of the key tools for regulating human-ecosystem relations, environmental conservation policies can promote ecological rehabilitation across a variety of spatiotemporal scales. However, quantifying the ecological effects of such policies at the regional level is difficult. A case study was conducted at the regional level in the ecologically vulnerable region of the Loess Plateau, China, through the use of several methods including the Universal Soil Loss Equation (USLE), hydrological modeling and multivariate analysis. An assessment of the changes over the period of 2000–2008 in four key ecosystem services was undertaken to determine the effects of the Chinese government's ecological rehabilitation initiatives implemented in 1999. These ecosystem services included water regulation, soil conservation, carbon sequestration and grain production. Significant conversions of farmland to woodland and grassland were found to have resulted in enhanced soil conservation and carbon sequestration, but decreased regional water yield under a warming and drying climate trend. The total grain production increased in spite of a significant decline in farmland acreage. These trends have been attributed to the strong socioeconomic incentives embedded in the ecological rehabilitation policy. Although some positive policy results have been achieved over the last decade, large uncertainty remains regarding long-term policy effects on the sustainability of ecological rehabilitation performance and ecosystem service enhancement. To reduce such uncertainty, this study calls for an adaptive management approach to regional ecological rehabilitation policy to be adopted, with a focus on the dynamic interactions between people and their environments in a changing world.

Editor: Ben Bond-Lamberty, DOE Pacific Northwest National Laboratory, United States of America

Funding: This work was financially supported by the National Basic Research Program of China (No. 2009CB421104), the National Natural Science Foundation of China (Nos. 40930528 and 40971065), the State Forestry Administration (No. 201004058), and the CAS/SAFEA International Partnership Program for Creative Research Teams of "Ecosystem Processes and Services." The funders had no role in study design, data collection and analysis, decision to publish, or preparation of the manuscript.

Competing Interests: The authors have declared that no competing interests exist.

* E-mail: bfu@rcees.ac.cn

Introduction

Ecosystem services are the benefits that people obtain from nature [1]. They are affected by a number of factors including changes in demographic, economic, sociopolitical, scientific and technological, cultural and religious, physical and biological conditions. The impacts of human activity on ecosystem services are most obviously reflected at the local and regional levels. Historically, natural, semi-natural, or managed ecosystems have been able to provide ecosystem services to meet the needs of social development. However, due to the accelerated growth of society, the gaps between the capacity of ecosystems to provide services and human needs are steadily widening. Over the last 50 years, 60% of worldwide ecosystem services have degraded due to increases in the global population and economic growth [2]. These human-ecosystem relationships have usually been governed by resource use and environmental conservation policies. However, policy issues have been under-evaluated in regards to their effects on improving ecosystem services and human-ecosystem relationships [3].

In China, widespread ecological degradation has constrained sustainable socioeconomic development in recent decades, particularly in the period before the end of 20th century. For instance, 23% of the land area in China suffered ecological degradation of which approximately 35% of the Chinese population depended upon for ecosystem services between the early 1980s and 2000s. This also led to a reduced capacity for carbon sequestration during this period [4]. The estimated economic costs of interrelated problems associated with this degradation, including resource depletion, environmental pollution and ecological damage, have amounted to over 13% of the national Gross Domestic Product [5]. In recognizing the serious environmental and ecological issues during economic booms, the Chinese government implemented a series of policies towards ecological restoration. For example, the Grain to Green Program (GTGP) launched in 1999 is the largest land retirement program in the developing world and uses a public payment scheme that directly engages millions of rural households as core agents of project implementation. This is distinct from China's other soil and water conservation and forestry programs because it is one of the first, and certainly the most ambitious, "payment for ecosystem services" program in China [6]. During the 1999–2008 period, the Chinese Central Government made a direct investment of 191.8 billion RMB (approximately 28.8 billion USD) in the implementation of GTGP. This has resulted in the involvement of 0.12 billion farmers in retiring and re-vegetating 9.27 million hectares of sloping croplands [7].

This paper quantitatively evaluates the effects of GTGP implementation on ecosystem services in the Loess Plateau region (Figure 1), which is prioritized as a pilot region for the GTGP. It is necessary to assess the spatial and temporal changes in ecosystem services following the implementation of the GTGP in order to quantify the performance of large-scale ecological rehabilitation efforts and mainstream ecosystem services for future science-based decision-making [8]. The objectives of this study are to: a) examine the land cover change in the Loess Plateau between 2000 and 2008; b) quantify the changes in ecosystem services in terms of water regulation, soil conservation, carbon sequestration and grain production; and c) examine the socioeconomic effects of the GTGP and policy impacts on human-ecosystem relationships.

Results

Land cover change between 2000 and 2008 and the broad climate regime

Prior to the GTGP implementation, the Loess Plateau was dominated by grasslands and farmlands. Between 2000 and 2008 the land cover patterns of the Loess Plateau changed remarkably. Woodland, grassland and residential land cover increased by 4.9%, 6.6% and 8.5%, respectively. Farmland decreased by 10.8% and desertification increased slightly, by 0.3% (Figure 2). The increases in grassland and woodland were distributed along a northeast to southwest land strip (Figure 3) and were mostly

converted from farmlands. This land cover change resulted in over 43% grassland, nearly 30% cropland, and about 16% woodland that dominated the Loess Plateau region in 2008.

The regional climate condition of the Loess Plateau region has exhibited a warming and drying trend. This climate trend was revealed from the analysis of time series data between 1951 and 2008, obtained from 85 weather stations located in the Loess Plateau region (Figure 4). Precipitation was found to decrease annually by an average of 0.97 mm and temperature was found to increase annually by an average of 0.02°C.

Hydrological regulation change

Regional water yield decreased after the implementation of the GTGP. Over half of the study area (northeast to southwest of the Loess Plateau) experienced a decrease in runoff (2–37 mm/year) with an average 10.3 mm/year decrease in runoff across the whole Loess Plateau over the 2002–2008 period (Figure 5). While, water yield increased in some local areas which accounted for less than 10% of the Loess Plateau region.

Soil conservation assessment

Soil conservation in the Loess Plateau, represented as a decrease in soil erosion, has improved since 2000 as a result of vegetation restoration (Figure 6). The annual average soil retention of the study area between 2000 and 2008 was found to be 3.44 billion tons (Table 1), equivalent to an annual average soil retention rate of 63.3% [Soil Retention Rate (%) = 1.2603Time (years since 2000)+56.556, R^2 = 0.3367 and P = 0.1. This linear relationship is not so significant statistically because of the large impacts from highly variable precipitations (Figure 4)]. The decreasing trend of soil loss per unit rainfall erosivity has also implied improvement on soil conservation service of the rehabilitated ecosystems (Table 1).

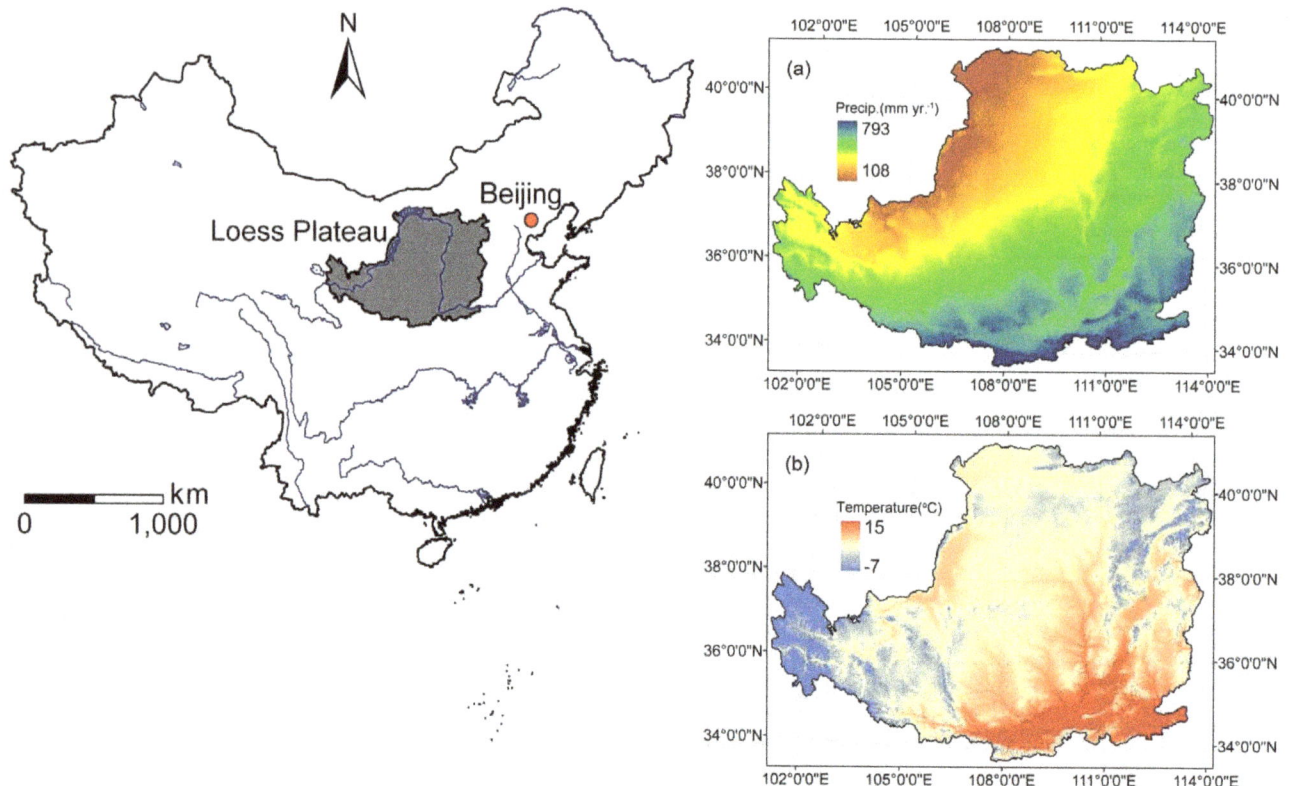

Figure 1. Location of the Loess Plateau and average climate conditions from 1999 to 2008. (a) Precipitation (b) Temperature.

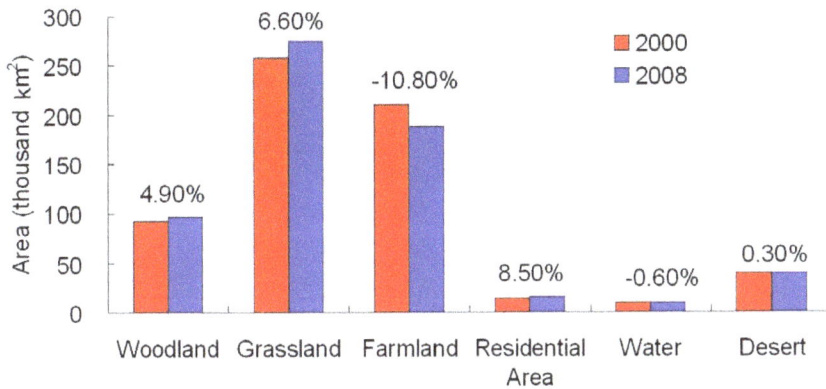

Figure 2. Coverage of each land cover type in the Loess Plateau, in 2000 and 2008. Numbers above bars indicate the change in area covered in 2008 as compared to 2000.

After vegetation restoration, 84.4% of soil retention occurred on hill slopes with a slope angle between 8°–35°. However, the mean soil erosion rate in areas with a slope gradient of over 8° was still greater than 4,260 t km^{-2} yr^{-1} in 2008, which is far beyond the tolerable erosion rate of 1,000 t km^{-2} yr^{-1} [9]. Soil erosion is thus still considered one of the most critical environmental issues in the Loess Plateau and requiring further ecological rehabilitation efforts.

Carbon sequestration assessment

Net carbon sequestration was estimated from vegetation and soil carbon change after re-vegetation was undertaken in 2000. The findings suggest that the ecological rehabilitation efforts have brought about significant positive impacts on carbon sequestration, with carbon levels in soil and rehabilitated vegetation found to be 11.54 Tg, and 23.76 Tg, respectively (Table 2). The spatial variation of carbon sequestration in the Loess Plateau is shown in Figure 7. The carbon sequestration is most evident from northeast to southwest including provinces of Shanxi, Shaanxi, Ningxia, and Qinghai, respectively.

Grain production

In the early stages of the GTGP implementation process (from 2000 to 2004), average grain productivity increased by approximately 1.3 times and then fluctuated around a productivity level of 3,614 kg/ha. As a result of this cropland productivity change, the gross grain production also increased by approximately 1.3 times between 2001 and 2006. The time and rate of the gross production change appeared to occur later and more slowly than the grain productivity change (Figure 8). Actual grain production increased across the whole of the Loess Plateau at a rate of 18% between 2000 and 2008.

Discussion

This study's results suggest that GTGP has resulted in ecosystem property and service change under unfavorable climate change conditions. Specifically, the following changes have been detected: 1) Significant expansion of grassland, woodland and residential areas, and shrinkage of farmland; 2) Reduction in regional water yield; 3) Significant improvement in regional soil conservation capacity, grain production and carbon sequestration. Complex relationships may exist between these changes, as well as between the biophysical and socioeconomic conditions.

Uncertainties involved in ecosystem service assessment

Several factors affected the accuracy of estimating annual water yield at the regional scale. Firstly, the complex terrain of the Loess Plateau presented a challenge for deriving the spatial distribution of annual precipitation that was interpolated from climate records at 172 weather stations in the Loess Plateau region. In addition,

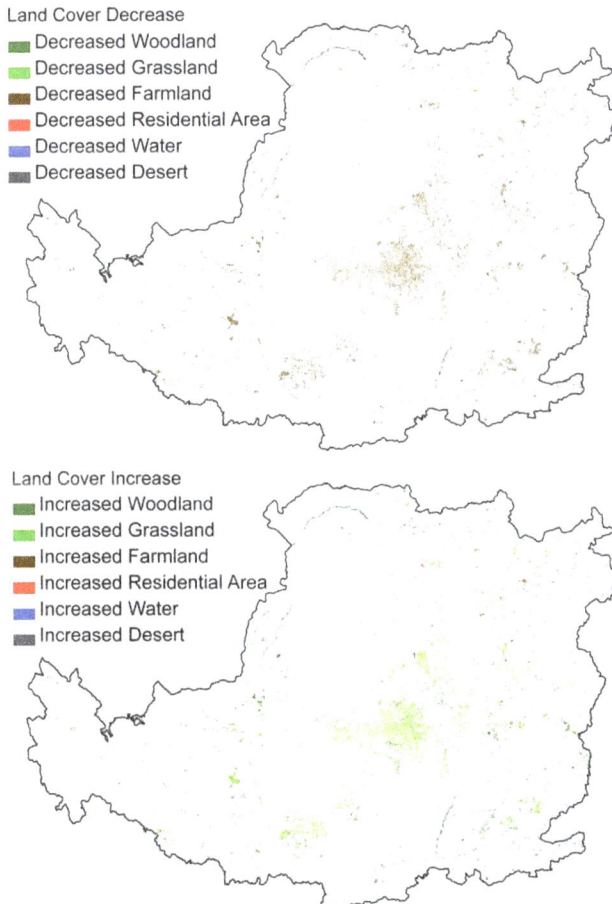

Figure 3. Decreased (above) and increased (below) land covers from 2000 to 2008.

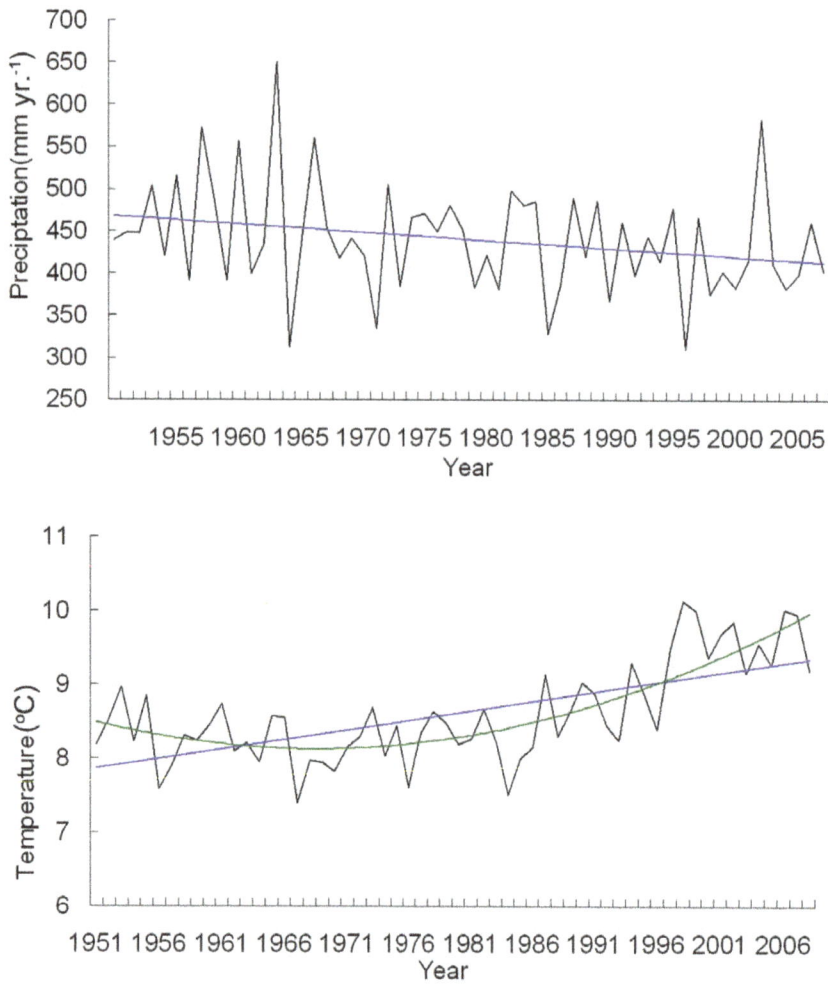

Figure 4. The trend towards a drier and warmer climate in the Loess Plateau region.

the seasonal and inter-annual variability of precipitation was considered to be high. The large spatial and temporal variability in precipitation thus made accurate mapping of precipitation distribution difficult at the 1-km resolution. Although the

evapotranspiration (ET) modeling results were believed to be much closer to reality than the results obtained from the remote sensing based product (MODIS-ET), uncertainty remained over

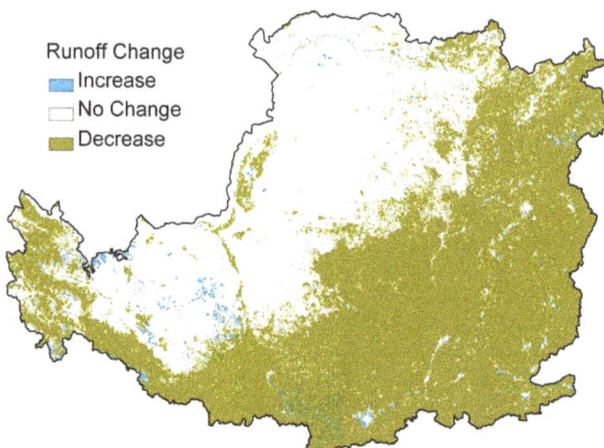

Figure 5. Average water yield change due to land cover change from 2000 to 2008.

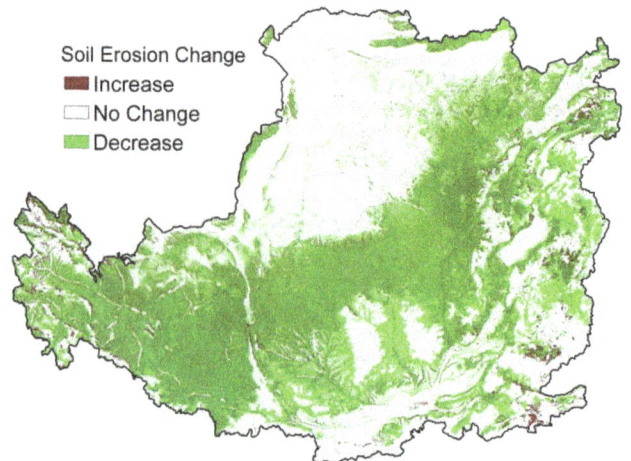

Figure 6. The change in soil erosion in the Loess Plateau region from 2000 to 2008.

Table 1. Rainfall erosivity and soil retention characteristics in the Loess Plateau region from 2000 to 2008.

Year	2000	2001	2002	2003	2004	2005	2006	2007	2008
Rainfall erosivity [megajoules·mm/ (ha·hour·yr)]	442.0	544.0	435.6	630. 8	487.8	408.7	456.3	539.0	434.4
Soil loss per unit rainfall erosivity (t)	0.048	0.044	0.044	0.040	0.046	0.045	0.038	0.031	0.035
Total soil retention(10^8 t)	34. 5	30.8	26.1	49.8	27.7	33.4	31.9	48.6	26.9

the seasonal distribution of ET by land cover type. Uncertainty also surrounded monthly ET estimates for two reasons: 1) Change in water storage may not have been negligible for certain wet years; and 2) Water resource use by communities and the impacts of soil conservation structures (e.g., check dams), were not considered. Anyway, water yield estimation is still an inaccurate science at this point in time [10], particularly at larger spatial scales. Overall, the method used in this paper may introduce systematic errors at a level of approximately 15% [11].

The estimation of soil conservation was undertaken through the application of the Universal Soil Loss Equation (USLE) together with remote sensing. The USLE is based on a statistical relationship established from a large number of plot scale rainfall-erosion experiments [12–13]. It estimates rill and inter-rill soil detachments on hill slopes from rainfall, soil and soil cover parameters, and management factors [14]. Therefore, it is a suitable method to estimate the effect of hill slope vegetation rehabilitation on soil conservation. However, this effect may have been overestimated in this research due to the omission of the local sediment deposition process [13]. Overestimation may have also occurred due to setting the control for soil conservation effects to a scenario of no vegetation cover or erosion control practice. Overestimation is evident after comparing soil conservation results to those from another similar soil conservation assessment using different methods, which reported an average soil conservation rate linked to vegetation restoration of 38.8% in the Zuli River basin of the Loess Plateau region [15]. These overestimations were made for the absolute values of spatial explicit annual soil conservation measurements but did not exclude the soundness of comparisons between annual soil conservation services brought about by vegetation rehabilitation. Uncertainties were also identified from the estimation of input parameters for the USLE [14]. Therefore, parameters established and experimentally verified in the Loess Plateau region were used for estimating the different factors in the USLE [16–19] to reduce this source of uncertainty.

For the assessment of carbon sequestration, only the effects on areas with land cover transitions from farmland to forest, shrub, or

grass were considered. However, evidence from the Loess Plateau suggests that significant carbon sequestration effects could also be detected in grassland and forestland from the process of ecological succession [20]. This research may therefore underestimate the carbon sequestration effects at the regional scale due to the exclusion of carbon sequestration effects associated with grassland, shrubland, and forest ecosystems that existed before implementation of the GTGP. Soil carbon sequestration effects at the sample point scale were also estimated for equal soil depths (20 cm) because of insufficient soil bulk density data. Furthermore, regional soil carbon sequestration effects were estimated through the upscaling of 103 samples collected from the Loess Plateau using a multi-regression method. The accuracy of the results from the CASA (Carnegie-Ames-Stanford Approach) model, which is fundamental to vegetation carbon sequestration assessment, is largely dependent on the resolution of remote sensing data. MODIS data at 1-km resolution and parameters transferred from national scale studies [21–22] were used for this research. All the above methods can introduce errors or uncertainties in the estimation of vegetation and soil carbon sequestration. These errors or uncertainties can be reduced with more soil data, higher resolution remote sensing data and localized model parameters. Given these uncertainties, the major characteristic of the carbon sequestration effects of the GTGP was revealed to be the dominance of vegetation carbon accumulation, which was found to be approximately twice the level of soil carbon sequestration in this study. This figure was 2.3 (vegetation carbon accumulation divided by soil carbon sequestration) in similar research conducted in Yunnan province of southwestern China [23].

Synergies and tradeoff between ecosystem services

The implementation of a large-scale vegetation rehabilitation program under a regional warming and drying climate (Figure 4) may contribute to the decrease of stream flow in the Loess Plateau region [24] because of the potential increase in vegetative water consumption. Vegetation cover in the Loess Plateau region has expanded due to a significant increase in grassland and woodland areas (Figure 3). The amount of vegetation cover improvement

Table 2. Area of cropland converted to forest (grassland) and the carbon sequestration by vegetation, soil and ecosystems in Loess Plateau between 2000 and 2008.

Types of conversion	Restoring to grassland	Restoring to shrub	Restoring to Broad-leaved forest	Restoring to coniferous forest	Total
Area of change (ha)	3.96×10^6	4.85×10^5	2.11×10^5	1.73×10^5	4.83×10^6
Soil carbon storage (Tg)	8.25	1.81	0.72	0.77	11.54
Vegetation carbon storage (Tg)	7.16	11.30	3.24	2.06	23.76
Total (Tg C)	15.41	13.11	3.96	2.83	35.30

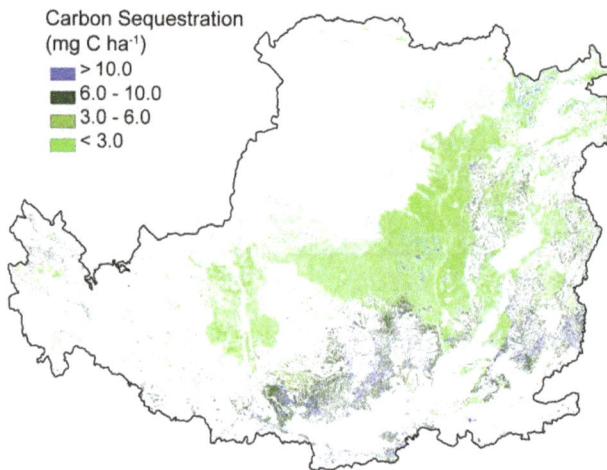

Figure 7. The spatial distribution of carbon sequestration.

achieved is higher than the national scale target of the GTGP to increase grassland and woodland coverage at a rate of 4.5%. This result was also supported by other research which estimated that vegetation cover in the whole Loess Plateau increased at an approximate rate of 6–8% during 2000–2006 [25] or 12.5% at local level in the central Loess Plateau during 1998–2005 [26]. The net primary production of regional ecosystems in the Loess Plateau that experienced a significant increase or remained stable between 1999 and 2008 accounted for 65.8% and 14.3% of the region, respectively [27]. Consequently, the trend of improving carbon sequestration, soil conservation and grain production may indicate that these key ecosystem services act in synergy. The decrease in regional water yield and the improvement in vegetation cover may be considered a tradeoff, as water resources and vegetation typically maintain an inverse relationship in semi-arid water limited environments under given climate conditions. However, both elements contribute significantly to the enhancement of soil conservation and carbon

sequestration (Figure 6–7 and Table 1). Due to the decline of regional water yield, nitrogen (influenced by population pressure and fertilizer use) and phosphorus (sourced from soil erosion in the Loess Plateau) transported in the lower reaches of the Yellow River have reduced significantly since the late 1990s [28]. The implementation of the GTGP vegetation rehabilitation program may therefore improve the water quality of the middle and lower reaches of the Yellow River, however, water shortage issues [29] may potentially be exacerbated. The significant improvement in cropland productivity were attributed to factors such as agricultural technological growth, the construction of high quality basic croplands (e.g., terrace croplands and check-dam derived croplands), the increase in resource input and farming management, and the improvement of extension services [30–31] as complementary or insurance measures for ecological rehabilitation.

Grain to Green Program and local empowerment

Under the GTGP, the government offered grain and cash to farmers annually as compensation (grain subsidy of 1500 kg/ha plus cash subsidy of RMB 300/ha) for their opportunity costs in discontinuing farming on sloping croplands [32]. The program has helped numerous farmers to gradually change their income structure by shifting from grain production to other income-generating activities [33]. Subsequently, the employment and sources of family income of farmers have been diversified due to the economic compensation obtained through the GTGP, which ranges from 10% to 30% of their total income [34–35].

The rural economic capacity of the Loess Plateau has also improved at both the regional and farmer household levels. Data from the National Bureau of Statistics of China indicates that the net per capita income of farmers in the Loess Plateau region increased annually from 1998 to 2007 at a rate of 8.6%, which could be actually reduced to 4.5% after subtracting the annual average inflation rate of 2.1% and the rural consumption price increasing rate of 2% in China during 2000–2008. The ratio of farmer respondents reporting significant increases in household income after the implementation of the GTGP varied with

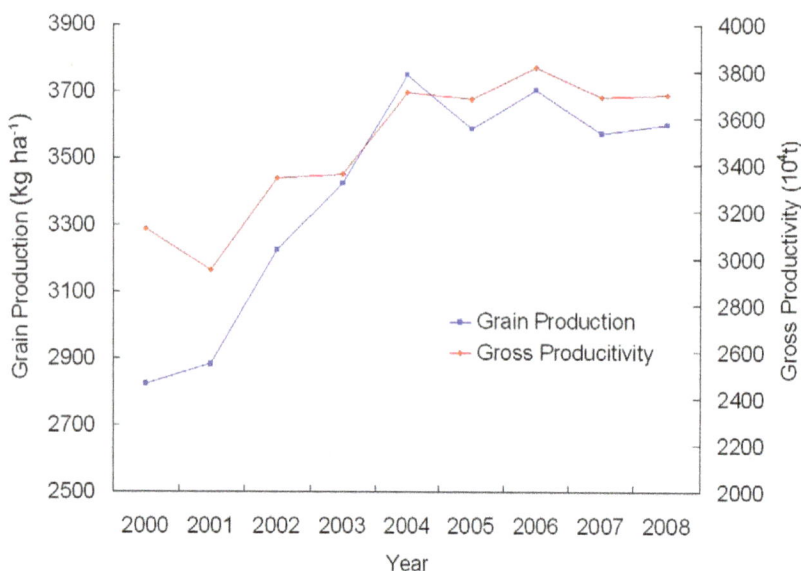

Figure 8. Grain and gross production change from 2000 to 2008 in the Loess Plateau region.

different study sites and ranged from 55% to over 90% [34,36]. During the implementation of the GTGP, local farmers developed a greater understanding of and support for ecological restoration programs [37]. However, the direct economic compensation from the GTGP has only been a minor contributor to farmers' income. The more significant effect of the GTGP has been to accelerate the socioeconomic transition from a food production-based rural community to a more active and profitable labor migration dominant rural economy (i.e. where the rural labor force can migrate to urban areas to earn a living or run a business).

Sustainability through adaptive management

This study suggests that the ecological rehabilitation policies of the GTGP and associated soil and water conservation measures implemented in the Loess Plateau tended to facilitate synergies on carbon sequestration, soil conservation, grain production, and farmers' economic welfare. These synergies are important goals that ecosystem management tries to reach.

The successful performance of ecological rehabilitation programs discussed above was largely due to the innovative ecosystem management systems and mechanisms. Close cooperation between local government and other stakeholders was found to be important for capitalizing on synergies between ecological rehabilitation initiatives, and for maximizing the outcomes of ecological management activities [38]. External funding other than government sources, such as private sectors, enterprises, and the World Bank, were also important in the success of restoration programs [39–40]. Project selections and designs have been increasingly informed by feasibility studies and demonstrations. Project planning has been taking a preliminary adaptive approach, informed by ongoing monitoring and evaluation as well as performance assessment [39,41].

Quantitative assessments of present ecological restoration policies have been increasingly available and the sustainability issues of regional ecological restoration programs have been recognized. For example, when non-native tree species were planted at a high density, soil drying was observed during re-vegetation which undermines the long-term capacity of soil to sustain ecosystems under a semi-arid climate [42]. Soil drying is at least partly due to the ecological rehabilitation policy that gives more weight to planting trees and less consideration of natural restoration that is more tailored to the local environment [43].

From a socioeconomic point of view, the sustainability of ecological rehabilitation depends largely on the economic incentives or benefits produced by the implementation of such activities. As the GTGP has been implemented in over 200 counties across seven provinces in the region, data insufficiency and uncertainty excluded a cost benefit analysis of the GTGP across the whole Loess Plateau. A local scale analysis in Dunhua county indicated that the net benefit (sometimes negative) varied widely according to geographical location (or environmental context), land productivity and discount rate [44]. Subsequently, the risk of re-cultivation of re-vegetated croplands will remain high if the policy-related economic compensation measures from the government are terminated [37,43,45].

Therefore, to improve the actual performance of regional ecological rehabilitation efforts, an adaptive management paradigm needs to be established to integrate the government-motivated "top-down" approach and the local stakeholder motivated "bottom-up" approach, with balanced considerations of the dynamics and sustainability requirements of the targeted ecological-socioeconomic coupled systems. Ecological rehabilitation is widely used in reversing environmental degradation and can contribute to the improvement of ecosystem services and

adaptability to climate change [46–47]. The success and sustainability of ecological rehabilitation efforts depend on the scientific understanding of the interactions between people and their surrounding ecosystems, rather than merely the ecosystems themselves [48]. Consequently, the way to secure sustainability in ecological rehabilitation and ecosystem service enhancement is to ensure net benefits, or at least, no net loss to the stakeholders and ecosystems involved in the adaptive management framework. An adaptive management approach allows for flexibility in human and financial resource allocation, an expansion of knowledge on the dynamic socioeconomic-ecological coupled systems, and efficacy of management operations [49]. The experience of ecological rehabilitation and the change in key ecosystem services in the Loess Plateau region exemplified the positive effects of environmental policies and the necessity of adopting an adaptive management approach.

Materials and Methods

Study area description

The Loess Plateau region is located in the middle reaches of the Yellow River basin in Northern China (Figure 1) and experiences arid and semi-arid climate condition over an area greater than 600,000 km^2. Precipitation occurs between June and September and accounts for 60–70% of the annual total in the form of high intensity rainstorms. The Loess Plateau is an ecologically vulnerable region and is well known for its high soil erosion rates and heavy sediment loads. The average erosion modulus is 5,000–10,000 t/km^2, with the highest rate up to 20,000–30,000 t/km^2 [9]. The areas characterized by slopes of 8–35 degrees are the main source areas for soil erosion and represent 45.63% of the whole Loess Plateau region. Therefore, restoring vegetation in these areas will play a key role in mitigating soil erosion.

The Loess Plateau comprises 6.67% of the territory in China and supports 8.5% of the Chinese population. By the end of 2007 the human population in the Loess Plateau region reached a magnitude of approximately 0.114 billion and a population density of 168 persons per square kilometer, a number four times that of the early 1910s. As a result, human pressure upon land resources has increased significantly in this region. Soil erosion has been accelerated by intensive land use (e.g., slope farming) and exploitive management for thousands of years, resulting in the loss of grassland and natural forest. Due to its great geographical magnitude, the Loess Plateau has diverse habitat conditions for different vegetation types which have shifted historically because of climate change. It can be inferred from literature that grassland and forest steppe were the dominant vegetation types across the whole Loess Plateau region. Forest was also dominant at a local scale in mountainous and valley areas in the Quaternary and particularly the Holocene periods [50–53]. In the last 2000 years, the vegetation in the Loess Plateau region has experienced significant degradation due to increasingly intensive human activities [52,54]. In 2000, woods (i.e. forests and shrubs) and grasses in the Loess Plateau Region covered areas of 77.3 and 252.8 thousand square kilometers, respectively (Figure 2). At present, the forest area in the Loess Plateau region accounts for only 7% of the total forest area in China [55].

Land cover change

Landsat TM/ETM images from 2000 were used to extract land cover data for the Loess Plateau. Prior to image interpretation, remote sensing data was geo-referenced through the use of 1:100,000 topographic maps. For each Landsat TM/ETM image a minimum of 30 evenly distributed sites were selected as Ground Control Points (GCPs). The Root Mean Squared Error of geometric rectification was less than 1 pixel (or 30 m). Land cover types were identified using ArcMap and based on the spectral reflectance and structure of objects. A total of 27 land cover subtypes in the study area were further grouped into six aggregated land cover types: woodland, grassland, farmland, residential areas, water bodies and desert. Based on the land cover map from 2000, the land cover classification of the Loess Plateau from 2008 was updated using China-Brazil Earth Resources Satellite (CBERS-2b) images. These images have a 20 m ground resolution and a similar amount of spectral bands as Landsat ETM images. To support image interpretation and validate the land cover map from 2008, a field survey was conducted to evaluate the classification accuracy. Field-measured land cover types and photos located with GPS coordinates were collected across the whole study area. Classification accuracy was measured as 95% at the level of the six aggregated land cover types.

Hydrological regulation

Water yield was used as an indicator of hydrological regulation. Water yield at the watershed-scale was modeled as precipitation minus evapotranspiration (ET), based on the assumption of negligible water storage change in the Loess Plateau region on an annual time scale. Monthly ET (mm) was estimated by **ET = 9.78+0.0072*PET*PPT+0.051*PPT*LAI**, where PET represents potential evapotranspiration (mm), PPT represents precipitation (mm), and LAI represents leaf area index (dimensionless) [11]. PET (mm) was calculated using the Hamon method [56]. The climatic parameters were obtained from the National Climatic Bureau and interpolated with ANUSPLIN [57]. LAI was derived from SPOT VEGETATION NDVI based on the relationship between NDVI and LAI for different types of land cover [58]. The monthly Loess Plateau ET model was calibrated and validated using runoff data from 46 basins in the region. This runoff data was retrieved from the web-based hydrological and sediment database of the Loess Plateau (http://www.loess.csdb.cn/hyd/user/index.jsp). The structure of the above monthly ET equation follows the empirical relationships established between monthly ET and the main influencing factors of 13 ecosystems with wide geographic distribution [11]. The present form of the ET equation has been established since calibration and validation was undertaken and is suitable for use in the Loess Plateau region.

Soil conservation

The soil conservation services of re-vegetation have been measured since 2000 by calculating the decrease in regional soil loss or regional soil retention on hill slopes. Soil retention is calculated as soil loss without vegetation cover and soil erosion control practices minus that under the current land use/land cover patterns and soil erosion control practices. The Universal Soil Loss Equation (USLE) is the most widely used method for soil erosion modeling and assessment [59] and was applied to quantify the amount of annual soil loss for the two situations described above. Soil retention can be expressed mathematically as: $\Delta A = A_0 - A_v = R \times K \times L \times S \times (1 - C_v \times P_v)$, where ΔA is the amount of soil conservation (t·ha^{-1}·yr^{-1}); A_0 is the potential soil erosion without vegetation cover (t·ha^{-1}·yr^{-1}); and A_v is the soil erosion under current land cover and management condition (t·ha^{-1}·yr^{-1}). R, K, L, and S represent rainfall erosivity [megajoules·mm/(ha·hour·yr)], soil erodibility [t ·ha· h/(ha·megajoules·mm)], slope length, and slope angle factors respectively. C_v and P_v

refer to current vegetation cover factors and erosion control practice factors, respectively. L, S, C_v, and P_v are all dimensionless factors.

Carbon sequestration

Soil Organic Carbon (SOC) sequestration for the GTGP in Loess Plateau was estimated by using a multiple regression approach. This approach included precipitation, the vegetation types converted from sloping croplands, and the time duration after conversion as the main independent variables. In the GTGP, sloping croplands (generally with a slope >15°) were converted principally into grassland, shrub, broad-leaved forest and coniferous forest. SOC sequestration for the GTGP in the Loess Plateau was estimated based on these four established ecosystems in two major climate zones, defined as zones with precipitation less than (north Loess Plateau) and greater than (south Loess Plateau) 550 mm. In each of the two climate zones, SOC sequestration under each plantation type was calculated by using the SOC sequestration rate derived from a multiple regression and the area of cropland involved in this plantation type. The total SOC sequestration across the whole Loess Plateau was determined from the sum of the SOC sequestration estimated in the four plantation type in the two climate zones. The multiple regression was undertaken based on SOC sequestration data collected from the top 20 cm soil layer collected from 103 samples across the Loess Plateau [$\log(Y) = 2.648 - 0.366$ P- $\alpha \times U + 0.023$ y ($R^2 = 0.256$, P<0.05, N = 103]. Y is the SOC sequestration rate (Mg C/ha) and P is a dummy variable representing precipitation. The value of P is set at 0 when precipitation is above 550 mm, while P is set at 1 when the precipitation is below 550 mm. U is also a dummy variable representing land use change. When cropland was converted into grassland, shrub, broad-leaved forest and coniferous forest, the value of α was set at 0.727, 0.533, 0.633 and 0, respectively. y is a variable representing plantation age]. Carbon sequestration in the vegetation of each plantation type was estimated from the carbon sink efficiency of the vegetation type and the NPP was calculated by using the CASA (Carnegie-Ames-Stanford Approach) model [21] [CSE = Cseq/NPP×100), where CSE is carbon sink efficiency; Cseq is the carbon sequestration in vegetation (MgC/ha/a); and NPP is Net Primary Productivity (gC/m^2/a)]. The value of CSE of grassland, shrub and forest (inclusive of broad-leaved forest and coniferous forest) was set at 0.015, 0.036 and 0.057, respectively according to Fang et al. [22].

Grain production

Data on grain production was obtained from provincial level Bureaus of Statistics in the Loess Plateau (287 counties in seven provinces).

Acknowledgments

We thank Changhong Su and Lang Wang for grain production data preparation, Professor Jianguo Liu from Michigan State University for valuable comments and suggestions on earlier versions of the manuscript, and Ms. Halinka Lamparski from the University of Western Australia for her help in improving the language of the revised manuscript.

Author Contributions

Conceived and designed the experiments: BJF YHL. Performed the experiments: BJF YHL XMF YZ YL RYC GS. Analyzed the data: XMF YZ YL RYC BFW. Contributed reagents/materials/analysis tools: BJF YHL GS BFW. Wrote the paper: YHL BJF.

References

1. Daily GC (1997) Nature's services: societal dependence on natural ecosystem. Washington, DC: Island Press.
2. Millennium Ecosystem Assessment (2005) Ecosystems and human well-being: synthesis. Washington, DC: Island Press.
3. Aronson J, Blignaut JN, Milton SJ, Le Maitre D, Esler KJ, et al. (2010) Are socioeconomic benefits of restoration Adequately quantified? a meta-analysis of recent papers (2000–2008) in Restoration Ecology and 12 other scientific journals. Restoration Ecology 18(2): 143–154.
4. Bai ZG, Dent D (2009) Recent land degradation and improvement in China. Ambio 38(3): 150–156.
5. Shi MJ, Ma GX, Shi Y (2011) How much real cost has China paid for its economic growth? Sustainability Science 6: 135–149.
6. Bennett MT (2008) China's sloping land conversion program: institutional innovation or business as usual? Ecological Economics 65: 699–711.
7. Li YC (2009) Grain for Green program is a great concrete action towards the ecological civilization in China: summary of a decade program implementation. Forestry Construction 27(5): 3–13 (in Chinese).
8. Yin RS, Yin GP, Li LY (2010) Assessing China's ecological restoration programs: what's been done and what remains to be done? Environmental Management 45: 442–453.
9. Cai QG (2001) Soil erosion and management on the Loess Plateau. Journal of Geographical Sciences 11(1): 53–70.
10. Allen RG (2008) Why do we care about ET? Southwest Hydrology 7: 18–19.
11. Sun G, Alstad K, Chen J, Chen S, Ford CR, et al. (2011) A general predictive model for estimating monthly ecosystem evapotranspiration. Ecohydrology 4: 245–255.
12. Ciesiolka CAA, Yu B, Rose CW, Ghadiri H, Lang D, et al. (2006) Improvement in soil loss estimation in USLE type experiments. Journal of Soil and Water Conservation 61: 223–229.
13. Kinnell PIA (2008) Discussion: Misrepresentation of the USLE in 'Is sediment delivery a fallacy?' Earth Surface Processes and Landforms 33: 1627–1629.
14. Tattari S, Barlund I (2001) The concept of sensitivity in sediment yield modelling. Physics and Chemistry of the Earth, Part B: Hydrology, Oceans and Atmosphere 26: 27–31.
15. Li C, Qi J, Feng Z, Yin R, Guo B, et al. (2010) Quantifying the Effect of Ecological Restoration on Soil Erosion in China's Loess Plateau Region: an Application of the MMF Approach. Environmental Management 45: 476–487.
16. Soil Survey Office of Shaanxi Province (1992) Soil of Shaanxi. Beijing: Science Press, China (in Chinese).
17. Wan YM, Guo PC, Cao WS (1994) A study on soil antierodibility in Loess Plateau. Journal of Soil and Water Conservation 8(4): 11–16 (in Chinese with English abstract).
18. Cai CF, Ding SW, Shi ZH, Huang L, Zhang GY (2000) Study of applying USLE and geographical information system IDRISI to predict soil erosion in small watershed. Journal of Soil and Water Conservation 14(2): 19–24 (in Chinese with English abstract).
19. Zhang K, Li S, Peng W, Yu B (2004) Erodibility of agricultural soils on the Loess Plateau of China. Soil and Tillage Research 76: 157–165.
20. Wang Y, Fu B, Lü Y, Chen L (2011) Effects of vegetation restoration on soil organic carbon sequestration at multiple scales in semi-arid Loess Plateau, China. Catena 85: 58–66.
21. Yu DY, Shi PJ, Shao HB, Zhu WQ, Pan YH (2009) Modelling net primary productivity of terrestrial ecosystems in East Asia based on an improved CASA ecosystem model. International Journal of Remote Sensing 30: 4851–4866.
22. Fang JY, Guo ZD, Piao SL, Chen AP (2007) The estimation of terrestrial vegetation carbon sink in china: 1981–2000. Science in China (Series D) 37: 804–812.
23. Chen XG, Zhang XQ, Zhang YP, Wan CB (2009) Carbon sequestration potential of the stands under the Grain for Green Program in Yunnan Province, China. Forest Ecology and Mamagement 258: 199–206.
24. Zhang XP, Zhang L, Zhao J, Rustomji P, Hairsine P (2008) Responses of streamflow to changes in climate and land use/cover in the Loess Plateau, China. Water Resource Research 44: W00A07. (doi:10.1029/2007WR006711).
25. Xin ZB, Xu JX, Zheng W (2008) Spatiotemporal variations of vegetation cover on the Chinese Loess Plateau (1981–2006): impacts of climate changes and human activities. Science in China Series D: Earth Sciences 51: 67–78.
26. Cao SX, Chen L, Yu XX (2009) Impact of China's Grain for Green Project on the landscape of vulnerable arid and semi-arid agricultural regions: a case study in northern Shaanxi Province. Journal of Applied Ecology 46: 536–543.
27. Lou XT, Du X, Sun YJ (2010) Analysis of remote sensing monitoring and dynamic variation of NPP in the Loess Plateau. Journal of Anhui Agricultural Science 38: 6057–6064 (in Chinese with English abstract).
28. Yu T, Meng W, Ongley E, Li Z, Chen J (2010) Long-term variations and causal factors in nitrogen and phosphorus transport in the Yellow River, China. Estuarine, Coastal and Shelf Science 86: 345–351.
29. Zhang Q, Xu CY, Yang T (2009) Variability of Water Resource in the Yellow River Basin of Past 50 Years, China. Water Resources Management 23: 1157–1170.
30. Li L, Tsunekawa A, Tsubo M, Koike A, Wang JJ (2010) Assessing total factor productivity and efficiency change for farms participating in Grain for Green program in China: A case study from Ansai, Loess Plateau. Journal of Food, Agriculture, and Environment 8: 1185–1192.
31. Yao SB, Li H (2010) Agricultural Productivity Changes Induced by the Sloping Land Conversion Program: An Analysis of Wuqi County in the Loess Plateau Region. Environmental Management 45: 541–550.
32. Xu ZG, Bennett M, Tao R, Xu JT (2004) China's sloping land conversion program four years on: current situation, pending issues. The International Forestry Review 6: 317–326.
33. Uchida E, Rozelle S, Xu JT (2009) Conservation payments, liquidity constraints, and off-farm labor: impact of the Grain-for-Green Program on rural households in China. American Journal of Agricultural Economics 91: 70–86.
34. Xie C, Liu JJ, Han Y, Yuan M (2009) 2008 conversion of cropland to forest project: farmer household effect analysis. Forestry Economics 31(9): 56–64 (in Chinese with English Abstract).
35. Li QR, Wang JJ (2010) Effects of Returning Farmland to Forest Project on Agricultural Eco-economic Succession in Mizhi County. Bulletin of Soil and Water Conservation 30: 206–210 (in Chinese with English Abstract).
36. Zhu QK, Lai YF (2009) Assessment on the comprehensive benefits of Converting Farmland to Forest Project in Loess Hilly Regions-a case study of Wuqi county, Shaanxi province. Journal of Northwest Forestry University 24: 219–223 (in Chinese with English Abstract).
37. Hu CX, Chen LD, Fu BJ, Gulinck H (2006) Farmer's attitude towards Grain for Green Program in the hilly area of the Loess Plateau. International Journal of Sustainable Development and World Ecology 13: 211–220.
38. Chang ZW (2002) Experiences from ecological rehabilitation at small watershed scale in Xinzhou prefecture. Soil and Water Conservation Science and Technology in Shanxi 29(1): 39–40 (in Chinese).
39. Liu ZR, Wang HZ (2006) The performance of and experiences from the World Bank loan project on soil and water conservation in Loess Plateau region. Soil and Water Conservation in China 27(10): 30–33 (in Chinese).
40. Liu GC, Lei YJ, Cheng JH (2007) The characteristics of integrated soil and water conservation practices in Qiushui River watershed. Soil and Water Conservation Science and Technology in Shanxi 34(4): 11–12 (in Chinese).
41. Zhou YL (2005) Experiences and inspirations of Loess Plateau watershed rehabilitation project. China Water Resources 25(12): 11–13 (in Chinese with English Abstract).
42. Li W, Wang QJ, Wei SP, Shao MA, Yi L (2008) Soil desiccation for Loess soils on natural and regrown areas. Forest Ecology and Management 255: 2467–2477.
43. Cao SX, Chen J, Chen L, Gao WS (2007) Impact of grain for green project to nature and society in North Shaanxi of China. Scientia Agricultura Sinica 40: 972–979 (in Chinese with English abstract).
44. Wang C, Ouyang H, Maclaren V, Yin Y, Shao B, et al. (2007) Evaluation of the economic and environmental impact of converting cropland to forest: a case study in Dunhua county, China. Journal of Environmental Management 85: 746–756.
45. Cao SX, Xu CG, Li C, Wang XQ (2009) Attitudes of farmers in China's northern Shaanxi Province towards the land-use changes required under the Grain for Green Project, and implications for the project's success. Land Use Policy 26: 1182–1194.
46. Benayas JMR, Newton AC, Diaz A, Bullock JM (2009) Enhancement of biodiversity and ecosystem services by ecological restoration: a meta-Analysis. Science 325: 1121–1124.
47. Harris JA, Hobbs RJ, Higgs E, Aronson J (2006) Ecological restoration and global climate change. Restoration Ecology 14: 170–176.
48. Halle S (2007) Science, art, or application - the "Karma" of restoration ecology. Restoration Ecology 15: 358–361.
49. Gilioli G, Baumgartner J (2007) Adaptive ecosocial system sustainability enhancement in Sub-Saharan Africa. Ecohealth 4: 428–444.
50. Feng ZD, Tang LY, Wang HB, Ma YZ, Liu KB (2006) Holocene vegetation variations and the associated environmental changes in the western part of the Chinese Loess Plateau. Palaeogeography Palaeoclimatology Palaeoecology 241: 440–456.
51. Shang X, Li XQ (2010) Holocene vegetation characteristics of the southern Loess Plateau in the Weihe River valley in China. Review of Palaeobotany and Palynology 160(3–4): 46–52.
52. Zhang K, Zhao Y, Zhou AF, Sun HL (2010) Late Holocene vegetation dynamic and human activities reconstructed from lake records in western Loess Plateau, China. Quaternary International 227: 38–45.
53. Sun AZ, Feng ZD, Ma YZ (2010) Vegetation and environmental changes in western Chinese Loess Plateau since 13.0 ka BP. Journal of Geographical Sciences 20: 177–192.
54. Wang L, Shao M, Wang QJ, Gale WJ (2006) Historical changes in the environment of the Chinese Loess Plateau. Environmental Science and Policy 9: 675–684.
55. Peng H, Coster J (2007) The Loess Plateau: Finding a place for forests. Journal of Forestry 105: 409–413.
56. Hamon WR (1963) Computation of Direct Runoff Amounts From Storm Rainfall. International Association of Hydrological Sciences Publication 63: 52–62.
57. Wahba G, Wendelberger J (1980) Some new mathematical methods for variational objective analysis using splines and cross-validation. Monthly Weather Review 108: 1122–1145.

58. Zhang WC, Zhong S, Hu SY (2008) Spatial scale transferring study on Leaf Area Index derived from remotely sensed data in the Heihe River Basin, China. Acta Ecologica Sinica 28: 2495–2503 (in Chinese with English abstract).

59. Kinnell PIA (2010) Event soil loss, runoff and the Universal Soil Loss Equation family of models: a review. Journal of Hydrology 385: 384–397.

15

Modeling the Impact of Soil and Water Conservation on Surface and Ground Water Based on the SCS and Visual Modflow

Hong Wang[1,3], Jian-en Gao[1,2,4,5]*, Shao-long Zhang[4], Meng-jie Zhang[5], Xing-hua Li[4]

1 Institute of Soil and Water Conservation, Chinese Academy of Sciences and Ministry of Water Resources, Yangling, Shaanxi Province, China, 2 Institute of Soil and Water Conservation, Northwest A&F University, Yangling, Shaanxi Province, China, 3 University of Chinese Academy of Sciences, Beijing, China, 4 College of Water Resources and Architectural Engineering, Northwest A&F University, Yangling, Shaanxi Province, China, 5 College of Natural Resources and Environment, Northwest A&F University, Yangling, Shaanxi Province, China

Abstract

Soil and water conservation measures can impact hydrological cycle, but quantitative analysis of this impact is still difficult in a watershed scale. To assess the effect quantitatively, a three-dimensional finite-difference groundwater flow model (MODFLOW) with a surface runoff model–the Soil Conservation Service (SCS) were calibrated and applied based on the artificial rainfall experiments. Then, three soil and water conservation scenarios were simulated on the sand-box model to assess the effect of bare slope changing to grass land and straw mulching on water volume, hydraulic head, runoff process of groundwater and surface water. Under the 120 mm rainfall, 60 mm/h rainfall intensity, 5 m^2 area, 3° slope conditions, the comparative results indicated that the trend was decrease in surface runoff and increase in subsurface runoff coincided with the land-use converted from bare slope to grass land and straw mulching. The simulated mean surface runoff modulus was 3.64×10^{-2} $m^3/m^2/h$ in the bare slope scenario, while the observed values were 1.54×10^{-2} $m^3/m^2/h$ and 0.12×10^{-2} $m^3/m^2/h$ in the lawn and straw mulching scenarios respectively. Compared to the bare slope, the benefits of surface water reduction were 57.8% and 92.4% correspondingly. At the end of simulation period (T = 396 min), the simulated mean groundwater runoff modulus was 2.82×10^{-2} $m^3/m^2/h$ in the bare slope scenario, while the observed volumes were 3.46×10^{-2} $m^3/m^2/h$ and 4.91×10^{-2} $m^3/m^2/h$ in the lawn and straw mulching scenarios respectively. So the benefits of groundwater increase were 22.7% and 60.4% correspondingly. It was concluded that the soil and water conservation played an important role in weakening the surface runoff and strengthening the underground runoff. Meanwhile the quantitative analysis using a modeling approach could provide a thought for the study in a watershed scale to help decision-makers manage water resources.

Editor: Jose Luis Balcazar, Catalan Institute for Water Research (ICRA), Spain

Funding: This paper was supported by National Technology Support Project (2011BAD31B05), by Natural Science Foundation of China (41371276), by the Subject of National Science and Technology Major Project (2009ZX07212-002-003-02) and by Knowledge Innovation Project of Institute of Soil and Water Conservation, CAS & MWR (Soil and Water Conservation Project) (A315021304). The funders had no role in study design, data collection and analysis, decision to publish, or preparation of the manuscript.

Competing Interests: The authors have declared that no competing interests exist.

* E-mail: gaojianen@126.com

Introduction

In recent years, with more and more officials and professionals paying attention to the problems such as soil erosion and water loss, the contradiction between water supply and demand, whether soil and water conservation measures on upper regions would have effect on the amount of water resources of the lower reaches is a scientific issue to be discussed and probed into urgently [1]. For many years, the work of soil and water conservation as same as the Yellow River management, not only obtains great achievement, but also brings some problems [2]. The biggest achievement was no more than soil erosion control in the watershed, and thereby reducing the amount of sediment flowed into the river. Took the Yellow River basin for example, changes in land use and vegetation in the Loess Plateau had a decisive impact on sediment transport of Yellow River [3], even about 60% sediment yield since 10 ka BP occurred during the last 1040 years of the period of

estrepement in Loess Plateau. The sediment monitoring at Sanmenxia hydrological station of Yellow River showed the annual average sediment discharge was 1040 million tons during 1950–2010, 589 million tons during 1987–2010, while it was only 198 and 351 million tons in 2009 and 2010 separately [4].

The reduction in the quantity of the sediment discharge by soil and water conservation is no doubt pleasing, but it also brings a new eco-environmental problem for its surface water reduction function. Some studies found that land use/land cover (LULC) change could alter hydrological cycles by affecting ecosystem evaportranspiration, soil infiltration capacity, surface and subsurface flow regimes [5–8]. Just as the study of the relation between control in Loess Plateau and no-flow in the Yellow River showed that the influence of water reduction on no-flow in the lower Yellow River for the comprehensive control changed the condition underlying surface and influenced the water cycle was nonnegligible [9]. Results of soil and water conservation method showed

Figure 1. The experimental flume.

that the mean flood reduced 0.5456 billion tons in He-Long section of middle Yellow River, Jinghe, Beiluohe and Weihe watershed from 1970 to 1996 [10]. The average runoff reduction benefits of soil and water conservation measures for Chabagou, Dalihe and Wudinghe basins in the 1970s were 14.47%, 20.22% and 20.78% respectively [11]. Jain and Mishra et al [12] evaluated the suitability to particular land use, soil type and combination a quantitative evaluation of the existing Soil Conservation Service Curve Number (SCS-CN) model using a large set of rainfall-runoff data from small to large watersheds of the U.S.A. Shi and Yi et al [13] evaluated the effect of land use/cover change on surface runoff by SCS model in Shenzhen region, China. Guo and Wang et al [14] calculated the volume of surface runoff during 5 rainfalls on 5 different kinds of land use types in sloping runoff plots by SCS model.

Soil and water conservation could reduce surface runoff and it is bound to affect groundwater recharge, which is the entry into the saturated zone of water made available at the table surface, together with the associated flow away from the water table within the saturated zone [15]. But less direct study has previously been undertaken to assess this effect quantitatively. It had been shown that groundwater recharge was closely related to land-use types very much. The impact of land-use on distributed groundwater recharge and discharge in the western Jilin, China, using MODFLOW, WetSpass, the Seepage packages, and ArcGIS showed that forest vegetation had the highest recharge, followed by agricultural farmlands and the recharge generally decreased when vegetated forests deteriorate to other landforms (bush, grassland or bare-land) [16]. A study by Cho and Barone using MODFLOW indicated that subsurface flow regimes could be negatively affected by urbanization due to increased withdrawal and reduced recharge [17]. A number of studies showed the effects of changes in forest cover on groundwater recharge [18–22]. These studies showed a range in the reduction of groundwater recharge beneath trees, from 15% to 90%, compared to that

under grass [18]. Thus the impacts of land-use on the atmospheric components of the hydrologic cycle (regional and global) are increasingly being recognized, though those on the subsurface components of the hydrologic cycle, particularly groundwater recharge are not equally known [23].

Therefore special attention should be given to the effect of soil and water conservation on the hydrologic cycle, especially recharge and discharge. Both monitoring and modeling approaches are used for conjunctive investigation of surface water and groundwater. The monitoring approach is expensive and time demanding yet measuring actual changes in stream and groundwater levels over time may lead to more direct estimates of the impact of land development on both surface and subsurface flows [17]. Through the model coupling approach, interaction between surface water and groundwater has become a trend [24,25]. Linkage between MODFLOW and existing surface models such as the SCS method to consider surface water and groundwater interactions could validate the outputs of the model.

The overall goal of this study is to assess the impact of soil and water conservation on the surface and subsurface flow by mathematical models combined with artificial rainfall simulation experiments and gives a quantitative analysis thought simultaneously.

Materials and Methods

SCS Curve Number Method

The runoff equation of the U.S. Soil Conservation Service, commonly called the curve number method, for estimating runoff from rainfall, came into common use in the mid - 1950s [26]. Although the SCS method for runoff estimation has changed little since the 1960s, its popularity has been maintained over the years [27]. The SCS-CN equation, in the typical form [28], is given as:

Table 1. Mechanical composition of experimental materials (%).

Types	1~0.5	0.5~0.25	0.25~0.05	0.05~0.01	0.01~0.005	0.005~0.001	<0.001
Lou soil	0.07	0.65	5.86	49.04	12.18	13.72	18.48
Mixed soil	1.83	24.34	30.91	25.29	4.82	5.53	7.27

$$\bar{Q}_{surf} = \frac{(P-I_a)^2}{(P-I_a+S)} \text{ for } P>I_a; \ \bar{Q}_{surf}=0 \text{ for } P\leq I_a \quad (1)$$

Where $\bar{Q}_{surf}$ is the surface runoff (mm), in this study it is the average runoff depth per hour; P is the precipitation (mm), in this study it is the precipitation per hour; S is the amount of water storage available in the soil profile or the maximum storage (mm), and I_a is the initial abstraction (mm). To reduce the numbers of variables, the empirical relationship $I_a = 0.2S$ was adopted, which then gives the most familiar form of the runoff equation [29]:

$$\bar{Q}_{surf} = \frac{(P-0.2S)^2}{(P+0.8S)} \text{ for } P>0.2S; \ \bar{Q}_{surf}=0 \text{ for } P\leq 0.2S \quad (2)$$

Table 2. Calibrated parameter value and the relative error (RE) (%).

Period	Calibration period	Verification period			
Rainfall Intensity (mm/h)	75	45	90	105	120
RE (%)	0.015	3.41	0.40	2.20	1.77

Because the range of the S value is too large to obtain a suitable value, the dimensionless parameter, curve number (CN), was introduced into this formula. The parameter is related to CN by the relationship:

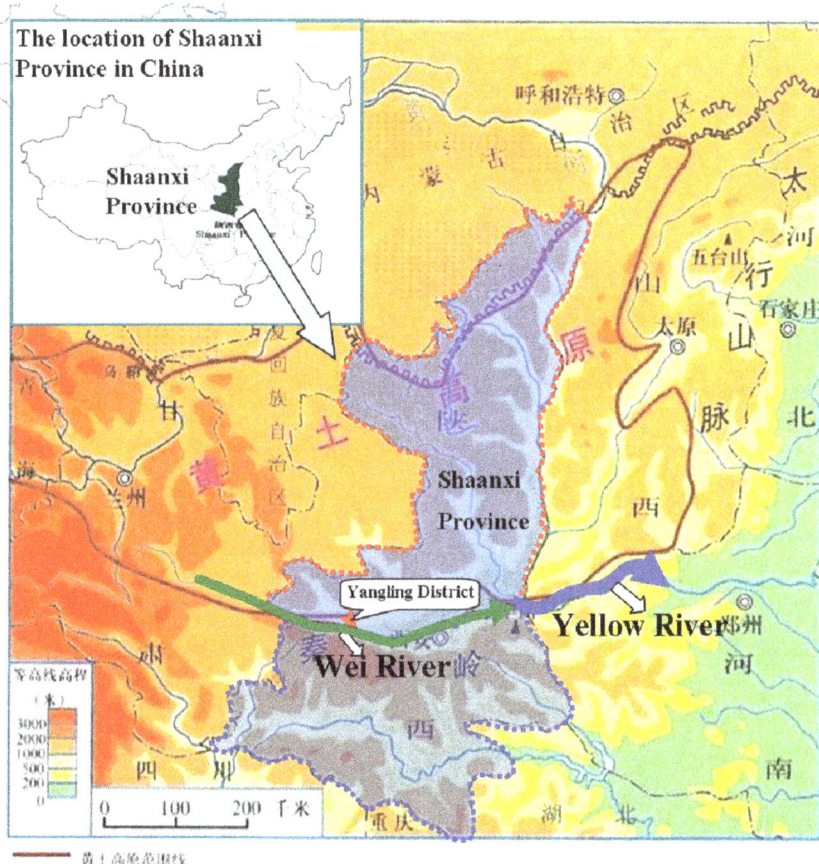

Figure 2. The schematic map of Wei River and Yangling District location.

Table 3. The main calibrated parameters of the MODFLOW model.

Parameters	Kx1(m/s)	Kz1(m/s)	Kx2(m/s)	Kz2(m/s)	Kx3(m/s)	Kz3(m/s)	Sy	Ss
Minimum	1.16×10^{-8}	1.16×10^{-9}	5.79×10^{-7}	5.79×10^{-8}	5.79×10^{-5}	5.79×10^{-6}	0.05	1×10^{-5}
Maximum	5.79×10^{-6}	5.79×10^{-6}	5.79×10^{-5}	5.79×10^{-5}	5.21×10^{-4}	5.21×10^{-4}	0.5	1×10^{-3}
Initial Value	2.90×10^{-6}	2.90×10^{-7}	3.18×10^{-5}	3.18×10^{-6}	2.2×10^{-4}	2.2×10^{-5}	0.2	1×10^{-4}
Calibrated Value	1.12×10^{-7}	2.52×10^{-9}	1.50×10^{-6}	7.47×10^{-7}	3.1×10^{-4}	7.60×10^{-5}	0.36	1.385×10^{-4}

$$CN = \frac{25400}{(S+254)}; \quad S = 25.4\left(\frac{1000}{CN} - 10\right) \qquad (3)$$

The CN is a comprehensive parameter which related to initial soil moisture, slope, vegetation, soil type and land use status etc [30]. It appeared that the curve numbers were used as a proxy for the retention parameter S in order to scale the curves to a convenient range between zero and 100 [29].

Meanwhile, the runoff amount could be calculated combined the rainfall duration by the relationship:

$$Q_{surf} = \bar{Q}_{surf} \times h \qquad (4)$$

Where Q_{surf} is the total runoff of rainfall (120 mm) (mm); h is the rainfall duration (h).

Relative error (RE) was selected as the statistical evaluation index of SCS and it was defined by the equation below:

$$RE = \frac{Q_{cal} - Q_{obs}}{Q_{cal}} \qquad (5)$$

Where Q_{cal} was the calculated runoff by SCS model (mm); Q_{obs} was the observed runoff obtained from the rainfall experiment (mm).

Visual Modflow Model

MODFLOW (Modular Three-dimensional Finite-difference Ground-water Flow Model)) is a modular finite-difference ground-water flow model published by the U.S. Geological Survey, which can simulate ground-water flow in a three-dimensional heterogeneous and anisotropic medium. Using the finite-difference method, the domain in which flow is to be simulated is divided into a rectilinear mesh of rows, columns, and layers [31–33]. Visual MODFLOW is also a three-dimensional groundwater flow and contaminant transport modeling application that integrates MODFLOW, MODPATH, MT3DMS, WinPEST, Zone Budget, and so on. Applications include well head capture zone delineation, pumping well optimization, aquifer storage and recovery, groundwater remediation design, simulating natural attenuation, and saltwater intrusion [32].

The three-dimensional movement of ground water of constant density through porous earth material may be described by the partial-differential equation [31]:

$$\frac{\partial}{\partial x}\left(Kxx-\frac{h}{x}\right) + \frac{\partial}{\partial y}\left(Kyy-\frac{h}{y}\right) + \frac{\partial}{\partial z}\left(Kzz-\frac{h}{z}\right) + W$$
$$= Ss\frac{\partial h}{\partial t} \qquad (6)$$

where Kxx, Kyy, and Kzz are values of hydraulic conductivity along the x, y, and z coordinate axes, which are assumed to be parallel to the major axes of hydraulic conductivity (m/s); h is the potentiometric head (m); W is a volumetric flux per unit volume representing sources and/or sinks of water, with W<0 for flow out of the ground water system, and W>0 for flow into the system; Ss is the specific storage of the porous material (1/m); and t is time (min).

The calibration statistics can be reported in the result plots and the statistical evaluation indexes, including Calibration Residual (CM), Residual Mean (RM), Absolute Residual Mean (ARM), Standard Error of the Estimate (SEE), Root Mean Squared (RMS), Normalized Root Mean Squared (NRMS) and Correlation Coefficient (Cor) [32]. Based on the principle of the modeling and simulated rainfall experiments, a transient three-dimensional groundwater flow model was built using Visual MODFLOW 4.1. The model domain was divided into a 50×265 array of 4×10^{-4} square meter cells uniformly. The model consisted of three individual layers with gradient of $3°$ and the thicknesses of layers were respectively 0.5, 1 and 98.5 cm from top to bottom. The transient flow simulation was selected as the flow type and the time unit for all parameters was minute. The edge of the model domain was modeled as a no-flow boundary. Actual quantities of groundwater abstraction from drain pipe were treated as flux boundary condition in form of pumping well. Groundwater recharge from rainfall was modeled as recharge package in MODFLOW. The initial ground water level was 0.39 m high and paralleled to the bottom of the model. The initial value and the range of hydraulic conductivity and storage coefficient values at different layers were assigned to each active grid cell by interpolation of discrete property data derived from water releasing test analysis and regional geology data.

Simulated Rainfall Experiments

Experimental conditions and equipment. The simulated rainfall experiments were carried out in the Rainfall Simulation Hall of the State Key Laboratory of Soil Erosion and Dryland Farming on the Loess Plateau during the period of June to October of the year 2012. The simulated rainfall system, with automatic simulation device of under sprinkler, could ensure the kinetic energy of simulated precipitation close to the natural rainfall for the mean fall-height of 18 meters. And the calibration tests showed that rainfall uniformity was greater than 85%. The experiments were conducted in the sand-box model [34,35] (figure 1), 5.3 m×1 m×1 m at the Rainfall Simulation Hall. The

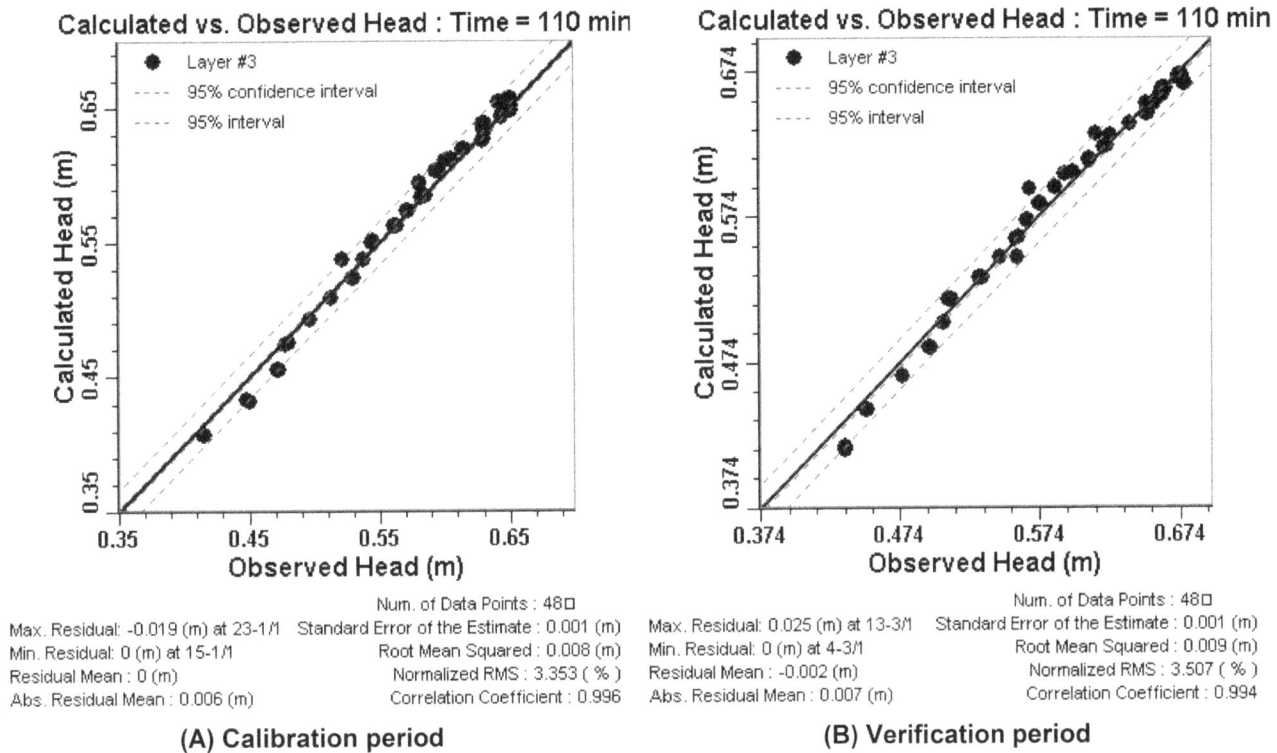

Num. of Data Points : 48☐
Max. Residual: -0.019 (m) at 23-1/1   Standard Error of the Estimate : 0.001 (m)
Min. Residual: 0 (m) at 15-1/1   Root Mean Squared : 0.008 (m)
Residual Mean : 0 (m)   Normalized RMS : 3.353 (%)
Abs. Residual Mean : 0.006 (m)   Correlation Coefficient : 0.996

(A) Calibration period

Num. of Data Points : 48☐
Max. Residual: 0.025 (m) at 13-3/1   Standard Error of the Estimate : 0.001 (m)
Min. Residual: 0 (m) at 4-3/1   Root Mean Squared : 0.009 (m)
Residual Mean : -0.002 (m)   Normalized RMS : 3.507 (%)
Abs. Residual Mean : 0.007 (m)   Correlation Coefficient : 0.994

(B) Verification period

Figure 3. The scatter graph of calculated vs. observed values.

slope of the model could be adjusted manually from 0° to 35°. There were two water tanks, 0.15 m×1 m×1 m in front and back of the model, for the regulation of the groundwater level. Above the front side of the water tank there was one surface water groove and there was one drainage pipe of groundwater in the 0.39 m high at the front side of the water tank.And one hundred and twenty 0.2 m×0.2 m sets of piezometric tubes of level observation were installed in left side of the sand-box model. And two neutron probe access tubes down to a depth of 0.9 m in experimental flume for soil moisture control.

The experimental flume was fixed at an angle of 3° in this study. Three water and soil conservation measures (bare slope, straw mulching and grass land) were considered and the soil surface condition should be roughen to keep the beginning condition consistent every time before experiment. The precipitation for a control to be equal was 120 mm. Six gradient rainfall intensities (45~120 mm/h) with uniform rainfall conditions were simulated for 160 to 60 min correspondingly. The initial ground water level was parallel to the bed of the flume and the distance between two lines was 0.39 m.

Experimental materials and monitoring methods. The test materials included riversand and Lou soil. The riversand

samples were dug from the middle and lower reaches of the Wei River bank in Yangling District and the Lou soil was also collected from Yangling District, Shaanxi Province, China. Figure 2 is the schematic map of Wei River and Yangling District location. The mechanical composition of soil particles was shown in table 1. The samples were air-dried for about ten days and sieved through a series of corresponding magnitude sieves. Then the test materials were packed into the flume layer by layer. The experimental flume consists of three individual layers and the thickness was respectively 0.5, 1 and 98.5 cm from top to bottom. The medium sand (0.25 mm ~ 0.5 mm) was paved in the third 98.5 cm deep layer and the soil bulk density was 1.4~1.5 g/cm^3. In the second layer the riversand particle size was less than 0.25 mm and the average soil bulk density was about 1.6 g/cm^3. On the top layer, the soil used was the composite sandy loam, composed of riversand (<0.25 mm) and Lou soil, the weight ratio of this composite soil was about 2:5 and the average soil bulk density was also about 1.6 g/cm^3. The setting of water and soil conservation measures as follows: for the straw mulching, the quantity of arid straw mulching was designed as 0.4 kg/m^2 uniformly each experiment with coverage of 85~90%; for the grassland, the grass was grew on the soil in the flume and its species was ophiopogon japonicus with coverage of 65%~70% and the planting structure was, 10 cm row spacing×5 cm plant spacing×8.2 cm average plant height.

The main monitoring items measured during the experiments were: the surface runoff amount, the surface runoff in the process, underground runoff and groundwater level etc. Surface runoff was collected by collecting buckets at the beginning of runoff yield. The surface runoff in the process was collected for 30 seconds at 5-min intervals during experiments and the runoff amount was collected during the whole process. After groundwater level adjustment, recorded the initial level first and the dynamic changes every 10 minutes with the changes begin for about 4

Table 4. Error indexes of calculated vs. observed values.

Evaluation Indexes	RM/m	ARM/m	SEE/m	RMS/m	NRMS/%	Cor
Calibration	0.0002	0.00727	0.0012	0.0097	4.146	0.996
Verification	0.000375	0.00843	0.00119	0.00981	4.259	0.994

Table 5. The benefits of water reduction in field controlled and uncontrolled small watersheds %.

Water System	Watershed Name	Area (km²)	Condition	Representative Type Area	Year	Benefits δ(%)
Kuye River	Mengjiagou	2.03	Controlled	Earth-rocky Mountainous Area	1959~1961	42.64
	Yangyagou	1.88	Uncontrolled			
Yuxi River	Qingcaogou	0.373	Controlled	Half sandstorm area	1959~1960	88.19
	Wangjiagou	0.434	Uncontrolled			
	Tiaogou	0.7677	Controlled		1959~1961	16.51
	Lijiagou	0.693	Uncontrolled			
Wuding River	Jiuyuangou	70.1	Controlled	Loess Hilly and Gully Region	1959~1969	16.88
	Peijiamaogou	41.2	Uncontrolled			
	Xiangtagou	0.454	Controlled		1958~1961	23.70
	Tuanyuangou	0.491	Uncontrolled			
	Wangmaozhuanggou	5.967	Controlled		1962~1963	43.70
	Lijiazhaigou	5.45	Uncontrolled			
Yanhe River	Dabiangou	3.7	Controlled		1963~1967	39.63
	Xiaobiangou	3.925	Uncontrolled			
Luohe River	Sigou	4.37	Controlled	gully region of loess plateau	1959~1961	57.98
	Nangou	5.11	Uncontrolled			
Juhe River	Guanzhuanggou	3.39	Controlled		1959~1961	57.98
	Yuanguzhuanggou	2.82	Uncontrolled			
Luohe River	Xingshugou	0.522	Controlled	Loess Hilly and Gully Region	1958~1961	83.62
	Beilougou	0.334	Uncontrolled			
Jinghe River	Fengyugou	1.176	Controlled		1958~1960	14.71
	Wangjiagou	0.874	Uncontrolled			
Chanhe River	Yaojiagou	7.815	Controlled	Terrace region of loess plateau	1960~1961	64.99
	Dicungou	4.4	Uncontrolled			
Average						44.09

hours. Meanwhile, underground runoff was collected by measuring cylinder per 2000 ml uninterruptedly for about 8 hours. Then according to the condition of the underground runoff, the monitoring interval was lengthened gradually until the water level dropped to the initial control water level. Soil moisture (neutron probe method) measurements were taken immediately before and after precipitation to ensure replicability of the initial soil surface conditions. And water temperature was also recorded during each simulation.

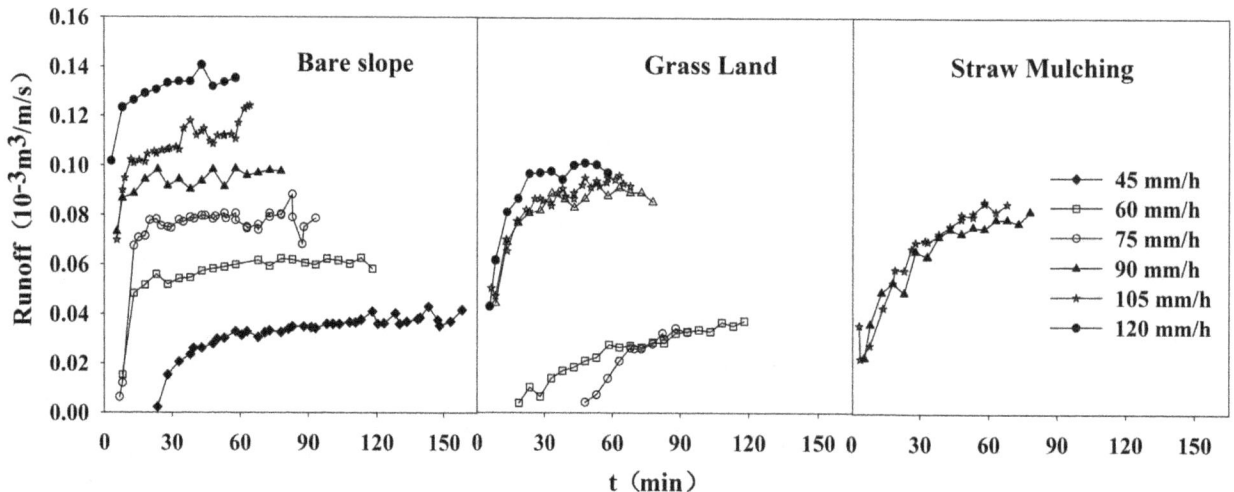

Figure 4. The time-series graphs of the surface runoff under three simulation scenarios.

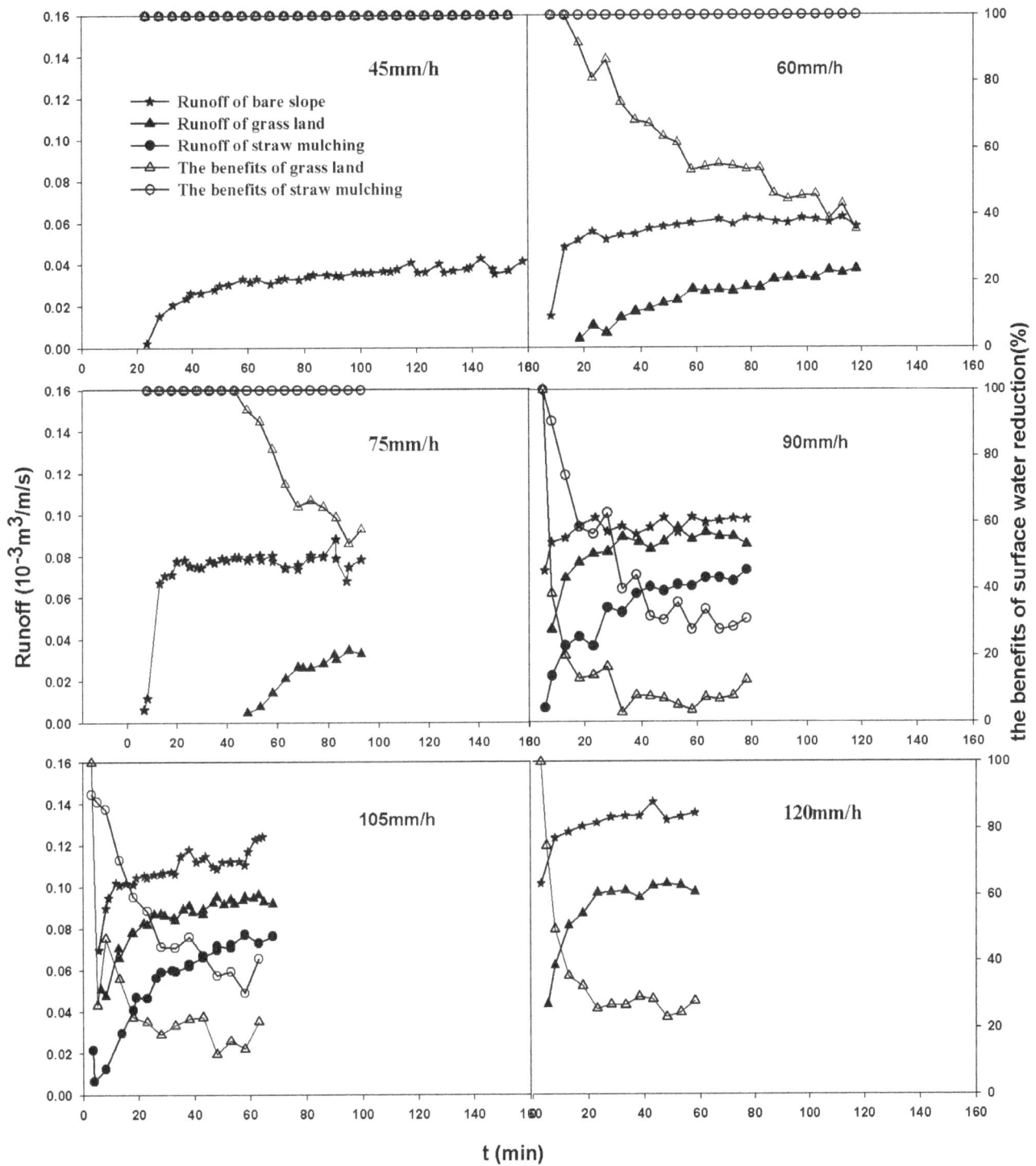

Figure 5. The time-series graphs of the surface runoff and the benefits of surface water reduction.

Ethics Statement

No specific permissions were required for these sampling activities because the location is not privately-owned or protected in any way and the field activities did not involve endangered or protected species.

The Benefits of Water Quantity Change for Soil and Water Conservation Measures

The quantity of surface water and groundwater could be changed by adoption of conservation management practices. Quantitative analysis of this change could be computed by the relationship:

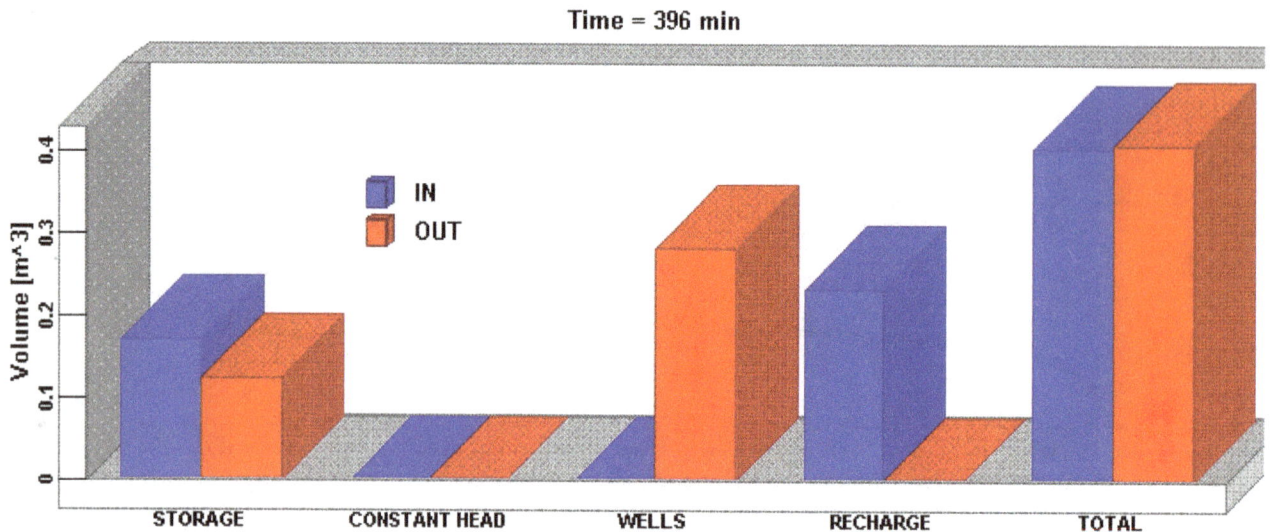

Figure 6. The flow mass balance graph.

$$\delta = \frac{Q_c - Q_{uc}}{Q_{uc}} \times 100\% \qquad (7)$$

Where δ is the benefit of water change contributed by soil and water conservation measures, %; Q_c is the runoff modulus in the scenario controlled by soil and water conservation; Q_{uc} is the runoff modulus in the bare slope scenario.

Results and Discussions

Model Calibration and Validation

Model calibration is the process whereby selected model input parameters are adjusted within reasonable limits to produce simulation results that best match the known or measured values [32]. It is the most critical process in building a model, because the quality of the calibration and validation inevitably determines the reliability of any conclusions and recommendations made using the simulation results.

Calibration and Validation of SCS Model

Where concurrent rainfall and runoff data are available, an 'optimal' curve number can be found by calibrating the curve number to the data [29]. Based on the rainfall runoff data obtained from simulated rainfall experiment for the rainfall intensity 75 mm/h of bare slope scenario, the SCS model was calibrated. Observed $\bar{Q}_{surf}$ and P were 50.01 mm and 75 mm respectively. According to eq. (2) and (3), S and CN values were calculated successively. The calibrated CN value was 90.33. Then based on precipitation data from rainfall intensity 45, 90, 105, 120 mm/h of bare slope scenarios respectively, the SCS model were verified by comparing the estimated runoff with in situ measured data.

Based on the statistical analysis of model results, the relative error (RE, eq. 5) in the estimated runoff ranged from 0.40 to 3.41% (table 2) during the verification period respectively. The calibration and verification statistics showed the model did a relatively good job for surface water predicting. Hence, the SCS model was available for the predicting in this study. So surface runoff of bare slope scenario could be simulated by the verified

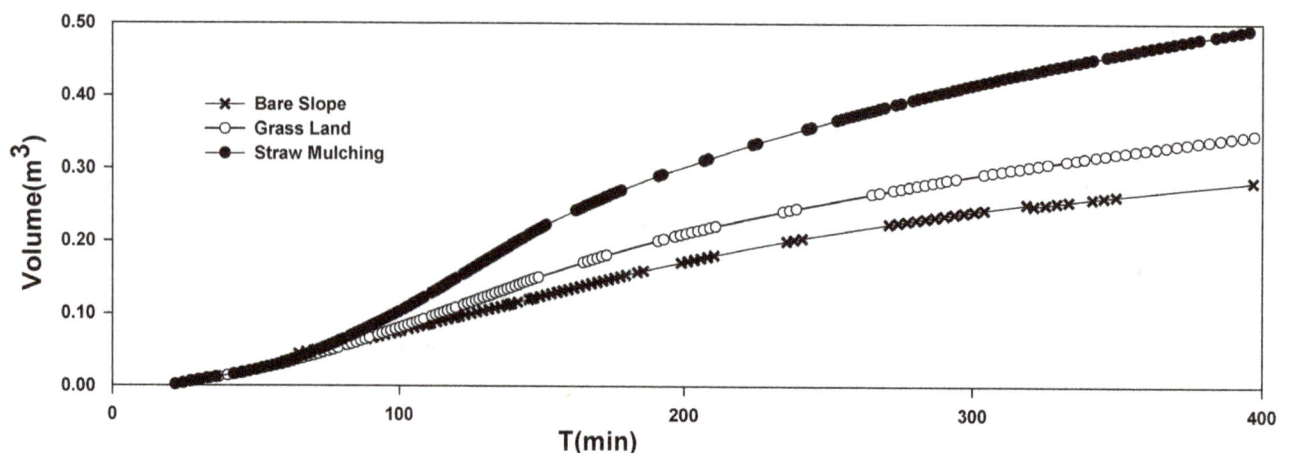

Figure 7. The time-series graphs of accumulated underground water volume under three simulation scenarios.

Table 6. Differences of calculated heads of bare slope vs. observed values of grass land/straw mulching.

Scenario	Average change(cm)	CM$_{max}$/cm	RM/cm	ARM/cm	RMS/cm	NRMS/%
Grass Land	3.91	−11.8	−3.6	4.1	4.5	14.1
Straw Mulching	7.63	−20.5	−7.3	7.4	7.8	23.4

SCS model and the calculated result was compared with the corresponding observed runoff of lawn and straw mulching experiments to study the impact of soil and water conservation on surface water.

Calibration and Validation of Visual MODFLOW Model

The main parameters of the model were: hydraulic conductivity (Kxx, Kyy, and Kzz), storage (specific storage (Ss), specific yield (Sy)). The initial values and the range of these parameters at different layers were provided by the water releasing test analysis combined with the regional geology data. These data were imported into the constructed MODFLOW model as the initial parameters and adjusted during the process of model calibration. Then based on the simulated rainfall experiment of the bare slope, which intensity was also 75 mm/h and rainfall was 120 mm, a uncertainty analysis was carried out after calibration to quantify the uncertainty in the calibrated model caused by uncertainty in the estimates of aquifer parameters, recharge boundary conditions, initial head conditions and river-aquifer interactions [18]. The uncertainty analysis showed that horizontal conductivity in the third layer (Kx3), specific yield (Sy) were the most sensitive parameters in this model. The main parameters calibrated were shown in table 3. Then to verify the model used another simulated

rainfall experiment data of the bare slope, which intensity was 45 mm/h and rainfall was also 120 mm.

The scatter graph of calculated vs. observed values was the default calibration graph. This graph represents a snap-shot in time of the comparison between the values calculated by the model (Y-axis), and the values observed or measured in the field (X-axis) [32]. For example, the scatter graphs at t = 110 min of the calibration period and the verification period (figure 3) showed that most of the data points intersect the 45 degree line on the graph where X = Y. These represented an ideal calibration that simulated hydraulic heads were consistent with the observed heads. The calibration statistics were reported in the footer of the calibration plots window when the calculated vs. observed scatter graph were displayed [32]. Based on these statistical analysis of model results, error indexes including the residual mean (RM), the absolute residual mean (ARM), the standard error of the estimate (SEE), the root mean squared error (RMS), the normalized root mean squared (NRMS) and the correlation coefficient (Cor) for groundwater levels at the location of 48 observation boreholes during the calibration period and the verification period respectively were shown in table 4.

Based upon comparisons the scatter graphs and the calibration statistics between the observed and simulated hydraulic head, we

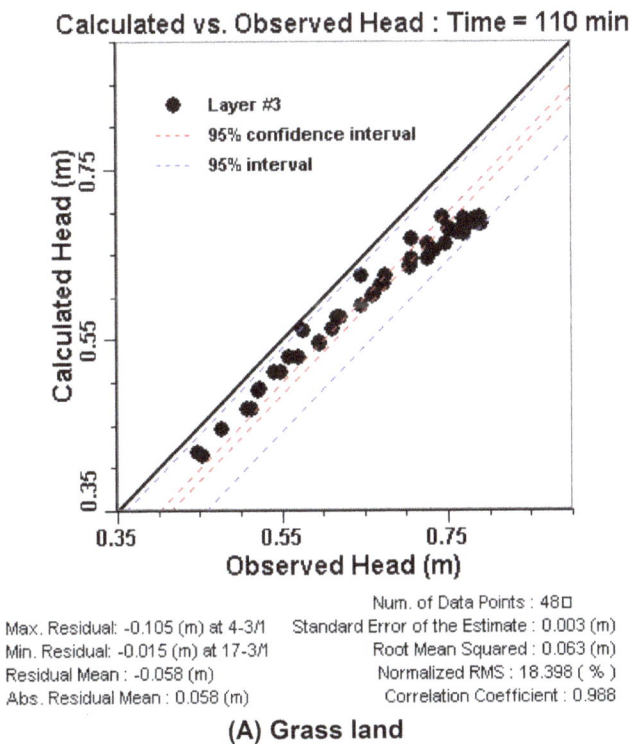

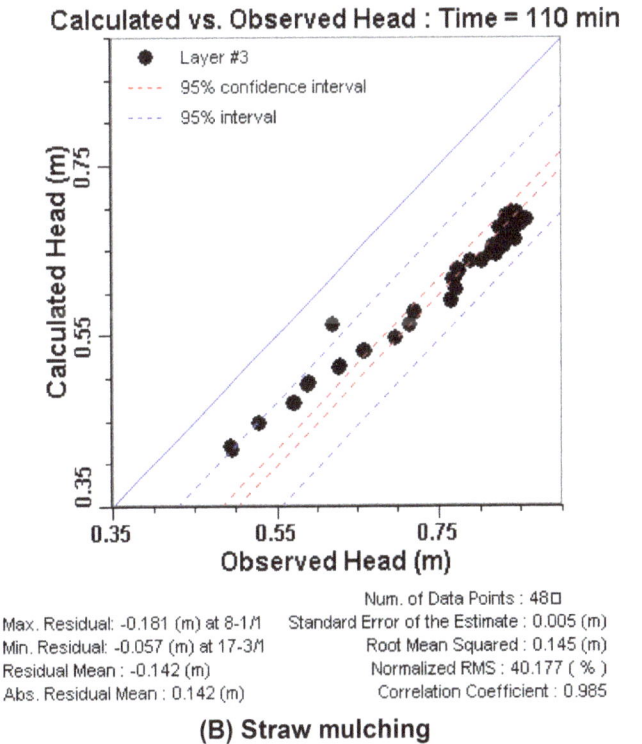

Calculated vs. Observed Head : Time = 110 min

Max. Residual: -0.105 (m) at 4-3/1   Num. of Data Points : 48□
Min. Residual: -0.015 (m) at 17-3/1   Standard Error of the Estimate : 0.003 (m)
Residual Mean : -0.058 (m)   Root Mean Squared : 0.063 (m)
Abs. Residual Mean : 0.058 (m)   Normalized RMS : 18.398 (%)
                                                       Correlation Coefficient : 0.988

(A) Grass land

Calculated vs. Observed Head : Time = 110 min

Max. Residual: -0.181 (m) at 8-1/1   Num. of Data Points : 48□
Min. Residual: -0.057 (m) at 17-3/1   Standard Error of the Estimate : 0.005 (m)
Residual Mean : -0.142 (m)   Root Mean Squared : 0.145 (m)
Abs. Residual Mean : 0.142 (m)   Normalized RMS : 40.177 (%)
                                                       Correlation Coefficient : 0.985

(B) Straw mulching

Figure 8. The scatter graph of calculated vs. observed values.

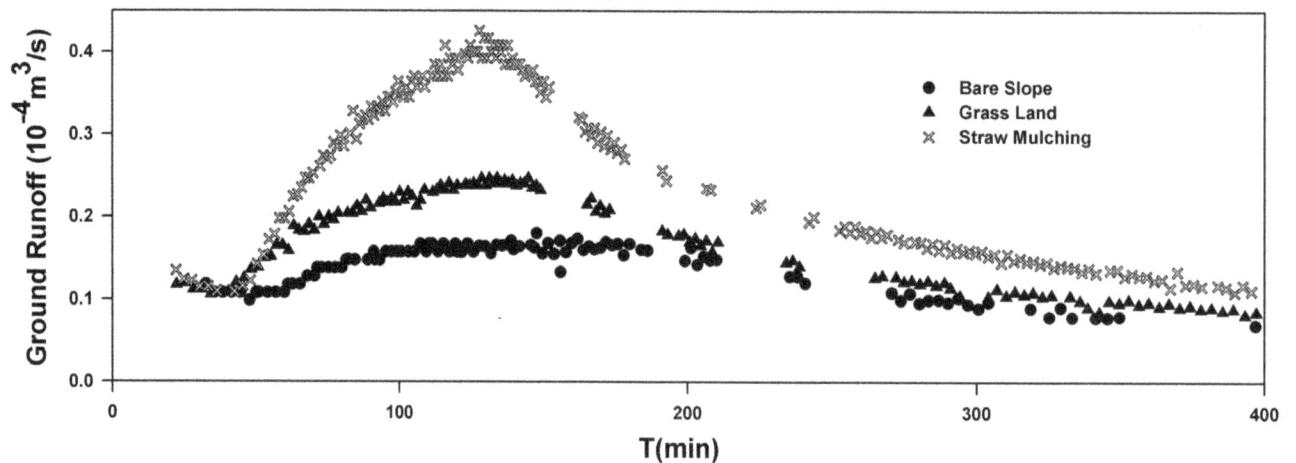

Figure 9. The groundwater runoff of bare slope vs. the discharge of lawn and straw mulching.

could see the groundwater levels simulated by the calibrated groundwater flow model were generally consistent with the physical system represented.

The Impact of Soil and Water Conservation on Surface and Ground Water

Then the calibrated and verified SCS and MODFLOW models were used to simulate rainfall intensity 60 mm/h of bare slope scenario to research the hydrologic impact of soil and water conservation. First, surface runoff amount was simulated by the SCS model and was compared with the corresponding results of lawn and straw mulching experiments to study the impact for the surface water. Second, the calibrated groundwater transient flow model with the input levels obtained from corresponding lawn and straw mulching experiments was used to assess the impact of land change on groundwater.

The Impact of Soil and Water Conservation on Surface Water

The surface runoff modulus amount calculated by SCS model was 0.364 m^3 in bare slope scenario, while the surface runoff observed from the rainfall experiments were 0.154 m^3 and 0.012 m^3 in the lawn and straw mulching scenarios respectively. Compared to the bare slope scenario, the benefits of surface water reduction by these two measures were 57.8% and 92.4% correspondingly. These results indicated that the soil and water conservation played an important role in the benefits of water reduction. These were consistent with the field observation results of the small watersheds. For example, by comparing the 12 groups (24) of controlled and uncontrolled small watersheds (table 5) in Shaanxi Province, China [36], we could see the benefits of surface water reduction by the comprehensive control of soil and water conservation varied from 14.71% to 88.19%, while the corresponding average value was 44.09%.

Not only the amount of surface runoff, but also the runoff process was important for studying the hydrologic impact of soil and water conservation. In this research, the SCS model could reliably estimate runoff amount, but it could not predict runoff process. So the influence for the process of surface runoff was analyzed using simulated rainfall experiments. The time-series graphs (figure 4) were used to evaluate and compare temporal trends in the runoff under three simulation scenarios. The process of runoff would appear to increase first and then stabilize gradually

with the passage of rainfall time under three simulation scenarios. This result was consistent with the research that rainfall infiltration rate was decreased first and then stabilize gradually with the passage of rainfall time [37].

Compared with the bare slope scenario, the reduced range of runoff was different in the lawn scenario and in the straw mulching scenario. When rainfall intensity increased from 75 mm/h to 90 mm/h, the latter average runoff was 7.3 times larger than that of the former in the lawn scenario and it started runoff in the straw mulch scenario. During the whole runoff process of the experiments, the benefits of surface water reduction by the grassland varied from 24.3% to 100% and its average value was 58.9%, while the range was 55.3% to 100% and the average value was 84.2% in the straw mulching scenario (figure 5).

The Impact of Soil and Water Conservation on Ground Water

As mentioned above, the calibrated and verified MODFLOW model was then applied to simulating the impact of soil and water conservation scenarios on groundwater recharge and level. The mass balance which is one of the key indicators of a successful simulation [32] was analyzed first. The flow mass balance graph (figure 6) showed the volume of water entering and leaving the system through the flow boundary conditions, and from aquifer storage at the end of simulation period. The total volume flow into the entire system was 0.4059 m^3 which included 0.1719 m^3 storage and 0.2345 m^3 recharge. The total volume flow out-of the model was 0.4070 m^3 which consisted of 0.1252 m^3 storage and 0.2818 m^3 well. The mass balance error for the simulation inflow and outflow was 1.47% <2%. The results of the simulation may generally be considered acceptable, provided the model was also calibrated [32].

Figure 7 compared the changes in accumulated underground runoff with scenarios converted from bare slope to grass land and straw mulching. As a result, the cumulative curve of straw mulching was the highest, the grassland took second place, and the bare slope was the lowest. Hence, there was a significant increase for straw mulching, contrasting to the changes for grassland. Then the impact of groundwater cumulant was analyzed quantitatively. At the end of simulation period, the accumulated underground runoff was 0.282 m^3 in the bare slope scenario, 0.346 m^3 in the lawn scenario, and 0.491 m^3 in the straw mulching scenario. Compared to the bare slope scenario, the amount and the benefit of groundwater recharge increased by 0.064 m^3 and 22.7% respectively in the lawn scenario. Accordingly, they were 0.209 m^3 and 60.4% in the straw mulching scenario. It was explicit that groundwater recharge very strongly depended upon land-use type [16].

The groundwater levels responded to the changes in groundwater recharge described as follows. Figure 8 compared the influence of soil and water conservation scenarios on groundwater levels at t = 110 min. Figure 8 (A) was the scatter plots of calculated levels of bare slope scenario vs. observed heads of grass land scenario. And figure 8 (B) was the graph of calculated levels of bare slope scenario vs. observed values of straw mulching scenario. As shown in the figures, the data points were under the X = Y line, so the calculated values were less than the observed values, especially in figure 8 (B). These indicated that groundwater levels increased with land-use converted from bare slope to grassland and even more with land-use converted to straw mulching. To quantify the impact on groundwater heads, the changes between the calculated heads of bare slope scenario and observed levels of grass land and straw mulching scenarios were showed in table 6.

It was explicit from the curve of figure 9 that the soil and water conservations had great influence upon the groundwater runoff. Straw mulching showed the higher runoff, followed by grassland. It was illustrated by further analysis that average flow was 0.14×10^{-4} m^3/s in the bare slope scenario, 0.17×10^{-4} m^3/s in the lawn scenario, and 0.26×10^{-4} m^3/s in the straw mulching scenario, during the simulation period. So the average runoff increased by 1.23 and 1.87 times through measures of lawn and straw mulching.

In summary, both grass and straw mulching scenarios played an important role in reducing the surface runoff and increasing the underground runoff. The effects of measures on redistribution of rainfall runoff was mainly due to the rainfall infiltration rata was changed. The grass and straw mulching dispersed the large raindrops into small raindrops, which reduced the actual rainfall intensity on the ground [38–42] and made more precipitation meet the condition of infiltration. But different soil and water conservation measures had different effects. In this study, straw mulching has more significant hydrological effects than lawn. The main difference between the two scenarios was that the straw mulching scenario had less bare land area compared to lawn scenario.

Conclusions

Based on the principle of water balance, a linked approach for SCS model and Visual MODFLOW was conducted to assess the impact of soil and water conservation on the surface water and groundwater runoff. The calibration results showed that the predicted results matched well with the observed data. Therefore,

three land use management scenarios were simulated on the sandbox model to assess the effect of bare slope converted to grass land and straw mulching on water volume, hydraulic head, runoff process of groundwater and surface water.

Soil and water conservation measures could reduce surface runoff effectively. Under the 120 mm rainfall, 60 mm/h rainfall intensity, 5 m^2 area, 3° slope conditions, the comparative results indicated that decrease in surface runoff and increase in subsurface runoff coincided with the land-use converted from bare slope to grass land and straw mulching. Compared to the bare slope, the benefits of surface water reduction by these two measures were 57.8% and 92.4% correspondingly. Not only the individual soil and water conservation measure but also the comprehensive management of the small watershed had a significant benefit of surface water reduction. The comparative results of 12 groups (24) of controlled and uncontrolled small watersheds showed that the benefits of surface water reduction varied from 14.71% to 88.19% while the corresponding average value was 44.09%.

Soil and water conservation measures could promote rainfall recharge groundwater. Under the same condition, compared to the bare slope, the amount and the benefit of groundwater recharge increased by 0.064 m^3 and 22.7% respectively in the lawn scenario. Accordingly, they were 0.209 m^3 and 60.4% in the straw mulching scenario. For the runoff, the average flow of straw mulching was highest, the grassland took second place, the bare slope was the lowest, and the average runoff increased by 1.23 and 1.87 times through measures of lawn and straw mulching.

It was concluded that the soil and water conservation played an important role in weakening the surface runoff and strengthening the underground runoff. Meanwhile the groundwater flow model coupled with the surface water model used in this study, when properly validated, could be used as a tool in the evaluation of soil and water conservation measures on surface water and groundwater resources, and this approach could provide a thought for the study in a watershed scale to help decision-makers manage groundwater resources.

Acknowledgments

We would like to thank Mr. Xiu-quan XU, Mr. Li-zhi JIA, and Mr. Tong ZHANG for their assistant in experiment processing and Miss. Chun-hong ZHAO, Mr. Hong-jie WANG and Mr. Yuan-xing ZHANG for their advices in writing the paper. At the same time, the authors are very grateful for the editors' and reviewer's hard work in reviewing the paper.

Author Contributions

Conceived and designed the experiments: JEG HW. Performed the experiments: HW SLZ MJZ XHL. Analyzed the data: HW JEG. Contributed reagents/materials/analysis tools: JEG HW. Wrote the paper: HW JEG.

References

1. Li ZJ, Zhou PX, Mao LH (2006) General Review on Effects of Soil and Water Conservation Measures on Water Resources in China. Progress in Geography 25: 49–57.
2. Zhu XM, Jiang DS, Zhou PH, Jin ZS (1984) Discussion on the strategic issues of soil and water conservation in the loess region. Bulletin of soil and water conservation 1: 15–18.
3. Ren ME (2006) Sediment Discharge of the Yellow River, China: Past,Present and Future–A Synthesis. Advances in earth science 6: 551–563.
4. Yellow River Conservancy Commission of MWR (2010) Yellow river sediment bulletin, http://www.yellowriver.gov.cn/nishagonggao/2010/index.html.Accessed 2013 March 6.
5. Kim J, Choi J, Choi C, Park S (2013) Impacts of changes in climate and land use/land cover under IPCC RCP scenarios on streamflow in the Hoeya River Basin, Korea. Science of The Total Environment 452: 181–195.

6. Skaggs RW, Amatya DM, Chescheir GM, Blanton CD, Gilliam JW (2006) Effect of drainage and management practices on hydrology of pine plantation. Notes.
7. Sun G, Riedel M, Jackson R, Kolka R, Amatya D, et al. (2004) Influences of management of Southern forests on water quantity and quality. Southern Forest Science, General Technical Report SRS-75 198.
8. Qi S, Sun G, Wang Y, McNulty SG, Moore Myers JA (2009) Streamflow response to climate and landuse changes in a coastal watershed in North Carolina. Trans ASABE 52: 739–749.
9. Li YS (1997) Relation Between Control in Loess Plateau and No-flow in the Yellow River. Bulletin of Soil and Water Conservation 17: 41–45.
10. Ran DC (2006) Water and Sediment Variation and Ecological Protection Measures in the Middle Reach of the Yellow River. Resources Science 28: 93–100.

11. Qi JY Cai QG, Cai L, Sun LY (2011) Scale Effect of Runoff and Sediment Reduction Effects of Soil and Water Conservation Measures in Chabagou, Dalihe and Wudinghe Basins. Progress in Geography 30: 95–102.

12. Jain MK, Mishra SK, Singh VP (2006) Evaluation of AMC-dependent SCS-CN-based models using watershed characteristics. Water Resources Management 20: 531–552.

13. Shi PJ, Yuan Y, Zheng J, Wang JA, GeY, etal. (2007) The effect of land use/cover change on surface runoff in Shenzhen region, China. Catena 69: 31–35.

14. Guo XJ, Wang DJ, Zhuang JQ (2010) Application of SCS model on simulation of the slope -runoff process in dry - hot valley. Science of Soil and Water Conservation 8: 14–18.

15. Freeze RA (1969) The Mechanism of Natural Ground - Water Recharge and Discharge: 1. One - dimensional, Vertical, Unsteady, Unsaturated Flow above a Recharging or Discharging Ground - Water Flow System. Water Resources Research 5: 153–171.

16. Paul MJ (2006) Impact of land-use patterns on distributed groundwater recharge and discharge. Chinese Geographical Science 16: 229–235.

17. Cho J, Barone VA, Mostaghimi S (2009) Simulation of land use impacts on groundwater levels and streamflow in a Virginia watershed. Agricultural water management 96: 1–11.

18. Zhang H, Hiscock KM (2010) Modelling the impact of forest cover on groundwater resources: A case study of the Sherwood Sandstone aquifer in the East Midlands, UK. Journal of Hydrology 392: 136–149.

19. Cooper JD (1980) Measurement of moisture fluxes in unsaturated soil in Thetford Forest. Research Report 66. Institute of Hydrology, Wallingford, UK.

20. Bosch JM, Hewlett JD (1982) A review of catchment experiments to determine the effect of vegetation changes on water yield and evapotranspiration. Journal of Hydrology 55: 3–23.

21. Dams J, Woldeamlak ST, Batelaan O (2008) Predicting land-use change and its impact on the groundwater system of the Kleine Nete catchment, Belgium. Hydrology and Earth System Sciences 12: 1369–1385.

22. Allen A, Chapman D (2001) Impacts of afforestation on groundwater resources and quality. Hydrogeology Journal 9: 390–400.

23. Pouget L, Escaler I, Guiu R, Mc Ennis S, Versini PA (2012) Global Change adaptation in water resources management: The Water Change project. Science of The Total Environment 440: 186–193.

24. Ferrer J, Pérez-Martin MA, Jiménez S, Estrela T, Andreu J (2012) GIS-based models for water quantity and quality assessment in the Júcar River Basin, Spain, including climate change effects. Science of The Total Environment 440: 42–59.

25. Candela L, Tamoh K, Olivares G, Gomez M (2012) Modelling impacts of climate change on water resources in ungauged and data-scarce watersheds. Application to the Siurana catchment (NE Spain). Science of The Total Environment 440: 253–260.

26. Boughton WC, Stone JJ (1985) Variation of runoff with watershed area in a semi-arid location. Journal of Arid Environments 9: 13–25.

27. Yu B (1998) Theoretical justification of SCS method for runoff estimation. Journal of irrigation and drainage engineering 124: 306–310.

28. Rallison RE (1980) Origin and evolution of the SCS runoff equation. Symposium on Watershed Management 1980. ASCE, New York. N, Y. II: 912–924.

29. Boughton WC (1989) A review of the USDA SCS curve number method. Soil Research 27: 511–523.

30. Wang ML, Guo SL, Yi Y (2004) SCS Monthly hydrological model building and its application. Journal of Water Resources Research 2: 002.

31. Harbaugh AW, Banta ER, Hill MC, McDonald MG (2000) MODFLOW-2000, the US Geological Survey modular ground-water model: User guide to modularization concepts and the ground-water flow process. Reston: US Geological Survey.

32. Hydrogeologic Water (2005) Visual MODFLOW v. 4.1 User's Manual. Waterloo, Ontario.

33. Harbaugh AW, McDonald MG (1996) User's documentation for MODFLOW-96, an update to the US Geological Survey modular finite-difference ground-water flow model. United States, Geological Survey.

34. Zhang WZ (1983) The Calculation of the Groundwater Unsteady Flow and Appraisal of Groundwater Resources. Beijing: Science Press.

35. Liang W, Liu L, PAN P (2007) Simulation of nitate-nitrogen transfer in trough scale. Lake Science 19: 710–717.

36. Shaanxi Provincial Bureau of Soil and Water Conservation (1976) Shaanxi Provincial runoff and sediment Data of soil and Water conservation. Xi'an: Shaanxi Provincial Bureau of Soil and Water Conservation.

37. Li YY, Shao MA (2004) Experimental study on characteristics of water transformation on slope land. Journal of hydraulic engineering 4: 48–53.

38. Laflen JM, Colvin TS (1981) Effect of crop residue on soil loss from continuous row cropping. Transactions of the ASAE 24: 1472–1475.

39. West LT, Miller WP, Langdale GW, Bruce RR, Laflen JM, et al. (1991) Cropping system effects on interrill soil loss in the Georgia Piedmont. Soil Science Society of America Journal 55: 460–466.

40. Zhu XM (1960) The impact of vegetation factors on soil and water erosion in Loess area. Acta pedologica sinica 8: 110–121.

41. Tang T, Hao MD, Shan FX (2008) Effects of Straw Mulch Application on Water Loss and Soil Erosion Under Simulated Rainfall. Research of Soil and Water Conservation 15: 9–11.

42. Li M, Yao WY, Li ZB (2005) Progress of the Effect of Grassland and Vegetation for Conserving soil and water on Loess Plateau. Advances in earth science 20: 74–80.

16

Adaptation of Pelage Color and Pigment Variations in Israeli Subterranean Blind Mole Rats, *Spalax Ehrenbergi*

Natarajan Singaravelan[1,2]*, **Shmuel Raz**[1], **Shay Tzur**[1], **Shirli Belifante**[1], **Tomas Pavlicek**[1], **Avigdor Beiles**[1], **Shosuke Ito**[3], **Kazumasa Wakamatsu**[3], **Eviatar Nevo**[1]

1 Institute of Evolution, University of Haifa, Mount Carmel, Haifa, Israel, 2 Bommanampalayam, Coimbatore, Tamil Nadu, India, 3 Department of Chemistry, Fujita Health University School of Health Sciences Toyoake, Aichi, Japan

Abstract

Background: Concealing coloration in rodents is well established. However, only a few studies examined how soil color, pelage color, hair-melanin content, and genetics (i.e., the causal chain) synergize to configure it. This study investigates the causal chain of dorsal coloration in Israeli subterranean blind mole rats, *Spalax ehrenbergi*.

Methods: We examined pelage coloration of 128 adult animals from 11 populations belonging to four species of *Spalax ehrenbergi* superspecies (*Spalax galili*, *Spalax golani*, *Spalax carmeli*, and *Spalax judaei*) and the corresponding coloration of soil samples from the collection sites using a digital colorimeter. Additionally, we quantified hair-melanin contents of 67 animals using HPLC and sequenced the *MC1R* gene in 68 individuals from all four mole rat species.

Results: Due to high variability of soil colors, the correlation between soil and pelage color coordinates was weak and significant only between soil hue and pelage lightness. Multiple stepwise forward regression revealed that soil lightness was significantly associated with all pelage color variables. Pelage color lightness among the four species increased with the higher southward aridity in accordance to Gloger's rule (darker in humid habitats and lighter in arid habitats). Darker and lighter pelage colors are associated with darker basalt and terra rossa, and lighter rendzina soils, respectively. Despite soil lightness varying significantly, pelage lightness and eumelanin converged among populations living in similar soil types. Partial sequencing of the *MC1R* gene identified three allelic variants, two of which were predominant in northern species (*S. galili* and *S. golani*), and the third was exclusive to southern species (*S. carmeli* and *S. judaei*), which might have caused the differences found in pheomelanin/eumelanin ratio.

Conclusion/Significance: Darker dorsal pelage in darker basalt and terra rossa soils in the north and lighter pelage in rendzina and loess soils in the south reflect the combined results of crypsis and thermoregulatory function following Gloger's rule.

Editor: Alexandre Roulin, University of Lausanne, Switzerland

Funding: EN acknowledges the financial support of the Ancell Teicher Research Foundation for Genetics and Molecular Evolution and the Israel Discount Bank Chair of Evolutionary Biology. KW and SI acknowledge the funding support by Japan Society for the Promotion of Science (JSPS) KAKENHI (No. 20591357, 21500358). NS acknowledges postdoctoral fellowship support by University of Haifa and the Council of Higher Education, Israel. The funders had no role in study design, data collection and analysis, decision to publish, or preparation of the manuscript.

Competing Interests: The authors have declared that no competing interests exist.

* E-mail: yoursings@gmail.com

Introduction

Mammalian pelage coloration plays an important role in crypsis, intra-specific signaling, thermoregulation, and ultraviolet screening [1–4]. Several studies demonstrated a strong positive correlation between rodents' coat color and background color of the environment in which they live, indicating that natural selection is operating [5–9]. Such adaptive coat color variations are caused by the switch between 'brown to black' eumelanin and 'yellow to red' pheomelanin [10–13]. This dual melanogenesis is controlled by the interaction of two proteins: melanocortin-1-receptor (MC1R) and agouti-signaling protein (ASIP) [14–16]. MC1R is a G protein-coupled receptor expressed highly in melanocytes involved in the production of eumelanin. Agouti is an antagonist of MC1R. The expression of ASIP suppresses the synthesis of eumelanin and triggers the production of pheomelanin.

A classic example of MC1R gene-driven coat color variation is demonstrated in pocket mice; individuals inhabiting dark volcanic lava have dark coats, and mice inhabiting light-colored granitic rocks exhibit light coats. This color polymorphism is considered as a cryptic adaptation to avoid predation [17,18]. Despite numerous studies on the dorsal coloration of rodents, how melanin contents are selected to form concealing coloration is least explored [19]. A clearer understanding is essential to elucidate the evolution of concealing coloration and the pigmental variation underlying it, including comparisons of: 'soil color vs. pelage color', 'pelage color vs. melanin contents', and 'pelage color vs. candidate gene', to suggest adaptation to soil color. Such comprehensive studies are still lacking. Hence, in the present study, we intend to address the above scenario in populations of the subterranean mole rats of the *Spalax ehrenbergi* superspecies in Israel.

Diversifying selection of pelage color occurs even in burrowing subterranean mammals that exhibit adaptation of pelage to the color of their background habitat (e.g., *Thomomys* [20] and *Geomys* [21]). Such observations were substantiated also in blind mole rats of the *Spalax ehrenbergi* superspecies whose populations live in the subterranean environment and are selected for different soil colors by differential predation (22). The pelage coloration of *Spalax ehrenbergi* varies and tends to match different soil colors [22]. Why and how selection works on the pelage color of these blind mole rats, restricted most of their lives to the underground ecotope, is important evolutionarily. There are four species of mole rats of the superspecies "*Spalax ehrenbergi*" in Israel; S. galili ($2n = 52$), S. golani ($2n = 54$), S. carmeli ($2n = 58$), and S. judaei ($2n = 60$), whose distribution in Israel in four distinct parapatric, climatically different areas is highly correlated with increasing aridity, both southward and eastward [23–25].

Pelage of blind mole rats is usually gray, but there are differences among the four species which enable them to be camouflaged above ground, especially at night [22]. The apical and sub-apical portion of the hairs of the northern mesic-species ($2n = 52, 54$) is reddish orange with darker pelage corresponding to reddish brown and dark tones of the terra rossa and basalt soils, respectively, which they mainly inhabit. The xeric southern species ($2n = 58$, and primarily $2n = 60$) tend to be yellowish with lighter pelage, except for those populations living in the alluvial soils around the coastal rivers. Such color variation between species of mole rats is hypothesized to be due to the selection pressures of predation and thermoregulation, even though they spend little time above ground (mostly at night). The correlation of pelage color with soil color certainly suggests that adaptation occurs and reflects an underlying genotypic variation. The study by Heth et al. [22] relied on Munsell color charts to determine both pelage and soil colors. This method is limited due to eye perception differences, illumination, and other factors that can affect color determination. In the current study, we measure the coloration of mole rats using a digital colorimeter. We test the working hypothesis that crypsis exists in mole rats when there is harmony in the causal chain (i.e., soil color, pelage color, hair melanin contents, and genetics) of pelage coloration. Complementarily, we explore the climatic cline on pelage color following the ecological Gloger's rule.

Methods

Ethics statement

Mole rats were trapped from various locations using special live traps by exposing the tunnels using a hoe. No specific permissions were required for these locations, which are wild habitats of mole rats; none belonged to protected areas or belonged to private property. The field studies did not involve endangered or protected species. Animals used in the study were adults. The experiments were approved by the Ethics Committee of the University of Haifa.

Animal collection and hair sampling

We measured the dorsal pelage color of 128 adult mole rats from 11 populations of the four *Spalax* species in Israel (Figure 1). Mole rats were captured in the field during 2002–2010 and were maintained in the animal facility at the Institute of Evolution, University of Haifa, under constant conditions ($22°C$ with relative humidity 70%; photoperiod 12L:12D) and were fed vegetables. Hair samples were excised with scissors to the full length obtained

from random locations over the animal's dorsal body across a 3×3 cm^2 area (~60 mg of hairs) from the different geographic regions. We made sure to excise uniform lengths (~95% of total length) of the pelage. This procedure was consistent for all individuals to standardize sample collections for melanin analysis. Animals were treated with care while removing hairs to prevent injury and/or suffering.

Color Measurement

All color measurements were made on adults. Note that the coloration changes during the course of ageing; pups and juveniles are lighter than fully matured adults. We measured the dorsal pelage color using a digital colorimeter (Spec boss 4000, JETI). Each animal was measured at least 12 times in random locations over the dorsal body. We used the L*a*b* color space model (CIE-LAB) under standard daylight illumination [26] to quantify different components of the measured color. This color space was selected because it is more appropriate for the biological aims of this study. The color space component 'L*' represents the level of lightness in color (L* estimates equivalent to 'brown to black' eumelanin), a positive value of 'a*' is represented in red/magenta, while a negative value of 'a*' is represented in green. The positive value of 'b*' is represented by the amount of purplish-red (magenta) yellow, while the negative value of 'b*' represents blue (a* and b* estimates are equivalent to the 'yellow to red' pheomelanin). Chroma ('C' = $\sqrt{a*^2 + b*^2}$) and hue ('h' = arc tan [a*/b*]) were derived from L*a*b* parameters and were also included in the analysis. For statistical analysis, we estimated the average value for each color component, for each individual, following elimination of the 20% percentile of the lowest and highest color measurements in order to avoid the inclusion of unreliable estimates.

Melanin Assay

Micro-analytical methods to quantify the amounts of eumelanin and pheomelanin were based on the formation of specific degradation products, pyrrole-2,3,5-tricarboxylic acid (PTCA) by alkaline H_2O_2 oxidation of eumelanin and 4-amino-3-hydroxy-phenylalanine (4-AHP) by reductive hydrolysis of pheomelanin with hydriodic acid (HI) [27,28]. Hair samples were homogenized with a Ten-Broeck glass homogenizer at a concentration of 10 mg/mL water.

Alkaline H_2O_2 oxidation was used to measure eumelanin (PTCA). A sample homogenate (100 μL) was taken in a 10-ml screw-capped conical test tube, to which 375 μL 1 mol/L K_2CO_3 and 25 μL 30% H_2O_2 (final concentration: 1.5%) were added. The mixture was mixed vigorously at $25°C \pm 1°C$ for 20 h. The residual H_2O_2 was decomposed by adding 50 μL 10% Na_2SO_3, and the mixture was then acidified with 140 μL 6 mol/L HCl (the generation of CO_2 occurs by adding HCl to the alkaline mixtures). After vortex-mixing, the reaction mixture was centrifuged at 4,000 g for 1 min, and an aliquot (80 μL) of the supernatant was directly injected into the HPLC system [28–30].

HI reductive hydrolysis was used to measure pheomelanin (4-AHP). A sample homogenate (100 μL) was taken in a 10-ml screw-capped conical test tube to which 20 μL 50% H_3PO_2 and 500 μL 57% HI were added. The tube was heated at $130°C$ for 20 h, after which the mixture was cooled. An aliquot (100 μL) of each hydrolysate was transferred to a test tube and evaporated until dried using a vacuum pump connected to a dry ice-cooled vacuum trap and two filter flasks containing NaOH pellets. The residue was dissolved in 200 μL 0.1 mol/L HCl. An aliquot (10 μL) of each solution was analyzed on the HPLC system [27].

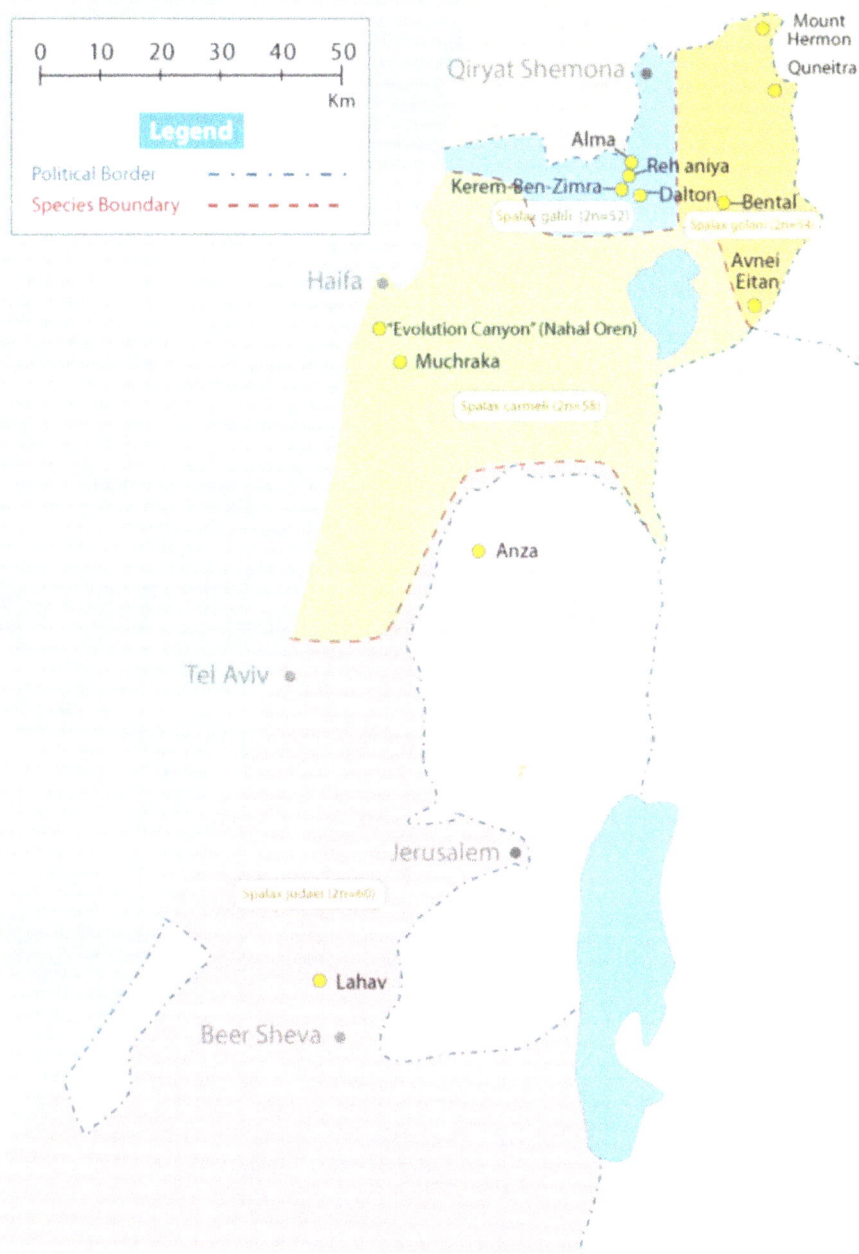

Figure 1. Map shows the studied populations of mole rats in Israel. Kerem-Ben-Zimra, Alma and Rehaniya populations are located in Galilee mountains; Quneitra A. Etan and Bental are in Golan heights. Muhraka and "Evolution Canyon" (Nahal Oren) populations are in Carmel mountains. Anza is located in West Bank, while Lahav is in Negev.

Soluene-350 solubilization was used to measure total melanin. A sample homogenate (100 μL) was taken in a 10-ml screw-capped conical test tube, to which 900 μL Soluene-350 (from Perkin-Elmer) was added. The tube was vortex-mixed and heated at 100°C (in a boiling water bath) for 15 min, after which the mixture was cooled. The tube was vortex-mixed and heated again at 100°C for an additional 15 min and then cooled. After vortex-mixing, the mixture was centrifuged at 4,000 g for 3 min, and the supernatant was analyzed for absorbance at 500 nm (A500). For a reference, a mixture of 100 μL water and 900 μL Soluene-350 was used after heating under the same conditions as for the samples. Background values (due to protein) of 0.019 at

500 nm and 0.001 at 650 nm for mouse hair samples were recorded [31].

HPLC analyses

H_2O_2 oxidation products were analyzed with the HPLC system consisting of a JASCO 880-PU liquid chromatograph (JASCO Co., Tokyo, Japan), a Shiseido C_{18} column (Shiseido Capcell Pak MG; 4.6×250 mm; 5 μm particle size), and a JASCO UV detector. The mobile phase was 0.1 mol/L potassium phosphate buffer (pH 2.1): methanol, 99: 1 (v/v). Analyses were performed at 45°C at a flow rate of 0.7 mL/min. Absorbance of the elute was monitored at 269 nm. A standard solution (80 µL) containing 1 µg each of PTCA (pyrrole-2,3,5-tricarboxylic acid), PDCA (pyrrole-2,3-dicarboxylic acid), TTCA (thiazole-2,4,5-tricarboxylic acid), and TDCA (thiazole-2,3-dicarboxylic acid) in 1 mL HPLC buffer was injected into the HPLC system every 10 samples.

HI reductive hydrolysis products were analyzed with an HPLC system consisting of a JASCO 880-PU liquid chromatograph, a JASCO C_{18} column (JASCO Catecholpak; 4.6×150 mm; 7 µm particle size), and an EICOM ECD-300 electrochemical detector. The mobile phase used for analysis of 4-AHP was 0.1 mol/L sodium citrate buffer, pH 3.0, containing 1 mmol/L sodium octanesulfonate and 0.1 mmol/L Na_2EDTA: methanol, 98: 2 (v/v). Analyses were performed at 35°C at a flow rate of 0.7 mL/min. The electrochemical detector was set at +500 mV versus an Ag/AgCl reference electrode. A standard solution (10 µL) containing 500 ng each of 4-AHP (4-amino-3-hydroxyphenylalanine) and 3-AHP (3-amino-4-hydroxyphenylalanine; 3-aminotyrosine from Sigma) in 1 mL 0.1 mol/L HCl was injected into the HPLC system every 10 samples.

Screening variations in the *MC1R* gene of mole rats

The *MC1R* gene was partially sequenced (905 bp out of 952 bp total length) in 68 mole rats from the four *Spalax* species, as presented in Supplemental Table S1. Specific PCR primers were used to amplify the gene fragment: S5F 'cagaagaggctgctggactc' and S3B 'gagctccgcatgacactcag'. PCR was performed in a 25-µL reaction volume containing 12.5 µL ReadyMixTM Taq PCR Reaction Mix with MgCl2 (Sigmàs product code P4600), 0.5 µL primer s5f 0.5 µL primer s3b, 10.5 µL water, and 1 µL DNA sample. PCR-amplification was done under the following conditions: 94°C for 5 min, then 39 cycles of 94°C for 30 s, 57°C for 30 s, and 72°C for 30 s, followed by 72°C for 7 min.

Statistical analysis

Numerical data were shown as means ± SD (standard deviation). The differences in pelage color coordinates and hair melanin contents between males and females, and between populations (where applicable) were examined using the Mann-Whitney test. The Kruskal Wallis test was employed to detect the significance of variation in soil and pelage color coordinates and in hair-melanin contents among populations. The association between soil and pelage color coordinates was detected by employing the Spearman correlation. We used multiple-regression, followed by stepwise forward-regression to reveal the effect of soil color variables (as independent variables) on pelage color variables (as dependent variables) of mole rats. We employed the chi-square test to detect variations in haplotype frequencies between northern and southern species of mole rats.

Results

Soil color variations

The soil colors of different populations were light gray, yellow, brown, and dark gray. *S. galili* inhabits a variety of soils including terra rossa, rendzina, and basalt. *S. golani* inhabits basalt soil, *S. carmeli* inhabits terra rossa, and *S. judaei* inhabits rendzina and loess soils. Basalt and terra rossa soils are dark due to the presence of oxide metals, while rendzina and loess soils are light due to the high content of chalk. These variations in soil types are clearly manifested in the measured values of soil color. The maximum lightness score ('L*') of soils in our samples is found in *S. judaei* (range L* =41.93–45.19) and *S. galili* in the Kerem Ben Zimra rendzina sample we tested (42.19–50.81), compared with *S. carmeli* (23.13–30.93) and *S. golani* (26.36–30.30). In addition, soils of *S.*

galili (a* = −0.84–10.47) and *S. golani* (a* = 5.16–8.08) tend to be more reddish, while *S. judaei* (b* = 13.58–15.94) and *S. galili* (b* = 9.32–16.56) tend to be more yellowish in our samples. As expected, basalt and terra rossa soils exhibited lower lightness, and rendzina and loess soils exhibited higher lightness scores indicating that the former two soil types are relatively darker than the latter, as is also clearly observed. Moreover, terra rossa soils are more reddish, whereas rendzina soils are more yellowish (Table 1). The rendzina soil in Kerem Ben Zimra (L* = 46.62±2.59) is the lightest, and the abutting basalt soil in Alma (26.31±5.73) is the darkest among populations investigated. Rihaniya soil is lighter than the Muhraka and "Evolution Canyon" (Nahal Oren) populations inhabiting terra rossa soils. Soil color parameters varied significantly among populations that live in similar soil types (see Table 1).

Variation between sexes

We did not find differences in pelage color between females and males (species and populations pooled data) both in the colorimeter measurements that were performed on 30 females and 13 males (Table 2), and in the comparison of the melanin content, which were performed on 47 females and 20 males (Table 3). These results firmly support the conclusion that mole rats do not show sex color variation in pelage color as expected from blind species.

Variation among populations in pelage color

Overall, pelage lightness (L*) varied significantly among 11 populations from the four species living in different soils ($H = 20.366$, $df = 10$, $P = 0.026$). Likewise, pelage color lightness of populations living in four soil types (basalt, terra rossa, rendzina, and loess) varied significantly ($H = 10.399$, $df = 3$, $P = 0.015$; populations' pooled data). Regardless of species and climatic divergence, the Anza population of *S. judaei* (39.89±3.63) from the south and Kerem-Ben-Zimra, KBZ population of *S. galili* (38.66±2.66) from the north, living 70 km apart, were the lightest, both inhabiting rendzina (light colored, chalky) soils. The KBZ population inhabits a significantly more humid region. The Rihaniya population of *S. galili* that lives in the darker terra rossa soil, 2 km apart from KBZ, was the darkest (32.32±8.50) (Table 4). This indicates how cryptic factors prevail over climate factors in nearby populations. Among basalt populations, Alma (38.11±4.18) exhibited the lightest pelage color and neighboring Dalton (34.49±6.38) exhibited the darkest. The Anza population showed lighter pelage than the Kerem-Ben-Zimra in the light rendzina soil, which may indicate the climatic determinant in Anza, which is drier than KBZ. Among populations that live in terra rossa soil, mole rats in "Evolution Canyon", Nahal Oren (38.07±3.50) exhibited the lightest pelage, and those living in Rihaniya (32.32±8.50) showed the darkest pelage (Table 4). Regardless of species, populations living in similar soil types did not vary significantly in lightness of pelage color (Table 4). However, populations living on basalt showed significant variations in color coordinates, a* ($H = 33.617$, $df = 4$, $P<0.001$), b* ($H = 29.466$, $df = 4$, $P<0.001$), chroma ($H = 30.521$, $df = 4$, $P<0.001$), and hue ($H = 17.739$, $df = 4$, $P = 0.001$). Populations that live in terra rossa soils exhibited significant variations only for pelage variables a* ($H = 12.974$, $df = 2$, $P = 0.002$) and h* ($H = 15.384$, $df = 2$, $P<0.001$); the rendzina populations did not show significant variation for any of the pelage color variables (Table 4). Population-wise variations are striking across 11 populations living in four soil types. But regardless of species, populations living in similar soil types showed convergence in pelage lightness (Table 4), again

Table 1. Soil color variation among populations living in different soil types.

Soil Type	Population (Species)	N	L*	a*	b*	C	h
Basalt	Alma (*Spalax galili*)	7	26.31±5.73	8.62±4.18	11.14±0.50	14.57±1.17	53.84±17.72
Basalt	Dalton (*Spalax galili*)	6	30.59±0.94	6.23±0.43	9.97±0.43	11.76±0.59	58.01±0.81
Basalt	Quneitra (*Spalax golani*)	7	27.20±0.57	5.68±0.27	10.39±0.86	11.84±0.87	61.27±1.20
	Kruskal-Wallis		*H = 13.972, P<0.001*	*H = 8.764, P = 0.013*	*H = 11.599, P = 0.003*	*H = 9.779, P = 0.008*	*H = 10.533, P = 0.005*
Terra rossa	Rihaniya (*Spalax galili*)	8	28.97±0.82	10.07±0.29	12.77±0.34	16.26±0.42	51.74±0.57
Terra rossa	Muhraka (*Spalax carmeli*)	9	28.89±1.41	7.32±0.63	11.08±0.87	13.29±0.93	56.51±2.32
Terra rossa	"Evolution Canyon" Nahal Oren (*Spalax carmeli*)	8	26.32±2.13	3.31±0.66	6.78±0.96	7.55±1.12	64.22±0.85
	Kruskal-Wallis		*H = 9.109, P = 0.011*	*H = 21.350, P = <0.001*	*H = 21.040, P = <0.001*	*H = 21.350, P = <0.001*	*H = 20.739, P = <0.001*
Rendzina	Kerem-Ben-Zimra (*Spalax galili*)	14	46.62±2.59	5.78±0.11	15.95±0.37	16.969±0.378	70.04±0.186
Loess	Lahav (*Spalax judaei*)	7	43.32±1.33	5.23±0.25	14.75±1.04	15.65±1.06	70.46±0.430

Footnote: 'L*' represents the level of lightness of the color, positive value of 'a*' represents red. Value of 'b*' represents the amount of purplish -red (magenta). Chroma ('C' = $\sqrt{a^{*2}+b^{*2}}$) and hue ('h' = arc tan [a*/b*]) were calculated from L*a*b* parameters.

indicating the soil determinant as an important variable in determining pelage color.

Hair-melanin contents

Similar to the measured pelage color, PTCA (eumelanin) contents of mole rat populations living in similar soil types did not vary in all three soil types (see Table 5): basalt ($U = 54.00$; $P = 0.791$), rendzina ($U = 33.00$; $P = 0.212$), and terra rossa ($U = 47.5$; $P = 0.672$). However, 4-AHP (pheomelanin) contents did vary among populations of mole rats living in rendzina ($U = 7.00$; $P = 0.001$) and terra rossa ($U = 18.0$; $P = 0.047$) but did not vary in basalt ($U = 116.00$; $P = 0.427$).

Pelage color vs soil color

The lightness of *Spalax* pelage is significantly correlated with the soils' hue (Spearman's r = 0.617, P = 0.043, N = 11). Multiple forward stepwise regression between soil color variables (as independent variables) and pelage color coordinates (as dependent variables) revealed that soil lightness (L*) influenced all pelage color variables (Table 6). The combined (i.e., average) R (0.4) of the multiple regressions signifies low to intermediate regression; yet pelage L* and h* are partially influenced by soil color (as the constant value is large in the equation: 34.42 for L* and 68.59 for h*). Populations living in darker basalt and terra rossa soils exhibited darker pelage, and lighter rendzina and loess soils exhibited lighter pelage. Among the four soil types in which the mole rats were studied, rendzina soil has significant impact on pelage coloration, which revealed similarities in scores of all color variables between populations of *S. galili* and *S. judaei* (Table 4),

despite being the distant populations (Figure 1), separated by dozens of kilometers. By contrast, abutting populations living on drastically different soil types (such as KBZ on rendzina and Dalton on basalt) differ in pelage color exemplifying cryptic coloration.

Hair-Melanin Content vs. Pelage and Soil Colors

Hair-melanin contents in the pelage of mole rats reflected the pelage coloration trend. As shown by pelage color scores (Table 4), populations inhabiting darker soils, like basalt or terra rossa, exhibited higher PTCA (eumelanin) contents; basalt-inhabiting populations, such as Alma (2050±584 ng/mg) and Quneitra (2040±500 ng/mg), and terra rossa-inhabiting populations, like Muhraka (2201±439 ng/mg) and N. Oren (2132±251 ng/mg), had more PTCA than the lighter rendzina populations [Kerem-Ben-Zimra (1873±459) and Anza (1641±371) populations (see Table 5)]. Likewise, 4-AHP (pheomelanin) is higher in the hairs of mole rat populations with higher pelage scores of a* (red magenta) and b* (magenta/yellow) – variables, which are likely to determine pheomelanin [e.g., Alma (74±48 ng/mg) and Kerem-Ben-Zimra (76±35 ng/mg)], whereas it is lower in populations with lower pelage scores a* and b* [e.g., Muhraka (32±10) and "Evolution Canyon", Nahal Oren (53±28)] (see Tables 4 and 5). Populations with higher reddish and yellowish shades to their pelage exhibit higher amounts of pheomelanin contents in their hairs.

Similarly, hair-melanin content appeared to be associated with soil color as well; for example, populations inhabiting darker soil (e.g., Alma, Quneitra and "Evolution Canyon" Nahal Oren)

Table 2. Pelage color variation between sexes (species and populations' pooled data).

Sex	N	L*	a*	b*	C	h
Female	30	37.60±2.53	0.80±0.53	4.25±1.52	4.34±1.58	81.45±5.74
Male	13	37.38±2.43	0.77±0.42	3.49±1.15	3.60±1.18	80.44±9.65
Mann-Whitney		*U = 196.00; P = 0.989*	*U = 189.00; P = 0.884*	*U = 258.00; P = 0.098*	*U = 255.00; P = 0.116*	*U = 232.00; P = 0.334*

Table 3. Hair-melanin content variation between sexes (species & populations' pooled data).

Sex	N	PTCA	4-AHP	TM
Female	47	2038±452	49±31	0.76±0.19
Male	20	1949±514	64±44	0.66±0.13
Mann-Whitney		U = 506.50; P = 0.622	U = 384.50; P = 0.244	U = 622.00; P = 0.038

exhibited higher eumelanin contents, and populations inhabiting darker basalt soils (Alma and Quneitra), with higher scores of a* and b*, exhibited higher pheomelanin and lower scores of a* and b* in pelage in "Evolution Canyon" (Nahal Oren) populations led to lower pheomelanin content (see Tables 1 and 5).

Variations in the *MC1R* gene of mole rats

The *MC1R* gene was sequenced in 68 samples. The sequences were compared to the consensus reference sequence. We identified three variants in this sample set: 1) a synonymous substitution in position c.228 C to T; 2) a non-synonymous substitution in position c.502 A to G, which changes the amino acid from methionine to valine (c.502A>G; p.168 M>V); and 3) a synonymous substitution in position c.592 C to T. The first and second mutations were in high linkage disequilibrium in all of the tested animals. We found three haplotypes: C-A-T (the consensus sequence), C-A-C, and T-G-C. C-A-C and C-A-T are almost restricted to northern species (*S. galili* and *S. golani*, see Figure 2A–C & Table S2); while C-A-C haplotype occurs in each of the southern species (*S. carmeli* and *S. judaei*), the latter is exclusive to *S. golani* (Supplemental Table S2), whereas T-G-C is restricted to southern species (*S. carmeli* and *S. judaei*, see Figure 2C and Supplemental Table S2). Thus, there is a clear separation

between the northern (*S. galili* and *S. golani*) and southern species (*S. carmeli* and *S. judaei*); $\chi^2(2) = 60.5$, $p<0.001$. Notably, northern populations with C-A-C haplotype exhibited significantly higher pheomelanin than southern populations with T-G-C haplotype ($U = 318.5$, $N_1 = 30$, $N_2 = 37$; $P = 0.003$), whereas eumelanin did not vary very much ($U = 605$, $N_1 = 30$, $N_2 = 37$; $P = 0.533$). Conceivably, the pheomelanin/eumelanin ratio of populations increased northward (Figure 3). But, the pelage lightness estimator exhibited a linear trend, and the association between *MC1R* haplotypes and pelage lightness (among populations) remains unclear. Nevertheless, the *MC1R* haplotypes completely diverged between the northern and the southern species of mole rats (Figure 2A–C), in correlation with pheomelanin concentrations, suggesting that it is subjected to climatic selection.

Discussion

Overview

Neither pelage coloration nor hair-melanin contents of mole rats vary between the visually identical males and females, and this confirms there is no sexual dichromatism that is conceivable, as *Spalax* is a completely blind mammal. Habitat soil coloration, pelage coloration, and hair-melanin contents varied among the populations. Soil lightness (L*) was a determinant of all pelage color variables. The pelage lightness increased with increasing aridity regionally across Israel. Among mesic north populations, KBZ displayed lighter pelage on lighter rendzina soil. How far do these results corroborate with the three functions of coat coloration: intraspecific communication, crypsis, and thermoregulation [8]? Intraspecific communication cannot be the case with *Spalax ehrenbergi* superspecies as the animal cannot use apparent coloration-based visual cues. Therefore, blending with the background environment to evade easy detection by predators and thermoregulation could be the only reasons behind coat color variation in mole rats.

Table 4. Pelage color variation among populations living in different soil types.

Soil Type	Population (Species)	N	L*	a*	b*	c	h
Basalt	Alma (*Spalax galili*)	21	38.11±4.18	2.20±1.02	6.68±2.29	7.05±2.47	72.14±4.60
Basalt	Dalton (*Spalax galili*)	13	34.49±6.38	2.01± 0.64	6.50±1.92	6.82±1.99	72.10±3.96
Basalt	Quneitra (*Spalax golani*)	15	35.91±2.06	0.67±0.33	2.94±1.03	3.04±1.06	80.50±9.46
Basalt	A. Etan (*Spalax golani*)	4	36.79±2.49	1.36±0.50	5.46±1.30	5.64±1.36	76.95±3.61
Basalt	Bental (*Spalax golani*)	5	36.89±3.25	1.10±0.32	4.59±1.04	4.72±1.06	77.08±3.00
	Kruskal-Wallis		H = 6.164, P = 0.187	H = 33.617, P<0.001	H = 29.466, P<0.001	H = 30.521, P<0.001	H = 17.739, P = 0.001
Rendzina	Kerem-Ben-Zimra (*Spalax galili*)	20	38.66±2.66	1.96±1.38	6.24±2.54	6.58±2.81	75.62±7.30
Rendzina	Anza (*Spalax judaei*)	13	39.89±3.63	1.56±1.02	5.74±2.17	5.99±2.31	79.31±12.23
	Mann-Whitney		U = 97, P = 0.231	U = 142, P = 0.672	U = 142,P = 0.672	U = 143, P = 0.645	U = 129, P = 0.985
Terra rossa	Rihaniya (*Spalax galili*)	8	32.32±8.50	1.90±0.90	5.86±2.78	6.18±2.90	71.06±5.18
Terra rossa	Muhraka (*Spalax carmeli*)	13	36.69 ±2.02	0.46±0.26	3.55±1.08	3.59±1.09	85.62±5.52
Terra rossa	"Evolution Canyon" Nahal Oren (*Spalax carmeli*)	8	38.07±3.50	0.90±0.92	4.31±2.02	4.44±2.15	82.95±9.77
	Kruskal-Wallis		H = 2.827, P = 0.243	H = 12.974, P = 0.002	H = 4.978, P = 0.083	H = 5.798, P = 0.055	H = 15.384, P<0.001
Loess	Lahav (*Spalax judaei*)	8	37.80±3.32	1.17±0.54	4.84±1.37	5.00±1.44	78.51±3.92

Table 5. Hair-melanin content variation among populations living in different soil types.

Soil Type	Population (Species)	N	PTCA	4-AHP	TM
Basalt	Alma (*Spalax galili*)	10	2050±584	74±48	0.57±0.14
Basalt	Quneitra (*Spalax golani*)	10	2040±500	60±48	0.67±0.12
	Mann-Whitney		U = 54.00; P = 0.791	U = 116.00; P = 0.427	U = 81.00; P = 0.076
Rendzina	Kerem-Ben-Zimra (*Spalax galili*)	10	1873±459	76±35	0.63±0.16
Rendzina	Anza (*Spalax judaei*)	10	1641±371	29±17	0.68±0.14
	Mann-Whitney		U = 33.00; P = 0.212	U = 7.00; P = 0.001	U = 61.00; P = 0.427
Terra rossa	Muhraka (*Spalax carmeli*)	12	2201±439	32±10	0.88±0.20
Terra rossa	"Evolution Canyon" Nahal Oren (*Spalax carmeli*)	7	2132±251	53±28	0.87±0.11
	Mann-Whitney		U = 47.5; P = 0.672	U = 18.0; P = 0.047	U = 42.00; P = 0.966
Loess	Lahav (*Spalax judaei*)	8	2178±423	55±22	0.82±0.08

Does pelage coloration exemplify 'crypsis'?

Driving forces – soil color or soil type? Darker basalt and terra rossa soils selected for darker pelage, whereas lighter rendzina selected for lighter pelage despite variation between populations and species. Pelage colors of populations living in heterogeneous soils varied; yet pelage lightness and eumelanin contents in the hairs of mole rat populations living in similar soil type did not vary much, despite soil lightness varying among these populations (Table 4). Though soil color is a strong evolutionary force driving pelage coloration [32,33], the soil type overrides soil color and selects for convergent evolution in pelage lightness and eumelanin of the *Spalax ehrenbergi* superspecies. These results support Hardy's [34] findings that soil types have significant influence on coat color and on the local distribution of mammals. Indeed pelage coloration of different populations approximated better with soil type than with microscale variation in soil color. Populations that inhabit lighter rendzina (Kerem-Ben-Zimra and Anza) and loess (Lahav) soils exhibited lighter pelage, and those

Table 6. Association between soil color on pelage color of mole rats (forward stepwise multiple regressions).

Dependent variables	Step	Independent determinants	R	R²	F	P
Pelage L*	0 (forced)	soil L*	0.099			
Pelage L*	1	soil L*, soil b*	0.277	0.077	4.649	0.011
Pelage a*	0 (forced)	soil a*	0.155			
Pelage a*	1	soil a*, L*	0.352	0.124	7.908	<0.001
Pelage a*	2	soil a*, L*, C*	0.495	0.245	12.033	<0.001
Pelage b*	0 (forced)	soil b*	0.091			
Pelage b*	1	soil b*, C*	0.283	0.081	4.893	0.009
Pelage b*	2	soil b*, C*, L*	0.476	0.226	10.815	<0.001
Pelage b*	3	soil b*, C*, L*, a*	0.492	0.242	8.793	<0.001
Pelage b*	4	soil b*, C*, L*, a*, h*	0.496	0.24	7.105	<0.001
Pelage C	0 (forced)	soil C*	0.186			
Pelage C	1	soil C*, b*	0.289	0.084	5.109	0.008
Pelage C	2	soil C*, b*, L*	0.481	0.231	11.111	<0.001
Pelage C	3	soil C*, b*, L*, a*	0.497	0.247	9.011	<0.001
Pelage C	4	soil C*, b*, L*, a*, h*	0.500	0.250	7.267	<0.001
Pelage h	0 (forced)	soil h*	0.220			
Pelage h	1	soil h*, L*	0.395	0.156	10.330	<0.001
Pelage h	2	soil h*, L*, b*	0.399	0.159	7.013	<0.001
Pelage h	3	soil h*, L*, b*, a*	0.420	0.176	5.880	<0.001
Pelage h	4	soil h*, L*, b*, a*, C*	0.428	0.183	5.891	<0.001

Footnote: 'L*' represents the level of lightness of the color, positive value of 'a*' represents red. Value of 'b*' represents the amount of purplish-red (magenta).
Dependent variables were the pelage color coordinates of mole rats and the independent variables consisted of soil color coordinates. Criterion for the variable to enter the regression F>0.300, to be removed F<0.100.
The program calculates the coefficient of the forced variable in the equation, and it is entered as R.
The F and P are calculated by ANOVA.

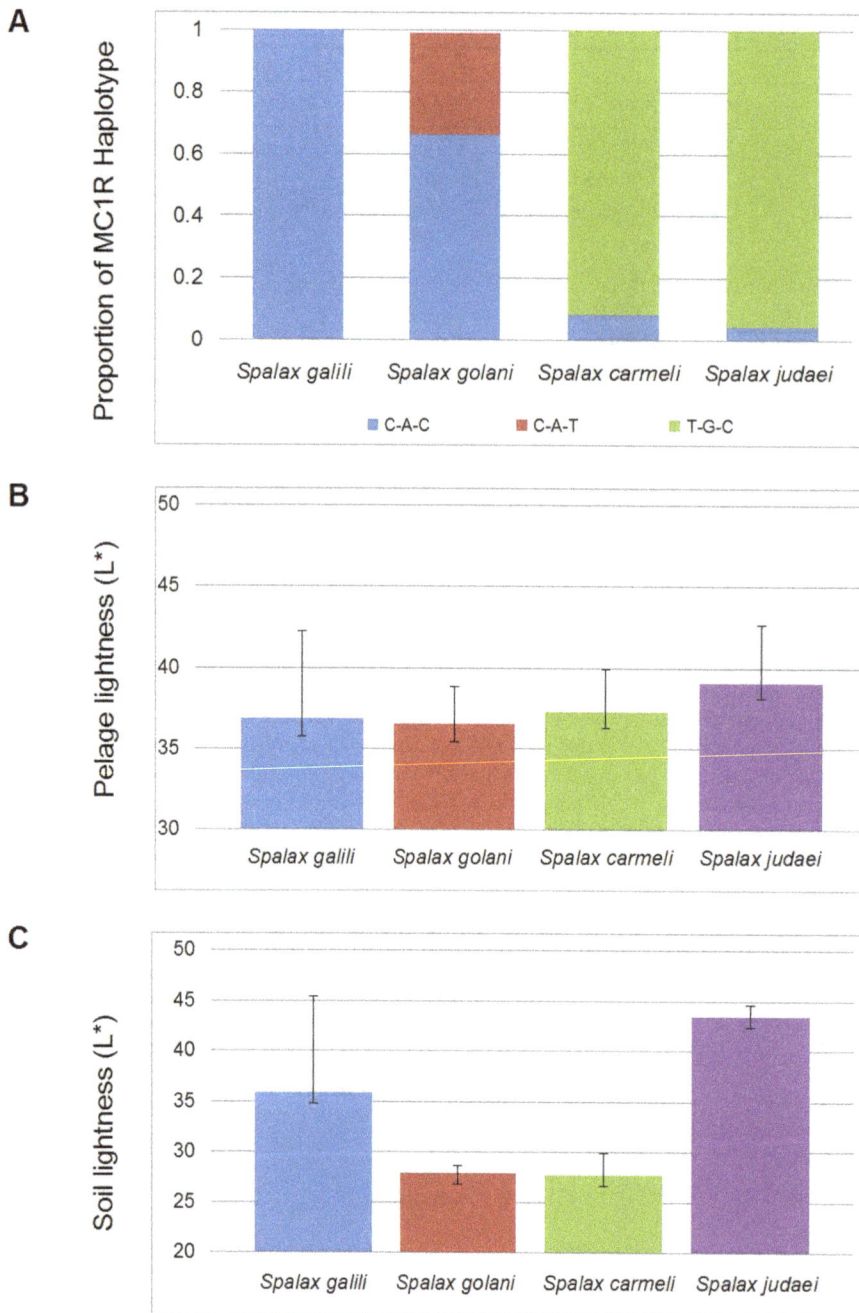

Figure 2. Patterns of variation across species of mole rats in Israel. A) genetic variation, B) phenotypic variation, and C) environmental variation.

that inhabit darker basalt (Dalton, Quneitra, A. Etan, and Bental) and terra rossa soils (Rihaniya and Muhraka) exhibited darker pelage (Table 4). Similarly, the populations inhabiting darker basalt (e.g., Alma, Quneitra) and terra rossa (e.g., Muhraka and "Evolution Canyon", Nahal Oren) soils exhibited higher amounts of PTCA (eumelanin) contents than populations living in lighter rendzina (Kerem-Ben-Zimra and Anza) that showed lower eumelanin contents (Table 5). Alma and Kerem-Ben-Zimra populations showed higher a* and b* scores (which would likely determine the prevalence of pheomelanin contents in pelage) in soil, resulting in higher pheomelanin contents (see Tables 1 & 5). Altogether, the results indicate that both coloration and melanin

contents of pelage are in accordance with soil type. The role of abrasive properties of the three soil types on both eumelanin and pheomelanin in the hairs (or pigment-type switching) is expected, but remains to be investigated.

In essence, the pelage lightness of mole rats show macro-geographic variations, but on a microscale, only the KBZ population exhibited lighter pelage on lighter soil color. Thus, the selection on pelage coloration towards crypsis is weak, but certainly exists. Nevertheless, the pelage coloration of a sympatric species, such as the spiny mouse *Acomys cahirinus*, in "Evolution Canyon" (Nahal Oren) responded well to variations in soil color even on a microscale [19]. This suggests that selection for crypsis

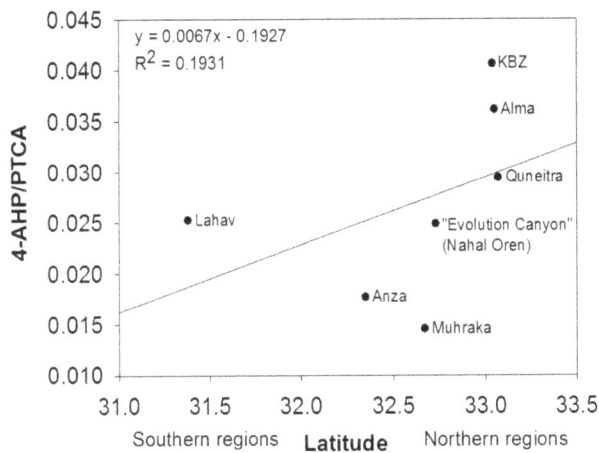

Figure 3. Pheomelanin/eumelanin ratio among *Spalax* populations across latitude.

might be stronger in such aboveground rodents than in subterranean mole rats, as the predation pressure on the former and latter varies drastically. Perhaps, aboveground vegetation, texture, and microclimatic differences among habitats might lead to the weak correlations between soil and pelage in Israeli mole rats (see also Rios et al. [35]).

Pelage coloration for thermoregulation

Climatic cline. Mole rats live in underground burrows that are microclimatically more or less stable [36]. However, soil moisture and temperature vary in their burrows across the climatically-divergent regions during seasonal changes [37]. Furthermore, mole rats seldom spend time aboveground during day time. Both soil moisture and aboveground activity together might influence the pelage coloration to regulate thermoregulation. Mole rats in Israel were distributed across a gradient of increasing aridity southward along the ranges of $2n = 52 \rightarrow 2n = 58 \rightarrow 2n = 60$ and eastward ($2n = 52 \rightarrow 54$) [24]. *S. carmeli* and *S. judaei* inhabit drier environments; therefore, their pelage is lighter. This could be in part to offset the hotter climate [22], in addition to its possible role in concealing coloration.

Though we don't have direct field-based evidence on whether *Spalax ehrenbergi* regulates body temperature during different seasons, a laboratory study by Haim et al. [38] shows that they might. Exposing cold-sensitive individuals to short photoperiods (8L:16D) at an ambient temperature (T_a) of 22°C, increased their thermoregulatory capacity under cold conditions (6°C for 6 h) when compared to individuals that were acclimated to 12L:12D at the same T_a. Conversely, acclimation of cold-resistant individuals to $T_a = 17°C$, but with a photoperiod of 16L:8D, decreased thermoregulatory capacity. It has been postulated that blind mole rats detect changes in photoperiods through its atrophied eyes or by other means involving the melatonin pathway [38]. Yet, how far mole rats could differentiate the already limited photic cues inside its burrow and then across seasons needs further research.

Soil and Climate – The dual factor

The coat coloration in rodents correlates both with the habitat gradients and climatic clines [39]. Animals in humid environments tend to be dark, whereas conspecifics in arid environments are light, a generalization recognized as Gloger's rule [40]. The pelage

lightness (L*) score across four *Spalax* species in Israel obeys this rule; *S. galili* and *S. golani* are darker and live relatively in humid environments, and *S. carmeli* and *S. judaei* (particularly the latter) are lighter and live in arid environments. Based on the comparisons of climatic data presented in Auffray et al. [41] and color scores of different variables in both Heth et al. [22] and the present study, some high-humid populations (e.g., KBZ population of *S. galili* and "Evolution Canyon", Nahal Oren population of *S. carmeli*) disagree with Gloger's rule, and did exhibit lighter pelage, and this is possibly to blend with lighter soils. Therefore, it is necessary to add the distinction between microgeographic (local) and macrogeographic (regional) scales to Gloger's rule. The northern, mesic but lighter KBZ population indicates that Gloger's rule is secondary to cryptic coloration.

Though Gloger's rule states that the color gradient of dark to light from humid to arid areas fulfills thermoregulatory needs, such a gradient also coincides with concealing coloration, as the soil color is darker in humid and lighter in arid areas. This is because the high density of vegetation is often correlated with high precipitation and humidity. This resulted in the darkening of the earth's surface in humid areas due to moisture and decomposing detritus, whereas arid areas became lighter [42]. Therefore, the selection forces behind concealing coloration and Gloger's rule are alike and obviously inseparable. Taken collectively, pelage coloration of mole rats characterizes both crypsis and Gloger's rule with almost equal exceptions to both.

Variation in *MC1R* and its implications on mole rats' dorsal coloration

In parallel to the phenotypic difference between northern and southern species, there is also a clear genetic differentiation between the *MC1R* gene variants. The haplotypes that carry valine in position 168 (c.502A>G variant) are exclusively restricted to the lighter southern species, which live in arid environments (*S. carmeli* and *S. judaei*), while the northern darker species, which live in humid environments (*S. galili* and *S. golani*), have methionine in the same position. Although we did not observe a correlation between this *MC1R* genotype and any of the pelage color parameters, there is a strong similarity in the divergence of the MC1R haplotypes distribution between the northern and southern species to the dramatic differences in pheomelanin concentrations in the same groups. This might suggest a functional effect of the MC1R haplotype on pheomelanin, in particular of the non-synonymous variant c.502A>G. However, only molecular biology experiments could prove that this variant is indeed the functional mutation that influences the pelage color. A similar example of MC1R effect was previously shown in a study of subspecies of the Gulf Coast beach mice where a strong association was found between *MC1R* genotypes, pigmentation and background sand brightness, consistent with local adaptation [43]. In pocket mice, four of nine *MC1R* non-synonymous variants were observed in the dark mice from the Pinacate locality. It was suggested that one or more of these four amino acid mutations are responsible for light/dark phenotypic differences seen in the Pinacate population [18]. In our study, the northern populations of mole rats with C-A-C haplotype exhibited higher pheomelanin than southern populations with T-G-C haplotype. Although in the absence of molecular biology studies, we could not confirm that the *MC1R* variants we found, were the causal variants for mole rats' pelage color and had clearly influenced pheomelanin in pelage, our results support that possibility.

The Pheomelanin/eumelanin ratio of *Spalax* populations increased with high latitudes indicating geographical variation. It is likely that higher pheomelanin (together with eumelanin) in northern populations caused darker pelage in *Spalax galili* and *Spalax golani*. We base this claim on report [13] that the darker coat of some rodent species is caused by the deposition of eumelanin and in other species by the deposition of pheomelanin. Geographical variation in melanin contents is known in owls [44], but yet this remains largely untested in rodents despite phenotypic studies often supporting Gloger's rule. However, it is puzzling why eumelanin didn't show such geographic variation in *Spalax*. It appears that eumelanin responds better to soil color variation and pheomelanin responds better to geographical variation in mole rats, but further studies are needed to test whether such a differential function of the two melanins exists and if so, how it optimizes the adaptive roles of dorsal coloration.

Conclusion

Multiple linear regressions revealed that soil lightness was a determinant of all pelage color coordinates. Darker pelage of mole rats coincides with darker basalt and terra rossa in humid regions, and lighter pelage coincides with lighter rendzina and loess soils in arid regions. Pelage color lightness and eumelanin content of different populations (of all four species) which inhabit the same soil type (in different climatic conditions) converge despite varied soil lightness. Therefore, it is hard to distinguish crypsis from the thermoregulatory function expressed by Gloger's rule. The pelage color differs distinctly between northern and southern species, and the mutations in the *MC1R* gene might therefore be involved. However, it should be noted that other genes in the blind subterranean mole rats genome might also affect the causal chain of its pelage coloration.

Supporting Information

Table S1　Populations and species of mole rats screened for Mc1r. The details of species, populations, soil type and number of animals sampled for each populations was given.

Table S2　MC1R haplotype frequencies in *Spalax* species. The three allelic variants in MC1R gene and the corresponding frequency of occurrence in each species are shown.

Acknowledgments

We thank Robin Permut for her help in language editing. We thank Aaron Avivi, Alma Joel, Assaf Malik, Yarin Hadid, and Ronit Younes for their help in molecular work and Noa Waitz for her assistance in graphics. We are indebted to the two reviewers (Tim Caro and Hynek Burda) and the academic editor for their insights and suggestions, which improved the quality and readability of the MS.

Author Contributions

Conceived and designed the experiments: NS SR ST SB KW EN. Performed the experiments: NS ST SR SB TP SI KW. Analyzed the data: NS ST AB KW. Contributed reagents/materials/analysis tools: KW EN. Wrote the paper: NS EN. Helped to draft the Manuscript: ST AB KW. Carried out soil and pelage color measurements: ST SR. Conducted the initial phase of molecular works and then mentored SB: NS. Conducted the major parts of molecular works: SB. Helped with the animals and hair sampling: TP ST SR. Directed the protocols for hair-sampling: NS. Conducted the melanin assay: KW SI. Analyzed and interpreted the results: NS ST AB KW EN. Critically analyzed the data: NS AB KW.

References

1. Burtt EH Jr (1981) The adaptiveness of colors. BioScience 31: 723–729.
2. Cloudsley-Thompson JL (1999). Multiple factors in the evolution of animal coloration. Naturwissenschaften 86: 123–132.
3. Majerus MEN (1998) Melanism: evolution in action. Oxford: Oxford University Press. 364 p.
4. Caro T (2005) The adaptive significances of animal colors. Bioscience 31, 723–729.
5. Sumner FB (1929) The analysis of a concrete case of intergradation between two subspecies. Proc Natl Acad Sci U S A 15: 110–120.
6. Dice LR, Blossom PM (1937) Studies of mammalian ecology in southwestern North America with special attention to the colors of desert mammals. Carnegie Inst Wash Publ No 485.
7. Cott HB (1940) Adaptive coloration in animals. London: Methuen. 508 p.
8. Endler JA (1978) A predator's view of animal color patterns. Evol Biol 11: 319–364.
9. Krupa JJ, Geluso KN (2000) Matching the color of excavated soil: cryptic coloration in the plains pocket gopher (*Geomys bursarius*). J Mammal 81: 86–96.
10. Cleffmann G (1963) Agouti pigment cells in situ and in vitro. Ann New York Acad Sci 100: 749–761.
11. Hoekstra HE, Hirschmann RJ, Bundey RA, Insel PA, Crossland JP (2006) A single amino acid mutation contributes to adaptive beach mouse color pattern. Science 313: 101–104.
12. Linnen CR, Kingsley EP, Jensen JD, Hoekstra HE (2009) On the origin and spread of an adaptive allele in deer mice. Science 325: 1095–1098.
13. Walker WP, Gunn TM (2010) Shades of meaning: the pigment-type switching system as a tool for discovery. Pigment Cell Res 23: 485–495.
14. Le Pape E, Wakamatsu K, Ito S, Wolber R, Hearing VJ (2008) Regulation of eumelanin and pheomelanin synthesis by MC1R ligands in melanocytes. Pigment Cell Res. 21: 477–486.
15. Kingsley EP, Manceau M, Wiley CD, Hoekstra HE (2009) Melanism in *Peromyscus* is caused by independent mutations in *agouti*. PLoS ONE 4, e6435.
16. Walker WP, Gunn TM (2010) Piecing together the pigment-type switching puzzle. Pigment Cell & Melanoma Res, 23: 4–6.
17. Hoekstra HE, Krenz JG, Nachman MW (2004) Local adaptation in the rock pocket mouse (*Chaetodipus intermedius*): natural selection and phylogenetic history of populations. Heredity 94: 217–228.
18. Nachman MW, Hoekstra HE, D'Agostino SL (2003) The genetic basis of adaptive melanism in pocket mice. Proc Natl Acad Sci USA 100: 5268–5273.
19. Singaravelan N, Pavlicek T, Beharav A, Wakamatsu K, Ito S, et al. (2010) Spiny mice modulate eumelanin to pheomelanin ratio to achieve cryptic coloration in "Evolution Canyon," Israel. PLoS ONE 5: e8708.
20. Ingles LG (1950) Pigmental variations in populations of pocket gophers. Evolution 4: 353–357.
21. Kennerly TE Jr (1954) Local differentiation in the pocket gopher (*Geomys personatus*) in Sourthern Texas. Texas J Sci 6: 297–329.
22. Heth G, Beiles A, Nevo E (1988) Adaptive variation of pelage color within and between species of the subterranean mole rat (*Spalax ehrenbergi*) in Israel. Oecologia 74: 617–622.
23. Wahrman J, Goitein R, Nevo E (1969) Mole rat *Spalax*: evolutionary significance of chromosome variation. Science 164: 82–84.
24. Nevo E, Filippucci MG, Redi CD, Korol AB, Beiles A (1994) Chromosomal speciation and adaptive radiation of mole rats in Asia Minor correlated with increased ecological stress. Proc Natl Acad Sci U S A 91: 8160–8164.
25. Nevo E, Ivanitskaya E, Beiles A (2001) Adaptive radiation of blind subterranean mole rats: naming and revisiting the four sibling species of the *Spalax ehrenbergi* superspecies in Israel: *Spalax galili* (2n = 52), *S. golani* (2n = 54), *S. carmeli* (2n = 58) and *S. judaei* (2n = 60). Leiden: Backhuys Publishers.
26. Hunter RS (1948) Photoelectric color difference meter. JOSA 38: 661.
27. Wakamatsu K, Ito S, Rees JL (2002) The usefulness of 4-amino-3-hydroxyphenylalanine as a specific marker of pheomelanin. Pigment Cell Res 15: 225–232.
28. Ito S, Nakanishi Y, Valenzuela RK, Brilliant MH, Kolbe L, et al. (2011) Usefulness of alkaline hydrogen peroxide oxidation to analyze eumelanin and pheomelanin in various tissue samples: application to chemical analysis of human hair melanins. Pigment Cell & Melanoma Res. 24: 605–613.
29. Ito S, Wakamatsu K (1998) Chemical degradation of melanins: application to identification of dopamine-melanin. Pigment Cell Res. 11: 120–126.
30. Wakamatsu K, Ohtara K, Ito S (2009) Chemical analysis of late stages of pheomelanogenesis: conversion of dihydrobenzothiazine to a benzothiazole structure. Pigment Cell & Melanoma Res 22: 474–486.
31. Ozeki H, Ito S, Wakamatsu K, Thody AJ (1996) Spectrophotometric characterization of eumelanin and pheomelanin in hair. Pigment Cell Res 9: 265–270.
32. Krupa JJ, Geluso KN (2000) Matching the color of excavated soil: cryptic coloration in the plains pocket gopher (*Geomys bursarius*). J Mamm 81: 86–96.
33. Lai YC, Shiroishi T, Moriwaki K, Motokawa M, Yu HT (2008) Variation of coat color in house mice throughout Asia. J Zool 274: 270–276.
34. Hardy R (1945) The influence of types of soil upon the local distribution of some mammals in southwestern Utah. Ecol Monogr 15: 71–108.
35. Rios E, Ticus S, Alvarez-Castaneda ST (2012) Pelage color variation in pocket gophers (Rodentia: Geomyidae) in relation to sex, age and differences in habitat. Mammal Biol 77: 160–165.

36. Nevo E, Guttman R, Haber M, Erez E (1979) Habitat selection in evolving mole rats, Ecologia 43: 125–138.

37. Nevo E (1985) Speciation in action and adaptation in subterranean mole rats: patterns and theory. Boll Zool 52: 65–95.

38. Haim A, Heth G, Pratt H, Nevo E (1983). Photoperiodic effects on thermoregulation in a 'blind' subterranean mammal. J Exp Biol 107: 59–64.

39. Sumner FB, Swarth H (1924) The supposed effects of the color tone of the background upon the coat color of mammals. J Mammal 5: 81–113.

40. Huxley J (1942) Evolution: the modern synthesis. London: Allen & Unwin. 576 p.

41. Auffray J-C, Renaud S, Alibert P, Nevo E (1999) Developmental stability and adaptive radiation in the *Spalax ehrenbergi* superspecies in the near-East. J Evol Biol 12: 207–221.

42. Cowles RB (1958) Possible origin of dermal temperature regulation. Evolution 12: 347–357.

43. Mullen LM, Vignieri SN, Gore JA, Hoekstra HE (2009) Adaptive basis of geographic variation: genetic, phenotypic and environmental differentiation among beach mouse populations. Proc Roy Soc B 276: 3809–3818.

44. Roulin A, Wink M, Salamin N (2009) Selection on eumelanic ornaments is stronger in the tropics than in the temperate zones in the worldwide-distributed barn owl. J Evol Biol 22: 345–354.

17

Comparison of Four Spatial Interpolation Methods for Estimating Soil Moisture in a Complex Terrain Catchment

Xueling Yao, Bojie Fu*, Yihe Lü, Feixiang Sun, Shuai Wang, Min Liu

State Key Laboratory of Urban and Regional Ecology, Research Center for Eco-Environmental Sciences, Chinese Academy of Sciences, Beijing, P. R. China

Abstract

Many spatial interpolation methods perform well for gentle terrains when producing spatially continuous surfaces based on ground point data. However, few interpolation methods perform satisfactorily for complex terrains. Our objective in the present study was to analyze the suitability of several popular interpolation methods for complex terrains and propose an optimal method. A data set of 153 soil water profiles (1 m) from the semiarid hilly gully Loess Plateau of China was used, generated under a wide range of land use types, vegetation types and topographic positions. Four spatial interpolation methods, including ordinary kriging, inverse distance weighting, linear regression and regression kriging were used for modeling, randomly partitioning the data set into 2/3 for model fit and 1/3 for independent testing. The performance of each method was assessed quantitatively in terms of mean-absolute-percentage-error, root-mean-square-error, and goodness-of-prediction statistic. The results showed that the prediction accuracy differed significantly between each method in complex terrain. The ordinary kriging and inverse distance weighted methods performed poorly due to the poor spatial autocorrelation of soil moisture at small catchment scale with complex terrain, where the environmental impact factors were discontinuous in space. The linear regression model was much more suitable to the complex terrain than the former two distance-based methods, but the predicted soil moisture changed too sharply near the boundary of the land use types and junction of the sunny (southern) and shady (northern) slopes, which was inconsistent with reality because soil moisture should change gradually in short distance due to its mobility in soil. The most optimal interpolation method in this study for the complex terrain was the hybrid regression kriging, which produced a detailed, reasonable prediction map with better accuracy and prediction effectiveness.

Editor: Guy J.-P. Schumann, NASA Jet Propulsion Laboratory, United States of America

Funding: This work was funded by the National Natural Science Foundation of China (No. 40930528), State Forestry Administration (No. 201004058) and the CAS/SAFEA International Partnership Program for Creative Research Teams of "Ecosystem Processes and Services". The funders had no role in study design, data collection and analysis, decision to publish, or preparation of the manuscript.

Competing Interests: The authors have declared that no competing interests exist.

* E-mail: bfu@rcees.ac.cn

Introduction

Soil moisture (SM) is of fundamental importance in meteorology, agriculture, and hydrology, among other scientific disciplines [1,2,3,4]. In hydrology, SM partitions rainfall into runoff and infiltration, thus impacting the surface and groundwater recharge, flood forecasting, and flow routing modeling [5,6]. Scientists usually need accurate, spatially continuous data across a region in order to make justified interpretations, but such data are usually not readily available and are often difficult and expensive to acquire. Remote sensing techniques have great potential for measuring spatially continuous SM data, but typically involve observing the average SM close to the ground surface and over large geographical areas with low resolution [7]. Many approaches combining remotely sensed data and auxiliary model to estimate deeper soil moisture have been developed, such as infiltration models [8] and knowledge based techniques that use prior information of hydrology [9] and so on. However, these approaches cannot meet the requirement of small catchment scale researches, which need finer spatial resolution of deep soil moisture data to study the exchange of water between different layers within the soil column or between the land surface and the atmosphere [10]. In situ measurements of soil moisture are still an important portion in recent researches [5,11,12]. The popular in situ techniques of measuring soil moisture content include gravimetric method, neutron probes, electromagnetic techniques, cosmic-ray neutrons and so on [13]. However, theses in situ techniques typically involve measuring SM in points. Spatially continuous SM data in deep soil with finer resolution are needed in many cases. Thus, estimating the values at unsampled sites using data from point observations is necessary, and spatial interpolation methods provide an essential tool to meet this need.

In previous research, the geostatistical method (ordinary kriging (OK), cokriging) [14,15,16], geometric method (inverse distance weighting (IDW), local polynomial), and statistical methods such as the linear regression model (LR) [17,18,19] have been the most commonly used interpolation technologies [1,20]. In addition, hybrid interpolation techniques, which combine two conceptually different approaches, have received increasing attention in recent years [21,22]. One of these techniques is known as regression kriging (RK) [23,24], and first uses regression on auxiliary information and then uses simple kriging (SK) with a known mean (0) to interpolate the residuals from the regression model [25]. Zhu [23] compared the performance of OK and RK for soil properties in different landscapes and indicated that when a strong relationship existed between the target soil properties and auxiliary variables and the terrain was relatively complex, RK was more accurate for interpolating soil properties. Li and Heap [26] investigated the performance and impact factors of popular interpolation methods in environmental sciences by accessing 53 comparative studies. The results indicated that the OK and IDW methods were the most frequently used. The performance of a spatial interpolation method depends not only on the features of the method itself, but also on factors such as data variation and sampling design. Most of the methods performed at an acceptable level for predicting soil properties in a gentle terrain [27,28], but few performed well in a complex terrain.

Our research was conducted in the semi-arid Loess Plateau, which has a complex hilly gully terrain. The SM in the deep soil was paid much attention because it significantly affects the growth of the planted vegetation as well as the success of the Grain for Green Project (a state campaign in China to restore an ecological balance to the country's western parts, by turning the low-yielding farmland back into forests and pasture) in this region [29,30]. Due to the limitation of remote sensing technology for directly obtaining deep SM with fine resolution [7,31], many studies in the Loess Plateau have been based on ground and point measurements [32,33,34,35]. However, considering the intensive labor consumption and destruction to the ground when conducting sampling, the sample density is usually insufficient at the catchment or region scale. Based on these practical challenges, we aimed to determine an optimal interpolation method that fits to the hilly gully terrain in the semi-arid Loess Plateau.

Two of the most popular interpolation methods, OK and IDW, were chosen in our research [26,33,36,37], as investigating their suitability in complex terrain is practically valuable to further research. Secondly, considering the SM in the Loess Plateau was strongly impacted by geographic factors such as land use type [38,39], soil properties [35], gradients, slope aspects [40], and so on, the LR model was chosen, in which the strong correlation between SM and geographic factors would be helpful to create an optimal regression function [26]. A hybrid RK model was also chosen for its theoretical suitability in our research and good performance in previous research [23,41,42,43]. The performance of each method was assessed in terms of mean-absolute-percentage-error (MAPE), root-mean-square-error (RMSE), and goodness-of-prediction statistic (G). The theoretical and practical advantages and disadvantages of each method for a complex terrain are discussed in detail at the end of this research.

Study Area

The study was carried out in the Yangjuangou catchment (36°42′N, 109°31′E), which is located in the center of the Loess Plateau near Yan An City in Shaanxi Province, China (Fig. 1). The catchment has a total area of 2.02 km^2 and the elevation ranges from 1,050 to 1,298 m. It is a typical gully and hilly area with a gully density of 2.74 km km^{-2} and the slope gradients range from 10° to 30°. The area has a semi-arid continental climate with an average annual rainfall of 535 mm. Rainfall events occur mainly between June and September with large inter-annual

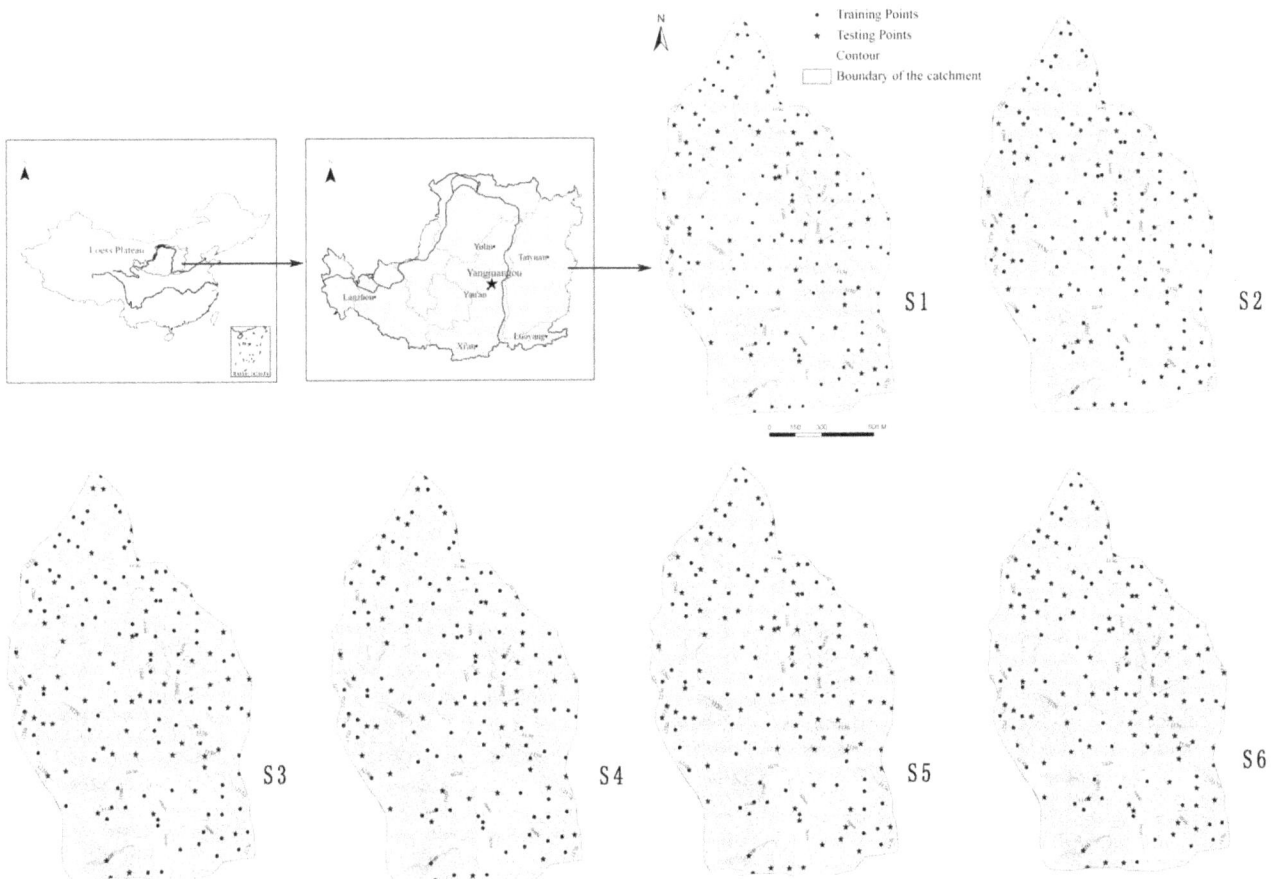

Figure 1. The location of the study catchment and the distribution of the samples.

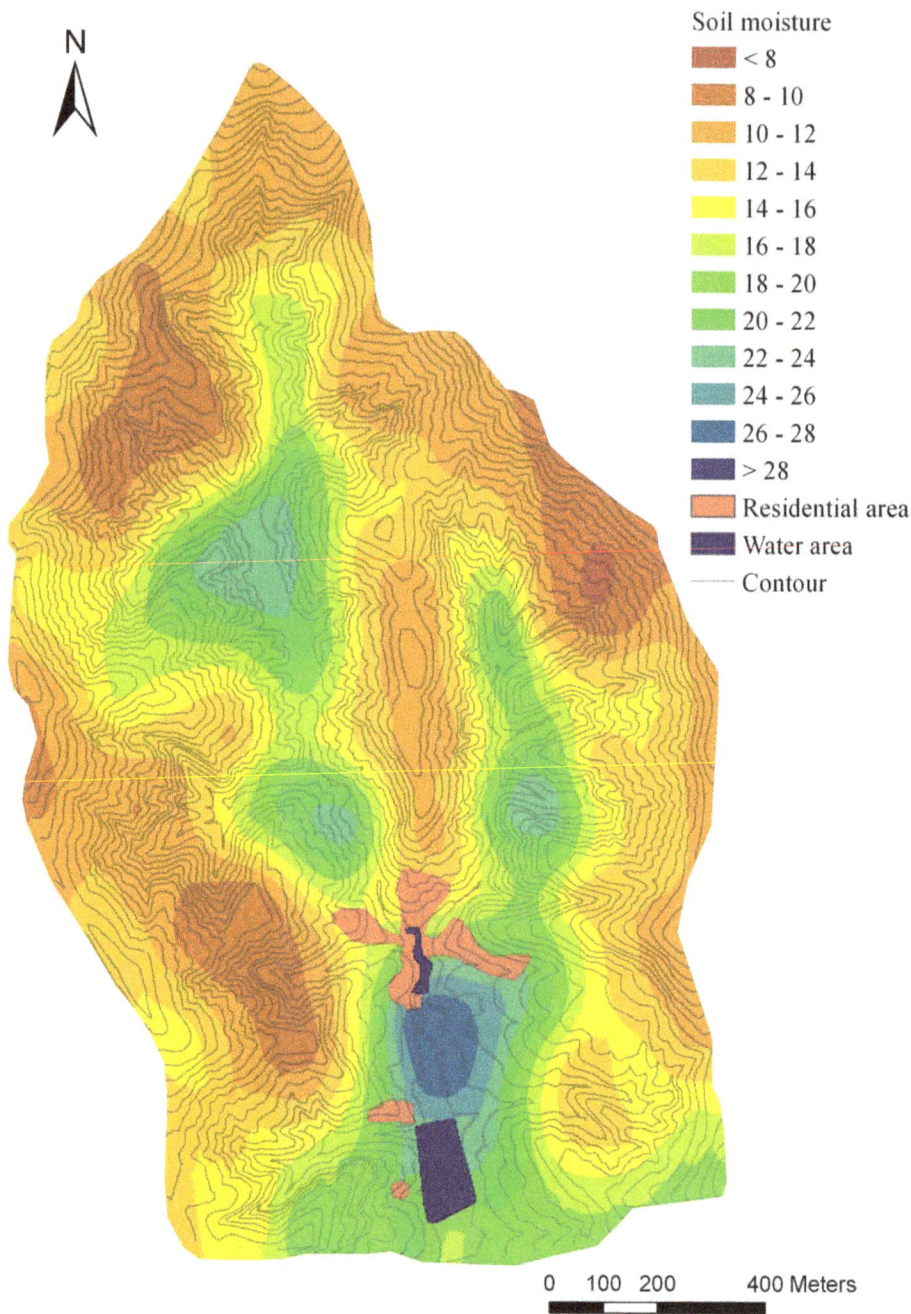

Figure 2. The soil moisture prediction map basing on ordinary kriging method.

variation. The soil in the study area was derived from loess with a maximum depth of approximately 200 m. The soil texture was rather homogeneous, containing mainly loessial sandy loam soil with the silt particles being 65–75% and bulk density being 1.2–1.4 g/cm^3 according to laboratory measurements. As result of the Grain for Green project that was launched in 1998, most of the cultivated lands on steep slopes were abandoned for natural or planted vegetation. Grasslands and forestlands now dominate the hillslopes, and shrubs are thriving at the bottom of the north-facing slopes. The main forest species in the Yangjuangou catchment is *acacia* (*Robinia pseudoacacia*), which was planted in the 1980s and after 1999. The dominant grass species are *Artemisia sacrorum*, *Stipa bungeana*, and *Artemisia scoparia*. The main shrub species are *Prunus*

armeniaca and *Hippophae rhamnoide*. A mosaic of patchy land cover is the typical landscape pattern in the Yangjuangou catchment as a result of human disturbances as well as climatic and topographical conditions [44].

Experimental Layout and Methods

Experimental Layout

The SM measurement was conducted in June 2010, with a total of 153 points being measured. The slope distances between neighboring points ranged from 50 m to 100 m and probably met the uniform distribution, which was required by Geostatistical interpolation methods. Besides, all the land use types and typical

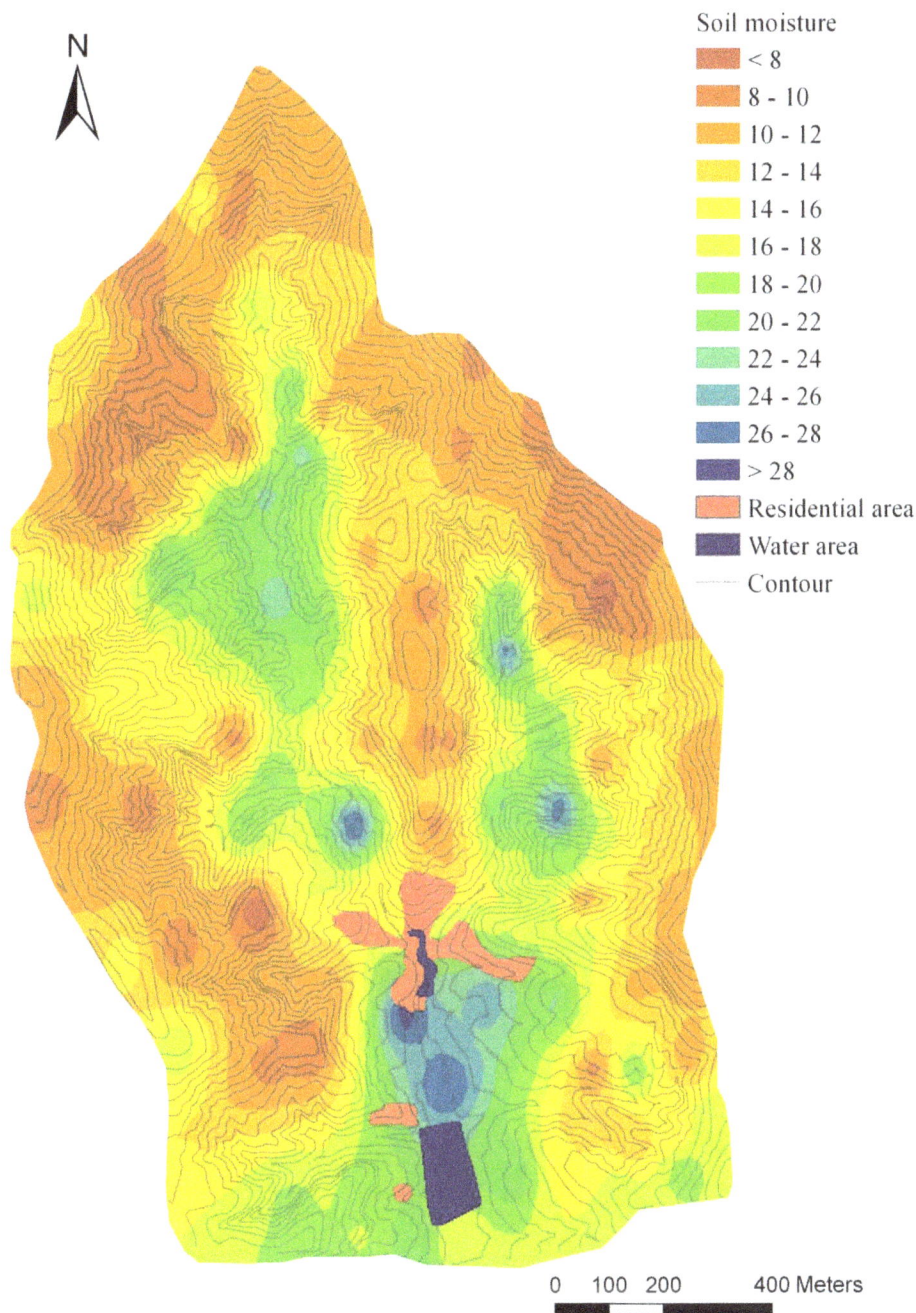

Figure 3. The soil moisture prediction map basing on inverse distance weighting method.

topography types in the catchment were involved and each land use/topography type contained at least ten sampling points to ensure the validity when conduct statistical analysis. The field sample collection lasted three days, and there was no rain during these days. The distribution of the points was shown in Figure 1. The soil samples in 10 cm, 20 cm, 40 cm, 60 cm, 80 cm and 100 cm depth were extracted using a soil auger at each point. Once extracted from the ground, the samples were placed in aluminum cans with tight-fitting lids, and the gravimetric water content was determined from the weight loss after oven drying at 105°C to a constant mass. The soil bulk density was measured synchronously in each plot using a ring cut. The volumetric water content was calculated by multiplying the gravimetric water

content and the soil bulk density and dropping the units. The soil moisture in 10–20 cm layers was averaged to present the upper soil layer for each point, while the 40–100 cm to present the deeper soil layer.

In the field, a GPS receiver with 5 m precision was used to obtain the altitude, longitude, and latitude, which were later imported into a geographic information system (ArcGIS 9.3) as Albers coordinates. The slope degree and aspect were measured with a geological compass. The land use types, primary soil types, and vegetation species and coverage (%) were estimated by observation.

To produce a spatially continuous surface and evaluate the performance of each interpolation method, the 153 sampling

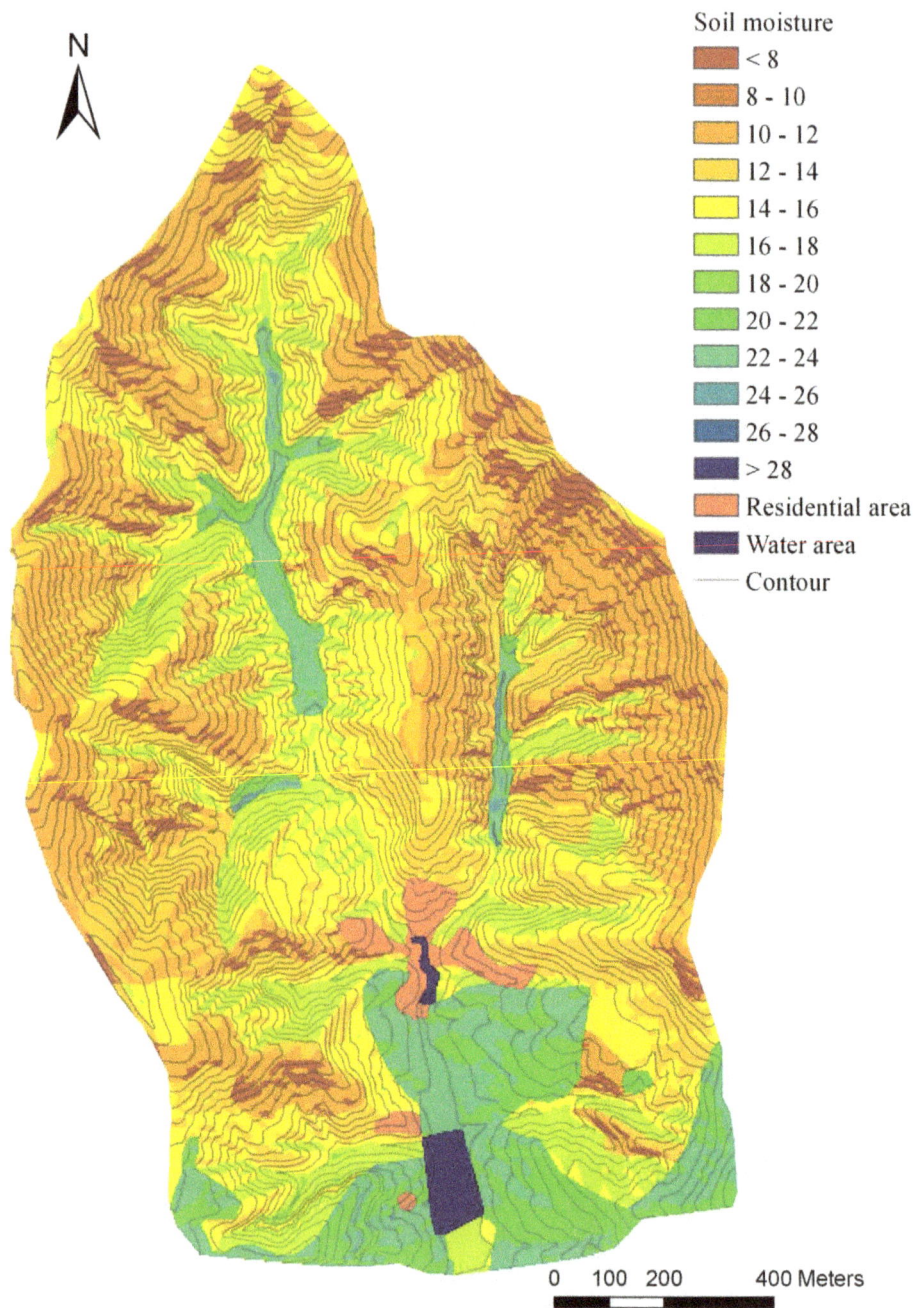

Figure 4. The soil moisture prediction map basing on linear regression method.

points were randomly divided into two sets by the "Create Subset" component of the Geostatistical Analyst extension in ArcGIS: the training data set and the test data set with 2:1 ratio. The training data set was used to create the model and the test data sets were used to assess the performance of the model. In the present study, the training data set included 102 points and the test data set included 51 points. Considering that the distribution of the training points may affect model performance, the subsetting was repeated six times (S1 to S6). For each subset, the four interpolation methods were used to produce the spatially continuous surface, and the performance was assessed accordingly.

Methods

For the IDW and OK interpolation methods, the value of variable Z at the unsampled location x_0, $Z^*(x_0)$ is estimated based on the data from the surrounding locations, $Z(x_i)$, as

$$Z^*(x_0) = \sum_{i=1}^{n} w_i Z(x_i) \qquad (1)$$

where w_i is the weight assigned to each $Z(x_i)$ value and n is the number of the closest neighboring sampled data points used for estimation. The weights for the IDW are usually proportional to the inverse of the squared distance between the prediction point

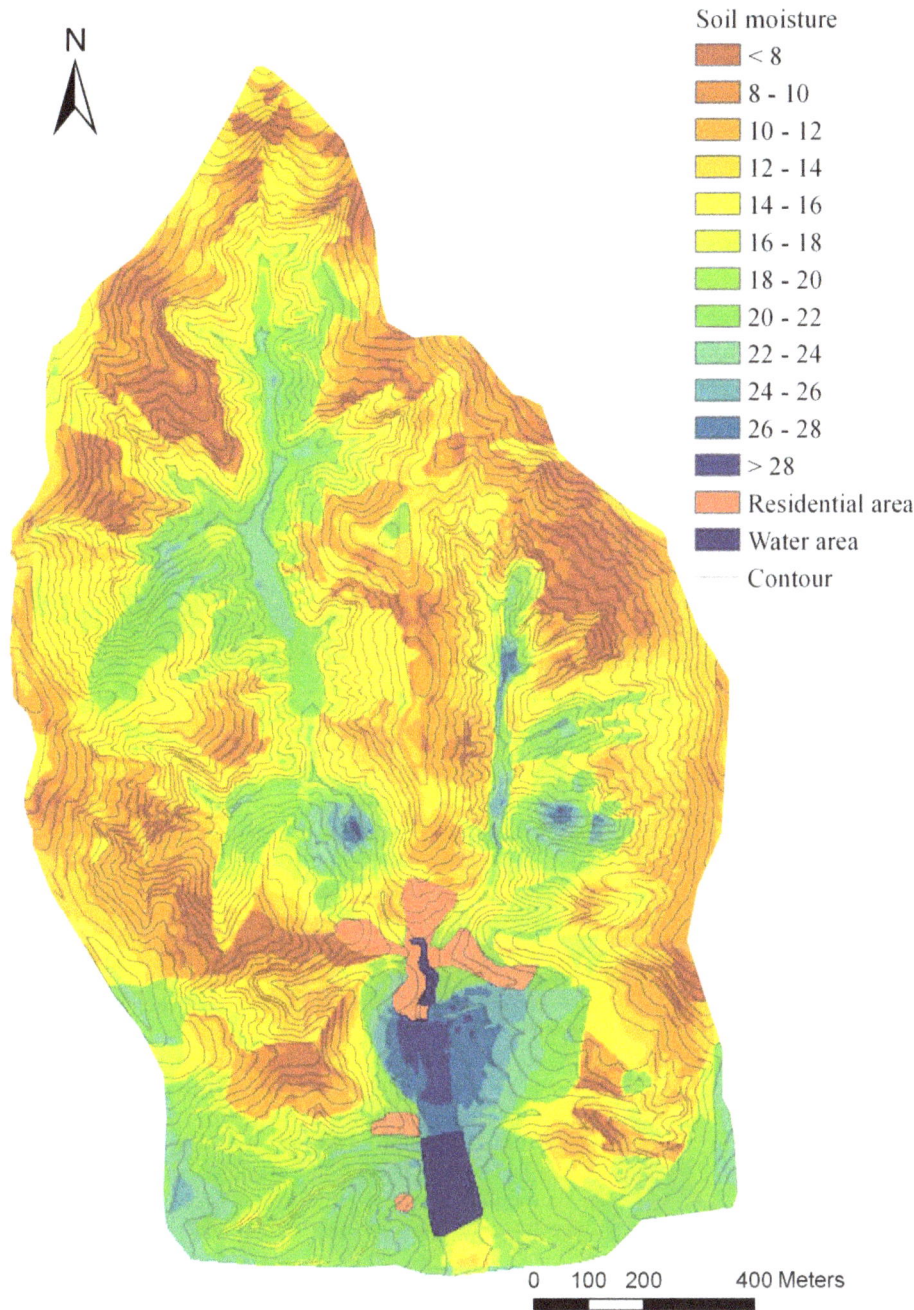

Figure 5. The soil moisture prediction map basing on regression kriging method.

and the observation points, and they sum to 1. That is,

$$w_i = \frac{1/d_i^2}{\sum_{i=1}^{n} 1/d_i^2} \tag{2}$$

where di is the distance between the estimated point and the observed point.

The number of the closest neighboring samples is an important factor affecting the precision of IDW. Considering the sample spatial density and the hilly gully terrain of our research area, the number of the closest neighboring samples should be small

because the samples taken on the other side of the hill should have little correlation with the predicted point in reality. Thus, the number of the closest neighboring samples we applied varied from 3 to 6. Cross-validation was used to compare the results obtained with a different number of the closest neighboring samples. The numbers of the closest neighboring samples producing the best agreement between the observed data and the estimates were chosen as the optimal IDW weighting parameters [20].

Kriging calculates the values of w_i by estimating the spatial structure of the variable's distribution represented by a sample variogram as

Table 1. The basic statistical properties of soil moisture of each data set.

	Count	Min	Max	Average	Std.dev	Skewness	Kurtosis	CV
					O/L	O/L	O/L	O/L
All	153	5.49	35.94	14.08	5.85/0.38	1.16/0.39	3.95/2.44	0.42/0.15
S1	102	5.49	28.59	13.96	5.57/0.38	0.82/0.25	2.67/2.12	0.40/0.15
S2	102	6.44	35.94	14.20	6.19/0.40	1.22/0.47	4.03/2.41	0.44/0.16
S3	102	6.60	35.94	14.03	5.96/0.39	1.21/0.53	4.02/2.36	0.42/0.15
S4	102	5.49	35.94	14.42	6.02/0.39	1.19/0.33	4.11/2.62	0.42/0.15
S5	102	5.49	29.16	13.70	5.47/0.37	1.16/0.43	3.70/2.55	0.40/0.15
S6	102	6.44	35.94	14.07	5.82/0.38	1.20/0.42	4.23/2.44	0.41/0.15

O. Statistical value from Ordinary dataset.
L. Statistical value from Log-transformed dataset.

$$\gamma(h) = \frac{1}{2n} \sum_{i=1}^{n} [Z(x_i) - Z(x_i + h)]^2 \quad (3)$$

$$y = \beta_0 + \sum_{i=1}^{n} \beta_i X_i \quad (4)$$

where x_i and x_i+h are sampling locations separated by a distance h, and $Z(x_i)$ and $Z(x_i+h)$ are the observed values of variable Z at the corresponding locations. The sample variogram is fitted with a variogram model and the adequacy of the chosen model is tested using cross-validation. In this study, we considered the spherical, Gaussian, and exponential models for the sample variogram fitting. The cross-validation was conducted with varying model parameter values and the numbers of the closest neighboring samples ranging from 3–10 until the highest estimation accuracy was reached.

Linear regression is a common forecasting tool for many research areas. It is a statistical tool for modeling the relationship between a dependent variable and one or more independent variables. In linear regression models, the dependent variable is a linear function of one or more independent variables, as shown in the equation below.

The parameters of the linear regression model are typically estimated using the least squares method, which results in a line that minimizes the sum of squared vertical distances from the observed data points to the line [19].

A large body of research indicates that the SM in the Loess Plateau is strongly affected by land use types, soil properties, and terrain [11,40,45,46,47]. Considering the soil properties were fairly homogeneous in our research area (mainly loessial sandy loam soil with the silt particles being 65–75% and bulk density being 1.2–1.4 g/cm^3 basing on laboratory detection), the independent variables were selected as land use types, slope, and annual average solar radiation, which were preliminary detected basing on correlation analysis and finally determined basing on the significance of regression coefficient (P<0.05) in the regression equation. The slope and annual average solar radiation were continuous variables, which were produced by digital elevation model (DEM, 3 m×3 m). The land use types were categorical variables and were converted into dummy variables before they were introduced into the regression analysis. When all of the parameters of the linear function were produced and satisfied the

Table 2. The correlation between the G-values and the sample pattern properties basing on correlation analysis.

	Std.dev		Skewness		Kurtosis		CV	
Method (soil depth, cm)	PCC	P	PCC	P	PCC	P	PCC	P
OK (10–20)	−0.75	0.08	−0.68	0.14	−0.73	0.10	−0.74	0.09
IDW (10–20)	−0.70	0.12	−0.45	0.37	−0.39	0.44	−0.85	0.03
LR (10–20)	−0.44	0.38	−0.32	0.54	−0.40	0.43	−0.44	0.38
RK (10–20)	−0.60	0.21	−0.54	0.27	−0.56	0.25	−0.68	0.14
OK (40–100)	−0.53	0.27	−0.12	0.82	−0.09	0.86	−0.67	0.15
IDW (40–100)	−0.59	0.22	−0.20	0.70	−0.14	0.79	−0.74	0.09
LR (40–100)	−0.76	0.08	−0.23	0.66	−0.28	0.59	−0.80	0.06
RK (40–100)	−0.67	0.14	−0.27	0.60	−0.30	0.57	−0.75	0.08

PCC. Pearson correlation coefficient.
P. Significance value (2-tailed).

Table 3. The performance assessment of the four interpolation methods for 10–20 cm soil layer.

	MAPE				RMSE				G-value			
	OK	IDW	LR	RK	OK	IDW	LR	RK	OK	IDW	LR	RK
S1	0.28	0.29	0.19	0.15	3.65	3.58	2.68	2.28	0.54	0.56	0.75	0.82
S2	0.36	0.37	0.26	0.24	3.77	3.86	2.64	2.99	0.17	0.13	0.59	0.56
S3	0.40	0.35	0.22	0.19	4.40	3.69	2.72	2.53	0.24	0.47	0.71	0.75
S4	0.28	0.25	0.21	0.19	3.61	3.27	2.13	2.29	0.42	0.53	0.80	0.83
S5	0.41	0.43	0.20	0.24	5.01	4.88	3.16	3.43	0.50	0.53	0.80	0.83
S6	0.39	0.29	0.27	0.22	3.77	3.15	2.85	2.79	0.20	0.44	0.54	0.56
Average	0.36	0.33	0.23	0.20	4.04	3.74	2.70	2.72	0.35	0.44	0.70	0.69

significance test ($P<0.05$) in SPSS 13.0 software, they were adopted to produce the prediction map in ArcGIS 9.3.

RK is a spatial interpolation technique that combines the regression of the dependent variable on auxiliary variables with the kriging of the regression residuals. The target variable SM, was fitted with each auxiliary data set using the linear regression. By detrending the regression predictions, the residuals were geostatistically analyzed and interpolated using SK, and finally the regression predictions and interpolated residuals were summed. Eventually, the SM predictions were back-transformed to normal SM values. In RK, the auxiliary data sets in the regression were the same as in the LR [25].

The IDW and OK method were conducted using ArcGIS 9.3. With the LR and RK methods, the linear regression function was established by SPSS 13.0 and the prediction of the continuous spatial surface was conducted by ArcGIS. Eventually, all of the predicted maps were laid out using ArcGIS with comparable design (Fig. 2, 3, 4, 5).

Assessment of Method Performance

The performance of the methods was assessed by identifying the error in the predictions. For each method, the prediction values and corresponding observed values in the test data set were compared, and the following evaluation indicators were calculated.

The accuracy was measured by MAPE, which is an accuracy measure based on percentage (or relative) errors and RMSE, which measures the average magnitude of the error [26]. The errors are squared before they are averaged, so the RMSE gives a relatively high weight to large errors. This means that RMSE is

most useful when large errors are particularly undesirable. Small MAPE and RMSE values indicate a model with few errors and more accurate predictions.

The MAPE is calculated as follows:

$$MAPE = \frac{1}{n} \sum_{i=1}^{n} |(p_i - o_i)/o_i| \qquad (5)$$

Where n is the number of validation points, p_i is the predicted value at point i, o_i is the observed value at point i.

The RMSE is calculated as follows:

$$RMSE = [\frac{1}{n} \sum_{i=1}^{n} (p_i - o_i)^2]^{1/2} \qquad (6)$$

Where n is the number of validation points, p_i is the predicted value at point i, o_i is the observed value at point i.

The effectiveness of the models was evaluated using a goodness-of-prediction statistic (G). The G-value measures how effective a prediction might be relative to that which could have been derived using the sample mean:

$$G = 1 - [\sum_{i=1}^{n} (p_i - o_i)^2 / \sum_{i=1}^{n} (o_i - \overline{o})^2] \qquad (7)$$

Table 4. The performance assessment of the four interpolation methods for 40–100 cm soil layer.

	MAPE				RMSE				G-value			
	OK	IDW	LR	RK	OK	IDW	LR	RK	OK	IDW	LR	RK
S1	0.36	0.33	0.26	0.25	5.82	5.44	4.72	4.50	0.28	0.37	0.53	0.57
S2	0.28	0.29	0.23	0.23	4.03	4.09	3.54	3.84	0.09	0.06	0.30	0.28
S3	0.36	0.34	0.23	0.23	4.99	4.93	3.80	3.50	0.26	0.28	0.57	0.64
S4	0.29	0.26	0.28	0.26	4.91	4.47	4.47	4.21	0.40	0.50	0.50	0.55
S5	0.41	0.40	0.26	0.27	5.91	5.49	4.40	4.18	0.37	0.45	0.65	0.68
S6	0.27	0.27	0.26	0.26	4.97	4.64	4.25	4.41	0.26	0.36	0.46	0.42
Average	0.33	0.32	0.25	0.25	5.10	4.84	4.20	4.11	0.28	0.34	0.50	0.51

Table 5. The means comparison of the G-value of the four interpolation methods for 10–20 cm soil layer.

Method	Sample Number	Sub-classification for $P=0.05$	
		1	2
OK	6	0.35	
IDW	6	0.44	
LR	6		0.70
RK	6		0.73
P		0.25	0.75

P. Significance value.

Table 6. The means comparison of the G-value of the four interpolation methods for 40–100 cm soil layer.

Method	Sample Number	Sub-classification for $P=0.05$	
		1	2
OK	6	0.28	
IDW	6	0.34	
LR	6		0.50
RK	6		0.52
P		0.45	0.06

P. Significance value.

Where n is the number of validation points, p_i is the predicted value at point i, o_i is the observed value at point i, and $\bar{o}$ is the sample arithmetic mean. A G-value equal to 1 indicates perfect prediction, a positive value indicates a more reliable model than if the sample mean had been used, a negative value indicates a less reliable model than if the sample mean had been used, and a value of zero indicates that the sample mean should be used [20].

Results

Basic Statistics of Soil Moisture

The SM of the overall samples ranged from 5.49% to 35.94%, with an average value of 14.08%. The SM spatial variability was significant, with a standard deviation (Std.dev) of 5.85 and coefficient of variation (CV) of 0.42 (Table 1). The skewness values of the data sets were positive and the asymmetry was obvious, with the value of 1.2. After the log-transformation, the corresponding skewness values were much smaller (0.39), which means that the data distribution was closer to normal. The log-transformed data sets were used in the Kriging interpolation method, which strictly demands normality of the data set [48]. The statistics values of each subset (S1 to S6) were also listed in Table 1.

Correlation between the Sample Pattern and the Performance of the Interpolation Methods

The Std.dev, skewness, kurtosis, and CV statistics were used to represent the pattern properties of each subset. The Pearson Correlation between the pattern properties and the G-values were used to analyze the effect of the sample pattern on the performance of each method. Table 2 shows that all of the correlations were poor and none of the correlation coefficients were significant in this study.

Performance of the Interpolation Methods

The MAPE (RMSE) were generally decreasing from OK to RK in Table 3 and Table 4, with the corresponding average values of 0.36 (4.04), 0.33 (3.74), 0.23 (2.70), 0.20 (2.72) for the 10–20 cm layer and 0.33 (5.10), 0.32 (4.84), 0.25 (4.20), 0.25 (4.11) for the 40–100 cm layer, which indicated a greater probability that errors occur in OK and IDW than in LR and RK.

The G-value reflects the prediction effectiveness of each method. For the S1 to S6 subsets, the G-values were ranked as OK<IDW<LR<RK, with corresponding average values being 0.35, 0.44, 0.70, 0.69 for the 10–20 cm soil layer and 0.28, 0.34, 0.50, 0.51 for the 40–100 cm soil layer. According to the multiple

mean comparison (Student-Newman-Keuls) of G-value of each method, the four interpolation methods were divided into two classes. The OK and IDW methods were classed into a same group because the G-values were not significantly different, while the LR and RK were classed into another group (Table 5, Table 6). The G-values difference between the two classes was significant ($P<0.05$), which means that the effectiveness of the LR and RK methods was significantly better than the distance-based OK and IDW methods. For OK and IDW, the G-value of data set S2 for 40–100 cm soil layer was close to 0, which indicates that the effectiveness of the prediction was not better than if the sample mean was used [20]. Although the relatively better performance of IDW than OK, both of them were not optimal in complex terrain because of their larger error and lower prediction effectiveness. Comparatively, the performances of LR and RK were much better, with the average G-values about 0.7 for 10–20 cm soil layer and 0.5 for 40–100 cm soil layer. Although the G-value of LR was higher than RK in several cases, the RK performance was generally better than LR in terms of all the three assessment indicators (Table 3, Table 4).

Discussion

Many researchers indicated that data normality [48] and CV [26] might affect the performance of spatial interpolation methods. In our research, the S1 to S6 subsets corresponded to six different sample patterns. For each interpolation method, the prediction errors between the six patterns obviously differed, indicating that the sample pattern may significantly affect the performance of the methods, which has been referred in many studies [49,50]. However, from the correlation analysis, we did not find any significant correlation between the sample pattern properties and the G-values (Table 2). The factors inducing the performance difference between each sample pattern is still not clear in this study and need to be analyzed in further research.

The spatial autocorrelation of the target variable is a basic assumption for the distance-based interpolation method. The spatial autocorrelation is present when the value of a variable at one location exerts an influence on the value of the same variable in neighboring locations [51]. Theoretically, the spatial autocorrelation of SM should exist at a small scale because of the water mobility in soil. However, SM may have little autocorrelation at small catchment scale because of the strong control of local geographical factors to SM [11]. Climate could be seen as homogeneous in small catchment scale, thus the local factors, such as land use types, vegetation, soil types and topography therefore stood out as the main factors dominating the soil moisture spatial

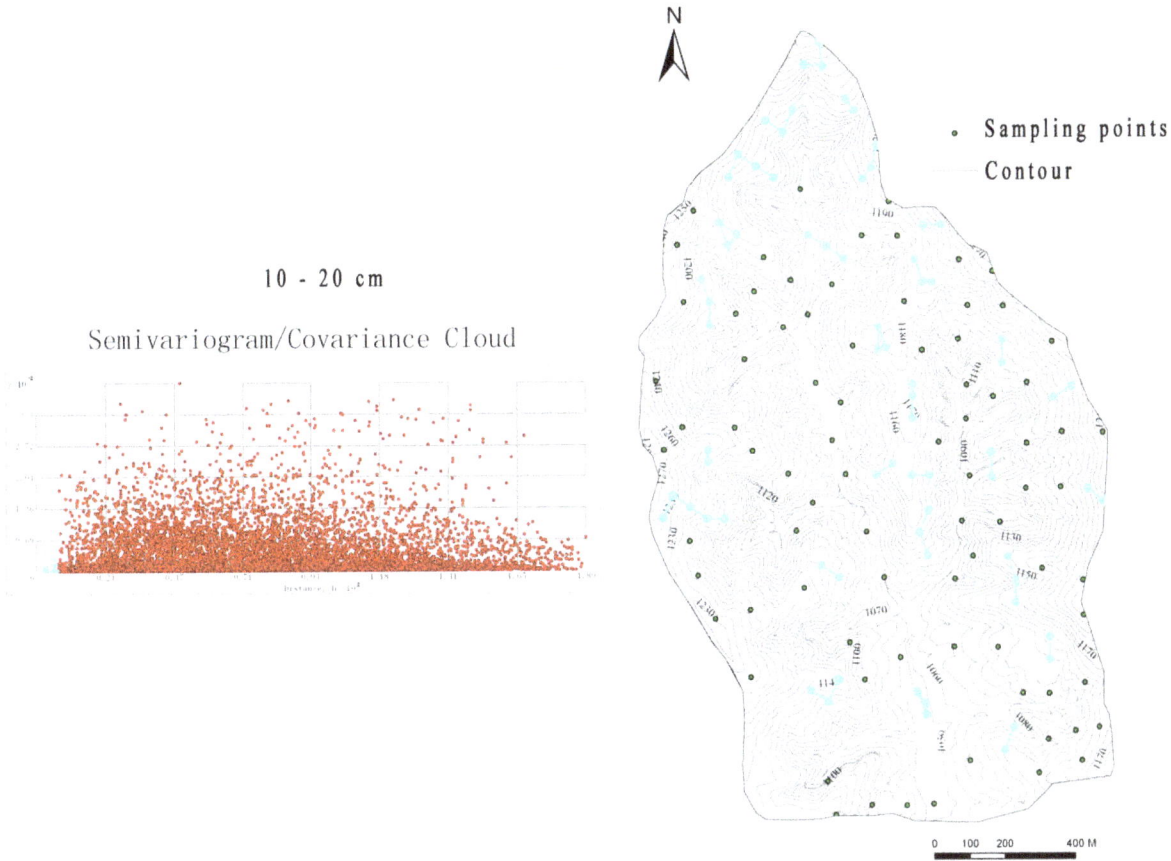

Figure 6. The autocorrelation range of the 10–20 cm soil moisture in complex terrains.

distribution. Because of the spatial fragmentation of these factors in hilly-gully area, the soil moisture may show poor spatial autocorrelation. The semivariogram cloud confirmed the short-distance autocorrelation of the SM in this complex terrain area (Fig. 6, Fig. 7), showing that spatial autocorrelation only exists at slope scale (<100 m) and no obvious autocorrelation exists at small catchment scale. One of the basic assumptions of ordinary kriging was that the observations have obvious spatial autocorrelation and the autocorrelation is a function of the distance between the observations [50]. Obviously, the short autocorrelation range of SM in this area was a lack of statistical effectiveness, which could hardly fit this assumption. Thus, it is not difficult to understand the poor performance of OK in the present research (Fig. 3). For IDW, the performance was slightly better than OK because it simply relies on the similarity of neighboring sample points to predict the unmeasured points. However, the IDW still did not perform its best because of the limitation of the low sampling density relating to the complex terrain (Fig. 4).

Although the SM in upper soil was more sensitive to ground impact factors (such as weather) and was expected to have better spatial autocorrelation at catchment scale, the results of the semivariogram analysis (Fig. 6) showed little difference with that in the deeper soil (Fig. 7). The performance of the four methods for the upper soil moisture generally appears better than that for the deeper soil moisture in terms of RMSE and G-value. However, the relative performance showed no difference, both with the accuracy ranking as OK<IDW<LR<RK.

In addition to the theoretical unsuitability, there are many practical problems if distance-based interpolation methods are used in a complex terrain area. Figure 8 shows two examples of these problems. In the first example, point A (located in the bottom of the valley) was covered with farmland and had high SM, which was similar to nearby points A1 and A2. The points B1 and B2 located on the top and the other side of the hill around the valley, covered with forest/shrub, usually had low SM. However, in the distance-based interpolation methods, the points A1 and A2 would not be chosen to predict the SM in point A because they are not the closest to point A based on the horizontal distance. In contrary, points B1 and B2, which had little similarity with point A, would be determined as the closest points to predict the SM at point A. Thus, a rather unreliable prediction would be produced. Similarly, in the second example, the SM at the bottom of the valley (point A) would be predicted based on the SM on the ridges (points B1, B2 and B3), where the SM condition is usually rather different from that in the valley. On the other hand, the points (point A1)similar to point A would be neglected or only given a very small weight (Fig. 8). Thus, it is difficult to produce a reliable prediction basing on the distance-based interpolation methods in this type of complex terrain, which was confirmed by the performance assessment (Table 3, Table 4).

The LR method showed more effectiveness than the distance-based methods, clearly expressing the impact of the environmental factors on SM, and the prediction was more detailed and accurate than OK and IDW. The variance of SM between different land use types, slopes, and slope aspects is fully displayed in the prediction map (Fig. 4). However, there are some disadvantages to this method. The SM change was fairly sharp at the boundary of the land use types and the junction of the sunny (southern) and

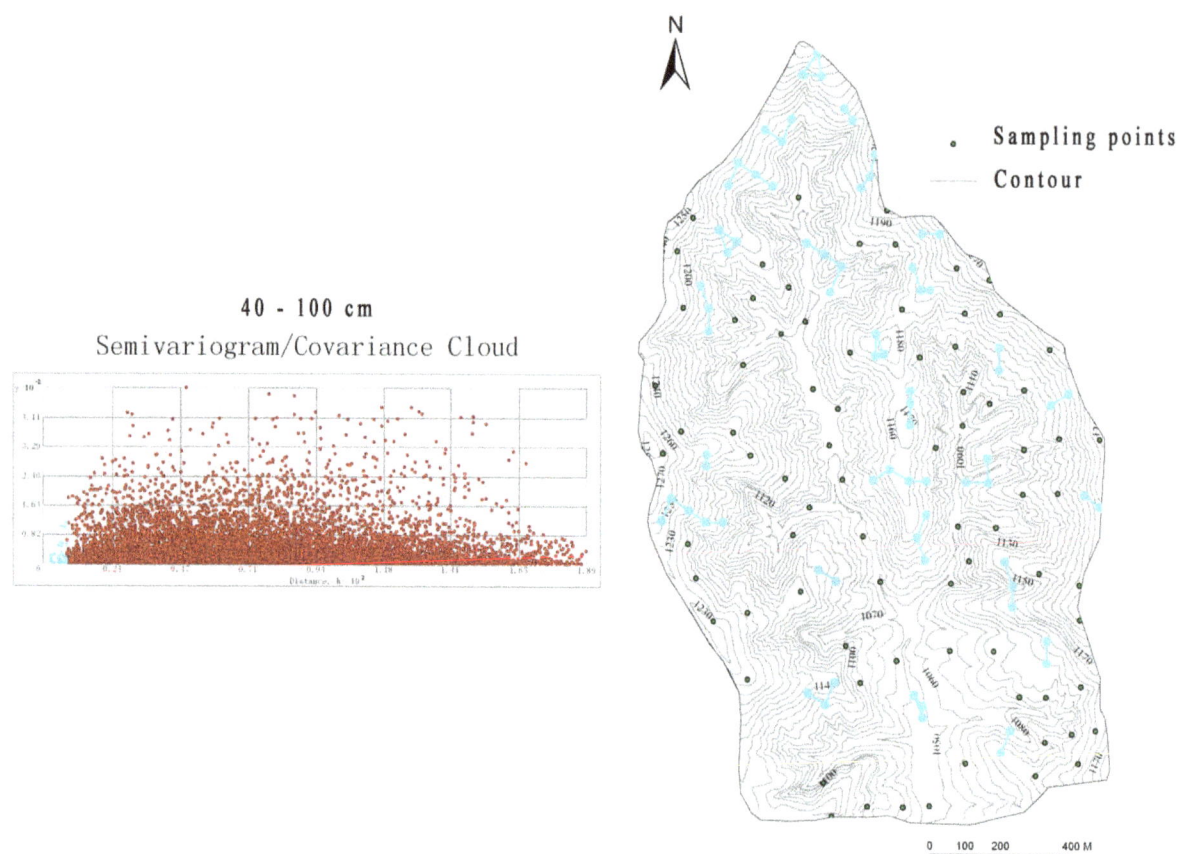

Figure 7. The autocorrelation range of the 40–100 cm soil moisture in complex terrains.

shady (northern) slopes. This is inconsistent with the reality, as water is mobile in soil and its spatial change should be gradual at a small scale.

The RK method effectively released the problems of LR, and combined the advantages of LR and kriging model. It not only reserved the impact of the environmental factors on SM obtained from the regression model, but also added the gradually changing property obtained from the kriging model. The prediction map displayed a detailed, reasonable continuous surface, which was more consistent with the reality (Fig. 5). The performance assessment evidently confirmed that this hybrid interpolation method was the most suitable and accurate to the complex terrain in our research.

In conclusion, we suggest that in an area with complex terrain, where the spatial autocorrelation of the interested variable exists only at a small scale and the target variable is significantly correlated with auxiliary variables, the hybrid RK model would perform much better than the distance-based methods in predicting the spatially continuous surface. This conclusion is consistent with the work of Zhu and Lin [23], whose study indicated that RK is more accurate for interpolating soil properties when a strong relationship exists between target soil properties and auxiliary variables as well as when the terrain is more complex.

Conclusions

The case study of SM spatial interpolation in the hilly gully Loess Plateau shows three main things. First, the distance-based OK and IDW methods performed poorly due to the poor spatial autocorrelation of soil moisture in complex terrain areas, where

the environmental impact factors were discontinuous in space at small catchment scale. Second, the LR model performed much better than OK and IDW, and adequately showed the SM difference with the variance of the impact factors. However, the predicted SM changed too sharply near the boundary of the land use types and at the junction of the sunny (southern) and shady (northern) slopes, which was inconsistent with the reality because the soil moisture should change gradually in short distance due to its mobility in soil. Third, the hybrid RK model has evident advantages over the three ordinary approaches for predicting SM in complex terrain area in terms of MAPE, RMSE, and G assessment, with the prediction map being more accurate and realistic.

Acknowledgments

We thank Mr. Changhong Su, Mr. Ruiying Chang, Mr. Guangyao Gao, and Mr. Yu Liu for field assistance. No specific permits were required for the described field studies. Our field studies did not involve endangered or protected species. The study location is not privately-owned or protected and no specific permissions were required for these locations because most land in the catchment had been abandoned for more than 10 years and the catchment has been a long-term experiment station of our institute.

Author Contributions

Conceived and designed the experiments: BJF XLY. Performed the experiments: XLY BJF YHL FXS SW. Analyzed the data: XLY FXS SW. Contributed reagents/materials/analysis tools: XLY FXS SW ML. Wrote the paper: XLY BJF YHL.

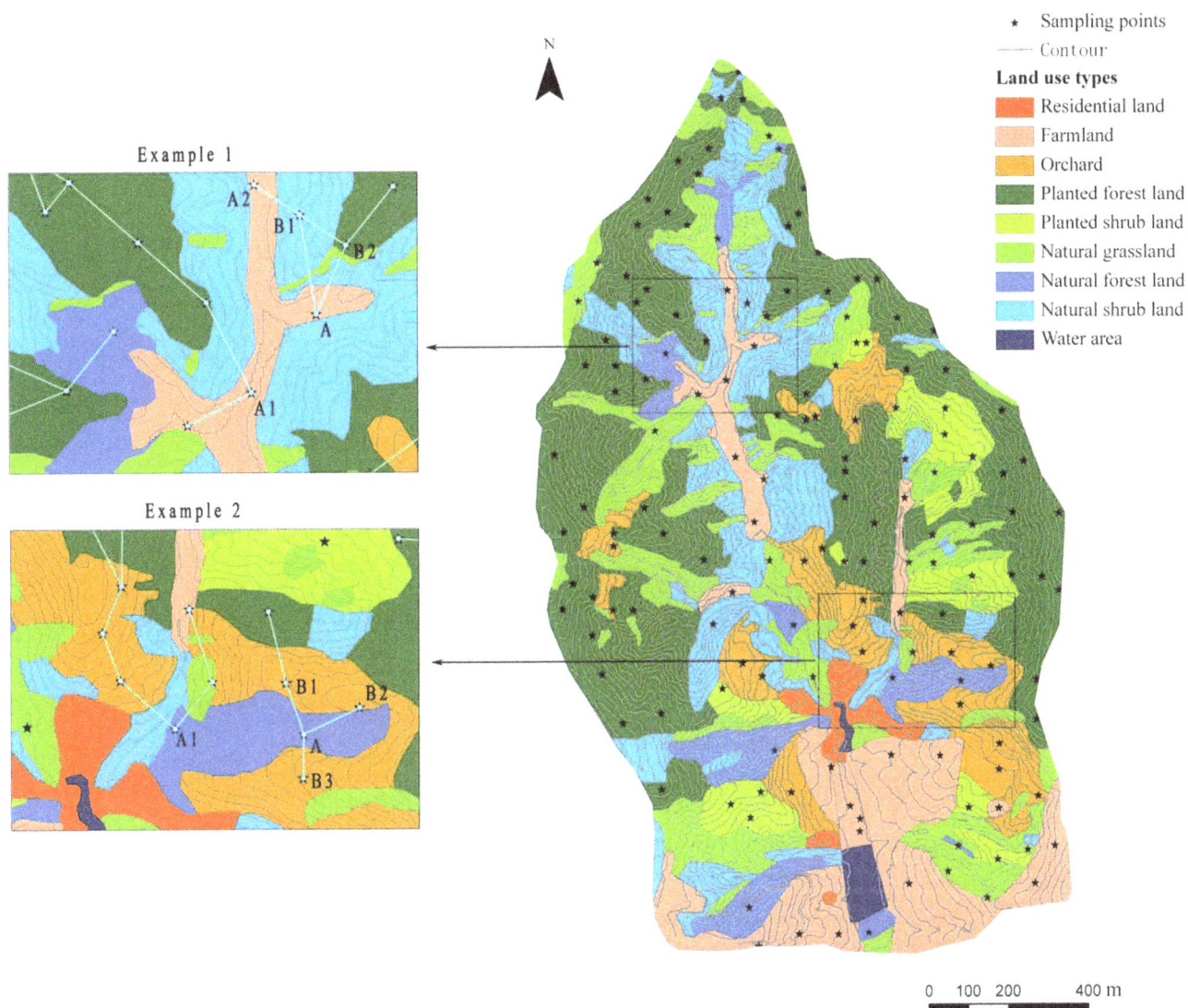

Figure 8. Two examples of the problems when distance-based methods were used in complex terrains.

References

1. Dripps WR, Bradbury KR (2007) A simple daily soil-water balance model for estimating the spatial and temporal distribution of groundwater recharge in temperate humid areas. Hydrogeology Journal 15: 433–444.

2. Toth E, Farkas C, Nagy V, Hagyo A, Stekauerova V (2008) Assessment of spatial variation of the soil water regime in the soil-plant system. Cereal Research Communications 36: 307–310.

3. Yu XX, Zha TS, Pang Z, Wu B, Wang XP, et al. (2011) Response of soil respiration to soil temperature and moisture in a 50-year-old oriental arborvitae plantation in China. Plos One 6.

4. Wang M, Shi S, Lin F, Hao ZQ, Jiang P, et al. (2012) Effects of soil water and nitrogen on growth and photosynthetic response of manchurian ash (fraxinus mandshurica) seedlings in northeastern China. Plos One 7.

5. Tramblay Y, Bouvier C, Martin C, Didon-Lescot JF, Todorovik D, et al. (2010) Assessment of initial soil moisture conditions for event-based rainfall-runoff modelling. Journal of Hydrology 387: 176–187.

6. Minet J, Laloy E, Lambot S, Vanclooster M (2011) Effect of high-resolution spatial soil moisture variability on simulated runoff response using a distributed hydrologic model. Hydrology and Earth System Sciences 15: 1323–1338.

7. Tischler M, Garcia M, Peters-Lidard C, Moran MS, Miller S, et al. (2007) A GIS framework for surface-layer soil moisture estimation combining satellite radar measurements and land surface modeling with soil physical property estimation. Environmental Modelling & Software 22: 891–898.

8. Wagner W, Lemoine G, Rott H (1999) A method for estimating soil moisture from ERS scatterometer and soil data. Remote Sensing of Environment 70: 191–207.

9. Moran MS, Peters-Lidard CD, Watts JM, McElroy S (2004) Estimating soil moisture at the watershed scale with satellite-based radar and land surface models. Canadian Journal of Remote Sensing 30: 805–826.

10. Albergel C, Rudiger C, Pellarin T, Calvet JC, Fritz N, et al. (2008) From near-surface to root-zone soil moisture using an exponential filter: an assessment of the method based on in-situ observations and model simulations. Hydrology and Earth System Sciences 12: 1323–1337.

11. Yao XL, Fu BJ, Lu YH, Chang RY, Wang S, et al. (2012) The multi-scale spatial variance of soil moisture in the semi-arid Loess Plateau of China. Journal of Soils and Sediments 12: 694~703.

12. Perry MA, Niemann JD (2008) Generation of soil moisture patterns at the catchment scale by EOF interpolation. Hydrology and Earth System Sciences 12: 39–53.

13. Dorigo WA, Wagner W, Hohensinn R, Hahn S, Paulik C, et al. (2011) The International Soil Moisture Network: a data hosting facility for global in situ soil moisture measurements. Hydrology and Earth System Sciences 15: 1675–1698.

14. Govaerts A, Vervoort A (2010) Geostatistical interpolation of soil properties in boom clay in Flanders. Geoenv Vii - Geostatistics for Environmental Applications 16: 219–230.

15. Baskan O, Erpul G, Dengiz O (2009) Comparing the efficiency of ordinary kriging and cokriging to estimate the Atterberg limits spatially using some soil physical properties. Clay Minerals 44: 181–193.
16. Fritsch C, Coeurdassier M, Giraudoux P, Raoul F, Douay F, et al. (2011) Spatially explicit analysis of metal transfer to biota: influence of soil contamination and landscape. Plos One 6.
17. Qiu Y, Fu BJ, Wang J, Chen LD (2003) Spatiotemporal prediction of soil moisture content using multiple-linear regression in a small catchment of the Loess Plateau, China. Catena 54: 173–195.
18. Lesch SM, Corwin DL (2008) Prediction of spatial soil property information from ancillary sensor data using ordinary linear regression: Model derivations, residual assumptions and model validation tests. Geoderma 148: 130–140.
19. Tabari H, Sabziparvar AA, Ahmadi M (2011) Comparison of artificial neural network and multivariate linear regression methods for estimation of daily soil temperature in an arid region. Meteorology and Atmospheric Physics 110: 135–142.
20. Kravchenko A, Bullock DG (1999) A comparative study of interpolation methods for mapping soil properties. Agronomy Journal 91: 393~400.
21. Minasny B, McBratney AB (2007) Spatial prediction of soil properties using EBLUP with the Matern covariance function. Geoderma 140: 324–336.
22. Lin GF, Chen LH (2004) A spatial interpolation method based on radial basis function networks incorporating a semivariogram model. Journal of Hydrology 288: 288–298.
23. Zhu Q, Lin HS (2010) Comparing ordinary kriging and regression kriging for soil properties in contrasting landscapes. Pedosphere 20: 594~606.
24. Sun W, Whelan BM, Minasny B, McBratney AB (2012) Evaluation of a local regression kriging approach for mapping apparent electrical conductivity of soil (ECa) at high resolution. Journal of Plant Nutrition and Soil Science 175: 212–220.
25. Hengl T, Heuvelink GBM, Rossiter DG (2007) About regression-kriging: From equations to case studies. Computers & Geosciences 33: 1301~1315.
26. Li J, Heap AD (2011) A review of comparative studies of spatial interpolation methods in environmental sciences: Performance and impact factors. Ecological Informatics 6: 228~241.
27. Pandey V, Pandey PK (2010) Spatial and temporal variability of soil moisture. International Journal of Geosciences 1: 87–98.
28. Sigua GC, Hudnall WH (2008) Kriging analysis of soil properties - Implication to landscape management and productivity improvement. Journal of Soils and Sediments 8: 193–202.
29. Chen HS, Shao MG, Li YY (2008) Soil desiccation in the Loess Plateau of China. Geoderma 143: 91–100.
30. Bai TL, Yang QK, Shen J (2009) Spatial variability of soil moisture vertical distribution and related affecting factors in hilly and gully watershed region of Loess Plateau. Chinese Journal of Ecology 28: 2508–2514.
31. Zribi M, Le Hetarat-Mascle S, Ottle C, Kammoun B, Guerin C (2003) Surface soil moisture estimation from the synergistic use of the (multi-incidence and multi-resolution) active microwave ERS Wind Scatterometer and SAR data. Remote Sensing of Environment 86: 30–41.
32. Liu W, Zhang XC, Dang T, Zhu O, Li Z, et al. (2010) Soil water dynamics and deep soil recharge in a record wet year in the southern Loess Plateau of China. Agricultural Water Management 97: 1133–1138.
33. Bi HX, Li XY, Liu X, Guo MX, Li J (2009) A case study of spatial heterogeneity of soil moisture in the Loess Plateau, western China: A geostatistical approach. International Journal of Sediment Research 24: 63–73.
34. Zhao PP, Shao MA (2010) Soil water spatial distribution in dam farmland on the Loess Plateau, China. Acta Agriculturae Scandinavica Section B-Soil and Plant Science 60: 117–125.
35. Wang YQ, Shao MA, Liu ZP (2010) Large-scale spatial variability of dried soil layers and related factors across the entire Loess Plateau of China. Geoderma 159: 99–108.
36. Zhang CL, Liu SQ, Fang JL, Tan KZ (2008) Research on the spatial variability of soil moisture based on GIS. Computer and Computing Technologies in Agriculture, Vol 1 258: 719–727.
37. Zhang CL, Liu SQ, Zhang XY, Tan KZ (2009) Research on the spatial variability of soil moisture Computer and Computing Technologies in Agriculture Ii, Vol 1 293: 285–292.
38. Zeng C, Shao MA, Wang QJ, Zhang J (2011) Effects of land use on temporal-spatial variability of soil water and soil-water conservation. Acta Agriculturae Scandinavica Section B-Soil and Plant Science 61: 1~13.
39. Chen LD, Huang ZL, Gong J, Fu BJ, Huang YL (2007) The effect of land cover/vegetation on soil water dynamic in the hilly area of the loess plateau, China. Catena 70: 200~208.
40. Bi HX, Zhang JJ, Zhu JZ, Lin LL, Guo CY, et al. (2008) Spatial dynamics of soil moisture in a complex terrain in the semi-arid Loess Plateau region, China. Journal of the American Water Resources Association 44: 1121~1131.
41. Carre F, Girard MC (2002) Quantitative mapping of soil types based on regression kriging of taxonomic distances with landform and land cover attributes. Geoderma 110: 241~263.
42. Stacey KF, Lark RM, Whitmore AP, Milne AE (2006) Using a process model and regression kriging to improve predictions of nitrous oxide emissions from soil. Geoderma 135: 107~117.
43. Li Y (2010) Can the spatial prediction of soil organic matter contents at various sampling scales be improved by using regression kriging with auxiliary information? Geoderma 159: 63–75.
44. Liu Y, Fu BJ, Lu YH, Wang Z, Gao GY (2012) Hydrological responses and soil erosion potential of abandoned cropland in the Loess Plateau, China. Geomorphology 138: 404~414.
45. Chen LD, Huang ZL, Gong J, Fu BJ, Huang YL (2007) The effect of land cover/vegetation on soil water dynamic in the hilly area of the loess plateau, China. Catena 70: 200–208.
46. Wang ZQ, Liu BY, Zhang Y (2009) Soil moisture of different vegetation types on the Loess Plateau. Journal of Geographical Sciences 19: 707–718.
47. Zeng C, Shao MA, Wang QJ, Zhang J (2011) Effects of land use on temporal-spatial variability of soil water and soil-water conservation. Acta Agriculturae Scandinavica Section B-Soil and Plant Science 61: 1–13.
48. Wu J, Norvell WA, Welch RM (2006) Kriging on highly skewed data for DTPA-extractable soil Zn with auxiliary information for pH and organic carbon. Geoderma 134: 187~199.
49. Li J, Heap AD (2011) A review of comparative studies of spatial interpolation methods in environmental sciences: Performance and impact factors. Ecological Informatics 6: 228~241.
50. Zimmerman D, Pavlik C, Ruggles A, Armstrong MP (1999) An experimental comparison of Ordinary and UK and Inverse Distance Weighting. Mathematical Geology 31: 375~390.
51. Bruno Gilbert KL (1997) Forest attributes and spatial autocorrelation and interpolation: effects of alternative sampling schemata in the boreal forest. Landscape and Urban Planning 37: 235–224.

18

Soil Carbon and Nitrogen Changes following Afforestation of Marginal Cropland across a Precipitation Gradient in Loess Plateau of China

Ruiying Chang[1,2**9**]**, Tiantian Jin**[2,3**9**]**, Yihe Lü**[2]**, Guohua Liu**[2]**, Bojie Fu**[2*]

1 Key Laboratory of Mountain Surface Processes and Ecological Regulation, Institute of Mountain Hazards and Environment, Chinese Academy of Sciences, Chengdu, Sichuan, China, **2** State Key Laboratory of Urban and Regional Ecology, Research Center for Eco-Environmental Sciences, Chinese Academy of Sciences, Beijing, China, **3** China Institute of Water Resources and Hydropower Research, Beijing, China

Abstract

Cropland afforestation has been widely found to increase soil organic carbon (SOC) and soil total nitrogen (STN); however, the magnitudes of SOC and STN accumulation and regulating factors are less studied in dry, marginal lands, and therein the interaction between soil carbon and nitrogen is not well understood. We examined the changes in SOC and STN in younger (5–9-year-old) and older (25–30-year-old) black locust (*Robinia pseudoacacia* L., an N-fixing species) plantations that were established on former cropland along a precipitation gradient (380 to 650 mm) in the semi-arid Loess Plateau of China. The SOC and STN stocks of cropland and plantations increased linearly with precipitation increase, respectively, accompanying an increase in the plantation net primary productivity and the soil clay content along the increasing precipitation gradient. The SOC stock of cropland decreased in younger plantations and increased in older plantations after afforestation, and the amount of the initial loss of SOC during the younger plantations' establishment increased with precipitation increasing. By contrast, the STN stock of cropland showed no decrease in the initial afforestation while tending to increase with plantation age, and the changes in STN were not related to precipitation. The changes in STN and SOC showed correlated and were precipitation-dependent following afforestation, displaying a higher relative gain of SOC to STN as precipitation decreased. Our results suggest that the afforestation of marginal cropland in Loess Plateau can have a significant effect on the accumulation of SOC and STN, and that precipitation has a significant effect on SOC accumulation but little effect on STN retention. The limitation effect of soil nitrogen on soil carbon accumulation is more limited in the drier area rather than in the wetter sites.

Editor: Xiujun Wang, University of Maryland, United States of America

Funding: This research was supported the National Natural Science Foundation of China (41230745), the Chinese Academy of Sciences (No. XDA05060000) and the International Partnership Program for Creative Research Teams of "Ecosystem Processes and Services." The funders had no role in study design, data collection and analysis, decision to publish, or preparation of the manuscript.

Competing Interests: The authors have declared that no competing interests exist.

* E-mail: bfu@rcees.ac.cn

9 These authors contributed equally to this work.

Introduction

Clearing trees to create cropland can reduce soil organic carbon (SOC) significantly by up to 50% of the SOC in the initial forest [1,2,3,4]. In contrast, the afforestation of previously arable land is generally found to sequester carbon (C) in the soil and play an important role in climate change mitigation [5,6,7], which was proposed as an effective method of C sequestration in Article 3.3 of the Kyoto Protocol in 1997. Subsequently, large-scale afforestation has been practiced worldwide in recent decades, especially in marginal land [8]. Thus, it is meaningful and necessary to study the process and mechanism of soil C changes and their causes during afforestation, especially when considering large-area plantation activities conducted in marginal land to mitigate CO_2 emissions.

Soil C sequestration following afforestation has been the subject of a substantial body of research, which suggests that the direction and magnitude of SOC changes are determined by many factors and processes, such as climate, stand age and soil depth [6,7,9,10,11]. However, the effects of these factors on SOC sequestration are still controversial and unclear. Some reviews suggest that climate factors have a weak effect on soil C accumulation on a global scale [6,11], whereas other meta-analyses indicate that climate plays an important role in influencing SOC during afforestation [9,12]. Along a precipitation gradient (650 to 1450 mm), Berthrong *et al.* [10] found that SOC sequestration was significantly related to precipitation. In arid areas, precipitation is the most important ecological limiting factor and determines the net primary productivity (NPP); it may therefore have a significant effect on SOC, but such evidence is limited. The changes in soil C with stand age can vary, particularly in the initial stage of afforestation (<10 years). Soil C can decrease [7,9,10,11], increase [10,11], show no trend [13,14], or produce a Covington curve [11] with stand age. It has been suggested that the different initial trend may depend on mean annual precipitation (MAP) [10], with soil C decreasing in wetter regions (MAP>1150 mm) but increasing in drier regions (MAP< 1150 mm).

In contrast to SOC, the changes in soil nitrogen (N) following afforestation are poorly documented, even though soil N is strongly correlated with SOC, and SOC sequestration is controlled by the accumulation of soil N as a result of a stoichiometric relationship [15,16]. In their meta-analysis, Li *et al.* [7] and Yang *et al.* [11] found a linear relationship between soil C and N following afforestation, and a 1 g N accretion was associated with a 7 to 13 g C accumulation. However, the interactions between soil C and N are less studied and less understood at the regional scale [17].

Plantations of N-fixing trees accumulate more soil C than non N-fixing forest [18,19,20]. An increase of 12 to 15 g C in the soils was found to be associated with an increase of a 1 g N in 19 case studies on N-fixing tree plantations [18]. The mechanisms behind the greater accumulation of soil C in N-fixing tree plantations may be the reduced decomposition of older soil C, increased formation of soil C, and a higher C input from vegetation, which is associated with the supply of N derived from fixation [18,19,20]. Yet the processes affecting soil C and N change once an N-fixing forest is established, and their relationship with stand age, remain unclear, especially in water-limited marginal lands. The N supply from decomposition is severely limited on dry land because of water limitations. As a result, increases in the external N supply from N-fixing tree plantations in arid areas may result in greater soil C than similar increases in moist regions.

Globally, most of the modern afforestation activities have occurred in China [8]. From 2000 to 2010, the area of afforestation and reforestation was approximately 4.9×10^6 ha globally, and over 60% of the new forest area was in China [8]. Large areas of new forest have expanded during the modern period in China due to a series of ecological restoration programs, such as the Natural Forest Conservation Program and the Grain for Green project (GFG). The GFG is one of the most ambitious programs, initiated in 1999 in some areas and expanded in 2000 to the whole country. It aims to convert marginal, low-yield cropland slopes into woodland and grassland. The Loess Plateau, located in northwestern China, has a semi-arid continental climate (MAP ranges from approximately 650 mm in the southern Loess Plateau to approximately 250 mm in the north) and is the most important target for the GFGP due to its tremendous soil erosion. Between 2000 and 2008, the total area of the marginal cropland involved in the project was approximately 4.83×10^6 ha, of which 3.84×10^5 ha was converted into forest [21]. An N-fixing tree species, black locust (*Robinia pseudoacacia* L.), was one of the most commonly planted trees. Such large-scale plantation activity provides suitable conditions for studying the effects of afforestation on SOC and soil total nitrogen (STN) in marginal lands at a regional scale. The objectives of this study are to (1) detect the effects of MAP on soil C and N changes in younger (<10 years) and older (approximately 30 years) black locust stands in marginal cropland and (2) analyze the interaction between SOC accumulation and soil N accretion at the regional scale.

Materials and Methods

Site Description

All our field activities were conducted with the permission of the farmers who owned the land we tested, and we confirmed that the field studies did not involve any endangered or protected species.

Six sites (S1 to S6) were selected across the Loess Plateau (Fig. S1). The MAP increases from 380 mm in the northern site (S1) to 650 mm in the southern site (S6), with most precipitation occurring between June and August at all sites. The mean annual temperature is similar among the sites (Table 1). In all sites, the landform is typical loess and hilly, and the soil is a calcareous loamy soil (classified as an Entisol in USA soil taxonomy) and well drained [22]. The clay content of the soils increased significantly and linearly from 21.3% in the driest site (S1) to 38.2% in the wettest site (S6), whereas the soil sand content decreased significantly and linearly from 27.2% to 8.5% along this gradient (Fig. S2).

In July and August of 2008, 2009, and 2010, a total of 17 younger (5 to 9 years at sample time) and 16 older (25 to 30 years at sample time) black locust plantation stands were chosen in the six sites, and in each site nearby, steep cropland stands also were chosen for comparison (Table 1). The aspect and angle of the cropland sites were similar to those of the corresponding black locust plantation stands. Most of the sloping croplands had been converted to other land use, especially since 1999 to 2000, so only a small amount of cropland existed, and only one or two cropland sites were selected in each location. Therefore, a few additional cropland data were cited and used from previous studies conducted at the same site (Table 1), which were believed to be reliable because the management history of cropland was similar and the soil C or N content of sloping cropland were found to be quite consistent in the same site in the Loess Plateau (e.g., Sun *et al.*, [23], Table 1).

We assumed there were no differences in the soil physicochemical properties between the black locust plantations and cropland sites before afforestation, and the changes in soil C and N afterward were primarily due to the plantation establishment. There are three lines of evidence to support these assumptions. First, all plantations were converted from long-term (at least 30 years) cultivated cropland, and soil C and N are generally thought to reach a constant value after long-term cultivation [3]. Second, the forest and cropland sites had similar histories and management before afforestation. Third, in each site, the soil characteristics, such as soil type (loess soil) and texture, were largely uniform [22,24]. Furthermore, our assumption that there were no differences in soil C and N between pre-plantation and cropland was able to be strongly supported by the evidence that the soil C and N content of sloping cropland were consistent in the same site as shown above.

In each forest stand, black locust trees were planted in a regular pattern, and the same pre-planting site preparation, including disc trenching, was used across all the sites. After plantation, but prior to 2000, the older stands were occasionally thinned or used for livestock-grazing activities. Since 2000, there has been little human activity in the older plantations, and there has never been much in the younger stands. None of the stands has ever been fertilized. The annual fertilization application to cropland (farm manure or urea) was approximately 300 to 500 kg ha^{-1} across all sites. There has never been irrigation in the plantations or cropland.

Field Measurements, Soil Sampling and Analysis

Generally, the area of the plantation stands was less than 0.5 ha, and only one plot (10 m×10 m or 20 m×10 m plot, depending on the area of the stand) was established in the center of each. The diameter at breast height (DBH), tree height, and tree density of the black locust trees in each plot were recorded (Table 1).

In each plot, locust trees were separated into three or four (dependent on the DBH values distribution) classes based on DBH, and one sample tree with the mean DBH value was selected per class. A 1 m×1 m subplot was established using the sample tree as a vertex, with three or four sample tree subplots in each plot. Sometimes, in addition to the sample tree subplots, another 1 m×1 m subplot (sometimes two) was randomly established in some plots. In total, three to six subplots were selected and were

Table 1. Basic Descriptive Statistics of Six Plantation Sites Along a Precipitation Gradient.

Sites	Longitude/Latitude	Temperature(°C)	Precipitation(mm)	aCategory	Plantation age (years)	Mean DBH (cm)	Mean tree height (m)	Tree density (trees·ha^{-1})	bFine roots	cDescription
S1	110.22°/37.47°	9.5	380	Cropland (4)	0					
				Cropland (3)	0					
				Y (3)	7	6.5	5.2	900		
				A (3)	30	15.0	8.0	350		
				A (4)	30	14.0	11.0	700		
S2	108.11°/37.11°	8.5	420	Cropland (3)	0					
				Y (3)	8	5.0	3.5	1000	Y (6)	
				Y (4)	8	7.8	7.0	625	Y (7)	
				A (3)	30	15.0	7.0	800		
				A (5)	25	10.6	6.0	625	Y (9)	
S3	109.30°/36.92°	9.0	460	Cropland	0					[62]
				Cropland	0					[63]
				Cropland (4)	0				Y (8)	
				Y (5)	5	2.4	3.0	1200	Y (16)	
				Y (4)	8	6.1	6.5	1950	Y (16)	
				Y (3)	9	4.2	2.5	1100		
				A (3)	30	13.5	8.0	400		
				A (3)	30	11.0	8.0	1800		
				A (5)	30	16.1	10.0	500	Y (14)	
S4	109.53°/36.52°	8.7	530	Cropland	0					[64]
				Cropland	0					[23]
				Cropland	0					[65]
				Y (3)	5	6.3	7.0	2200		
				Y (3)	9	5.0	6.2	2700		
				Y (3)	9	5.1	5.8	2300		
				A (3)	25	9.0	6.8	1800		
				A (3)	30	10.5	6.5	2000		
				A (3)	30	18.0	7.0	1500		
S5	109.18°/36.07°	9.2	580	Cropland	0					[66]
				Cropland (5)	0					
				Y (3)	5	2.0	2.5	1400	Y (10)	
				Y (3)	6	5.0	7.0	5200	Y (8)	
				Y (3)	8	-	-	-		Tree census not detected
				Y (4)	8	5.6	7.0	2700	Y (16)	
				A (3)	26	12.0	12.0	1000		

Table 1. Cont.

Sites	Longitude/Latitude	Temperature(°C)	Precipitation(mm)	[a]Category	Plantation age (years)	Mean DBH (cm)	Mean tree height (m)	Tree density (trees·ha⁻¹)	[b]Fine roots	[c]Description
				A (4)	28	13.2	15.0	1300		
				A (5)	30	7.1	7.5	1700	Y (17)	
S6	109.12°/35.33°	8.5	650	Cropland (4)	0					
				Cropland (5)	0				Y (10)	
				Y (3)	5	5.3	6.0	3000		
				Y (3)	6	5.2	6.5	3800		
				Y (6)	8	5.6	8.0	3200	Y (24)	
				Y (3)	9	5.8	8.0	4000		
				A (3)	25	9.2	12.0	2200		
				A (3)	30	14.3	13.0	1000		
				A (5)	30	13.8	13.0	1550	Y (20)	

[a]Y and A indicate younger and older plantations, respectively. The number in the parentheses is the soil sample size in each stand.

[b]Y (Yes) indicates that the fine root biomass was collected in this stand. The number in the parentheses is the sample size.

[c]The data on soil organic carbon, total nitrogen content, and soil bulk density of some croplands were cited from other studies in the same site, and the data in the other cropland and forest stands were collected and analyzed in this paper. There was no tree census of the S5 forest stand, but the soil properties were measured.

evenly distributed in each plot, and the sample size of each stand is shown in Table 1. In each subplot, an auger (3.3 cm diameter, 85 cm³) was used to collect soil samples for SOC and STN concentrations, and another soil auger (5.0 cm diameter, 100 cm³) was used to collect the soil samples for bulk density (BD). In all, there were three to six soil samples for SOC, STN, and BD taken from each plot, and all the samples were collected in increments of 0–10 cm and 10–20 cm. In some of the plots, soil cores (3.3 cm diameter, 85 cm³) were also collected to investigate the fine root (≤2 mm in diameter) biomass in each subplot, for a total of 6 to 24 soil cores in each plot (Table 1). Fine root samples were also partitioned into 0–10 cm and 10–20 cm increments. A detailed description of the fine root collection protocol is provided in Chang et al. [25].

In each cropland site, a 30 to 50 m transect was established at the center of the stand. Three to five 1 m × 1 m subplots were established at approximately 10 m intervals along each transect. In each subplot, one soil core (3.3 cm diameter, 85 cm³) was collected to analyze SOC and STN concentrations. Another soil core was collected using a 100 cm³ cylindrical auger (5.0 cm diameter) for soil BD. In some of the cropland stands, two soil samples were collected in each subplot to assess the fine root biomass (Table 1). The soil samples for both physicochemical properties and fine root biomass were partitioned into 0–10 cm and 10–20 cm increments.

The SOC was measured using potassium-dichromate oxidation (titration of dichromate with ammonium ferrous sulfate), and the C content of the soil organic matter was assumed to be 0.58 [26]. The STN concentration was determined by the dry combustion method (1150°C) using a CN analyzer (Elementar Vario EL)(principle with thermal-conductivity detection for N_2). Fine root samples were washed to collect the fine roots. Roots that were less than 2 mm in diameter were dried at 65°C to a constant mass in order to estimate the biomass. The soil samples for soil BD were dried at 105°C to a constant mass and weighed to determine the soil BD.

Data and Statistical Analysis

Plantation aboveground biomass and fine root biomass. The aboveground biomass at the stand level in the plantations was calculated by multiplying the mean stem biomass by the tree density. The mean stem biomass was determined from the mean DBH based on the allometric relationship (between stem biomass and DBH) established in a previous study on black locust plantations [27]. The fine root biomass at stand level was calculated using the sample value.

SOC and STN stock estimation. The equivalent mass method (SOC stock calculated based on a reference soil mass) was not used in this study because there was a significant difference in soil BD in only a few plantation stands (four stands, 12.9% of total stands) and the corresponding cropland across the six sites. Thus the SOC or STN stock of each stand was calculated at a fixed depth as follows:

$$SCT_i = SC_i \times BD_i \times H_i \times 10 \times (1 - \eta_i) \qquad (1)$$

where SCT_i represents either the SOC or STN stock in layer i (in the top 10 cm layer or the 10–20 cm layer, Mg C·ha⁻¹); SC_i represents the SOC or STN concentration (g·kg⁻¹) in layer i; BD_i is the soil bulk density (g·cm⁻³) in layer i; and H_i is the thickness (cm) of layer i. The term η_i represents the volumetric percentage of particles with a size fraction >2 mm (rock fragments, %) in layer i, and this term was zero in each soil layer. The parameter 10 is a conversion factor.

SOC and STN stock changes. For each site, the mean SOC and STN of the cropland were calculated, and the absolute changes in SOC and STN stock (Mg C·ha^{-1}) after planting were estimated as the difference between the SOC or STN stock in the tree plantation soil and the mean value of cropland soil, as in the following:

$$Absolute\ change = SCT_{forest} - SCT_{cropland} \qquad (2)$$

where SCT_{forest} is the SOC or STN stock in the plantation (Mg C·ha^{-1}), and $SCT_{cropland}$ is the mean SOC or STN stock of the corresponding cropland for each site (Mg C·ha^{-1}).

The relative SOC or STN stock change (%) after the establishment of tree plantations was calculated as follows:

$$\begin{aligned} Relative\ change \\ = (SCT_{forest} - SCT_{cropland})/SCT_{cropland} \times 100 \end{aligned} \qquad (3)$$

where the mean SCT_{forest} and $SCT_{cropland}$ are shown above.

The C:N ratio change was used to represent the relative gain of SOC to STN following afforestation, and it was calculated as follows:

$$\begin{aligned} C:N\ ratio\ change = C:N\ ratio\ of\ plantations \\ /C:N\ ratio\ of\ cropland \end{aligned} \qquad (4)$$

where C:N ratio means SOC:STN.

Statistical Analysis

A simple linear regression was used to test whether the tree density, tree height, and the plantation aboveground biomass increased with MAP and MAP had a significant effect on the SOC and STN stock and the C:N ratio (SOC to STN ratio) of the cropland and plantation stands and whether the MAP were significant predictors of the effects of afforestation on the SOC stock, STN stock and C:N ratio. In addition, the relationships between the SOC and STN stock changes during afforestation were tested using linear regression. The difference in the slope of the linear regression for different soil layers was tested (Z-test) using the following formula proposed in Paternoster *et al.* [28]:

$$Z = (b_1 - b_2)/\sqrt{SEb_1^2 + SEb_2^2} \qquad (5)$$

where b_1 and b_2 are the regression coefficient, and SEb_1 and SEb_2 are the coefficient variances of the two regressions, respectively.

A one-way ANOVA was used (Scheffé's test was used if the data had equal variances and, if not, Tamhane's T2 test was used) to compare the fine root biomass among the cropland and the younger and older plantations within each site, as well as among the sites for younger and older plantations. All analyses were performed in SPSS 11.0, based on a significance level of 0.05.

Results

Tree Census along the Precipitation Gradient

Tree density increased significantly linearly with MAP increasing ($R^2 = 0.42$, $p<0.001$), and tree height tended to increase along the gradient but with a low R^2 ($R^2 = 0.12$, $p = 0.054$). As the trend in tree density and height increased, the black locust plantation aboveground biomass also increased along the MAP gradient ($R^2 = 0.18$, $p = 0.015$). In contrast, the fine root biomass decreased

from the drier sites to wetter sites for younger and older plantations (Fig. 1).

With stand age, the fine root biomass of the younger plantations increased significantly in all four sites (Fig. 1), although the average age difference between younger and older plantations was approximately 22 years. In comparison to cropland, the fine root biomass of the younger plantations was significantly greater in the two drier sites (S2, MAP = 420 mm and S3, MAP = 460 mm), but not in the two wetter sites (S5, MAP = 580 mm and S6, MAP = 650 mm), and the fine root biomass of the older plantations was pronounced greater in all sites (Fig. 1).

SOC, STN Stocks and C:N Ratio along the Precipitation Gradient

With an increase in MAP, the SOC stock of the top 20 cm layer of the cropland increased linearly from 5.2 Mg C·ha^{-1} in the driest site (S1) to 19.6 Mg C·ha^{-1} in the wettest site (S6; about a fourfold increase; $R^2 = 0.92$, $p<0.001$), and the STN stock increased linearly from 0.8 Mg N·ha^{-1} at S1 to 2.2 Mg N·ha^{-1} at S6 (about a threefold increase; $R^2 = 0.83$, $p<0.001$, Fig. 2). The greater sensitivity of SOC to MAP (compared with STN) led to a positive relationship between the C:N ratio of cropland and MAP ($R^2 = 0.50$, $p = 0.01$, Fig. 2). For all plantations, the SOC and STN stocks in the top 20 cm layers also increased linearly with MAP ($R^2 = 0.75$, $p<0.001$ for SOC of younger plantation; $R^2 = 0.80$, $p<0.001$ for STN of younger plantation; $R^2 = 0.44$, $p = 0.005$ for SOC of older plantation; and $R^2 = 0.48$, $p = 0.004$ for STN of older plantation; Fig. 2). However, the regression slope was similar between SOC and STN for both younger and older plantations, resulting in a poor relationship between the C:N ratio of the plantations and MAP (Fig. 2).

SOC and STN Stock Changes Following Afforestation along the Precipitation Gradient

The changes in SOC stock following the afforestation of cropland were dependent on soil depth and plantation age (Table 2, Fig. 3). For all the sites, the average absolute and relative SOC remained unchanged in the uppermost 10 cm and for the total 0–20 cm layer following afforestation, but decreased significantly (by 1.45 Mg C·ha^{-1} or 20.9%) in the lower 10–20 cm of soil ($p<0.05$, Table 2). In addition, the size of the decrease in absolute SOC in the 10–20 cm soil layer had a positive relationship with MAP ($R^2 = 0.39$, $p<0.01$, Fig. 3), indicating that SOC stock decreased more in the wetter sites relative to the drier sites in the first plantation period.

With an increase in plantation age, absolute and relative SOC increased significantly, by 6.70 Mg C ha^{-1} or 161.1% in the top 10 cm layer, and by 7.57 Mg C ha^{-1} or 93.8% in the total 0–20 cm layer ($p<0.05$ for both, Table 2). However, the absolute SOC showed little increase in the lower 10–20 cm soil layer ($p>0.05$, Table 2), suggesting that the initial decrease of SOC stock in this layer recovered to the level of cropland approximately 30 years after planting. For older plantations, the SOC stock in both the uppermost 10 cm layer and lower 10–20 cm increased more quickly in the drier sites than in the wetter sites, as shown by the negative linear relationship between MAP and the relative SOC changes ($R^2 = 0.57$, $p<0.01$ for the top 10 cm, and $R^2 = 0.35$, $p = 0.02$ for the 10–20 cm layer, Fig. 3). Moreover, the slope of the linear regression was greater for the top 10 cm than for the 10–20 cm layer ($p<0.01$, Fig. 3), implying that the relative SOC increased more rapidly in surface layers (0–10 cm) than in the subsurface layer (10–20 cm) as precipitation decreased.

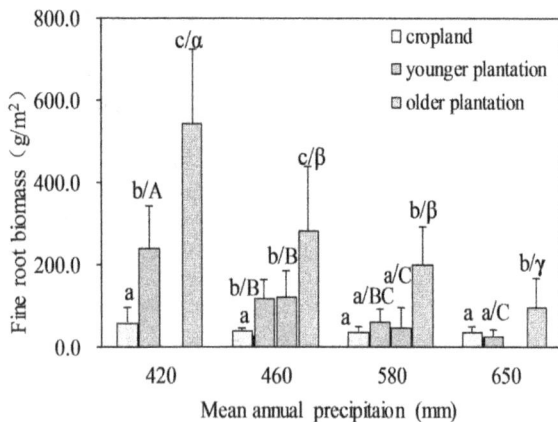

Figure 1. Comparison of fine root biomass among cropland and younger and older plantations in each site as well as among the younger and older plantations in four sites. Younger plantations are 5- to 9-year- old and older plantations are 25- to 30-year-old forests, respectively. The error bars represent the standard deviations of the means. The sample numbers are shown in Table 1. A different lowercase English letter is used for cropland and for younger and older plantations in the same site, and a different uppercase English letter represents younger plantations across the sites, whereas a different Greek letter is used for older plantations to indicate a significant difference at the 0.05 level.

In contrast to SOC, neither absolute nor relative STN were related to MAP at any depth (Fig. 4). For all values of MAP, the absolute STN increased in the top 10 layer and remained constant in the lower 10–20 cm following afforestation, whereas there was no trend or a decrease in the absolute SOC in the upper and lower soil layers, respectively (Table 2). In older plantations, the absolute and relative STN had increased significantly in each soil layer, compared with the lack of change in absolute SOC in the lower 10–20 cm in younger plantations (Table 2). These results suggested that STN increased more quickly than SOC following cropland afforestation.

Relationship between SOC and STN Changes Following Afforestation

Absolute and relative STN were found to increase significantly and linearly with increasing SOC (Figs. 5A, 5B), indicating that there is a close relationship between SOC and STN during afforestation at the regional scale. According to the relationship between absolute SOC and STN, an average gain of 1 g N was associated with a 9.1 g gain in C during afforestation (Fig. 5A). The slope (1.27) of the relationship between the relative SOC and STN was significantly greater than one (Fig. 5B), which implies that the relative STN exceeded SOC across the sites, confirming the relationship in average gains. However, the changes in SOC and STN during afforestation were dependent on precipitation (Figs. 5C, 5D). The negative linear relationship between the C:N ratio and MAP suggests that the relative gain of SOC to STN following afforestation was higher in drier lands and lower in wetter areas.

Discussion

SOC and STN Stocks along the Precipitation in the Loess Plateau

The SOC and STN stocks of younger and older plantations increased linearly as MAP increased from the north to the south of

the Loess Plateau. These findings were consistent with the measured trend of SOC content in *Pinus tabulaeformis* plantations along the increasing precipitation gradient of the Loess Plateau [29] and the SOC stock's tendency in the Nordic forest soils [30]. The positive relationship between precipitation and SOC and STN stocks could be related to an increase in NPP with precipitation [30,31]. Also, the accompanying increase in tree density with precipitation is generally found to lead to a higher aboveground NPP and total belowground carbon allocation in forest [32]; however, the tree density has a nonsignificant effect on SOC stocks [32] because of the synchronous increase in soil CO_2 efflux with tree density [33]. In addition, the greater SOC and STN stocks in the wetter Loess Plateau is associated with the higher clay content, which, it is suggested, plays a positive role in accumulating SOC and STN resulting from the chemical protection of SOC by bonds with clay surfaces and from physical protection through occlusion with aggregates [34,35].

SOC Changes Following Afforestation

Soil C is generally found to decrease in the first few years (>10 years) following afforestation, after which it increases with plantation age to pre-afforestation levels and then to net gains, which is indicated as a Covington curve [7,9,11]. Our findings showed a similar temporal pattern in absolute SOC, with an initial decrease in 8-year-old plantations followed by a return to the initial cropland SOC level in 30-year-old plantations in the 10–20 cm soil layer. The balance between soil C input and output determined the soil C pool. Accordingly, the initial decrease in SOC following afforestation indicates that the C input was too low to match the decomposition of SOC in younger plantations. A study using a natural ^{13}C abundance approach indicated that when cane (a C_4 photosynthetic pathway) is planted with *Eucalyptus* (a C_3 photosynthetic pathway), the new SOC_3 (derived from the *Eucalyptus* planting) in the 10–20 cm layer increased by 1.8 Mg C/ha, whereas the older SOC_4 (derived from cane) was lost at a rate of 3.2 Mg C/ha, resulting in a net C loss of 1.4 Mg C/ha in 10 to 13 years [14].

The initial accelerated SOC decomposition may be due in part to the intensive pre-planting site preparation, which included disc trenching in our study sites. In two reviews, Johnson [36] suggested that site preparation results in a net loss of soil C, although there is an accompanying increase in productivity, and Laganière *et al.* [6] suggested that low-intensity preparation (e.g., hand planting) may lead to a greater accumulation of SOC, and that high-intensity preparation (e.g., disc trenching) stops affecting the SOC level only 21 years after afforestation of cropland. Besides, the decomposition of older SOC is frequently found to be stimulated by the input of new liable C to soil in a short time scale, which is called a priming effect (PE) [37,38,39]. PE is suggested to be a global phenomenal, including in terrestrial and aquatic ecosystems [40], and the rhizosphere priming effect on SOM decomposition is from −70% to more than 380% [37,39]. Thus the priming effect associated with the new C input, especially derived from fine roots during forest establishment, may in part induce the initial loss of old C from the arable soil.

The lose in absolute SOC in the 10–20 cm layer in younger stands increased with rising MAP, even though the NPP of vegetation is generally found to increase along the MAP gradient across the Loess Plateau [41]. This result provided evidence supporting the idea that the initial SOC loss during afforestation is more likely to occur in wetter soils other than in drier soils [10]. This condition may be ascribed to four causes. First, as predicted by the functional balance theory [42,43], the fine root biomass of plantation stands was found to be greater in the drier sites where

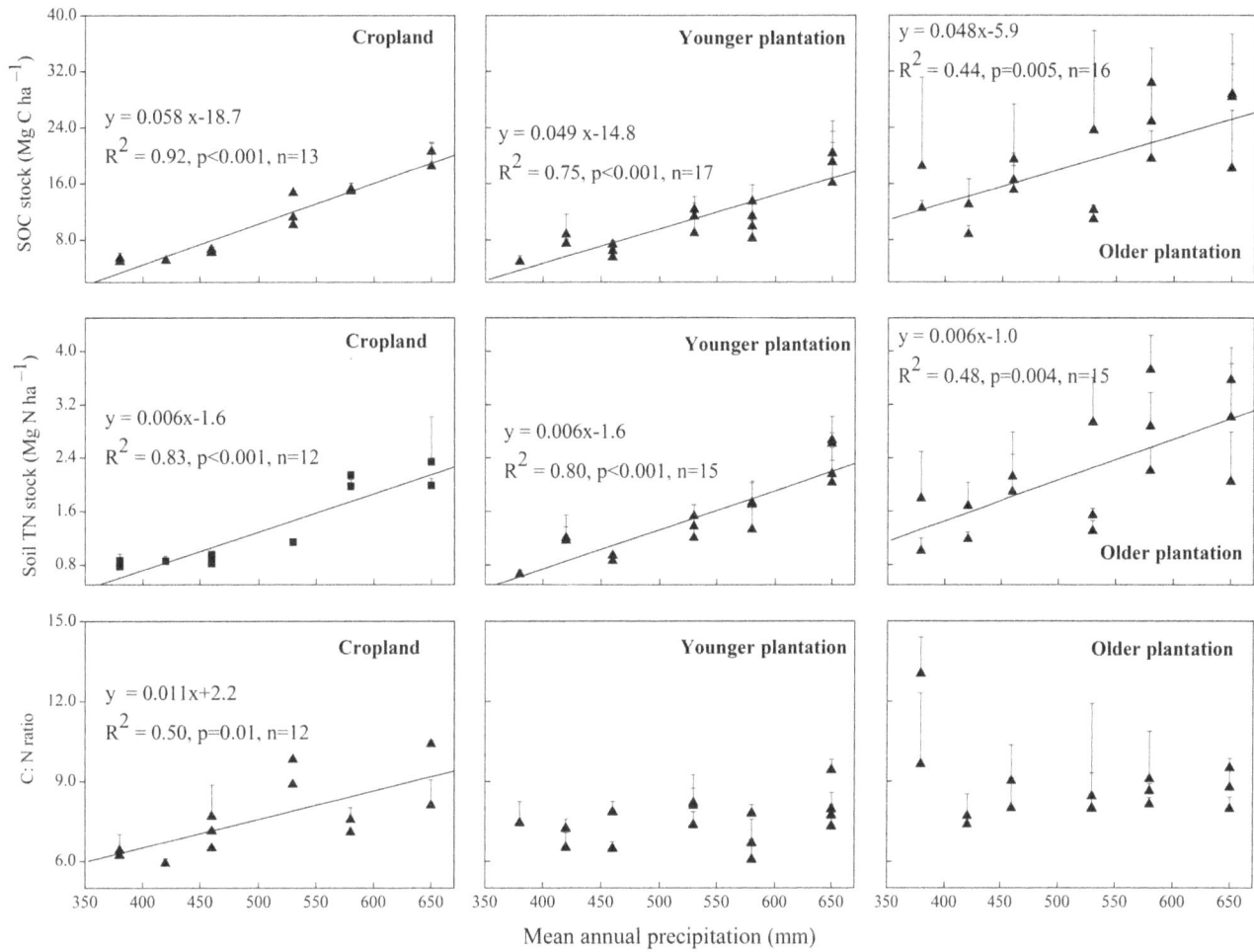

Figure 2. Relationship between mean annual precipitation and soil organic carbon (SOC), soil total nitrogen (STN) stock, and C:N ratio in cropland and younger and older plantations. Younger and older plantations are 5- to 9-year-old and 25- to 30-year-old forests, respectively. The error bars represent the standard deviations of the means (only shown in positive, n = three to six for each stand). The data for some cropland stands were cited from other studies (Table 1). Regressions were conducted for the top 20 cm of soil, and only significant regression models are displayed.

Table 2. Changes in Soil Organic Carbon (SOC) and Soil Total Nitrogen (STN) Stocks in Younger (n = 17 and 15 for SOC and STN, Respectively) and Older Plantations (n = 16 and 15 for SOC and STN, Respectively) Established on Croplands.

Afforestation stages	Absolute SOC change (Mg C·ha⁻¹)			Relative SOC change (%)			Absolute STN change (Mg N·ha⁻¹)			Relative STN change (%)		
	0–10 cm	10–20 cm	0–20 cm	0–10 cm	10–20 cm	0–20 cm	0–10 cm	10–20 cm	0–20 cm	0–10 cm	10–20 cm	0–20 cm
Younger plantation	0.43	−1.45**	−1.01	16.6	−20.9**	−2.3	0.12**	−0.08	0.04	21.5**	−9.8**	5.5
	(−0.44/1.31)	(−2.07/−0.82)	(−2.41/0.38)	(−4.7/37.8)	(−30.9/−10.9)	(−16.8/12.3)	(0.01/0.23)	(−0.17/0.01)	(−0.16/0.23)	(3.8/39.1)	(−19.5/−0.2)	(−7.1/18.2)
Older plantation	6.70**	0.85	7.57**	161.1**	27.1**	93.8**	0.66**	0.11**	0.77**	109.4**	18.4**	63.5**
	(4.79/8.65)	(−0.23/1.93)	(4.85/10.29)	(97.5/224.4)	(3.9/50.4)	(52.3/135.4)	(0.41/0.91)	(0.01/0.21)	(0.45/1.10)	(63.8/154.9)	(5.0/31.8)	(35.8/91.2)

**Significant at 0.05 level compared with the value of zero (one-sample T test).
The numbers in parentheses indicate the 95% confidence interval.
Younger and older plantations mean 5- to 9-year-old and 25- to 30-year-old plantations, respectively.

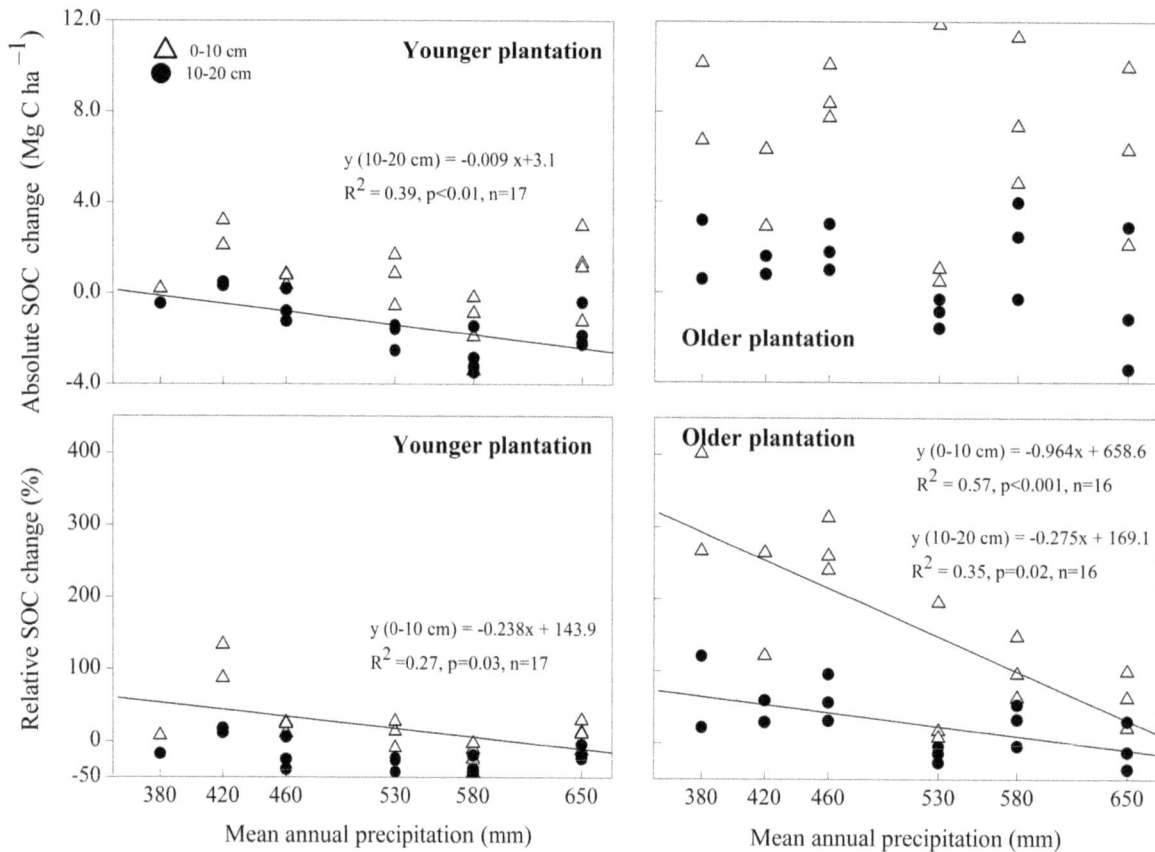

Figure 3. Relationship between mean annual precipitation and the absolute and relative change in soil organic carbon (SOC) in younger and older plantations. Younger and older plantations are 5- to 9-year-old and 25- to 30-year-old forests, respectively. Regressions were conducted separately for different soil layers, and only significant regression models are displayed. Positive values indicate an increase in the SOC stock due to afforestation, and negative values indicate a decrease.

soil water availability was limited, unlike aboveground biomass. Fine roots have been suggested to play a more important role in soil C accumulation than the aboveground parts [44,45]; therefore, the increased fine root biomass along the increasing MAP gradient may drive the trend in SOC change across the Loess Plateau. Second, the tree density increased with increasing MAP. Although tree density is suggested to have a limited effect on SOC changes during afforestation [6], higher tree density could entail an increase in the associated plantation disturbance. Thus the SOC loss resulting from the pre-planting preparations may be higher in the wetter sites. Third, microbial activity is suggested to be limited by the amount of soil water available [46], and soil respiration is generally found to be positively correlated with precipitation [47]. Thus the loss of soil C derived from decomposition would be less in the drier sites. Fourth, in the drier, infertile sites, the PE may also be constrained by limited soil water availability and decomposition substrate [38,48], which also could reduce the SOC loss. Therefore, our results suggest that less SOC decomposition and higher belowground C input determined the initial SOC changes along the decreasing MAP gradient, rather than the NPP of afforestation.

After the initial decrease in SOC in younger plantations, soil C is usually found to increase with plantation age and to reach net gains in approximately 30 years [6,7,9,10]. Based on global afforestation meta-data, Laganière et al. [6] found that the SOC stock decreased by 5.6% in young plantations (<10 years) but increased by 6.1% and 18.6% in medium-aged (10 to 30 years)

and older (>30 years) plantations, respectively. Our findings showed a greater gain in SOC stock (161.1% and 93.8% in the top 10 cm and total 20 cm layers, respectively) in the 30-year-old plantations, which may be because the plantations were primarily N-fixing tree species.

In older plantations, there was a weak relationship between the change in absolute SOC stock and MAP. However, the negative relationship between the relative SOC changes and MAP suggests that the relative SOC gain was higher in drier sites than in the wetter sites. Berthrong et al. [10] also showed that more SOC accumulated in drier lands (MAP<1150 mm) than in wetter lands (MAP>1150 mm) following the afforestation of grassland. This current finding, in contrast to the trend of SOC during forest cultivation, indicates that SOC loss is greater in wetter regions than in drier regions [49,50]. The greater relative increase in SOC in the drier sites was associated with a lower initial value of soil C in cropland. Thus, afforestation can increase SOC more rapidly and improve soil quality more in infertile arable lands than in fertile land, although it may have a similar effect on the accumulation capacity for SOC across the sites.

STN Changes Following Afforestation

As with SOC, STN following afforestation also can show various temporal patterns [7,10,11,13,51]. In a global meta-analysis, STN stocks were generally found to follow a Covington curve, similar to the general trends in SOC, although the duration of the initial decrease in STN (50 years) is longer than that of SOC

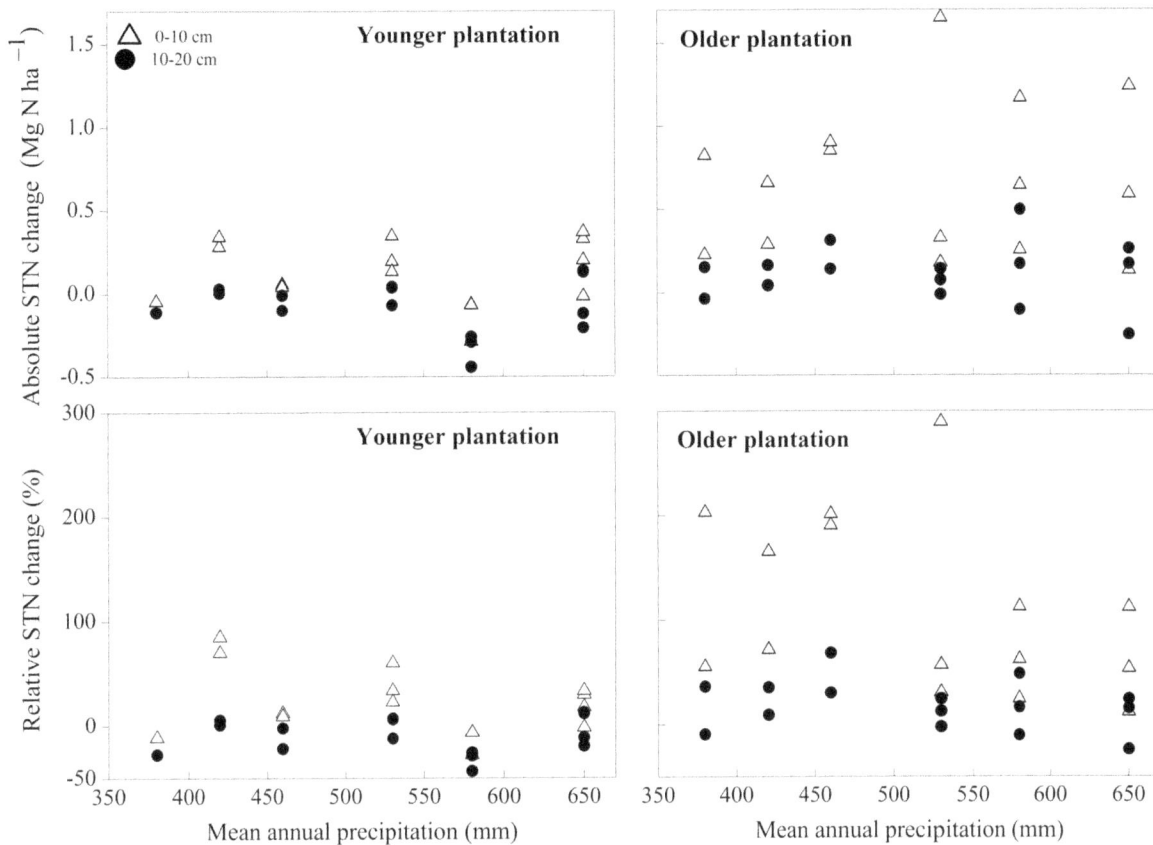

Figure 4. Association of absolute and relative change in soil total nitrogen (STN) in younger and older plantations with mean annual precipitation. Younger and older plantations mean 5- to 9-year-old and 25- to 30-year-old forests, respectively. Positive values indicate an increase in the STN stock due to afforestation, and negative values indicate a decrease. No significant association was found.

(30 years) [7]. However, our results showed no significant decrease in STN during early afforestation, whereas STN stocks tended to increase with plantation stand age. These findings were consistent with other studies of black locust plantations in the Loess Plateau [52,53]. This increase in STN may, to a large extent, be due to the effect of biological nitrogen fixation by the black locust. Boring and Swank [54] reported that a 4-year-old black locust stand fixed 30 kg N/ha/yr during a one-year study period, and Rice et al. [55] found that there was 86 kg N/ha/yr of leaf litter returned to the soil by a 20- to 35-year-old black locust stand. These estimates exceed the average observed annual N accretion in our sites (4 to 9 kg N/ha/yr for younger stands and 26 to 31 kg N/ha/yr for older stands, respectively). Other N sources and mechanisms also may affect STN accumulation. For example, atmospheric N deposition (including dry and wet deposition) is often used to explain increasing STN during afforestation [7,11]. However, few studies have examined the difference in atmospheric N deposition between plantation and cropland, and the N retained in the soil through deposition has rarely been quantified after afforestation. In addition, plantation areas may be less susceptible to soil erosion than croplands in the Loess Plateau [56]; therefore, the lower N loss from soil erosion may result in higher soil N after afforestation.

N mainly enters soil in three ways: atmospheric deposition, fertilizer, and N fixation. There was no N input from fertilizer in the plantations of the Loess Plateau. N inputs from atmospheric deposition are generally found to increase with MAP increase from

the northern to southern areas of the Loess Plateau [57]. Meanwhile, biological N fixation is predicted to increase with tree density increase from the drier to wetter sites, and N-fixation is also suggested to be limited by drought conditions in the drier areas [58,59]. Nevertheless, the changes in STN stock after afforestation were not found to be correlated with MAP in any age stand, which is in contrast to Berthrong et al.'s [10] finding that STN changes following afforestation were significantly related to MAP. This condition may result from higher N loss from plant uptake and mineralization and from less N gain from soil-erosion control in the wetter southern Loess Plateau. First, the above-ground biomass of plantation stands was positively correlated with MAP, and this higher aboveground biomass in the wetter sites may consume more soil N. Evidence shows that available inorganic soil N (NH_4^+ and NO_3^-) is lower with higher tree density (higher aboveground biomass) [32] and which may offset, in part, the benefit of greater N-fixation resulting from higher tree density. Second, stronger microbial activity and a more pro-nounced PE in the wetter environment (as discussed above) may lead to higher SOM mineralization and consequently higher N loss from gas (e.g., NH_3, N_2O). However, McCulley et al. [60] suggested the N cycling is tighter in the wetter grassland sites along a precipitation gradient in the Central Great Plains, and the net soil N mineralization did not increase along the precipitation gradient. Thus a more precise measurement of N mineralization in the forests across the Loess Plateau is needed to understand the

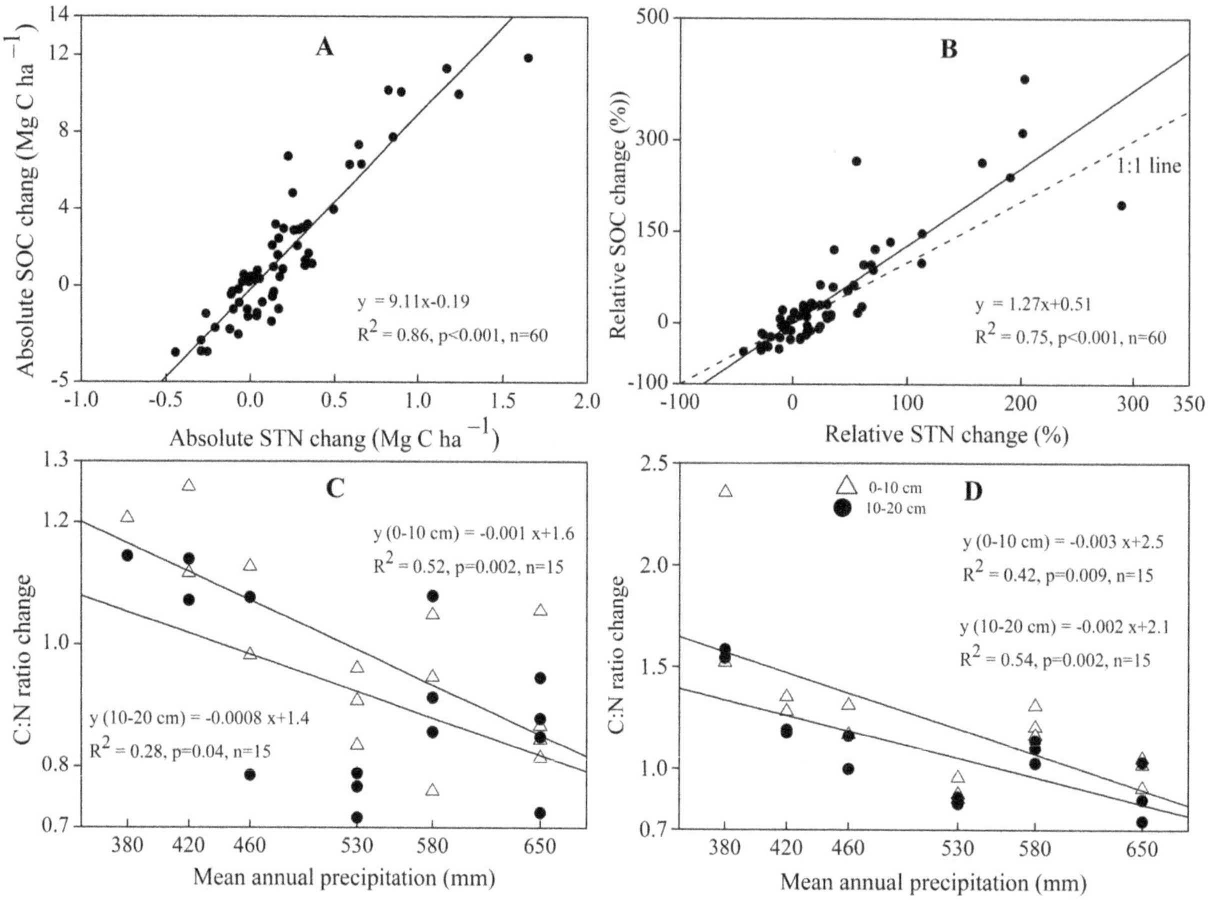

Figure 5. Relationship of changes in absolute soil organic carbon (SOC) and soil total nitrogen (STN) (A), changes in relative SOC and STN (B), and changes in the C:N ratio in younger (C) and older (D) plantations with mean annual precipitation. Younger and older plantations mean 5- to 9-year-old and 25- to 30-year-old forests, respectively. The dashed line in panel B indicates the 1:1 line. Values greater than 1.0 in panels C and D indicate a relative gain of SOC to STN following afforestation, and lower values indicate the inverse.

changes in soil N. Third, soil erosion in cropland is more serious in the northern, drier sites relative to the southern, wetter areas [61], and the N gain derived from the reduced soil erosion in afforested sites may be more pronounced in the northern, more eroded sites.

Soil C and N Interaction and the Implications

The loss of SOC is greater than that of STN when forests were converted into cropland, resulting in a lower C:N ratio in the cropland [1,4]. In contrast, the accumulation rate of SOC is suggested to be higher than that of soil N after cropland afforestation [4,7,51]. In the present study, however, STN was found to increase more quickly than SOC following afforestation, as indicated by higher relative changes in STN. Nevertheless, the relative gains of SOC and STN also were precipitation dependent, with higher C gain (compared with N gain) in the drier lands and lower C gain in the wetter sites. Furthermore, SOC changes following afforestation were related to MAP, whereas STN changes were not. Therefore, C and N were more likely to be decoupled following marginal cropland afforestation in this semi-arid environment, and the changes in soil C and N were not synchronous at the regional scale, which is a proxy for precipitation. This finding implied that the mechanisms and

factors controlling the accumulation rates were different for C and N, as suggested by McLauchlan [4].

As a result of the stoichiometric relationship between C and N, the terrestrial C (including soil C) sequestration is suggested to be determined by the availability of N and may be down-regulated as the supply of N is limited, as argued in the progressive N limitation theory [15]. In our study, however, the relative soil C gain to N retention increased when precipitation decreased, indicating a limited effect of N limitation on SOC accumulation in the drier area, at least in a relatively short plantation stage (during about 30 years). It seemed that an increase in soil N derived from N-fixing tree plantation in such N-limited arid lands could have a significant effect on soil C accumulation, and this effect was greater in the drier areas rather than in the wetter ones. Nevertheless, these results are based on a relatively short period of afforestation (30 years), and they may not hold up in the long term, although some studies suggest that forest ecosystems have an intrinsic ability to regulate the soil C and N accumulation and reduce N limitation in a relatively longer time [11]. So further investigations focused on the older stands in such arid areas are needed.

Conclusion

Our study suggests that afforestation of marginal cropland could lead to a significant increase in SOC and STN stocks in a relatively short time (about 30 years) and underlines the effect of precipitation on SOC dynamics during afforestation in an arid environment. The results show that, along the increasing precipitation gradient, the initial absolute loss of SOC increased but the relative increase in SOC in the older ages decreased. This phenomenon is suggested to be associated with the greater soil C decomposition loss and the less belowground C input in the wetter sites, rather than the NPP of afforestation.

Unlike the SOC dynamics, there was no significant loss in STN during initial afforestation, and the changes in STN were not correlated with precipitation. This decoupled relationship between STN and SOC implies that a mechanism may have been regulating the effect of soil N retention on soil C accumulation to prevent progressive N limitation during forest development.

Supporting Information

Figure S1　Geographic locations of the study sites and croplands involved in Grain for Green project in China's Loess Plateau.

Figure S2　Relationship between soil clay content of croplands and mean annual precipitation. The error bars represent the standard deviations of the means (only shown in positive, n = three to five for each stand). The datum of one stand in S4 (with MAP of 530 mm) was cited from Li *et al.* [66].

Author Contributions

Conceived and designed the experiments: RC TJ BF. Performed the experiments: RC TJ YL. Analyzed the data: RC. Contributed reagents/materials/analysis tools: RC TJ GL. Wrote the paper: RC YL BF.

References

1. Murty D, Kirschbaum MUF, McMurtrie RE, McGilvray H (2002) Does conversion of forest to agricultural land change soil carbon and nitrogen? A review of the literature. Global Change Biology 8: 105–123.
2. Liu SG, Bliss N, Sundquist E, Huntington TG (2003) Modeling carbon dynamics in vegetation and soil under the impact of soil erosion and deposition. Global Biogeochemical Cycles 17: 1074.
3. Davidson E, Ackerman I (1993) Changes in soil carbon inventories following cultivation of previously untilled soils. Biogeochemistry 20: 161–193.
4. McLauchlan K (2006) The nature and longevity of agricultural impacts on soil carbon and nutrients: a review. Ecosystems 9: 1364–1382.
5. Post WM, Kwon KC (2000) Soil carbon sequestration and land use change: processes and potential. Global Change Biology 6: 317–327.
6. Laganière J, Angers DA, Par D (2010) Carbon accumulation in agricultural soils after afforestation: a meta analysis. Global Change Biology 16: 439–453.
7. Li D, Niu S, Luo Y (2012) Global patterns of the dynamics of soil carbon and nitrogen stocks following afforestation: a meta-analysis. New Phytologist 195: 172–181.
8. FAO (2010) The state of forest resources – a regional analysis. In: FAO, editor. The State of the World's Forests 2011.
9. Paul KI, Polglase PJ, Nyakuengama JG, Khanna PK (2002) Change in soil carbon following afforestation. Forest ecology and management 168: 241–257.
10. Berthrong ST, Piñeiro G, Jobbágy EG, Jackson RB (2012) Soil C and N changes with afforestation of grasslands across gradients of precipitation and plantation age. Ecological Applications 22: 76–86.
11. Yang Y, Luo Y, Finzi AC (2011) Carbon and nitrogen dynamics during forest stand development: a global synthesis. New Phytologist 190: 977–989.
12. Guo LB, Gifford RM (2002) Soil carbon stocks and land use change: a meta analysis. Global Change Biology 8: 345–360.
13. Sartori F, Lal R, Ebinger MH, Eaton JA (2007) Changes in soil carbon and nutrient pools along a chronosequence of poplar plantations in the Columbia Plateau, Oregon, USA. Agriculture, Ecosystems & Environment 122: 325–339.
14. Bashkin MA, Binkley D (1998) Changes in soil carbon following afforestation in Hawaii. Ecology 79: 828–833.
15. Luo Y, Su BO, Currie WS, Dukes JS, Finzi A, et al. (2004) Progressive Nitrogen Limitation of Ecosystem Responses to Rising Atmospheric Carbon Dioxide. BioScience 54: 731–739.
16. Kirkby CA, Richardson AE, Wade LJ, Batten GD, Blanchard C, et al. (2013) Carbon-nutrient stoichiometry to increase soil carbon sequestration. Soil Biology and Biochemistry 60: 77–86.
17. Gärdenäs AI, Ågren GI, Bird JA, Clarholm M, Hallin S, et al. (2011) Knowledgegaps in soilcarbon and nitrogeninteractions – From molecular to global scale to global scale. Soil Biology and Biochemistry 43: 702–717.
18. Binkley D (2005) How nitrogen-fixing trees change soil carbon. In: Binkley D, Menyailo O, editors. Tree species effects on soils: implications for global change. Dordrecht: Kluwer Academic Publishers. pp. 155–164.
19. Forrester DI, Pares A, O'Hara C, Khanna PK, Bauhus J (2013) Soil Organic Carbon is Increased in Mixed-Species Plantations of Eucalyptus and Nitrogen-Fixing Acacia. Ecosystems 16: 123–132.
20. Resh SC, Binkley D, Parrotta JA (2002) Greater soil carbon sequestration under nitrogen-fixing trees compared with Eucalyptus species. Ecosystems 5: 217–231.
21. Lü YH, Fu BJ, Feng XM, Zeng Y, Liu Y, et al. (2012) A Policy-Driven Large Scale Ecological Restoration: Quantifying Ecosystem Services Changes in the Loess Plateau of China. PLoS ONE 7: e31782.
22. Wen QZ, editor (1989) Geochemistry in Chinese Loess. Beijing: Science Press.
23. Sun WY, Guo SL, Zhou XG (2010) Effects of Topographies and Land Uses on Soil Organic Carbon in Subsurface in Hilly Region of Loess Plateau. Environmental Science 31: 2740–2747.
24. Guo ZY (1992) Shaanxi Soils. Beijing: Science Press.
25. Chang RY, Fu BJ, Liu GH, Yao XL, Wang S (2012) Effects of soil physicochemical properties and stand age on fine root biomass and vertical distribution of plantation forests in the Loess Plateau of China. Ecological Research 27: 827–836.
26. Lu R (1999) Analytical Methods of Soil Agrochemistry. Beijing: Science and Technology Press.
27. Tian QF, Du LH, Li XJ (1997) Study on Biomass of *RobiniaPseudoacacia* Plantation in the Beijing Xishan National Forest Park. Journal of Beijing Forestry University 19: 104–107.
28. Paternoster R, Brame R, Mazerolle P, Piquero A (1998) Using the correct statistical test for the equality of regression coefficients. Criminology 36: 859–866.
29. Wei XR, Shao MA, Fu XL, Horton R, Li Y, et al. (2009) Distribution of soil organic C, N and P in three adjacent land use patterns in the northern Loess Plateau, China. Biogeochemistry 96: 149–162.
30. Callesen I, Liski J, Raulund-Rasmussen K, Olsson MT, Tau-Strand L, et al. (2003) Soil carbon stores in Nordic well-drained forest soils–relationships with climate and texture class. Global Change Biology 9: 358–370.
31. Schimel DS, Braswell BH, Holland EA, McKeown R, Ojima DS, et al. (1994) Climatic, edaphic, and biotic controls over storage and turnover of carbon in soils. Global Biogeochemical Cycles 8: 279–293.
32. Litton CM, Ryan MG, Knight DH (2004) Effects of tree density and stand age on carbon allocation patterns in postfire lodgepole pine. Ecological Applications 14: 460–475.
33. Litton CM, Ryan MG, Knight DH, Stahl PD (2003) Soil-surface carbon dioxide efflux and microbial biomass in relation to tree density 13 years after a stand replacing fire in a lodgepole pine ecosystem. Global Change Biology 9: 680–696.
34. Six J, Conant RT, Paul EA, Paustian K (2002) Stabilization mechanisms of soil organic matter: Implications for C-saturation of soils. Plant and Soil 241: 155–176.
35. Krull ES, Baldock JA, Skjemstad JO (2003) Importance of mechanisms and processes of the stabilisation of soil organic matter for modelling carbon turnover. Functional Plant Biology 30: 207–222.
36. Johnson DW (1992) Effects of forest management on soil carbon storage. Water, Air, & Soil Pollution 64: 83–120.
37. Cheng WX (2009) Rhizosphere priming effect: Its functional relationships with microbial turnover, evapotranspiration, and C–N budgets. Soil Biology and Biochemistry 41: 1795–1801.
38. Kuzyakov Y, Friedel JK, Stahr K (2000) Review of mechanisms and quantification of priming effects. Soil Biology and Biochemistry 32: 1485–1498.
39. Cheng WX, Kuzyakov Y (2005) Root effects on soil organic matter decomposition. In: Zobel RW, Wright, SF., editor. Roots and Soil Management: Interactions between Roots and the Soil. Madison Wisconsin: ASA-SSSA.
40. Guenet B, Danger M, Abbadie L, Lacroix G (2010) Priming effect: bridging the gap between terrestrial and aquatic ecology. Ecology 91: 2850–2861.
41. Li Z, Yan F, Fan X. The variability of NDVI over Northwest China and its relation to temperature and precipitation; 2003. IEEE. pp. 2275–2277.
42. Hendricks JJ, Hendrick RL, Wilson CA, Mitchell RJ, Pecot SD, et al. (2006) Assessing the patterns and controls of fine root dynamics: an empirical test and methodological review. Journal of Ecology 94: 40–57.
43. Farrar JF, Jones DL (2000) The control of carbon acquisition by roots. New Phytologist 147: 43–53.
44. Rasse DP, Rumpel C, Dignac MF (2005) Is soil carbon mostly root carbon? Mechanisms for a specific stabilisation. Plant and Soil 269: 341–356.

45. Kuzyakov Y, Schneckenberger K (2004) Review of estimation of plant rhizodeposition and their contribution to soil organic matter formation. Archives of Agronomy and Soil Science 50: 115–132.

46. Broughton LC, Gross KL (2000) Patterns of diversity in plant and soil microbial communities along a productivity gradient in a Michigan old-field. Oecologia 125: 420–427.

47. Raich JW, Schlesinger WH (1992) The global carbon dioxide flux in soil respiration and its relationship to vegetation and climate. Tellus B 44: 81–99.

48. Dijkstra FA, Cheng W (2007) Moisture modulates rhizosphere effects on C decomposition in two different soil types. Soil Biology and Biochemistry 39: 2264–2274.

49. Ogle SM, Breidt FJ, Paustian K (2005) Agricultural management impacts on soil organic carbon storage under moist and dry climatic conditions of temperate and tropical regions. Biogeochemistry 72: 87–121.

50. Lugo A, Sanchez M, Brown S (1986) Land use and organic carbon content of some subtropical soils. Plant and Soil 96: 185–196.

51. Hooker TD, Compton JE (2003) Forest ecosystem carbon and nitrogen accumulation during the first century after agricultural abandonment. Ecological Applications 13: 299–313.

52. Wang B, Liu G, Xue S (2012) Effect of black locust (Robinia pseudoacacia) on soil chemical and microbiological properties in the eroded hilly area of China's Loess Plateau. Environmental Earth Sciences 65: 597–607.

53. Liu D, Huang YM, An SS (2012) Changes in Soil Nitrogen and Microbial Activity During *Robinia Pseudoacacia* Recovery Period in the Loess Hilly-Gully Region. Chinese Journal of Eco-Agriculture 20: 322–329.

54. Boring LR, Swank WT (1984) Symbiotic Nitrogen Fixation in Regenerating Black Locust (Robinia pseudoacacia L.) Stands. Forest Science 30: 528–537.

55. Rice S, Westerman B, Federici R (2004) Impacts of the exotic, nitrogen-fixing black locust (Robinia pseudoacacia) on nitrogen-cycling in a pine–oak ecosystem. Plant Ecology 174: 97–107.

56. Fu BJ, Wang YF, Lu YH, He CS, Chen LD, et al. (2009) The effects of land-use combinations on soil erosion: a case study in the Loess Plateau of China. Progress in Physical Geography 33: 793–804.

57. Peng L, Yu CZ, Wang JZ (1995) The Nutrient Content and Nutrient Supply of Dry-farming Soil in Loess Plateau. Journal of Northwest University (Natural Science Edition) 25: 117–122.

58. Zahran HH (1999) Rhizobium-Legume Symbiosis and Nitrogen Fixation under Severe Conditions and in an Arid Climate. Microbiology and Molecular Biology Reviews 63: 968–989.

59. Gálvez L, González EM, Arrese-Igor C (2005) Evidence for carbon flux shortage and strong carbon/nitrogen interactions in pea nodules at early stages of water stress. Journal of Experimental Botany 56: 2551–2561.

60. McCulley R, Burke I, Lauenroth W (2009) Conservation of nitrogen increases with precipitation across a major grassland gradient in the Central Great Plains of North America. Oecologia 159: 571–581.

61. Wang WZ, Jiao JY (2002) Temporal and Spatial Variation Features of Sediment Yield Intensity on Loess Plateau. Acta Geographica Sinica 57: 210–217.

62. Wang B, Liu GB, Xue S, Zhu BB (2011) Changes in soil physico-chemical and microbiological properties during natural succession on abandoned farmland in the Loess Plateau. Environmental Earth Sciences 62: 915–925.

63. Jiao JY, Zhang ZG, Bai WJ, Jia YF, Wang N (2010) Assessing the Ecological Success of Restoration by Afforestation on the Chinese Loess Plateau. Restoration Ecology 20: 240–249.

64. Wang XL, Guo SL, Ma YH, Huang GY, Wu JS (2007) Effects of land use type on soilorganic C and total N in a small watershed in loess hilly-gully region. Chinese Journal of Applied Ecology 18: 1281–1285.

65. Liu Y, Zheng FL, An SS, Guo M (2007) The responses of soil organic carbon, total nitrogen and enzymatic activity to vegetation restoration in abandoned lands in Yangou watershed. Agricultural Research in the Arid Areas 25: 220–225.

66. Li YY, Shao MA, Zheng JY, Zhang XC (2005) Spatial-temporal changes of soil organic carbon during vegetation recovery at Ziwuling, China. Pedosphere 15: 601–610.

19

Identification of Paleo-Events Recorded in the Yellow Sea Sediments by Sorting Coefficient of Grain Size

Liguang Sun[1][*][9], **Xin Zhou**[1][9], **Yuhong Wang**[2], **Wenhan Cheng**[1], **Nan Jia**[1]

1 Institute of Polar Environment and School of Earth and Space Sciences, University of Science and Technology of China, Hefei, China, **2** Advanced Management Research Center, Ningbo University, Ningbo, China

Abstract

Identification of natural and anthropogenic events in the past is important for studying their patterns and mechanisms; and sensitive proxies in marine sediments are more reliable for identifying these events than those in terrestrial sediments, which are usually disturbed by human activities. Since the main source materials for the sediments in the Northern Yellow Sea Mud are transported by the Yellow River, sedimentary characteristics can be used to reconstruct the historical events that occurred in the Yellow River Valley. In the present study, by analyzing sorting coefficient of grain size in a 250-year sediment core from the Northern Yellow Sea Mud, we identified several major historical events: the Haiyuan Earthquake in AD 1920 and several times of relocation of the Yellow River estuary. The proxy has the potential of detecting and reconstructing historical events; in combination with historical archives, they also provide an accurate dating method.

Editor: Vanesa Magar, Plymouth University, United Kingdom

Funding: This study is supported by the National Basic Research Program of China (2010CB428902). The funders had no role in study design, data collection and analysis, decision to publish, or preparation of the manuscript.

Competing Interests: The authors have declared that no competing interests exist.

* E-mail: slg@ustc.edu.cn

9 These authors contributed equally to this work.

Introduction

Identification of natural and anthropogenic events in the past is important for studying their patterns and mechanisms; and sensitive proxies in marine sediments are more reliable for identifying these events than those in terrestrial sediments, which are usually disturbed by human activities.

In China's Near Sea, suspended fine-grained materials are carried by rivers and deposited on the inner shelves to form the mud sediments [1]. Grain size of these sediments is subject to the influence of natural and anthropogenic events in the relevant regions and thus sensitive to various climatic factors. For example, a coarse sand layer found in the muddy sediments is a good indication of storm surge [2]. Grain size of these sediments has been widely used as paleoclimate proxies [2,3]; however, the historical events recorded in these sediments have attracted much less interests.

The major Yellow River estuary relocation events have been recorded in historical archives [4] and in the sediments transported by the Yellow River. The AD 1855 relocation, one of the largest, has been detected by elements and organic matter $\delta^{13}C$ changes in the sediments from the southern Yellow Sea [5]. Some other major relocation events have been detected by using sediment facies and total organic carbon / total nitrogen (TOC/ TN) changes in a long, but imprecisely dated sediment core from the northern Yellow River [6]. To date, however, no systematic analysis has been performed to identify these events by using sedimentary proxy.

In the present study, we analyzed sedimentary characteristics, including parameters for grain size, magnetic susceptibility (MS)

and SiO_2/Al_2O_3 ratio to identify paleo-events recorded in the Northern Yellow Sea Mud (NYSM).

Materials and Methods

NYSM is located in the north of Shandong Peninsula of China, and its main source materials come from the Yellow River and are transported across the Bohai Bay along the Shandong Peninsula. The Yellow River can transport materials directly to the NYSM region, as shown in the satellite image [7]. In this study, we collected a 34 cm long sediment core, labeled as M38002, from Station 38002 (122°30.21′ E, 37°59.92′ N, water depth 49.2 m; Figure 1) of NYSM by box-corer in 2009 during "The Offshore Sea Opening Research Cruise (Autumn)" on the scientific survey ship "Kexue 1", Institute of Oceanology, Chinese Academy of Sciences. The permits, from the Institute of Oceanology of Chinese Academy of Sciences were obtained for the described field studies.

The sediment core was divided at 0.5 cm intervals to collect 68 subsamples, which were analyzed for magnetic susceptibility, grain size, and levels of Si and Al. Chronology of the core is determined by ^{210}Pb-^{137}Cs dating method (Figure 2).

The samples were pre-treated for grain size analysis. 10–20 ml H_2O_2 solution (30%) was added, and the mixture was heated to 100°C for 0.5 h to remove organic matter and bathed in 10 ml HCL solution (10%) for 48 h to remove calcareous cement and shell materials. All samples were fully desalted and dispersed by adding 10 ml $(NaPO_3)_6$ (10%) and by ultrasonic treatment of 10 minutes before measurement.

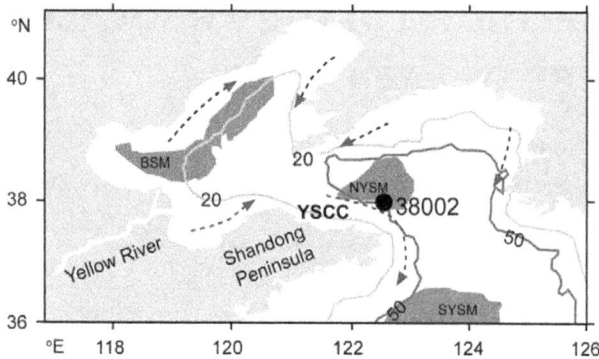

Figure 1. Location of the sampling site (map is modified from [24]). Topographic lines are shown in light grey, mud areas are shown in dark grey, and coastal currents are marked by dashed lines and arrows. BSM: Bohai Sea Mud; NYSM: Northern Yellow Sea Mud; SYSM: Southern Yellow Sea Mud; YSCC: Yellow Sea Coastal Current.

Grain size was measured using Mastersizer 2000 (Malvern Instruments) at the Lab of Soil and Environmental Changes, Taishan University, Taian, China. The measurement range of the instrument is $0.02-2000$ μm, the resolution is 0.01 Φ, and the repeated measurement error is less than 2%. Measurements of low frequency MS were carried out using the Bartington MS2 susceptibility meter in Nanjing University, Nanjing, China. The absolute contents of SiO_2 and Al_2O_3 were measured using XRF-1800 (X-Ray Fluorescence) Spectrometry (Shimadzu Corporation) in Physical and Chemical Science Experimentation Center of the University of Science and Technology (USTC), Hefei, China.

Radioactivity was measured by germanium detector manufactured by AMETEK Company in the Institute of Polar Environment, USTC, Hefei, China. The samples were dried to constant weight at a temperature of 50 °C, homogenized using a mortar and pestle, and passed through a 120-μm sieve. Samples (between 5–10 g) were then packed into standard counting geometries for gamma analyses and sealed and stored for about one week to allow

radioactive equilibration between ^{226}Ra and its daughter product ^{214}Pb. Spectra were continuously measured for 24 h to obtain enough counts. The resulted spectrum files showed ^{210}Pb activity with a peak at 46.5 keV.

Results and Discussion

Chronology

The whole core consists of muddy sediments. The upper 23 cm is a light yellow oxide layer, and the lower part dark grey. Some layers contain shell fragments. Because the excess ^{210}Pb activity showed a simple exponential relation with depth (Figure 2), typical of the decay curve, indicating a nearly constant sedimentation rate. The ^{210}Pb chronology was thus constructed by using a Constant Initial Concentration (CIC) model [8]. The age at 8.5 cm depth, calculated from the ^{210}Pb profile, is AD 1961, close to the age of AD 1963 as indicated by the peak value of ^{137}Cs profile; thus our ^{210}Pb age model appears to be correct. The average sedimentation rate was 0.13 cm·yr^{-1}, consistent with the earlier results [9,10]. The time span of the core is about 254 years (AD 1755–2009), as estimated by extrapolation of the average sedimentation rate.

Grain size distributions

Four typical samples were selected from the core M38002 at depths of 3.5, 11.5, 15, and 25cm to study the grain size distributions of the sediments. The frequency distributions (Figure 3) of these four samples are close to log-normal distribution with corresponding skewness of 0.09, 0.08, 0.09, and −0.01, and they are similar to that of the loess [11].

Identification of events by sorting coefficient for grain size

Median grain size, MS, and SiO_2/Al_2O_3 ratio are usually used as proxies of sedimentary characteristics. The profiles of median grain size and SiO_2/Al_2O_3 ratio in the sediment core M38002 show similar trends during the past 254 years. They have large oscillations without clear trend before AD 1820, gradually rise

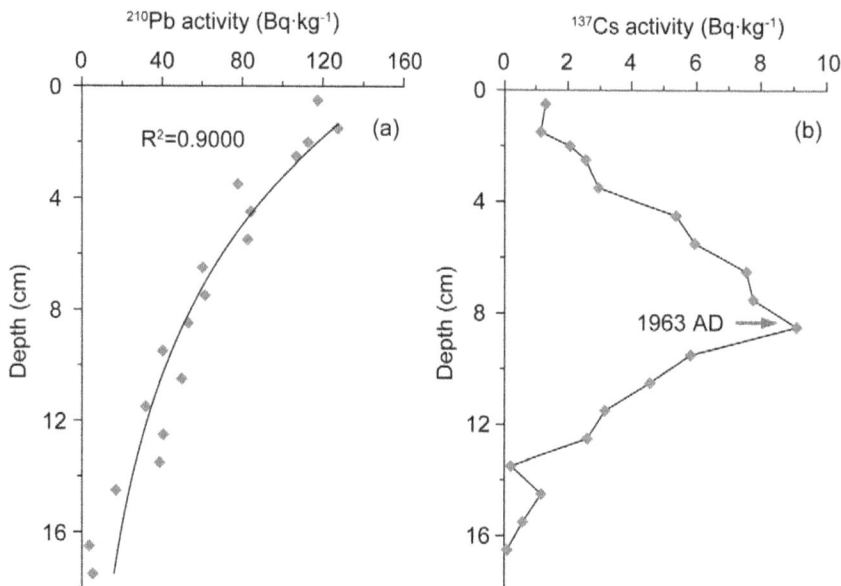

Figure 2. Activity profiles of ^{210}Pb (a) and ^{137}Cs (b) for the sediment core M38002.

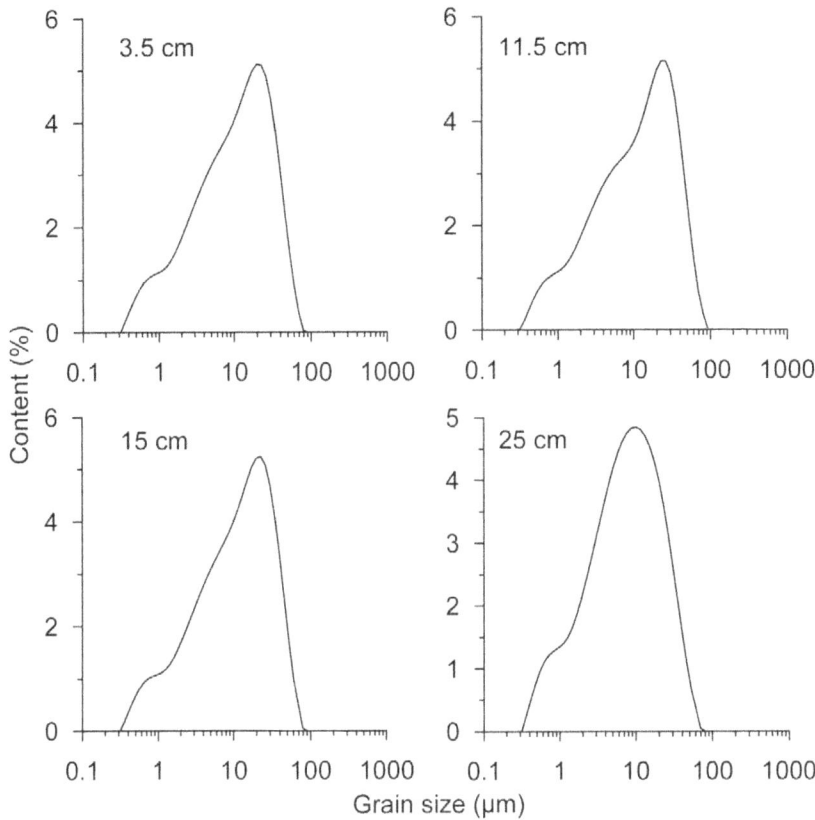

Figure 3. Grain size distributions of four samples selected from the sediment core M38002 at depths of 3.5, 11.5, 15, and 25cm.

between AD 1820 and 1960, and then fall abruptly after AD 1960. The MS profile shows an opposite trend (Figure 4).

The main source materials of the sediments from the NYSM were originated from the Loess Plateau, and transported by the Yellow River [12]. As expected, the sedimentary characteristics of the core M38002, as shown by the plot of median grain size versus MS and SiO_2/Al_2O_3 ratio (Figure 5), are consistent with those of the loess [13,14]. Great events, such as relocations of the Yellow River estuary, for the past 254 years have been well documented in historical archives. These events, especially those occurred on the Loess Plateau, could be recorded in the sediments transported by the Yellow River. However, no obvious, corresponding changes in the median grain size, the MS and the SiO_2/Al_2O_3 ratio of the core M38002 could be identified (Figure 4), likely due to relatively homogeneous sedimentary characteristics of the loess.

To search for a proxy that could identify these well-documented events, we examined the link between these events and sorting coefficient (SC). SC, one of the parameters for grain size, has been successfully used as a proxy for hydrodynamic and material source changes, and it can be calculated by two different methods:

$$SC1 = \sqrt{\frac{Mm75}{Mm25}} \qquad (1)$$

Where Mm_{25} and Mm_{75} are the reaches of the 25th and 75th percentiles using millimeter values [15], respectively, and

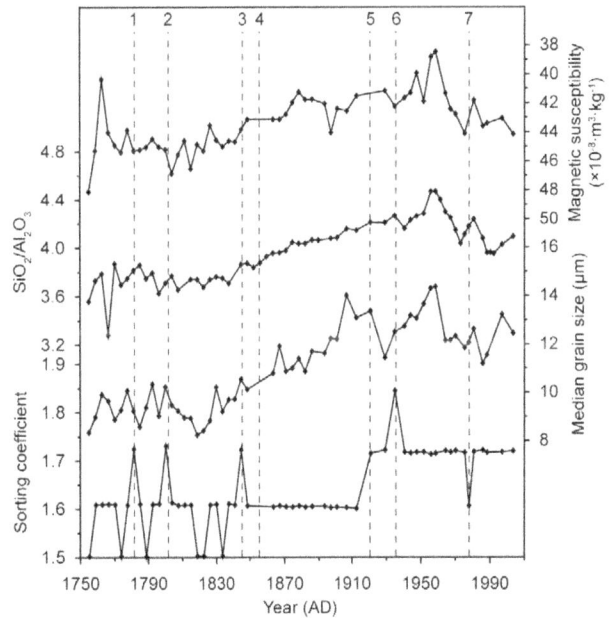

Figure 4. Comparison of sedimentary characteristics of the sediment core M38002 in NYSM with paleo-events. The numbers "1", "2" and "3" mark the extra-large flood around the Shandong Peninsula in AD 1781, the extra-large flood of Hai and Luan River in AD 1801 and large flood of Liao River in AD 1846, respectively; "4", "6" and "7" the events of the Yellow river relocation in AD 1855, 1938 and 1976, respectively; "5" the AD 1920 Haiyuan Earthquake.

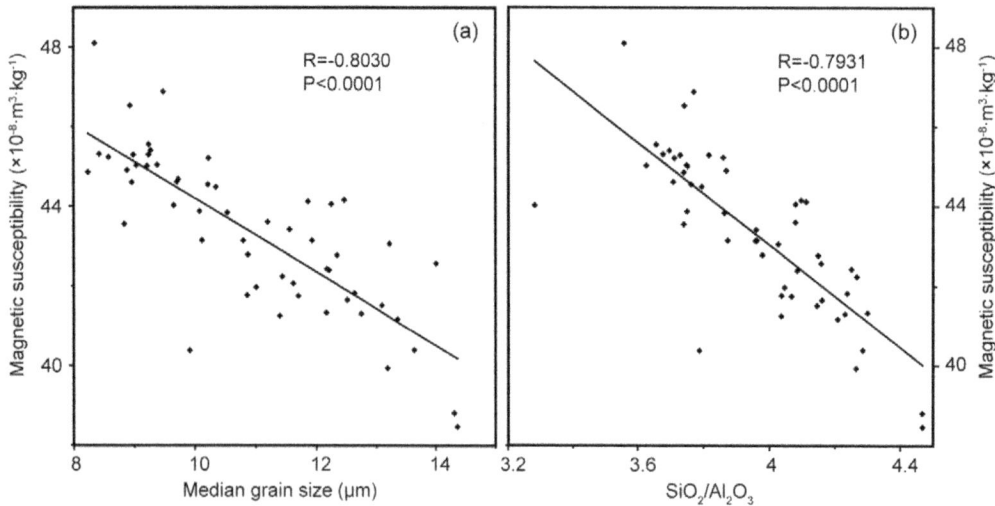

Figure 5. Correlations between MS and the median grain size and the SiO₂/Al₂O₃ ratio of sediment core M38002.

$$SC2 = \frac{\phi 84 - \phi 16}{4} + \frac{\phi 95 - \phi 5}{6.6} \qquad (2)$$

Where ϕi is the reach of percentile of i by using a ϕ scale [16].

SC_1 represents changes in the central part of the grain size distribution; therefore it is considered to be insensitive to environmental changes, which are expressed in the tails. SC_2 is frequently used as a sedimentary characteristic because it covers the tails of a grain size distribution. In our studied region, the source materials are relatively homogeneous, and transportation processes could have significant impacts on the tails of the grain size distribution. The good correlation between SC_2 and concentrations of >63 μm contents ($R = 0.87$, $P<0.0001$; Figure 6) confirmed the impact of transportation processes on SC_2. Thus, SC_1, being less sensitive to transportation processes and less noisy, provides a better and more robust proxy of the events occurred in the Yellow River Valley.

The calculated profile of SC_1 (called SC below) showed three major stages (Figure 4). Before AD 1855, SC varies between 1.50 and 1.73. From AD 1855 to 1920, SC stays stable around 1.60, indicating constant material sources and hydrodynamics. Around AD 1920, SC increases abruptly to around 1.71 and then keeps relatively stable with one peak value of 1.84 around AD 1935 and one bottom value of 1.61 around AD 1977.

These changes in SC values correspond well to the events documented in historical archives (Figure 4). The Yellow River discharged into the southern Yellow Sea between AD 1128 and 1855 (Figure 7), and then changed to the Bohai Bay in AD 1855 [4]. The Hai River, the Luan River, the Liao River, and the rivers on the Shandong Peninsula provided the main material sources for the sediments in the NYSM before the relocation of the Yellow River estuary from Southern Yellow Sea to the Bohai Bay; thus the large oscillations of SC between AD 1128 and 1855 reflect the hydrodynamic changes of these rivers. Three higher values during these periods coincide with the great flood in the Shandong Peninsula in AD 1781 [17], in the Hai and Luan River regions in AD 1801 and in the Liao River region in AD 1846 [18] (Figure 4).

Large amounts of soil materials from the Loess Plateau are carried by the Yellow River to the sea; the average annual sediment discharge between AD 1952 and 2005 is 7.78 billion tons [19]. After the relocation of Yellow River estuary in AD 1855, the

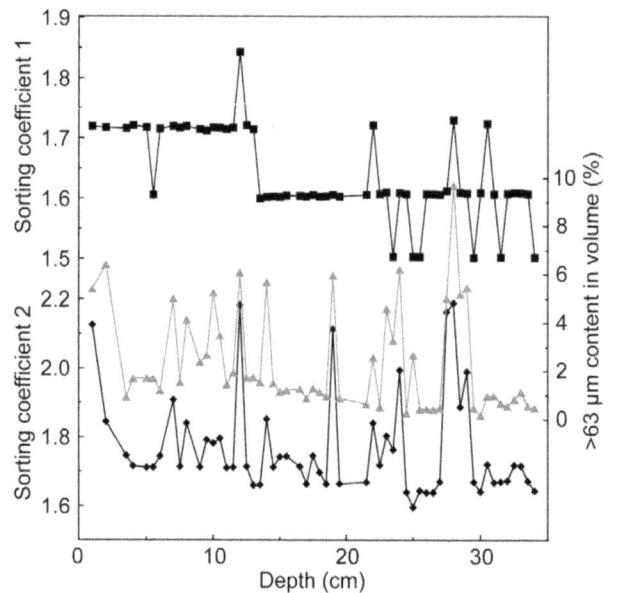

Figure 6. Comparison of sorting coefficients calculated using both Folk & Ward's (SC₂) [16] and Trask's method (SC₁) [15]. The concentrations of >63 μm contents are given in volume.

dominant material source of the core M38002 is the Loess Plateau. The abrupt changes of SC around AD 1855 are very likely caused by the relocation of the Yellow River estuary, and the stability of the SC values after that indicates a stable material source supply.

The abrupt increase of SC after AD 1920 concurred with the great Haiyuan Earthquake (Figure 5). This earthquake had a magnitude of 8.5 on the Richter scale, was felt throughout China, caused more than 200,000 deaths, induced many major landslides, changed the topography on the Loess Plateau, and ravined the "Yuan" of loess [20,21]. Loess is undiagenesised, erosion-prone eolian sediment, so a large amount of loess could be loosed by the great earthquake and transported by the Yellow River into NYSM to cause the increase of SC in the studied sediment core. For the past several decades, human activities may

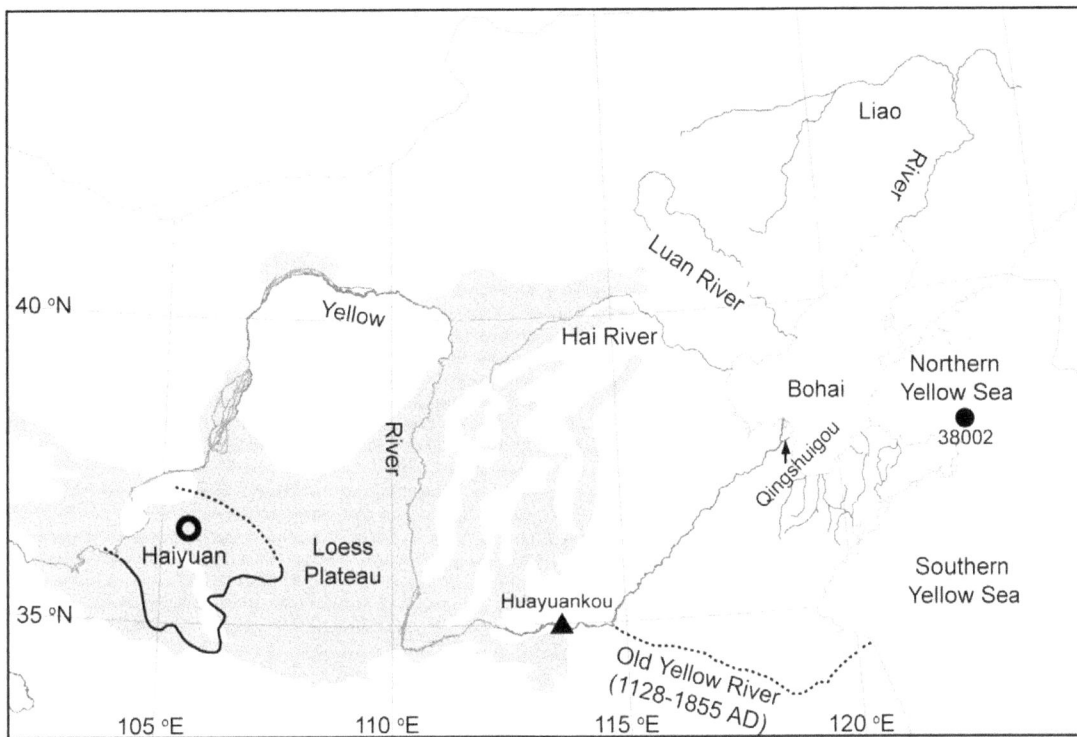

Figure 7. Locations of the Haiyuan County, the Huayuankou, the Old Yellow River, and the Loess Plateau. Areas with seismic intensity of 1920 Haiyuan Earthquake greater than seven degrees on the Chinese Seismic Intensity Scale (modified from [21]) are marked by solid and dashed lines around Haiyuan County; range of the Loess Plateau is modified from [28].

also contribute to the high silt content in the Yellow River [22] and thus the observed SC increase. Additionally, both the East Asian summer and winter monsoons likely had abrupt changes around AD 1920; but they are unlikely the cause for abrupt increase of SC after AD 1920 since they have opposite impacts on the hydrodynamics at the 38002 Station and their strength have been oscillating even since [23,24].

The peak of SC around AD 1940 coincided with the blast of the Yellow River Dyke in AD 1938 and the great flood in AD 1939. During the China's Anti-Japanese War, the Chinese army blew up the dyke at Huayuankou (Figure 7) in AD 1938, causing the relocation of the Yellow River estuary to the southern Yellow Sea during AD 1938 to 1946 [4]. The relocation of the Yellow River estuary changed the main source materials of M38002 from the Yellow River to other rivers. The great flood (the largest flood since AD 1801) at Hai River Basin (Figure 4) in AD 1939 [25] dramatically increased the soil material supply to these rivers and very likely contributed to the SC increase.

The sharp drop of SC around AD 1978 is very likely linked to the relocation of the Yellow River outflow to Qingshuigou River in AD 1976. The event resulted in a large and rapid accumulation of muddy/sand material in the subaqueous delta near the estuary until AD 1980 [26]. After AD 1980, the accumulation rate decreased abruptly and then remained stable up to now [27], so the SC bounded back and stayed stable.

Conclusions

In the present study, by analyzing grain size, MS and SiO_2/Al_2O_3 ratio in a 254-year sediment core from NYSM, we identified several historical events: the Haiyuan Earthquake in AD 1920 and several times of relocation of the Yellow River estuary. Different proxies have different sensitivities to different geological and environmental events. SC seems to be less noisy and more robust, and as a new method, it has a good potential of detecting earlier events. Furthermore, in combination with accurate historical archives, the SC proxy from the sediments of relatively homogenous material sources can provide additional dating control for sediment cores.

Acknowledgments

Thanks are expressed to the anonymous reviewer for his/her useful comments and good suggestions.

Author Contributions

Conceived and designed the experiments: LS XZ. Performed the experiments: XZ WC NJ. Analyzed the data: LS XZ YW WC NJ. Contributed reagents/materials/analysis tools: WC. Wrote the paper: XZ LS YW.

References

1. Shen S, Li A, Yuan W (1996) Low energy environment of the central South Yellow Sea. Oceanol Limnol Sin 27: 518–523.
2. Xiang R, Yang Z, Saito Y, Guo Z, Fan D, et al. (2006) East Asia Winter Monsoon changes inferred from environmentally sensitive grain-size component

records during the last 2300 years in mud area southwest off Cheju Island, ECS. Sci China Ser D 49: 604–614.
3. Xiao S, Li A, Jiang F, Li T, Huang P, et al. (2005) Recent 2000-year geological records of mud in the inner shelf of the East China Sea and their climatic implications. Chinese Sci Bull 50: 466–471.

4. Shi C, Zhang DD (2005) A sediment budget of the lower Yellow River, China, over the period from 1855 to 1968. Geogr Ann A 87: 461–471.

5. Yang W, Chen M, Li G, Cao J, Guo Z, et al. (2009) Relocation of the Yellow River as revealed by sedimentary isotopic and elemental signals in the East China Sea. Mar Pollut Bull 58: 923–927.

6. Qiao S, Shi X, Saito Y, Li X, Yu Y, et al. (2011) Sedimentary records of natural and artificial Huanghe (YellowRiver) channel shifts during the Holocene in the southern Bohai Sea. Cont Shelf Res 31: 1336–1342.

7. Yang ZS, Liu JP (2007) A unique Yellow River-derived distal subaqueous delta in the Yellow Sea. Mar Geol 240: 169–176.

8. Appleby RG (2001) Chronostratigraphic techniques in recent sediments. In: Last WM, Smol JP, editors. Tracking Environmental Change using Lake Sediments Basin Analysis, Coring and Chronological Techniques. Dordrecht: Kluwer Academic. pp. 171–203.

9. Li FY, Gao S, Jia JJ, Zhao YY (2002) Contemporary deposition rates of fine-grained sediment in the Bohai and Yellow Seas. Oceanol Limnol Sin 33: 364–369.

10. Qi J, Li FY, Song JM, Gao S, Wang G, et al. (2004) Sedimentation rate and flux of the North Yellow Sea. Mar Geol & Quaternary Geol 24: 9–14.

11. Sun Y, Lu H, An Z (2006) Grain size of loess, palaeosol and Red Clay deposits on the Chinese Loess Plateau: Significance for understanding pedogenic alteration and palaeomonsoon evolution. Palaeogeogr, Palaeoclimatol, Palaeoecol 241: 129–138.

12. Zhang J, Huang WW, Shi MC (1990) Huanghe (Yellow River) and its estuary: Sediment origin, transport and deposition. J Hydrol 120: 203–223.

13. Liu T, Ding Z (1998) Chinese loess and the paleomonsoon. Annu Rev Earth Pl Sc 26: 111–145.

14. Peng S, Guo Z (2001) Geochemical indicator of original eolian grain size and implications on winter monsoon evolution. Sci China Ser D 44 supp.: 261–266.

15. Trask PD (1932) Origin and Environment of Source Sediments of Petroleum. Texas: Houston. 323 p.

16. Folk RL, Ward WC (1957) Brazos River Bar: A study in the significance of grain size parameters. J Sediment Petrol 27: 3–26.

17. Academy of Meteorological Science (1981) Yearly Charts of Dryness/Wetness in China for the Last 500-year Period. Beijing: Cartographic Publishing House. 332 p.

18. Luo C, Le J (1996) Floods in China. Beijing: Chinese Bookstore. 434p.

19. The Ministry of Water Resources of the People's Republic of China (2010) Specifications of River Sediment Bulletin (2009 AD). Beijing: China Water-Power Press. 58 p.

20. Close U, McCormick E (1922) Where the mountains walked. National Geographic 41: 445–464.

21. Zhang Z, Wang L (1995) Geological disasters in loess areas during the 1920 Haiyuan Earthquake, China. GeoJournal 36: 269–274.

22. Tang K, Wang B, Zheng F, Zhang S, Ci L, et al. (1994) Effects of human activities on soil erosion in the Loess Plateau. Yellow River: 13–17.

23. Guo Q, Cai J, Shao X, Sha W (2004) Studies on the variations of East-Asian summer monsoon during AD 1873~2000. Chinese J Atmos Sci 28: 206–215.

24. D'Arrigo R, Jacoby G, Wilson R, Panagiotopoulos F (2005) A reconstructed Siberian High index since A.D. 1599 from Eurasian and North American tree rings. Geophys Res Lett 32, L05705, doi: 10.1029/2004GL022271.

25. Xie Z, Li Y (2011) On the Floods of Haihe River Basin in 1939. J Hohai Univ (Philos Soc Sci) 13: 15–19.

26. Gao W, Zhang G, Jiang M, Han F (1997) Fluvial process for the Qingshuigou river course of the Yellow River estuary. J Sediment Res: 1–7.

27. Liu J, Qin H, Kong X, Li J (2007) Comparative researches on the magnetic properties of muddy sediments from the Yellow Sea and East China Sea shelves and the Korea Strait. Quaternary Sci 27: 1031–1039.

28. Kukla G (1987) Loess stratigraphy in central China. Quaternary Sci Rev 6: 191–219.

20

Spatial Analysis of Soil Organic Carbon in Zhifanggou Catchment of the Loess Plateau

Mingming Li[1,2], **Xingchang Zhang**[2,3], **Qing Zhen**[1,2], **Fengpeng Han**[2,3]*

1 Institute of Soil and Water Conservation, Chinese Academy of Sciences and Ministry of Water Resources, Yangling, PR China, **2** University of Chinese Academy of Sciences, Beijing, PR China, **3** State Key Laboratory of Soil Erosion and Dryland Farming on the Loess Plateau, Institute of Soil and Water Conservation,Northwest A & F University, Yangling, PR China

Abstract

Soil organic carbon (SOC) reflects soil quality and plays a critical role in soil protection, food safety, and global climate changes. This study involved grid sampling at different depths (6 layers) between 0 and 100 cm in a catchment. A total of 1282 soil samples were collected from 215 plots over 8.27 km². A combination of conventional analytical methods and geostatistical methods were used to analyze the data for spatial variability and soil carbon content patterns. The mean SOC content in the 1282 samples from the study field was 3.08 g·kg⁻¹. The SOC content of each layer decreased with increasing soil depth by a power function relationship. The SOC content of each layer was moderately variable and followed a lognormal distribution. The semi-variograms of the SOC contents of the six different layers were fit with the following models: exponential, spherical, exponential, Gaussian, exponential, and exponential, respectively. A moderate spatial dependence was observed in the 0–10 and 10–20 cm layers, which resulted from stochastic and structural factors. The spatial distribution of SOC content in the four layers between 20 and 100 cm exhibit were mainly restricted by structural factors. Correlations within each layer were observed between 234 and 562 m. A classical Kriging interpolation was used to directly visualize the spatial distribution of SOC in the catchment. The variability in spatial distribution was related to topography, land use type, and human activity. Finally, the vertical distribution of SOC decreased. Our results suggest that the ordinary Kriging interpolation can directly reveal the spatial distribution of SOC and the sample distance about this study is sufficient for interpolation or plotting. More research is needed, however, to clarify the spatial variability on the bigger scale and better understand the factors controlling spatial variability of soil carbon in the Loess Plateau region.

Editor: Ben Bond-Lamberty, DOE Pacific Northwest National Laboratory, United States of America

Funding: This study was financially supported by the National Natural Science Foundation (41101528), China Clean Development Mechanism Fund grant project (2012027-1), and the State Key Laboratory of Soil Erosion and Dry-land Farming on the Loess Plateau. The funders had no role in study design, data collection and analysis, decision to publish, or preparation of the manuscript.

Competing Interests: The authors have declared that no competing interests exist.

* E-mail: hanfp@ms.iswc.ac.cn

Introduction

Soil organic carbon (SOC) is an important aspect of soil quality and plays an important role in soil productivity, environmental protection, and food safety [1]. Because SOC is the biggest part of the terrestrial carbon cycle and carbon-based greenhouse gas balance research [2], slight changes in SOC can greatly impact atmospheric CO_2 concentrations and global climate change. Therefore, SOC has become a core topic in global climate change research. Considerable attention has focused on SOC in relation to climate change and greenhouse gas emissions [3,4].

The SOC has a strong spatial heterogeneity which can be expressed by a function [5,6]. A precise understanding of SOC spatial characteristics can improve the accuracy of SOC stock estimations and contribute to the development and implementation of effective carbon sequestration methods. Recently, a series of studies regarding SOC spatial distribution and stock were conducted by international researchers. ie., in some European countries [7,8], the United States [9], India [10], Brazil [11], and other countries. These studies indicated that the spatial variability of SOC characteristics was affected by multiple factors, including

land use, soil parent material, topography, vegetation, climate, and agricultural use [12–15].

The Loess Plateau of China is located in an ecologically vulnerable semi-arid region that is affected by one of the most serious soil erosion problems in the world. In the past decade, large-scale vegetation recovery and ecosystem improvement (to a certain extent) have occurred as a result of the "Grain for Green Project" implemented by the Chinese government [16]. Due to its complex and broken topography and hilly and gully landforms, spatial heterogeneity in the Loess Plateau region is relatively high [17]. Although many studies have been conducted, the data in these studies were mainly collected at slope and [18,19] ecosystem scales [20,21] and from shallow soil layers [22–24]. In addition, SOC spatial variability studies at a catchment scale have mainly focused on the environmental features that resulted from different land uses and soil types [25–27]. These SOC measurements were rarely related to the depth of the soil layers. Generally, only small amounts of data were used in these analyses, due to the considerable effort required to obtain data in this complex terrain. Many of the studies mentioned above are associated with significant uncertainty. This uncertainty results from the unavail-

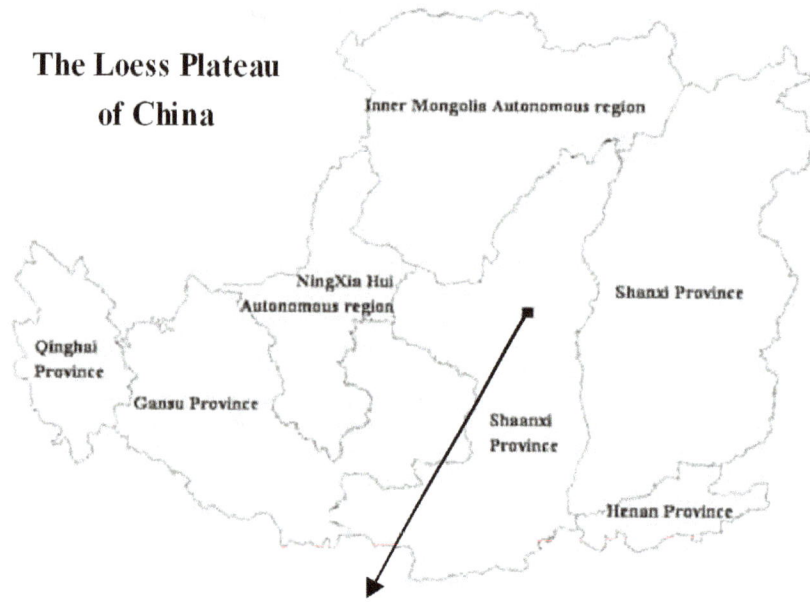

Figure 1. The location of the catchment.

ability of complete data sets, the diversity of the data sources, and the inherent spatial heterogeneity of the SOC [28].

Two objectives were addressed in this study: 1) obtaining the vertical distribution of SOC in a typical Loess Plateau small catchment; 2) elucidating the spatial variability and distribution of SOC at different depths within the catchment.

Materials and Methods

Study area

The Zhifanggou catchment is a typical small catchment on the Loess Plateau. which is located in Ansai County, Shaanxi Province, China (longitude 108°51′44″−109°26′18″, latitude 36°30′45″−37°9′31″, altitude 1,010″−1,1431 m, 8.27 km²) (Fig. 1). The geomorphology of this catchment is extremely

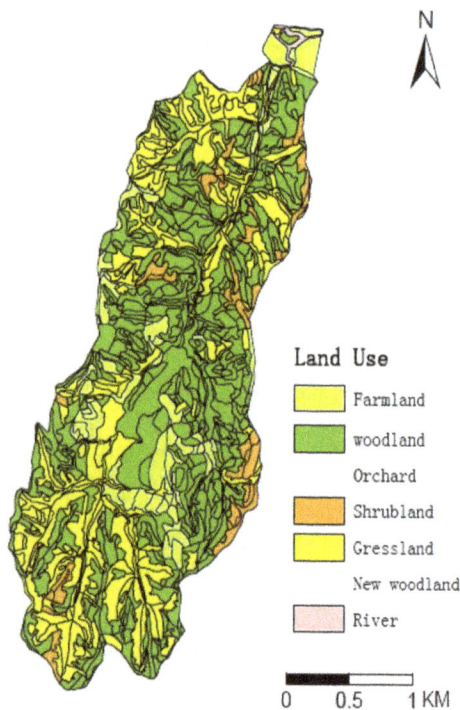

Figure 2. The land use types.

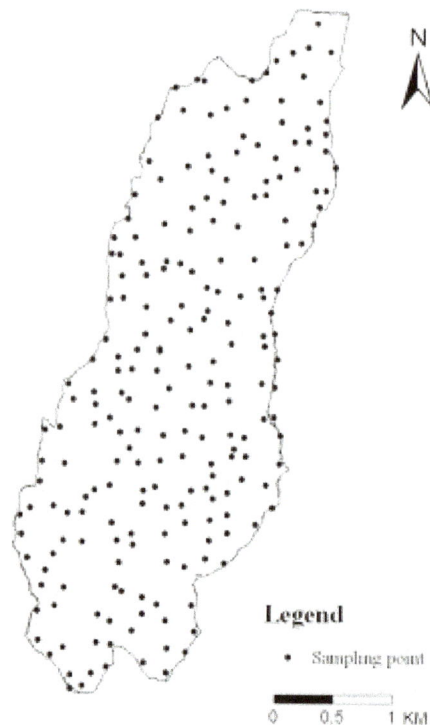

Figure 3. The locations of the sampling.

Table 1. Summary statistics from the classical analyses of soil organic carbon (SOC) content.

Soil Depth(cm)	N	Mean (g·kg⁻¹)	Median (g·kg⁻¹)	Min (g·kg⁻¹)	Max (g·kg⁻¹)	Std.D.	C.V.(%)	Skewness	Kurtosis	Distribution type
0–10	215	6.36^a	5.09	1.30	30.22	3.96	62	2.30	7.40	NN
								0.51	0.54	N^{log} *
10–20	215	4.43^b	3.84	1.33	14.87	2.12	48	1.62	3.21	n
								0.36	0.12	N^{log} *
20–40	215	2.99^c	2.58	0.95	8.78	1.41	47	1.74	3.42	n
								0.50	0.34	N^{log} *
40–60	215	2.49^d	2.17	1.00	6.69	1.10	44	1.70	3.14	n
								0.61	0.23	N^{log} *
60–80	213	2.29^d	2.01	0.66	11.69	1.14	50	1.96	5.42	NN
								0.51	0.93	N^{log} *
80–100	209	2.20^e	1.96	0.25	8.51	1.03	47	1.50	3.64	n
								0.17	0.95	N^{log} *

Notes: N., Number of samples; C.V., Coefficient of Variation; Std. D., Standard Deviation.
a, b, c, d, e, Different lowercase letters represent a significant difference between the layers (P<0.05).
*, Natural logarithm transformation with the corresponding skewness and kurtosis values.
N, Normal distribution; n, Near Normal Distribution; NN, Non-Normal Distribution; N^{log}, Log-Normal Distribution.

broken and exhibits the characteristics of a valley. The soils are predominantly loess and uniform in texture. The sand, silt, and clay contents of the soil are 65, 24, and 11%, respectively. The average annual precipitation in the catchment is 541.2 mm. In addition, 75% percent of the annual rainfall in this region occurs between July and September. During these months, the rainfall is intense and causes extensive erosion. The study area is under four main land use types that cover woodland (54%), grassland (32%), farmland (8%)and shrubland (6%). (Fig. 2). The main land uses (and vegetation species) are shrubland(r), woodland *(Populous simonii Carr., Fruit trees)* grassland *(Medicago sativa L., Artemisia gmelinii, Stipa bungeana, Artemisia scoparia,)* and farmland *(Triticum aestivum, Zea mays, Glycine max)* [29].

Sampling method

The grid method was used to collect soil samples. All of the designated sample sites were arranged on a 1:10,000 scale topographic map. A grid interval of 200×200 m was used, and each grid was considered an independent study unit. A portable GPS was used to locate each sample site. Each site was divided into 6 depths between 0–100 cm as follows: 0–10, 10–20, 20–40, 40–60, 60–80 and 80–100 cm. All samples were collected with a 5-cm-diameter hand auger. 215 soil sampling sites including farmland 28, shrubland 33, woodland 77 and grassland 77. A total of 1,282 soil samples were collected from 215 soil sampling sites (Fig. 3). Soil samples were air dried before passage through a 0.25 mm sieve for laboratory analysis. The SOC content of each sample was determined in duplicate with the dichromate oxidation (external heat applied) method [30]. The samples were collected in November 2010.

Data processing and analysis

The geostatistical method is a spatial analysis method that was developed from classical statistics. Based on the theory of regionalized variables, this method effectively uses semi-variogram and Kriging interpolations to determine the spatial distribution, variability, and related characteristics of the various random structural variables [31]. The semi-variance function was fit based

on the coefficient of determination R^2 and the residual sum of squares (RSS) to obtain an optimal theoretical mode [32].

The Kriging interpolation method was used to estimate the values of the unmeasured sites x_0 by assuming that z' (x_0) equals the linear sum of the known measured values. This process is expressed by the following equation [33]:

$$Z'(x_0) = \sum_{i=1}^{N} \lambda_i Z(X_i) \tag{1}$$

where Z (x_0) is the predicted value at position x_0, $Z(X_i)$ is the known value at sampling site X_i. λ_i is the weighting coefficient of the measured site. and N is the number of sites within the neighborhood searched for the interpolation.

The data that were used in this study were analyzed with classical statistical methods in the program SPSS 18.0. Analysis of variance (ANOVA) was performed with the least significant difference (LSD) method to compare the impacts of different soil depths on SOC content (P<0.05). The K-S (Kolmogorov-Smirnov) test was used to determine if the data were normally distributed. Logarithmic or other transformations were performed on data that were not normally distributed to obtain a normal distribution. The use of non-normally distributed data would increase the estimation of error. Therefore, it was necessary to transform these non-normally distributed data. The test results indicated that the SOC distributions were skewed at soil depths of 10–20, 20–40, 40–60, and 80–100 cm and were normal at soil depths of 0–10 and 60–80 cm. However, the normally distributed SOC contents were highly skewed and had a high kurtosis. Thus, a logarithmic conversion of the SOC contents of the six soil layers was performed. The kurtosis and skewness of the SOC content decreased in each soil layer and were normally distributed. After logarithmic conversion, the normally distributed data were imported into the software GS +9.0 for semi-variance fitting., GS +9.0 software was used to obtain semi-variance fits and an optimal theoretical model. The ArcGIS9.3 software was used for

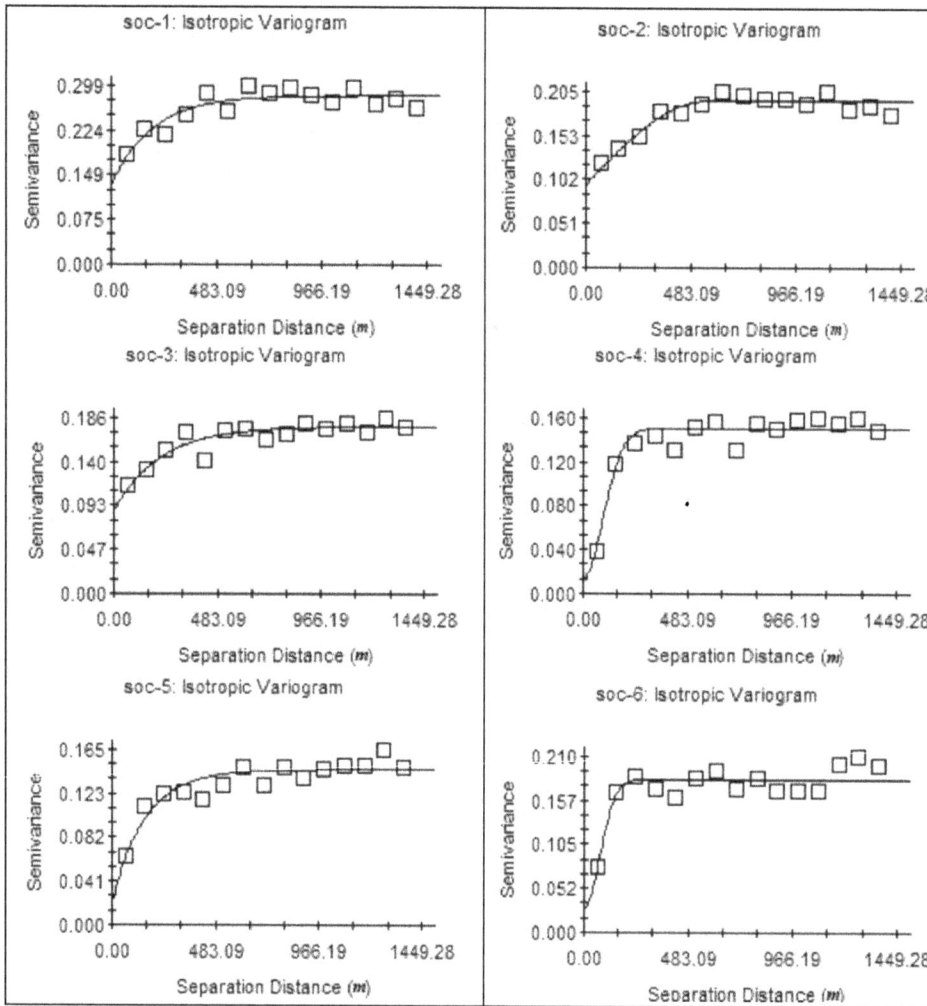

Figure 4. Semi-variance charts of soil organic carbon (SOC) under different soil depths.

the classical Kriging interpolation and for plotting the spatial distribution.

Results and Discussion

Descriptive statistics of the SOC content

The descriptive statistics obtained from SOC in the study area are presented in Table 1. The mean SOC content was 3.08 g·kg^{-1} in the study area, well below the average SOC level in China [34].

The mean SOC of all soil layers was between 2.20 and 6.36 g·kg^{-1}. The highest SOC content in the study area was observed in the 0–10 cm layer. The SOC content decreased with increasing soil depth. As anticipated, the lowest SOC content was observed at a depth of 80–100 cm because SOC is mainly formed by the decomposition of animal and plant residues that are primarily distributed in the soil surface and decrease with depth. The higher SOC in the surface soils indicates that the surface soil

Table 2. Geostatistical parameters for soil organic carbon(SOC) content.

Soil Depth(cm)	Model	C_0	C_0+C	Proportion (C_0/C_0+C)	Range (m)	R^2	RSS
0–10	Exponential	0.1306	0.2832	0.461	552	0.822	3.036E-03
10–20	Spherical	0.0968	0.1946	0.497	562	0.907	8.505E-04
20–40	Exponential	0.0314	0.1788	0.176	234	0.682	1.121E-03
40–60	Gaussian	0.0201	0.1472	0.137	233	0.915	1.240E-03
60–80	Exponential	0.0249	0.1348	0.185	254	0.533	1.354E-03
80–100	Gaussian	0.0204	0.1358	0.150	264	0.774	2.733E-04

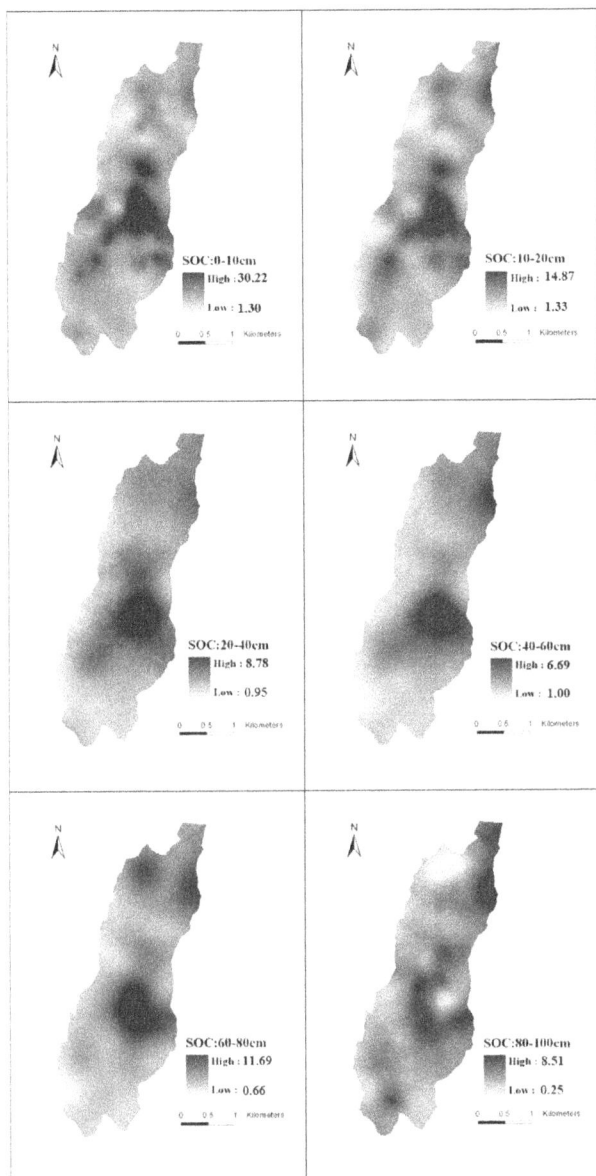

Figure 5. The spatial distribution of the soil organic carbon (SOC)under different soil depths in the catchment.

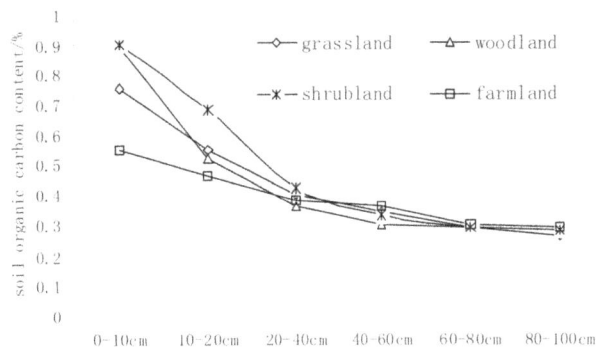

Figure 6. The soil organic content in the different land use types.

The average SOC contents were significantly different for each soil layer. This finding indicated that the central tendency of the SOC distribution was likely affected by anomalous values that led to a non-normal distribution.

Geostatistical analysis of the SOC contents

A table of SOC variability characteristics was generated from semi-variance fitting (Table 2). In Table 2, C_0 is the nugget variance, C is the structural variance, and C_0+C is the sill. $C_0/$ C_0+C represents the degree of spatial variability, which is affected by both structural and stochastic factors. Higher ratios indicate that the spatial variability is primarily caused by stochastic factors, such as fertilization, farming measures, cropping systems, and other human activities. By contrast, a lower ratio suggests that structural factors, such as climate, parent material, topography, soil texture, soil type, and other natural factors, play a significant role in spatial variability. In addition, a proportion less than 25% indicates a strong spatial correlation in the system, a proportion between 25% and 75% indicates a moderate spatial correlation, and a proportion larger than 75% indicates a weak spatial correlation. If the proportion is near 1, then the variable is constant at all scales [36].

As shown in Table 2, the C_0/C_0+C values for SOC were 0.461, 0.497, 0.176, 0.137, 0.185, and 0.150, respectively, in the six different soil layers. The proportion was between 25 and 75% at soil depths of 0–10 and 10–20 cm, indicating a moderate spatial correlation. This correlation was apparent in the 552 and 562 m ranges, respectively, and was subjected to the impacts of stochastic and structural factors. The C_0/C_0+C was less than 25% in the four layers at a depth of 20 to 100 cm, indicating a strong spatial correlation. This spatial correlation was apparent in the 414, 234, 534, and 264 m ranges and was affected by structural factors. The spatial correlation ranges are different from Han [17] and Liu [15] that are caused by the different study area.

The variability range determines the spatial autocorrelation. When a variable is within the range values, it is spatially autocorrelated, and when it is outside of the range values, it is not. This determination provides guidelines for effectively designing sampling schemes [26]. In this study, large range variations occurred between 234 and 562 m that is to say during the large range the data have the spatial autocorrelation. In general, the sampling distances that are outside of the range are invalid for interpolation or plotting [37]. The average sampling grid interval was 200 m in this study. This sampling grid was smaller than the minimal range of 234 m, which indicates that the sampling interval in the study area met the requirements for spatial variability analysis.

actively participates in "carbon sequestration". The relationship between SOC content (unit: g/kg) and soil depth (unit: cm) is expressed by the following power function: y = 17.501x−0.462, R2 = 0.9889, p<0.001. These results are similar to those reported by [17].

The SOC coefficients of variation in the six layers were 62, 48, 47, 44, 50, and 47%, respectively. According to the classification system proposed by Nielson and Bouma [35], the variable is considered to have weak variability if the coefficient of variation (CV) is less than 10% and moderate variability if the CV is between 10% and 100%; otherwise, the variable has strong variability. Therefore, these values all correspond to moderate variability. The highest coefficient of variation was 62%, at a soil depth of 0–10 cm. The lowest coefficient of variation was 44%, at a soil depth of 40–60 cm. This low coefficient of variation resulted from the influence of multiple factors on the soil surface, including human intervention, vegetation type, land use, and topography.

The semi-variance function model fitting curve for each soil layer was obtained using the semi-variance function. The semi-variance function of the SOC in the soil layers displayed the same trend (Fig. 4). The function values gradually increased with increasing spatial distance before stabilizing. The semi-variogram of the SOC contents at depths of 0–10, 10–20, 20–40, 40–60, 60–80, and 80–100 cm corresponded with the following models: exponential, spherical, exponential, Gaussian, exponential, and Gaussian, respectively. All six layers had coefficients of determination R2 of 0.682 to 0.915 and a small RSS. These results indicate that the theoretical model was an adequate representation of the spatial structural characteristics of the SOC contents in the soil layers. In addition, the curve fit for each layer was optimized.

Spatial distribution of the SOC content

To visualize directly the spatial distribution of SOC content in this catchment (according to the obtained semi-variogram model), the ordinary Kriging interpolation method from geostatistics was adopted to interpolate each layer in the study area and to generate a spatial distribution diagram of SOC content (Fig. 5).

As shown in Fig. 5, the overall spatial distribution of SOC density in each layer was observed in patches or speckles. Previous studies have shown that the distributions of SOC contents in soils result from the combined effects of soil parent material, climate, topography, landscape, and human intervention [38]. In this study, the catchment area was small with a uniform climate, soil parent material, and soil type. Consequently, the SOC content variations were only related to the landscape and human activities.

Figures 2 and 5 depict areas with significantly high SOC content in each layer in the mid-east and north regions of the catchment. These areas are mainly covered by woodland and fruit trees. The gully channels also contained high concentrations of SOC. The SOC content of the peripheral areas of the catchment was lower due to their higher elevation. From the vertical direction, the 0–10 and 10–20 cm depths had smaller spot areas

with dispersed distributions, indicating strong variability. The highest-content spots occurred in the woodlands and shrub lands. By contrast, the lowest-content spots occurred in the grasslands and farmlands. That is to say the woodland and shrub lands can increase the soil organic carbon content. Form Fig 6 we can know that the SOC content in the 0–10 and 10–20 cm depths was shrub lands > woodlands > grasslands > farmlands. No significant variations in SOC spatial distributions were observed in the other four soil layers that had concentrated and high-content areas. The 80–100 cm depth had loosely distributed spots in which the low-content spots corresponded with grasslands and farmlands. Therefore, topographical factors, land use, and human activities were the major causes of spatial variability in SOC distribution. In addition, the ordinary Kriging interpolation directly reflected the spatial distribution of SOC in this catchment.

Conclusions

This study showed that the overall spatial distribution of the SOC density in each layer of the study area was observed in patches or speckles and the coefficient of variation of the SOC content in each layer was moderate variability. Correlations within each layer were observed between 234 and 562 m. Our results suggest that the ordinary Kriging interpolation can directly reveal the spatial distribution of SOC and the sample distance about this study is sufficient for interpolation or plotting. More research is needed, however, to clarify the spatial variability on the bigger scale and better understand the factors controlling spatial variability of soil carbon in the Loess Plateau region.

Author Contributions

Conceived and designed the experiments: ML XZ FH. Performed the experiments: ML FH. Analyzed the data: ML QZ FH. Contributed reagents/materials/analysis tools: ML FH. Wrote the paper: ML FH.

References

1. Gregorich EG, Carter MR, Angers DA, Monreal CM, Ellert BH (1994) Toward minimum data set to assess soil organic-matter quality in agricultural soils. Canadian Journal of Soil Science, 74, 885–901.
2. Kern JS (1994) Spatial patterns of soil organic carbon in the contiguous United States. Soil Sci Soc Am J, 58,439–455.
3. Lal R (2004) Soil C sequestration impacts on global climatic change and food security. Science, 304, 1623–1627.
4. Heimann M, Reichstein M (2008) Terrestrial ecosystem carbon dynamics and climate feedbacks. Nature, 451, 289–292.
5. Wang HQ, Hall CAS, Cornell JD, Hall MHP (2002a) Spatial dependence and the relationship of soil organic carbon and soil moisture in the Luquillo, Experimental Forest, Puerto Rico. Landscape Ecol,17: 671–684.
6. Walter C, Viscarra RA, McBratney AB (2003) Spatio-temporal simulation of the field-scale evolution of organic carbon over the landscape. Soil Sci. Soc. Am. J., 67,1477–1486.
7. Batjes NH (2002) Carbon and nitrogen stocks in the soils of central and Eastern Europe. Soil Use Manage, 18, 324–329.
8. Krogh L, Noergaard A, Hermanen M, Hunlekrog GM, Balstroem T, et al. (2003) Preliminary estimates of contemporary soil organic carbon stocks in Denmark, using multiple datasets and four scaling-up methods. Agric. Ecosyst. Environ, 96, 19–28.
9. Batjes NH (2000) Effect of mapped variation in soil conditions on estimates of soil carbon and nitrogen stocks for South America. Geoderma, 97, 135–144.
10. Bhattacharyya T, Pal DK, Mandal C, Velayutham M. (2000) Organic carbon stock in Indian soils and their geographical distribution. Curr. Sci, 79, 655–660.
11. Bernoux M, Carvalho S, Volkoff B, Cerri CC (2002) Brazil's soil carbon stocks. Soil Sci. Soc. Am. J, 66, 888–896.
12. Tan ZX., Lal R (2005) Carbon sequestration potential estimates with changes in land use and tillage practice in Ohio, USA. Agr. Ecosyst. Environ,111,140–152.
13. Liu DW, Wang ZM, Zhang B (2006) Spatial distribution of soil organic carbon and analysis of related factors in croplands of the black soil region, Northeast China. Agr.Spatial Variability Of Soil Organic Carbon 495 Ecosyst. Environ. 113,73–81.

14. Su ZY, Xiong YM, Zhu JY, Ye YC (2006) Soil organic carbon content and distribution in a small landscape of Dongguan, South China. Pedosphere, 16,10–17.
15. Liu ZP, Shao MA (2012) Large-scale spatial variability and distribution of soil organic carbon across the entire Loess Plateau, China. Soil Research, 50, 114–124.
16. Fu BJ, Chen DX., Qiu Y, Wang J, Meng QH (2002) Land Use Structure and Ecological Processes in the Loess Hilly Area, China. Commercial Press, Beijing, 1–50.
17. Han FP, Hu W (2010) Spatial variability of soil organic carbon in a catchment of the Loess Plateau. Acta Agriculturae Scandinavica Section B Soil and Plant Science, 60,136–143.
18. Wang J, Fu BJ, Qiu Y (2001) Soil nutrients in relation to land use and landscape position in the semi-arid small catchment on the loess plateau in China. Journal of Arid Environments, 48, 537–550.
19. Fu BJ, Chen LD, Ma KM, Zhou HF, Wang J (2003) The relationships between land use and soil conditions in the hilly area of the Loess Plateau in northern Shaanxi, China. Catena, 39,69–78.
20. Yang XM, Cheng JM, Meng L (2010) Study on soil organic carbon pool at forest-steppe zone of loess plateau. Pratacultural Science, 27(02),18–23.
21. Li JF, Cheng JM, Liu W, Gu XL(2010) Distribution of Soil Organic Carbon and Total Nitrogen of Grassland in Yun-Wu Mountain of Loess Plateau. Acta Agrestia-Sinica, 18(5),661–668.
22. Wang XL, Duan JJ, Guo SL (2007) Organic carbon density of topsoil and its spatial distribution of small watershed in hilly region of loess plateau. Journal of Northwest A & F University(Nat. Sci. Ed.),35(10),98–109.
23. Wang Y, Fu B, Yi HL, Song C, Luan Y (2010) Local-scale spatial variability of soil organic carbon and its stock in the hilly area of the Loess Plateau. China Quaternary Research, 73(1),70–76.
24. Fang X., Xue ZJ (2012) Soil organic carbon distribution in relation to land use and its storage in a small watershed of the Loess Plateau, China. Catena 88, 6–13.
25. Wang YQ, Zhang XC, Huang CQ (2009) Spatial Variability of Soil Organic Carbon in a Watershed on the Loess Plateau. Pedosphere,19(4),486–495.

26. Wang J, Fu BJ, Qiu Y (2002) Spatial Heterogeneity of Soil Nutrients in a Small Catchment of the Loess Plateau. Acta Ecologica Sinica, 22(8),1173–1178.
27. Jia YP, Su ZZ, Duan JN (2004) Spatial Variability of Soil Organic Carbon at Small Watershed in Gully Region of Loess Plateau. Journal of Soil and Water Conservation, 18(1): 31–34.
28. Xie XL, Sun B, Zhou HZ, Li Z, Li AB (2004) Organic carbon density and stock in soils of China and spatial analysis. Acta Pedologica Sinica 41(1), 35–43.
29. Wang GL, Liu GB, Xu MX (2002) Effect of vegetation restoration on soil nutrient changes in Zhifanggou Watershed of Loess Hilly region. Bulletin of Soil and Water Conservation, 22(1): 1–5.
30. Nelson DM, Sommer LE (1975) A rapid and accurate method for estimating organic carbon in soil. Proceeding Indiana Academic Science,84,456–462.
31. Trangmar BB, Yost RS, Uehara G (1985) Application of geostatistics to spatial studies of soil properties. Advances in Agronomy, 38,45–94.
32. Journel AG, Huijbregts CJ (1978) Mining Geostatistics. London: Academic Press, 600.
33. Wang ZQ (1999) Geostatistics and its applications in ecology. Beijing: Science Press.
34. Wang SQ, Zhou CH (2000) Analysis on spatial distribution characteristics of soil organic carbon reservoir in China. Acta geographica sinica, 55(5),533–544.
35. Nielsen DR. and Bouma J (1985) Soil Spatial Variability. Proceedings of a Workshop of the ISSS and the SSSA,Las Vegas, USA. 30th November to 1st December, 1984. Pudoc, Wageningen.
36. Cambardella CA, Moorman TB, Novak JM (1994) Field-scale variability of soil properties in central low a soils. Soil Sci. Soc. Am. J., (58), 1501–1511.
37. Zhou HZ, Gong ZT, Lamp L (1996) Study on soil spatial variability.Acta Pedologica Sinica 33(3),232–241.
38. Goovaerts P (1999)Geostatistics in soil science state of the art and perspectives. Geoderma. 89, 1–45.

21

Assessment of the Soil Organic Carbon Sink in a Project for the Conversion of Farmland to Forestland: A Case Study in Zichang County, Shaanxi, China

Lan Mu[1,2]**, Yinli Liang**[1,2]*****, Ruilian Han**[1,2]

1 Institute of Soil and Water Conservation, Northwest A&F University, Yangling Shaanxi Province, China, **2** Institute of Soil and Water Conservation, Chinese Academy of Sciences and Ministry of Water Resources, Yangling, Shaanxi Province, China

Abstract

The conversion of farmland to forestland not only changes the ecological environment but also enriches the soil with organic matter and affects the global carbon cycle. This paper reviews the influence of land use changes on the soil organic carbon sink to determine whether the Chinese "Grain-for-Green" (conversion of farmland to forestland) project increased the rate of SOC content during its implementation between 1999 and 2010 in the hilly and gully areas of the Loess Plateau in north-central China. The carbon sink was quantified, and the effects of the main species were assessed. The carbon sink increased from 2.26×10^6 kg in 1999 to 8.32×10^6 kg in 2010 with the sustainable growth of the converted areas. The black locust (*Robinia pseudoacacia L.*) and alfalfa (*Medicago sativa L.*) soil increased SOC content in the top soil (0–100 cm) in the initial 7-yr period, while the sequestration occurred later (>7 yr) in the 100–120 cm layer after the "Grain-for-Green" project was implemented. The carbon sink function measured for the afforested land provides evidence that the Grain-for-Green project has successfully excavated the carbon sink potential of the Shaanxi province and served as an important milestone for establishing an effective organic carbon management program.

Editor: Wen-Xiong Lin, Agroecological Institute, China

Funding: This work was financially supported by the National Science and Technology Support Program (2011BAD31B05) and by the National Science and Technology Support Program (2014BAD14B006). The funders had no role in study design, data collection and analysis, decision to publish, or preparation of the manuscript.

Competing Interests: The authors have declared that no competing interests exist.

* E-mail: liangyl@ms.iswc.ac.cn

Introduction

Soil organic carbon (SOC) is an important component of soil that plays a key role in the functions of both natural and agricultural ecosystems. In ecosystem services, SOC is critical for ensuring sustainable food production owing to its nutrient retention function and water-holding capacity [1,2]. The global SOC stock has been estimated to be 1400–1500 Pg C in the upper 100 cm soil layer [3–5], which is approximately twice the amount of C in the atmosphere and three times the amount stored in terrestrial vegetation [6]. Todd-Brown et al [7] reported that the present-day global SOC stocks range from 514 to 3046 Pg C among 11 earth system models (ESMs). Thus, slight reductions in SOC contents due to changes in land-use, soil management, or rates of soil erosion, could significantly raise the CO_2 in the atmosphere. Due to natural drought conditions, intensive human disturbance and severe soil erosion, the hilly-gully area of the Loess Plateau has the lowest soil organic carbon density (SOCD) in China [8]. However, it is possible to increase the organic carbon content and carbon sequestration capacity in the soils of this region through appropriate reforestation of degraded sloping croplands and other ecosystems, whose resilience capacity is intact [1,9].

Changes in land use may alter the land cover patterns in ways that can impact the biomass and soil carbon stocks [10] and alter the rate of C input and output to the soil, ultimately changing the soil C content [11]. Soil C sinks are not permanent and often persist only as long as appropriate management practices (conservation tillage and erosion control) are maintained. When a land-management or land-use change is reversed, the C accumulated as a result of the change is lost, usually more rapidly than it was accumulated [12]. And greater attention to the possibility of encouraging the growth of the forests as a means of removing carbon dioxide (CO_2) from the atmosphere [13]. Several studies have addressed the effects of the conversion of farmland to forestland on the SOC. In a long-term experiment, Guo and Gifford [14] showed that the conversion of forestland or grassland to farmland resulted in significant SOC reduction; the conversion of forestland to both farmland and grassland did not lead to SOC reduction in all cases, but always resulted in significant carbon sequestration from the atmosphere. Ostle et al [15] also found that soil carbon losses occur when grasslands, managed forest lands or native ecosystems are converted to croplands, and soil carbon gains are made when croplands are converted to grasslands, forest lands or native ecosystems. Carbon stored in forest ecosystems represents a substantial portion of the global C stock; worldwide, forests contain ~70% of all plant C and ~20% of all soil C [16]. The conversion of farmland to forestland is therefore an effective method for preventing soil erosion, which also has a large net sink effect on the SOC storage in the hilly-gully areas of China. However, the balance between inputs of organic matter, primarily from vegetation, and losses, as a result of decomposition, leaching and erosion, determines the magnitude of the carbon reservoir of the hilly-gully areas of China. The net CO_2 released via soil and water loss is calculated to be 8.4 g C m^{-2} yr^{-1} [17].

The hilly-gully areas of the Chinese Loess Plateau are known for their agricultural history and severe soil erosion [18]. Since the 1950s, the Chinese government has made substantially efforts to control the soil erosion and restore vegetation to the region. For example, an extensive tree-planting project was undertaken in the 1970s, and integrated soil erosion control was performed on the watershed scale in the 1980s and 1990s. Despite these efforts, the soil erosion remained unchecked, and vegetation had not been substantially restored to the region by the late 1990s [10]. In 1999, an extensive ecological rehabilitation program known as "Grain-for-Green" (conversion of farmland to forestland) was initiated by the Chinese government; this program is currently the largest land retirement project undertaken in the developing world. The Grain-for-Green program was initially intended to reduce soil erosion and desertification and increase the forest cover in China by retiring steep sloping and marginal lands from agricultural production in the hilly-gully area of the Loess Plateau [19]. The program has now been operating for about 10 years and the natural environment in parts of the Loess Plateau is improving as annual crops are replaced by perennial plants. However, limited attention has been paid to the effect of land use conversion on soil carbon sequestration in the Loess Plateau [20,21]. The Grain-for-Green program has potential for C gain through the improved land use conversions [[19] and we hypothesized that the conversion of farmland to forestland is one method that may increase SOC storage in soils in this study region in Loess Plateau.

Therefore, the objective of this paper was to assess the changes in the soil organic carbon content and carbon sink benefit following the implementation of the Grain-for-Green project on the Loess Plateau between 1999 and 2010. The paper also discussed the variation of the soil organic carbon content with the type of vegetation and soil age in the study area and assesses the effect of the land cover change project.

Material and Methods

Research region

Zichang County was located in the northern Shaanxi Province of the Loess Plateau in China within latitudes 36°59′–37°30′N and longitudes 109°11′–110°01′E. The region had a semiarid continental monsoon climate with a mean annual temperature of 9.1°C and mean annual precipitation of 514.7 mm. Over 70% of the precipitation occurred between June and September. The landform had the typical hilly-gully topology of the Loess Plateau, and the valleys and ravines accounted for approximately 94.6% of the total area, in which the gully density was 8.1 km km^{-2}. Soils in the research area originated from parent material of calcareous loess, with a low SOC concentration of less than 10 g kg^{-1}. These belong to the Calcic Cambisol group according to the Food and Agriculture Organization and United Nations Education Scientific and Cultural Organization (FAOUNESCO) soil classification system [10]. The soil type was loessal soil with a pH value of 8.6. The soil bulk density was 1.4 g cm^{-3} and total N content was 0.4 g kg^{-1}. The area of the croplands was 11.08×10^4 ha, accounting for 46.24% of the total land area. In this area, agricultural land occupied 39.5% of the area on slopes exceeding 25° (Table 1); this cultivated steep carried soil load of approximately 229.4×10^4 t yr^{-1} and a water loss of up to 857.8×10^4 m^3 yr^{-1}. The natural vegetation had been largely destroyed by

Table 1. The area of the cultivated slope land classification of the ZiChang county.

slope (degree)	area	percentage
	hm²	%
<5	17927.87	16.9
5~10	3050.33	2.9
10~15	9950.33	9.4
15~25	33247.4	31.3
25~35[a]	32532.93	30.7
>35	9362.93	8.8

[a]The slopes of greater than 25° were cultivated for agriculture occupied 39.5%.

deforestation and cultivation. Slope land reclamation and cultivation along the slope was the primary cause of soil and water loss.

Land degradation, desertification, soil erosion and declining soil fertility had degraded the environment and severely limited the crop yield. In addition, the SOC concentrations in these areas were less than 10 g C kg^{-1}. Consequently, an extensive ecological rehabilitation program known as "Grain-for-Green" was initiated in 1999 in the Loess Plateau region. Zichang was the demonstration county for the larger Shaanxi project. The program in the county had now been implemented for approximately 10 yr (since 1999), and the natural environment was improving as annual crops were replaced by perennial plants and as the vegetation cover increases. The vegetation coverage rate had increased from 6.16% to 32.8%, and the greening rate in the county had reached 62% (Figure 1). And the vegetation reduced the soil and water lost with the amount of approximately 27.5×10^4 t yr^{-1} and 102.9×10^4 m^3 yr^{-1}, respectively (the data come from the Bureau of Soil and Water conservation, Zichang county). These changes had significantly improved the ecological environment, reduced soil and water loss and produced a carbon sink effect in the area [22].

Land cover changes between 1998 and 2010

The main land use change in the Zichang region was the ongoing conversion of cropland to forestland between 1999 and 2010, with the afforestation area occupying up to 6.2×10^4 ha. Figure 2 showed the area over which the land cover was altered during the 10-yr period. 2 sites were selected for the study on the regional scale and a multi-year field study was conducted in the experimental sites at the Zichang experimental station, the institute of soil and water conservation, Chinese Academy of Science. The black locust (*Robinia pseudoacacia* L.) and alfalfa (*Medicago sativa* L.) were the main plantation species during this period.

Site A: alfalfa. Alfalfa is a perennial flowering plant in the pea family Fabaceae cultivated as an important forage crop in many countries around the world. Its root nodules contain bacteria, *Sinorhizobium meliloti* with the ability to fix nitrogen, producing a high-protein feed regardless of available nitrogen in the soil. It has been widely cultivated for restoration in Loess Plateau because of its nitrogen-fixing ability and livestock fodder. This site was situated in a hilly-gully slope which faced north, with a slope of 15°. The planting density was 150 plants /m^2. Three different ages of alfalfa (3, 7 and 12 yr old) were selected for detailed investigation to analyze the effect on the SOC content. The three afforestation chronosequences of alfalfa selected in the

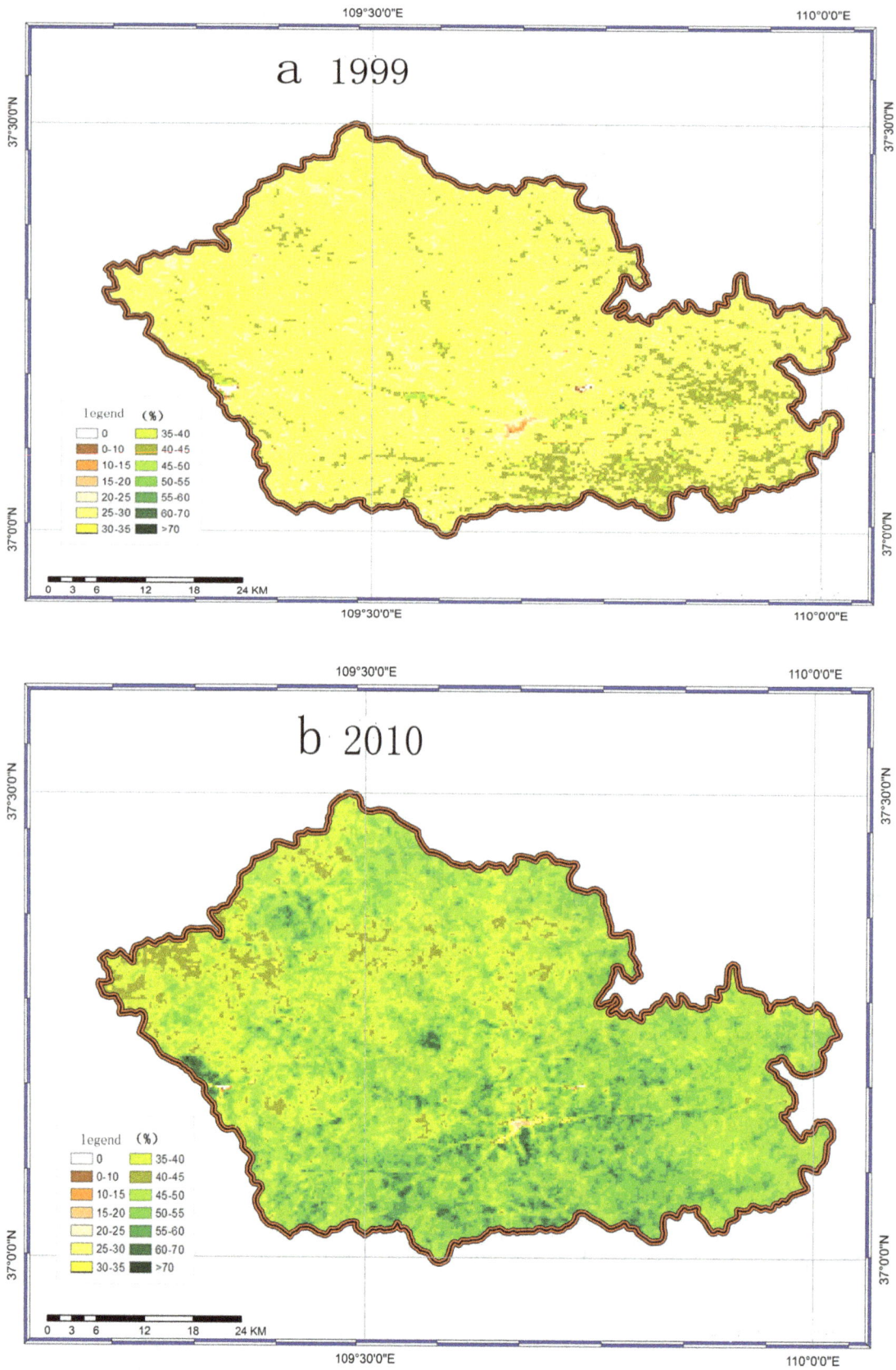

Figure 1. The images of the differences in the remote sensing vegetation coverage between 1999 and 2010 of ZiChang county.

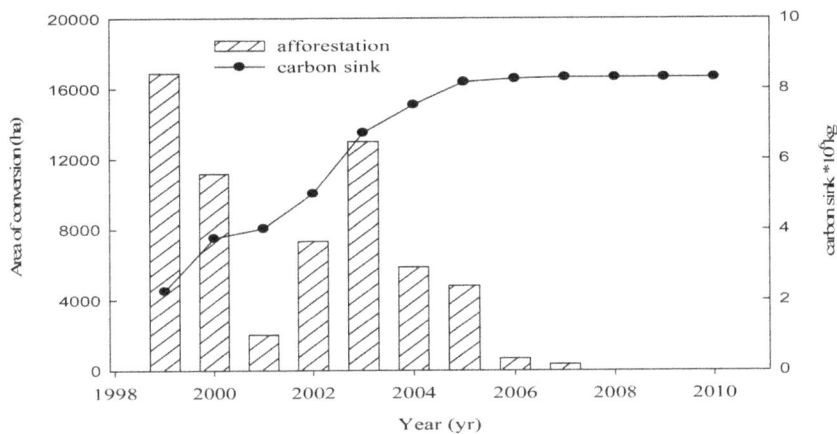

Figure 2. The areas of land cover change (farmland to forestland) and the carbon storage from 1998 to 2010.

same site and the only difference between them is the time when they had been afforested. Each chronosequences of alfalfa had three replications and the total were 3 ages×3 replications = 9 plots.

Site B: black locust. Black locust is a tree of the genus Robinia in the subfamily Faboideae of the pea family Fabaceae which is an exotic nitrogen-fixing tree native to Southeastern North America. It has been widely cultivated for restoration because of its drought resistance, high survival rate, ability to improve the soil nutrient status and remarkable growth rate [23]. At the present, it is the most widely cultivated species in the region. The black locust site was situated in a hilly - gully slope which faced north, with a slope of 19°. And the stand density of black locust was 1667 stands/hm². Four afforestation chronosequences of black locust (3, 7, 12 and 15 yr old) were selected for detailed investigation to analyze the effect on the SOC content. The four afforestation chronosequences of black locust selected in the same site and the only difference between them is the time when they had been afforested. Each chronosequences of black locust had three replications and the total were 4 ages×3 replications = 12 plots.

Calculation of forest carbon sink

Plenty of calculation methods for forest carbon sink have been evolved by experts and scholars such as carbon density method, carbon balance model F-CARBON and so on, which are precise but trivial and fall into the category of traditional nature science. In this research, the carbon sink was calculated on the basis of calculation method as forest storage extension suggested by Xi and Li [24] considering of the practicality and maneuverability [25].

Carbon sink was calculated using the following equation:

$$CF = \sum (Sx \times Cy) + \alpha \sum (Sx \times Cy) + \beta \sum (Sx \times Cy), \ Cy$$
$$= V \times \delta \times \rho \times \gamma$$

Where S_x is the area of forest in the research region; C_y is the carbon density of the forest in the research region; V is the volume per unit area of forest; α is the carbon transfer coefficient of undergrowth plants, which is 0.195 in this research; β is the carbon transfer coefficient of forestland, which is 1.244 in this research; δ is the biomass expanding coefficient, which is 0.5 in this research; ρ is volumetric coefficient, which is 1.90 in this research; Υ is

carbon content rate, which is 0.5 in this research. Values of each conversion coefficient in calculations of forest carbon sink potential in this region are taken as default values prescribed by intergovernmental Panel on Climate Change [26].

Soil sample collection and analysis

In November 2010, the soil samples were collected with a 5-cm diameter soil auger and were extracted in 20-cm incremental subsamples, which were subsequently mixed by hand. Soil samples were collected from each soil layer (0–10 cm, 10–20 cm, 20–40 cm, 40–60 cm, 60–80 cm, 80–100 cm and 100–120 cm) at five different locations selected within a 10 m radius surrounding each plot to analyze the soil organic carbon content. The five samples collected for each layer were subsequently mixed by hand, yielding one representative sample for each layer at each site. The total carbon stock for multiple soil layers was calculated by summing the soil carbon stocks of the layers.

All of the soil samples were air-dried, sealed in airtight bags and transported to the Institute for Soil and Water Conservation of the Chinese Academy of Science in Yangling, Shaanxi province to determine the SOC contents. The SOC was determined using the oil bath-K_2CrO_7 titration method after digestion, and the soil bulk density (BD) was determined using the ring tube method suggested by the Chinese Editorial Committee for Soil Analysis [27]. Total N was measured by the Kjeldahl procedure and pH was measured by electronic pH-meter [28].

Statistical analysis

Each SOC contents in composite samples for each plantation species of various depths were averaged at the same ages following afforestation to perform statistical analysis. Analysis of variance was performed using the SPSS 16.0 software. Means were compared by least significant difference (LSD) at $p<0.05$ or $p<0.01$ level. A one-way analysis of variance (ANOVA) was used to analyze the effect of each factor (land use, soil depths, ages and species) on the SOC concentration.

Results

Carbon sink effects following conversion of farmland to forestland

The forest carbon storage increased continuously as the farmland to forestland conversion project was implemented between 1999 and 2010. The carbon sink increased from

2.3×10^6 kg in 1999 to 8.3×10^6 kg in 2010 with the sustainable growth of the converted areas (Figure 2). The total carbon sink increased by 268% compared to that in the first yr of implementation of the project. This result established that the sequestration of carbon (C) is another important environmental effect of afforestation. The land use policy affected the type, distribution, productivity and turnover of vegetation. This policy was therefore a key determinant of whether the land surface was a sink or source of carbon. Forestland had great carbon sequestration potential, especially compared to farmland. The carbon storage induced by the conversion of farmland to forestland in Zichang County was very large.

SOC content for alfalfa and black locust based on the afforestation chronosequence

In the alfalfa soil, there was a significant increase in the SOC content in the 0–20 cm and 40–100 cm layers with increasing age ($p < 0.01$; Figure 3a) from 3 yr to 7 yr. Subsequently, the SOC content decreased slightly with increasing age (from 5 to 4.5 g kg^{-1} and from 3.2 to 2.4 g kg^{-1} in the last 5 yr). The SOC did not change appreciably in the 20–40 cm layer (Figure 3a). In contrast with the soil samples from <100 cm depth, the changes in the soil C in the 100–120 cm layer were significantly correlated with plantation age; the SOC decreased continuously with age from 2.7 to 2.0 g kg^{-1}.

The SOC exhibited a different trend with age in the black locust soil compared to the alfalfa soil (Figure 3b). The SOC content decreased in the first 7 yr at depths below 100 cm, from 3.0 to 1.7 g kg^{-1} (for yr 4–7, the concentration was consistently lower than that measured for the first 3 yr). Interestingly, the C content then exhibits a gradually increasing trend between 7 and 15 yr (Figure 3b) reaching 3.8 g kg^{-1} at an age of 15 yr. In contrast, the ecosystem C stores in the 100–120 cm layer of the old plantation were clearly higher than those in the younger afforestation sites, which had increased by 31.3% 15 yr after afforestation. The SOC content in the two afforestation plantations therefore provide evidence for a carbon sink once the farmland was converted to forestland.

A comparison of the SOC contents for plantation species of various depths at the same ages following afforestation

The statistical comparisons in the ANOVA demonstrate that the dependence of the SOC on depth in the areas converted to forestland is significant at a level of $p < 0.05$ (Table 2). Between ages of 7 and 12 yr, the SOC content in the alfalfa soil decreases with increasing depth, and the SOC content in the 0–60 cm layer exceeds that at depths above 60 cm. The soil C content is greatest (5.5 and 4.5 g kg^{-1}) in the 0–10 cm layer. However, in the first 3 yr, the SOC decreases significantly ($p < 0.01$, Table 2) with increasing depths of 0–80 cm and then exhibits a smaller increase at depths of 80–100 cm. For black locust, the SOC decreases continuously at depths of 0–100 cm and then increases slightly in the 100–120 cm layer in the 3-, 7-, 12- and 15-yr-old forests. Although the C concentration trend based on depth was different for the two plantations, the SOC content in the topsoil was significantly greater ($p < 0.01$, Table 2) than that in the subsoil for both plantations.

To inform organic carbon management strategies, the dependence of the SOC content on the type of vegetation is also quantified (Figure 4). The SOC concentration in the alfalfa soil exceeded that for the black locust soil at a given age in the 0–100 cm and 100–120 cm layers. The soil C for the two plantations shows a highly significant difference between alfalfa and black locust ($p < 0.05$ or $p < 0.01$, Figure 4). The SOC in the 3-yr-old alfalfa was approximately 36.9% higher than that in the black locust in the 0–100 cm layer. The largest difference between the two species was measured for the 7-yr-old plantation; the SOC contents for this plantation exhibit highly significant differences between the two species at each depth ($p < 0.01$, Figure 4). The soil C is similar for both species on the 12-yr-old plantation and exhibits a significant difference ($p < 0.05$, Figure 4) compared to the 7-yr-old plantation at each depth.

Relationship between the SOC content, depth and other factors as a function of age following afforestation

A change in land use from farmland to forestland and grassland implies that the annual cycle of cultivating and harvesting crops is replaced by the much longer forest cycle. Therefore, many factors influence the SOC in the alfalfa and black locust soils after the land cover changes. The relationships between the SOC content, soil depth and age are summarized in Figure 5. For the 3-yr-old afforested land, there is a strong correlation ($R^2 = 0.93$, $R^2 = 0.99$, $p < 0.01$; Figure 5a, b) between SOC content and depth. As shown in Fig 5, the same trend is obtained for 7- (Figure 5c, d) and 12-yr-old (Figure 5e, f) soil. The SOC is related to the afforestation yr, with the SOC decreasing gradually with increasing age (Figure 5).

Discussion

Effect of the farmland-to-forestland conversion project on the SOC stocks

Land use change can cause a change in land cover and an associated change in the carbon stocks [29]. Each soil has a given carbon carrying capacity, i.e., an equilibrium carbon content depending on the nature of the vegetation, precipitation and temperature [30]. The equilibrium carbon stock arises from a balance between the inflow and outflow to the carbon pool [31]. The equilibrium between the carbon inflow and outflow to the soil is disturbed by any land change until a new equilibrium is reached in the new ecosystem. In our study, we have found that in the farmland-to-forestland conversion project, the soil acts as a carbon sink (from 2.26×10^6 kg in 1999 to 8.32×10^6 kg in 2010). Other studied has got the same results. It has also been estimated that UK forestry and grassland sequester 110 ± 4 kg and 240 ± 200 kg of carbon per hectare yr, respectively, whereas croplands lose on average 140 ± 100 kg of carbon per hectare per yr [32]. Ostle et al [15] indicated that soil carbon accumulates more slowly (decadal) but gains can be made when croplands are converted to grasslands, plantation forest or native woodland. The results of this study partially support the hypothesis that the afforestation of formerly arable land leads to increased C storage in the soil over the short term (approximately 30 yr).

The balance between the inputs (primarily from vegetation) and losses of organic matter (as a result of decomposition, leaching and erosion) determines the magnitude of the area's land carbon reservoir. The net effect of the afforestation of cultivated slope land on the soil C content depends not only on the new C gained, but also on the C lost as a result of the previous management. Although the vegetation carbon stock in the hilly-gully areas are relatively small, plant matter is the single most important source of carbon input to the soil [33]. Therefore, the higher SOC content in the two plantations in our study could be a result of primarily the greater annual litter input and fine root biomass compared to the cropland.

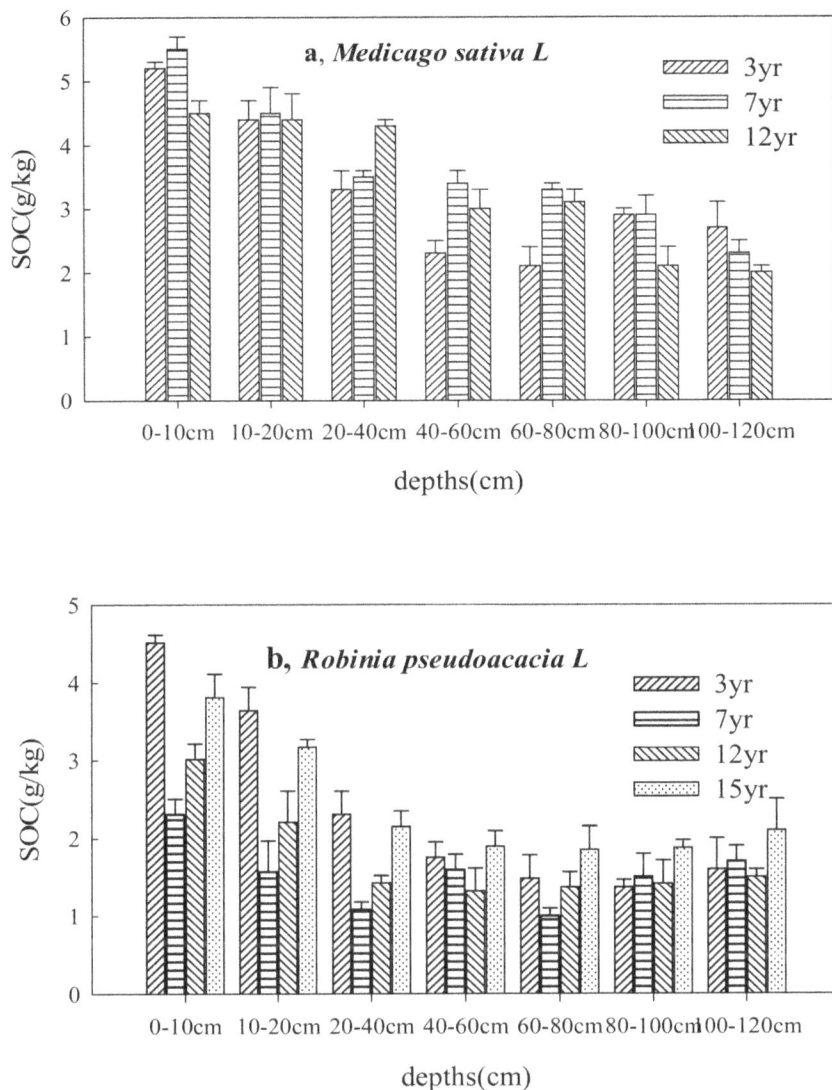

Figure 3. Changes of SOC sequestrations at different ages under alfalfa (*Medicago sativa L*) and black locust (*Robinia pseudoacacia L*) for each depth.

Table 2. SOC[a] at the different depth of the alfalfa (*Medicago sativa L*) and black locust (*Robinia pseudoacacia L*) following the afforestation.

Depth (cm)	alfalfa (*Medicago sativa L*) (g kg^{-1})				black locust (*Robinia pseudoacacia L*) (g kg^{-1})				
	3 yr	7 yr	12 yr	mean	3 yr	7 yr	12 yr	15 yr	mean
0–10 cm	5.2a	5.5a	4.5a	5.1a	4.5a	2.3a	3.0a	3.8a	3.0a
10–20 cm	4.4b	4.5b	4.4a	4.5a	3.6b	1.6b	2.2ab	3.2b	2.3b
20–40 cm	3.3c	3.5c	4.3a	3.8b	2.3c	1.1c	1.4bc	2.1c	1.5c
40–60 cm	2.3ef	3.4cd	3.0b	2.9c	1.8d	1.6b	1.3c	1.9c	1.6c
60–80 cm	2.01f	3.3cd	3.1b	2.8c	1.5d	1.0c	1.4bc	1.9c	1.4c
80–100 cm	2.9cd	2.9d	2.1c	2.6c	1.4d	1.5b	1.4bc	1.9c	1.6c
100–120 cm	2.7de	2.3e	2.0c	2.3c	1.6d	1.7b	1.5bc	2.1c	1.8cb

[a]Data in the column are mean values (n = 3 for alfalfa and black locust), which are compared among different depths within each ages and are not different at the 5% level of significance if followed by the same letter.

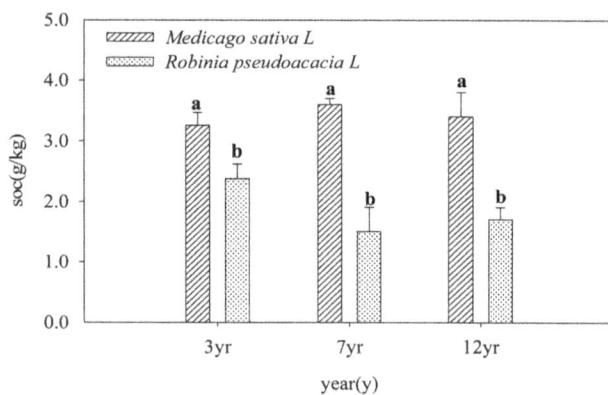

Figure 4. The difference of SOC between alfalfa (*Medicago sativa L*) and black locust (*Robinia pseudoacacia L*) at the same depths under the different ages at the 0–120 cm depths under the 3,7 and 12 ages after the conversion of farmland to forestland.

SOC accumulation for different depths and ages following the afforestation of the cropland

The soil organic carbon content increase in the topsoil (0–40 cm) for both plantations (from 3.8 to 5.1 g kg^{-1} and 1.5 to 3.0 g kg−1) after the conversion of cropland into the forest, is in consistent with other findings for the Loess Plateau [34,10] and other areas of the world [14,35–37] suggesting that the SOC may be highly sensitive to variations in the soil moisture, soil temperature and/or litter input (all of which depend strongly on depth). The increase in the topsoil SOC may be associated with the higher carbon input and lower rates of SOC decomposition and soil erosion associated with alfalfa plantations and black locust forest. It is well established that the top soil is the part of the soil profile that is most susceptible to land use change and disturbance, such as plowing and drainage. The increase in SOC decomposition resulting from tillage disturbance is a key reason for the decrease in SOC following the cultivation of perennial ecosystems [38]. Within limits, soil C increase with increasing soil water and decreasing temperature and the top soil of the increasing water within temperature zones can increase plant production and increase plant litter and root production [39]. In contrast, the cessation of tillage during the establishment of forest on cropland can reduce the SOC decomposition rate, thus increasing the SOC content. Furthermore, the addition of nitrogen derived from nitrogen-fixing species (in this case, alfalfa and black locust) has been well documented and can lower the decomposition of old and new carbon [40], hence favoring SOC sequestration. Moreover, the forest is widely found to be less susceptible to soil erosion compared to the cropland on the Loess Plateau [41], which may contribute to the higher topsoil SOC content following the afforestation of the cropland. And there is also an increase in the carbon input from aboveground litter and fine root biomass in the alfalfa and black locust, which may lead to SOC accumulation during the afforestation.

Understanding the dependence of the SOC content on the plantation age is important for the development of improved biological soil management practices on both local and national scales. This understanding will also help to identify different species that are ecologically compatible and economically useful. The SOC stores in the younger plantation were clearly higher than those in the old afforestation sites, which exhibited decreases of 15.6% and 50% for alfalfa and black locust 12 yr after afforestation. However, for black locust, the SOC continues to

increase gradually 15 yr after the conversion of farmland to forestland. A fundamental assumption regarding the effect of land change on the carbon sink is that younger, growing forests sequester carbon more rapidly than older forests. For several yr following a disturbance, the land becomes a carbon source, but as it matures, it becomes a large carbon sink, which slowly declines with age during late succession [42,43]. This trend may owe to the past application of organic fertilizer when the land was used as cropland, producing a higher initial soil C content just before afforestation. When agricultural land is no longer used for cultivation and is allowed to revert to natural vegetation, the balance of C input and output in the agricultural system is broken, as no fertilizer is added to the afforested land.

Factors affecting the SOC sequestration in the "Grain-for-Green" project

The contribution of afforestation to the SOC improvement is controlled by several factors, including the severity of the land degradation, plant species and composition, climate, duration, land use history and management. The effect of the Grain for Green project on SOC sequestration was mainly influenced by ages and depths, which explained 43.2% and 35.6% (Table 3) of the variation of SOC. vegetation species explained 21.2% (Table 3) of the variation in SOC. Xu et al [44] found that the effect of the "Grain-for-Green" project on the SOC sequestration was influenced primarily by the land use type and age, which explained 55.6% and 24.1% of the variation in the SOC, respectively. The SOC content therefore varies across plantations because the C sequestration is affected by the varying land management. Our findings on the carbon sink induced by the project and the relationship between the SOC content and the ages following afforestation supported the conclusion of Xu et al [44]. Topographic factors such as the slope aspect and position were also found to explain 8.5% of the variation in the SOC [44]. Reay ea al [45] indicated that by increasing the productivity, nitrogen enrichment of some natural ecosystems may enhance their capacity to sequester carbon. Clearly, we demonstrate that afforestation has a positive influence on the SOC concentration in our study, depending on the species planted (alfalfa has a higher SOC content than black locust). The species-induced difference in the SOC content and other soil properties may owe to a difference in biomass production and nutrient cycling via litter fall and root turnover [46].

Conclusion

We examined the changes in the carbon sink and SOC content of the topsoil and subsoil following the "Grain-for-Green" (farmland-to-forestland conversion) project. The afforested land (alfalfa and black locust) soil displayed remarkable SOC sequestration compared to the sloping croplands. The two plantations studied exhibited SOC sequestration potential in the top soil (0–100 cm) in the initial 7-yr period, while the sequestration occurred later in the 100–120 cm layer (>7 yr after the "Grain-for-Green" project was implemented). Furthermore, our analyses provide evidence that SOC can be influenced by afforestation in semi-arid region such as the Loess Plateau, China. From a regional perspective, type of land use, age and species all had significant effects on the SOC sequestration in the "Grain-for-Green" project. The implementation of the program has further enhanced the vegetation restoration and ecological conservation and strengthened the management of the Loess Plateau ecosystem.

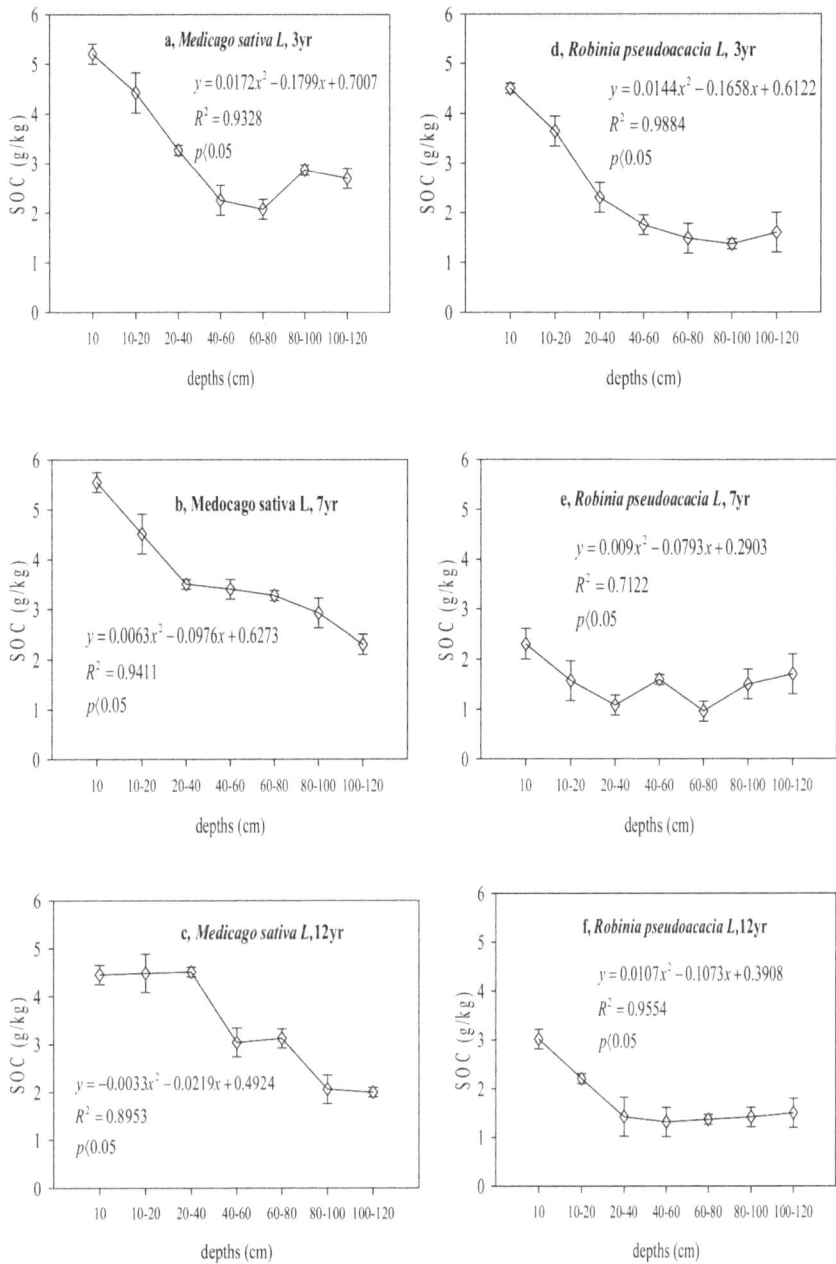

Figure 5. Relationship between the SOC and depths of the alfalfa (*Medicago sativa L*) and the black locust (*Robinia pseduoacacia L*) under 3, 7 and 12 years old following afforestation.

Table 3. Contribution of different factors to the variation of SOC content (variance components, n = 45).

SOC content	Variance Source		
	Afforestation ages	Depths	Plantation Species
Variance	16.62	13.69	8.15
Variance percentage to total variance (%)	43.20	35.60	21.20

Author Contributions

Conceived and designed the experiments: YLL LM. Performed the experiments: LM RLH. Analyzed the data: YLL LM. Contributed reagents/materials/analysis tools: YLL LM. Wrote the paper: LM YLL.

References

1. Lal R (2004) Soil carbon sequestration to mitigate climate change. Geoderma 123: 1–22.

2. Lal R (2010) Beyond Copenhagen: mitigating climate change and achieving food security through soil carbon sequestration. Food Secur 4: 1035–1064.

3. Post WM, Emanuel WR, Zinke PJ, Stangenberger AG (1982) Soil carbon pools and world life zones. Nature 298: 156–159.

4. Batjes NH (1996) Total carbon and nitrogen in the soils of the world. Eur.J. Soil Sci 47: 151–163.

5. Tarnocai C, Canadell JG, Schuur EAG, Kuhry P, Mazhitova G, et al. (2009) Soil organic carbon pools in the northern circumpolar permafrost region. Global Biogeochemical Cycles 23: 1–11.

6. Smith P (2004) Soils as carbon sinks- The global context. Soil Use Manage 20: 212–218.

7. Todd-Brown KEO, Randerson JT, Post WM, Hoffman FM, Tarnocai C, et al. (2013) Cause of variation in soil carbon simulations from CMIP5 Earth system models and comparison with observations. Biogeosciences 10: 1717–1736.

8. Yu DS, Shi XZ, Wang HJ, Sun WX, Chen JM, et al. (2007) Regional patterns of soil organic carbon stocks in China. J.Environ.Manage 85: 680–689.

9. Pan GX (2008) Soil organic carbon stock. Dynamics and climate change mitigation of China. Adv.Clim.Change Res 4: 282–289.

10. Chen LD, Gong J, Fu BJ, Huang ZL (2007) Effect of land use conversion on soil organic carbon sequestration in the loess hilly area, loess plateau of china. Ecol Res 22: 641–648.

11. Entry JA, Sojka RE (2002) Management of irrigated agriculture to increase organic carbon storage in soils. Soil Sci Soc Am. J 66: 1957–1964.

12. Smith P, Powlson DS, Glendining MJ (1996) Establishing a European soil organic matter network (SOMNET). In: Powlson DS, Smith P, Smith JU, editors. Evaluation of soil organic matter models using existing, long-term datasets. UK: Harpenden. pp: 429.

13. Lubowski RN, Plantinga AJ, Stavins RN (2005) Land-use change and carbon sinks: Econometric Estimation of the Carbon Sequestration Supply Function. Belfer Centerprograms or projects: Environment and Natural resources, Harvard Project on Climate Agreements 5: 135–152.

14. Guo LB, Gifford RM (2002) Soil carbon stocks and land use change: a meta analysis. Global Change Biol 8: 345–360.

15. Ostle NJ, Levy PE, Evans CD, Smith P (2009) UK land use and soil carbon sequestration. Land use polic 26s: S274–S283.

16. DeGryze S, Six J, Paustian K, Morris SJ, Merckx R, et al. (2004) Soil organic carbon pool changes following land-use conversions. Global Change Biology 10: 1120–1132.

17. Schlesinger WM (1999). Carbon sequestration in soils. Science 284: 2095.

18. Shi H, Shao MA (2000) Soil and Water loss from the Loess Plateau in China. J Arid Environment 45: 9–20.

19. Liu ZP, Shao MA, Wang YQ (2011) Effect of environment factors on regional soil organic carbon stocks across the Loess Plateau region, China. Agricultural, Ecosystemsand Environment 142: 184–194.

20. Xu XL, Zhang KL, Peng WY (2003) Spatial distribution and estimating of soil organic carbon on Loess Plateau. J Soil Water Conservation 17: 13–15.

21. Jia SW, He XB, Chen YM (2004) Effect of land abandonment on soil organic carbon sequestration in loess hilly areas. J Soil WaterConservation 18: 78–80.

22. Chang RY, Fu BJ, Liu GH, Wang SH, Yao XL (2012) The effects of afforestation on soil organic and inorganic carbon: A case study of the Loess Plateau of China. Catena 95:145–152.

23. Jin TT, Fu BJ, Liu GH, Wang Z (2011) Hydrologic feasibility of artificial forestation in the semi-arid Loess Plateau of China. Hydrol.Earth Syst.Sci 15:2519–2530.

24. Xi TT, Li SL (2006) Analysis of forestry carbon mitigation potential in Heilong jiang Province. Problems of Forestry Economics 26: 519–523.

25. Xu YR, Wang DT, Feng ZW (2003) Effect of carbon storage in several artificial vegetations on sea-beach salinity soil in Tianjin, China. Journal of Huazhong Agricultural University 22: 603–607.

26. IPCC (2000) Summary for Policymakers. In: Nakicenovic N, Swart R, editors. Special Report of the Intergovernmental Panel on Climate Change. Cambridge: Cambridge university. pp. 377

27. Chinese Editorial Committee of Soil Analysis (1996) Soil physical and chemical analysis and description of soil profile (in Chinese). China State Standards Press, Beijing.

28. ISS (1978) Physical and chemical analysis methods of soils. Institute of Soil Science. Shanghai: Shanghai Sci Technol Press. pp: 7–59.

29. Bolin B, Sukumar R (2000) Global perspective In: Watson RT, Noble IR, Bolin B, Racindranath NH, Verardo DJ, DokkenDJ, editors. Land use, land use change, and forestry. Cambridge: Cambridge university. pp. 23–51.

30. Gupta RK, Rao DLN (1994) Potential of wastelands for sequestering carbon by reforestation. Current science 66: 378–380.

31. Fearnside PM, Barbosa RI (1998). Soil carbon changes from conversion of forest to pasture in Brazilian Amazonia. Forest ecology and management 108: 147–166.

32. Dawson JJC, Smith P (2007) Carbon losses from soil and its consequences for land-use management. Science of the total environment 382: 165–190.

33. Leake MC, Jennifer HC, George HW, Fan B, Richard M, et al. (2006) Stoichiometry and turnover in single, functioning membrane protein complexes. Nature 443: 355–358.

34. Chang RY, Fu BJ, Liu GH, Liu SG (2011) Soil carbon sequestration potential for "grain for green" project in Loess plateau, China. Environmental management 48: 1158–1172.

35. Laganière J, Angers DA, Par D (2010) Carbon accumulation in agricultural soils after afforestation: a meta analysis. Global change biology 16: 439–453.

36. Paul KI, Polglase PJ, Nyakuengama JG, Khanna PK (2002) Change in soil carbon following afforestation. Forest ecology and management 168: 241–257.

37. Post WM, Kwon KC (2000) Soil carbon sequestration and land use change: processes and potential. Global change biology 6: 317–327.

38. Balesdent J, Chenu C, Balabane M (2000) Relationship of soil organic matter dynamics to physical protection and tillage. Soil and tillage research 53: 215–230.

39. Liski J, Iivesniemi A, Makela CJ (1999) CO_2 emission from soil in response to climate warming are overestimated-the decomposition of old soil organic matter is tolerantof temperature. Ambio 28: 171–174.

40. Binkley D (2005) How nitrogen-fixing trees change soil carbon. In: Binkley D, Menyailo O, editors. Tree species effects on soils: implications for global change. Dordrecht: Kluweracademic publishers. pp. 155–164.

41. Fu BJ, Wang YF, Lu YH, He CS (2009) The effects of land-use combinations on soil erosion: a case study in the loess plateau of China. Progress in physical geography 33: 793–804.

42. Law BE, Sun OJ, Campbell J, Vantuyl S, Thornton PE (2003) Changes in carbon storage and fluxes in a chronosequence of pondersosa pine. Global change biology 9: 510–524.

43. Turner DP, Guzy M (2003) Effects of land use and fine-scale environmental heterogeneity on net ecosystem production over a temperate coniferous forest landscape. Tellus 558: 657–668.

44. Xu MX, Wang Z, Zhang J, Liu GB (2012) Response of soil organic carbon sequestration to the "Grain for Green Project" in the hilly loess plateau regain. Acta Ecologicasinica 32: 5405–5415.

45. Reay DS, Dentener F, Smith P, Grace J, Feely R (2007) Global nitrogen deposition and carbon sinks. Nature Geoscience 1: 430–437.

46. Girmay G, Singh BR, Mitiku H (2008) Carbon stocks in Ethiopian soils in relation to land use and soil management. Land degradation and development 19: 351–367.

22

Modeling the Soil Water Retention Curves of Soil-Gravel Mixtures with Regression Method on the Loess Plateau of China

Huifang Wang[1,2,3], Bo Xiao[1,3]*, Mingyu Wang[2], Ming'an Shao[3]

1 Beijing Research & Development Center for Grass and Environment, Beijing Academy of Agriculture and Forestry Sciences, Beijing, China, **2** Center for Water System Security, Graduate University of Chinese Academy of Sciences, Beijing, China, **3** State Key Laboratory of Soil Erosion and Dryland Farming on the Loess Plateau, Institute of Soil and Water Conservation, Chinese Academy of Sciences, Yangling, Shaanxi, China

Abstract

Soil water retention parameters are critical to quantify flow and solute transport in vadose zone, while the presence of rock fragments remarkably increases their variability. Therefore a novel method for determining water retention parameters of soil-gravel mixtures is required. The procedure to generate such a model is based firstly on the determination of the quantitative relationship between the content of rock fragments and the effective saturation of soil-gravel mixtures, and then on the integration of this relationship with former analytical equations of water retention curves (WRCs). In order to find such relationships, laboratory experiments were conducted to determine WRCs of soil-gravel mixtures obtained with a clay loam soil mixed with shale clasts or pebbles in three size groups with various gravel contents. Data showed that the effective saturation of the soil-gravel mixtures with the same kind of gravels within one size group had a linear relation with gravel contents, and had a power relation with the bulk density of samples at any pressure head. Revised formulas for water retention properties of the soil-gravel mixtures are proposed to establish the water retention curved surface models of the power-linear functions and power functions. The analysis of the parameters obtained by regression and validation of the empirical models showed that they were acceptable by using either the measured data of separate gravel size group or those of all the three gravel size groups having a large size range. Furthermore, the regression parameters of the curved surfaces for the soil-gravel mixtures with a large range of gravel content could be determined from the water retention data of the soil-gravel mixtures with two representative gravel contents or bulk densities. Such revised water retention models are potentially applicable in regional or large scale field investigations of significantly heterogeneous media, where various gravel sizes and different gravel contents are present.

Editor: Han Y.H. Chen, Lakehead University, Canada

Funding: This study was funded by Natural Science Foundation of China (No. 50479063 and 41001156), by the Program of Hundreds of Talents of Chinese Academy of Sciences (99T3005WA2), and by The Key National Science & Technology Water Project (No. 2009ZX07212-003). The funders had no role in study design, data collection and analysis, decision to publish, or preparation of the manuscript.

Competing Interests: The authors have declared that no competing interests exist.

* E-mail: xiaoboxb@gmail.com

Introduction

A large proportion of soils containing rock fragments are present in the world due to soil evolution and erosion [1,2]. The highly variable gravel content or size in soilscape greatly increases the variability of the soil properties [3–5]. Knowledge of soil water retention curves (WRCs) is a prerequisite for modeling the fluxes of water and solutes in the vadose zone and consequently it is necessary to determine their spatial variability [6–8]. Since direct field measurements of WRCs are time-consuming and expensive, laboratory measurements continue to be the most frequent means of characterizing the vadose zone [9,10]. Soil water retention data are typically obtained in laboratory for fine soils (<2 mm) using pressure cells, pressure-plate extractors, and centrifuge methods [11–13]. Several reports noted that water held by gravel cannot be neglected in the determination of water retention properties due to the significant porosity of gravel and the changed pore-size distribution [14,15]. For example, the ironstone gravel contained a large amount of available water ranging from 0.03 cm^3 cm^{-3} to

0.15 cm^3 cm^{-3} [16], while the sandstone fragments and shale fragments held 0.11 cm^3 cm^{-3} and 0.23 cm^3 cm^{-3} available water, respectively [17]. For soil-glass mixtures, the volume of coarse lacunar pore increased with glass content when glass content was less than 50% [18]. As a large number of water retention curves for fine soils have been obtained in laboratory, gravel corrections for moisture retention in soil-gravel mixtures have been developed on the basis of WRCs of fine soils. For example, Gardner [19] used mass-based approach while Bouwer and Rice [20] used volume-based approach to make gravel corrections for water retention of soil-gravel mixtures [20–22]. Correction procedures are available to determine water retention properties for soil-gravel mixtures, but they have limited utility for soil-gravel mixtures with weathered gravels especially on high-suction range [23]. Pedotransfer functions for predicting WRCs have been developed from more easily measurable and more readily available soil properties [24–27]. Scheinost et al. [6] predicted WRCs for soils with a wide range of particles which included 2–67 mm gravel with a new pedotransfer function

Table 1. Mineral composition of soil and rock fragments sampled in this study.

Fine soil (<2 mm)				Rock fragments (2–10 mm)			
Texture	Grain (µm) content in volume (%)			Petrology	Mean density (g cm^{-3})	Shape	Weathering degree
Clay loam soil	0.02–2	2–20	20–2000	Shale clasts (S)	2.09±0.04	Flake, block	High
	17.9	39.4	42.7	Pebbles (P)	2.49±0.15	Sphere, ellipsoid	Low

including more textural fractions, but this did not greatly improve the prediction precision [25,28].

In order to quantify the effects of content and size of rock fragments on soil water retention and finally develop a new method for determining the WRCs of soil-gravel mixtures, especially for mixtures with weathered gravels and for WRCs at high-suction range, we investigated WRCs of a loess soil and gravels mixtures with various gravel contents and gravel sizes. The following constraints were soon evident in developing this new method: (1) To extend the practical use of the revised model, the water retention data of soil-gravel mixtures with weathered gravel should be obtained and the range of water potential should be extended to include very low values (high-suction). (2) To develop revised water retention models, the effects of gravel contents and gravel size on the effective degree of saturation (Se) of soil-gravel mixtures should be analyzed based on the measured WRCs data. The relationships between Se and gravel contents or bulk density of soil-gravel mixtures should be combined with a closed-form analytical equation such as the model of Brooks and Corey (BC-function) [29] or the van Genuchten equation (VG-function) [30]. (3) To validate the effectiveness of revised water retention models and simplify the procedure of parameter-obtaining, the way that the gravel size affects the shape parameters of revised models should be analyzed, and the shape parameters of the revised models should be obtained from the representative WRCs data of soil-gravel mixtures, such as that with two extremes values of gravel content range. Those parameters and the method of parameter-obtaining may give references for determining WRCs of other soil-gravel mixtures.

Materials and Methods

Ethics statement

No specific permissions were required for these sampling activities because the location is not privately-owned or protected in any way and the field activities did not involve endangered or protected species.

The samples were excavated from the soil profiles in 30 cm depth of Yaoxianliang in Tongchuan (108.93 E, 35.28 N, at 1570 m altitude) and Weihe river bank in Yangling (108.10 E, 34.25 N, at 437 m altitude), Shaanxi province, China. The soil in Yaoxianliang is recognized as Aric Regosol (FAO). The rock fragments in these soils are schists and shales (S) with brownish green in color and sheet-like shape. The other type of rock fragments sampled in Weihe river bank was pebble (P) from alluvial sediment. The rock fragments in sizes ranging from 2 to 10 mm were sieved and washed to remove the soil particles from their surface, and then sieved into three different diameter classes: 2–3 mm, 3–5 mm, and 5–10 mm. The weathering degree of the rock fragments was defined by observing their weathering characteristics in the field and observing the surface fissures and coarseness under a magnifier in the laboratory according to the Geotechnical Engineering Handbook [31]. The gravel weathering degree is described in Table 1.

The texture of the air-dried fine soil (<2 mm), from which the rock fragments (> 2 mm) and plant residues were removed, was measured by the Laser Diffraction Particle Size Analyzer (Mastersizer 2000, Malvern Instruments Ltd. in England). According to the International Soil Classification System, the disturbed soil sample was a clay loam soil (labeled with CL). The mean density of the rock fragments (ρ_r) in 2–10 mm, determined

Table 2. Packed bulk densities (g cm^{-3}) of clay loam soil and gravel mixtures.

Gravel content (%)	Shale clasts			Pebbles		
	2–3 mm	3–5 mm	5–10 mm	2–3 mm	3–5 mm	5–10 mm
0	1.28	1.28	1.28	1.28	1.28	1.28
10	1.31	1.31	1.32	1.35	1.35	1.36
15	1.34	1.35	1.35	1.39	1.39	1.39
25	1.39	1.40	1.40	1.47	1.47	1.48
35	1.45	1.46	1.45	1.56	1.56	1.56
45	1.50	1.50	1.50	1.67	1.66	1.67
55	1.58	1.58	1.58	1.79	1.77	1.78
65	1.63	1.66	1.66	1.93	1.91	1.91
100	1.36	1.39	1.36	1.60	1.61	1.61

Table 3. Rotation speeds and equilibrium times corresponding to the tested pressure heads in the centrifuge method.

Pressure head (cm)	Rotation speed (r s^{-1})	Equilibrium time (min)
102	16.3	26
204	23.1	36
408	32.7	45
612	40	51
816	46.2	55
1020	51.7	58
2040	73.1	68
4080	103.4	77
6120	126.6	83
10200	163.4	90

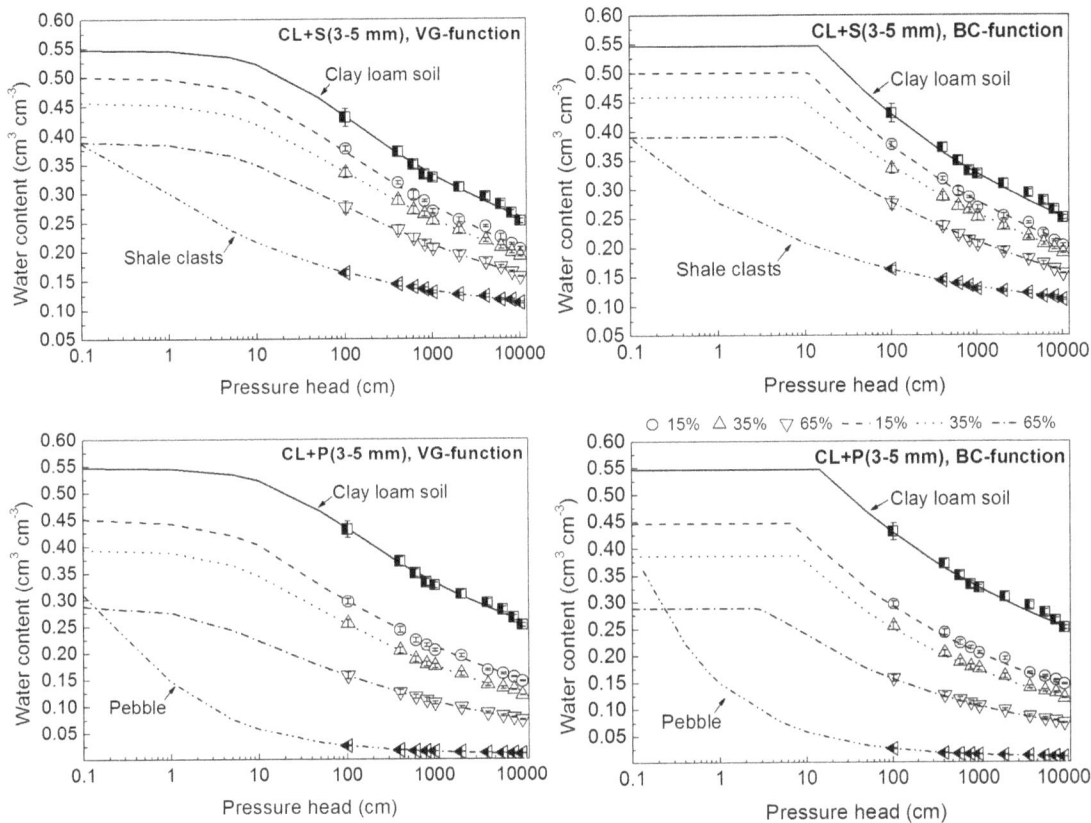

Figure 1. The measured data and fitted water retention curves for soil-gravel mixtures with 3–5 mm gravels. Point: Measured data; Line: Fitted data. CL+S stands for the mixtures of clay loam soil and shale clasts with varied gravel content; CL+P stands for the mixtures of clay loam soil and pebble with varied gravel content. The pressure head in the axe is the absolute value of the actual pressure head which is negative in the measurement.

using dividing masses by the corresponding volume of water, ranged from 2.09 to 2.49 g cm^{-3} (Table 1). It should be noted that the rock fragments were saturated in water before determining the water volume substituted by the rock fragments in a container with scale when measuring ρ_r. The saturated water contents (θ_s) of the gravels soaked into water for three days were measured by oven drying at 105°C for 24 h and the measurements of each kind of gravels were replicated five times. The results showed that θ_s of shale clasts was 0.09 ± 0.02 g g^{-1}, while that of pebbles was almost zero. In the sample saturation process, the air bubbles enclosed in the gravels possibly caused the gravel in incomplete saturated and subsequently the underestimation of θ_s in this study.

The clay loam soil was mixed with air-dried shale clasts or pebbles at the ratios of 10%, 15%, 25%, 35%, 45%, 55%, and 65% on a total mass basis in each of the three size classifications (2–3 mm, 3–5 mm, and 5–10 mm). We totally prepared 45 samples including one soil sample, two gravel samples, and 42 soil-gravel mixture samples. The soil-gravel mixtures were packed uniformly by hand into soil containers (98.2 cm^3 with 5 cm high and 5 cm inner diameter) and the bulk density of clay loam soil without gravel predetermined 1.28 g cm^{-3} for all the soil and soil-gravel mixture samples. Then the soil containers were posited inside the centrifuge rotor chamber (Hitachi-CR21G, Hitachi Ltd. in Japan) under dry conditions (i.e., water content was less than 0.5%). The bulk density of soil-gravel mixtures is presented in Table 2; it was calculated from the measured dry mass and the volume of the samples after finishing centrifuge rolling.

The WRCs of soil, gravels, and soil-gravel mixtures were measured separately by the centrifuge method [32], and each treatment had three replications. The tested suction head in the WRCs measurement was in the range of 102 cm to 10200 cm. The Hitachi-CR21G centrifuge was equipped to maintain air temperature constantly at 20°C. For the centrifugation method, soil water desorption was accomplished by applying a high gravity field (centrifugal force) to saturated soil samples. The suction heads were obtained by sequentially increasing the angular velocity or rotation speed of the centrifuge [33]. The rotation speeds and equilibrium times corresponding to the tested pressure heads are shown in Table 3. Water retention properties of the samples near saturation were not considered because the water retention of the artificially packed samples was conditioned by the samples packing and the geometry of soil samples, especially in the low matric suction region [34].

For those θ-h data, BC-function and VG-function parameters describing WRCs of soil, gravels, and soil-gravel mixtures were determined by the computer program RETC.FOR [35] using the Marquardt nonlinear least-squares optimization algorithm [36]. BC-function and VG-function are presented as Eq. (1) and Eq. (2):

$$S_e = \frac{\theta - \theta_r}{\theta_s - \theta_r} = \left(\frac{h}{h_a}\right)^{-\lambda} = (\alpha h)^{-\lambda} = A h^{-\lambda} (\alpha h > 1) \qquad (1)$$

Table 4. BC-function parameters describing the water retention characteristics of soils, soil-gravel mixtures, and gravels.

Samples	Gravel content (%)	Gravel size in 2–3 mm					Gravel size in 3–5 mm					Gravel size in 5–10 mm				
		θ_s^*	θ_r	$1/\alpha$ (cm^{-1})	λ	RMSE (×10^{-3})	θ_s	θ_r	$1/\alpha$ (cm^{-1})	λ	RMSE (×10^{-3})	θ_s	θ_r	$1/\alpha$ (cm^{-1})	λ	RMSE (×10^{-3})
CL	0	0.55	0.06	14.15	0.14	17.61	-	-	-	-	-	-	-	-	-	-
CL+S	10	0.52	0.00	12.68	0.11	13.42	0.53	0.00	11.11	0.12	8.37	0.54	0.07	18.62	0.17	12.25
	15	0.53	0.00	10.16	0.11	14.49	0.50	0.03	10.89	0.14	12.25	0.51	0.04	15.27	0.15	11.83
	25	0.51	0.00	10.15	0.11	12.25	0.49	0.00	8.58	0.13	12.65	0.48	0.02	13.69	0.14	5.48
	35	0.48	0.00	8.41	0.11	8.94	0.46	0.00	8.19	0.12	5.48	0.46	0.00	14.65	0.14	13.04
	45	0.46	0.00	8.25	0.12	10.49	0.43	0.00	8.95	0.12	7.75	0.43	0.00	13.98	0.14	10.49
	55	0.44	0.00	6.48	0.12	10.00	0.42	0.02	6.51	0.13	7.75	0.41	0.00	9.78	0.13	8.94
	65	0.44	0.02	5.74	0.13	10.00	0.39	0.00	5.93	0.12	8.37	0.36	0.05	16.74	0.20	9.49
	100	0.41	0.11	34.71	0.80	7.07	0.41	0.08	0.07	0.19	5.48	0.41	0.00	0.44	0.18	15.49
CL+P	10	0.48	0.09	21.39	0.27	5.48	0.49	0.09	17.77	0.29	9.49	0.47	0.08	16.77	0.28	7.75
	15	0.46	0.05	12.55	0.21	13.04	0.45	0.01	6.48	0.16	9.49	0.45	0.02	7.38	0.18	6.32
	25	0.44	0.02	7.00	0.18	4.47	0.43	0.00	4.37	0.15	8.94	0.43	0.00	4.61	0.16	7.75
	35	0.44	0.04	6.55	0.21	4.47	0.39	0.03	7.95	0.19	7.07	0.40	0.00	3.84	0.16	6.32
	45	0.43	0.02	4.07	0.19	7.07	0.37	0.04	6.58	0.22	4.47	0.38	0.03	6.53	0.20	4.47
	55	0.34	0.04	5.42	0.21	6.32	0.31	0.02	4.78	0.19	4.47	0.33	0.00	3.47	0.15	5.48
	65	0.33	0.02	4.26	0.21	11.83	0.29	0.01	2.83	0.18	3.16	0.28	0.03	7.76	0.24	5.48
	100	0.36	0.03	24.21	1.11	4.47	0.36	0.01	0.13	0.46	0.00	0.36	0.00	0.00	0.24	0.00

*θ_s for soil and soil-gravel samples were the measured value, and θ_s for gravel samples were calculated from density and bulk density. *RMSE*: square root of residual sum of squares values.

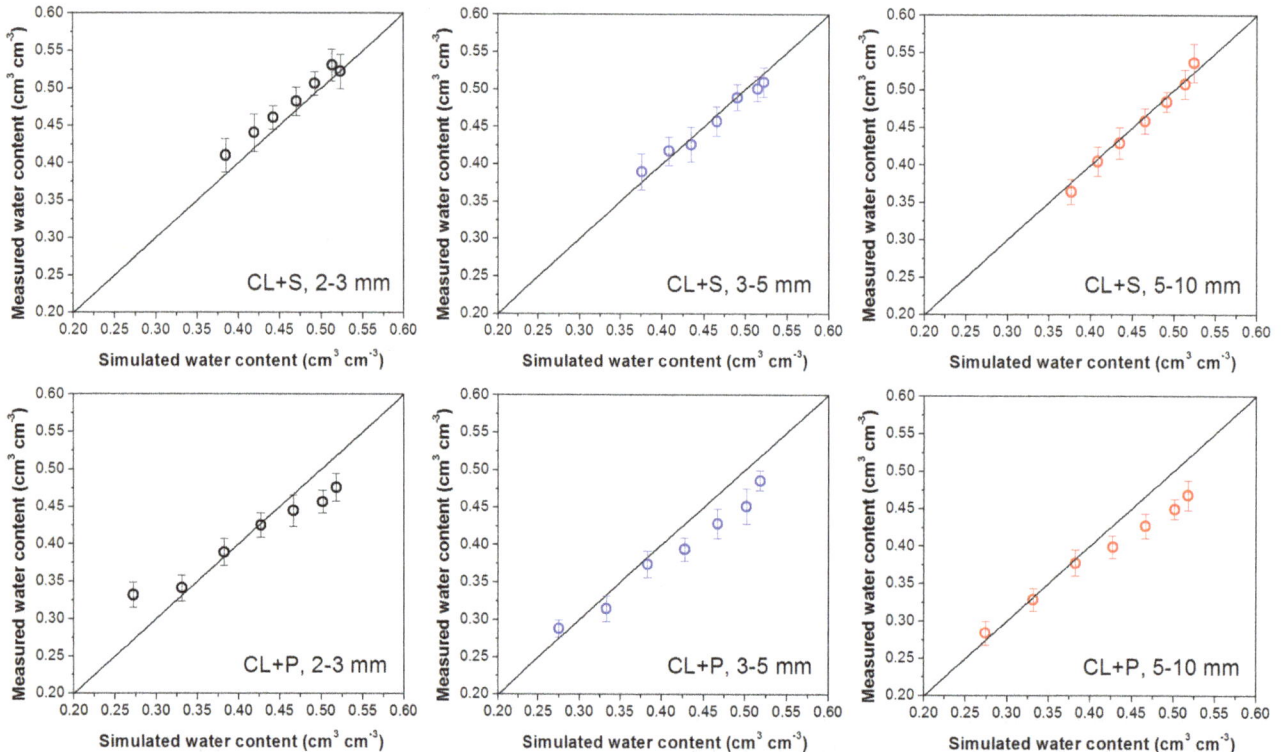

Figure 2. The measured and simulated saturated water contents for soil-gravel mixtures with varied gravel sizes. CL+S stands for the mixtures of clay loam soil and shale clasts; CL+P stands for the mixtures of clay loam soil and pebble.

Table 5. Regression empirical parameters for curved surface functions with M_r obtained by fitting all the measured water retention curves.

Samples	$S_e = A_M h^{-\lambda} M_r^{-\beta}$						$S_e = A_M' h^{-\lambda}(B_M' M_r + 1)$					
	A_M	λ	β	R2	SEE	REMS	A_M'	λ	B_M'	R2	SEE	REMS
CL+S (2–3)	1.18	0.12	0.07	0.97	0.01	0.13	1.39	0.12	0.23	0.98	0.01	0.10
CL+S (3–5)	1.25	0.12	0.03	0.98	0.01	0.10	1.33	0.12	0.09	0.98	0.01	0.10
CL+S (5–10)	1.36	0.14	0.03	0.97	0.01	0.12	1.48	0.14	0.13	0.98	0.01	0.11
CL+S (2–10)	1.26	0.12	0.04	0.96	0.02	0.29	1.40	0.12	0.15	0.97	0.02	0.28
CL+P (2–3)	1.12	0.16	0.13	0.98	0.02	0.12	1.52	0.16	0.39	0.98	0.02	0.12
CL+P (3–5)	1.15	0.16	0.08	0.97	0.02	0.15	1.41	0.16	0.28	0.98	0.01	0.11
CL+P (5–10)	1.24	0.16	0.04	0.97	0.01	0.11	1.36	0.16	0.14	0.98	0.01	0.11
CL+P (2–10)	1.17	0.16	0.08	0.96	0.02	0.27	1.43	0.16	0.27	0.97	0.02	0.25

R2, SEE, and REMS stand for coefficient of determination, estimated standard error, and square root of the residual sum of squares, respectively.

$$S_e = \frac{\theta - \theta_r}{\theta_s - \theta_r} = \frac{1}{[1 + (\alpha h)^n]^m} \quad (2)$$

where S_e is the effective degree of saturation, also called reduced water content; θ_s is the saturated water content ($cm^3 \, cm^{-3}$); θ_r is the residual water content ($cm^3 \, cm^{-3}$); h is the matric suction (cm); α is the empirical parameter whose inverse equals to h_a and is referred as the air entry value or bubbling pressure (cm^{-1}); and λ is the pore-size distribution parameter; n, m, $A = \alpha^{-\lambda}$ are empirical parameters affecting the shape of the retention curve, and $m = 1 - 1/n$.

Results and Discussion

Effects of rock fragments on WRCs

Soil matric suction (h) is a function of water content (θ) in an unsaturated soil and WRCs express the relationship between them. The VG-function resolves the coherence problem of WRCs near soil saturation and therefore it has become one of best choices for the analytical model for $\theta(h)$, especially for undisturbed field soil and many fine-texture soils [37,38]. While BC-function, leads to an air-entry value in the WRC above which soil is assumed to be saturated, has the ability of more accurate description for coarse texture soil with structural deterioration or compaction, especially in the dry end of WRCs [39]. However, in this study, they gave similar fine fitting performance (Fig. 1 and Table 4).

The maximum relative standard error (RSEm) for the saturated water contents of each sample was lower than 2.5% (n = 3), and the RSEm for the unsaturated water contents of them at various pressure heads was lower than 2.2% (n = 3), which showed a good representation for each sample; the measured WRCs data by the centrifuge method had acceptable accuracy for soil-gravel mixtures. While, the measured data of water retention for the soil-gravel samples might be underestimated due to the experimental set up, in which the gravels was possibly not moistured completely under normal air pressure conditions. Parameter sensitivity at one pressure head for WRCs was evaluated by the average value of the total ratios of the relative changes of θ to the relative changes of values of one parameter. The results of sensitivity analysis showed that the absolute value of sensitivity for parameter α in BC-function increased with soil suction till it reached $1/\alpha$, and then kept at that maximum value. Correspondingly, that value in VG-function increased with soil suction till it reached the dry end of curve ($h = 10200$ cm, except for the mixture samples containing 55% gravels, it reached at $h = 407$ cm). Parameter n became most sensitive at $h = 10200$ cm for both BC- and VG-function. The increasing sensitivity of α and n to BC-

Table 6. Regression empirical parameters for curved surface functions with ρ obtained by fitting all the measured water retention curves.

Samples	$S_e = A_\rho h^{-\lambda} \rho^{-\beta'}$						$S_e = A_\rho' h^{-\lambda}(B_\rho' \rho + 1)$					
	A_ρ	λ	β'	R2	SEE	REMS	A_ρ'	λ	B_ρ'	R2	SEE	REMS
CL+S (2–3)	1.61	0.12	0.62	0.98	0.01	0.10	2.07	0.12	0.26	0.98	0.01	0.10
CL+S (3–5)	1.41	0.12	0.13	0.98	0.01	0.08	1.59	0.12	0.08	0.98	0.01	0.10
CL+S (5–10)	1.60	0.14	0.32	0.98	0.01	0.10	1.87	0.14	0.17	0.97	0.01	0.12
CL+S (2–10)	1.54	0.12	0.40	0.97	0.02	0.26	1.85	0.12	0.19	0.97	0.02	0.27
CL+P (2–3)	1.81	0.16	0.71	0.98	0.02	0.12	2.20	0.16	0.25	0.98	0.02	0.14
CL+P (3–5)	1.59	0.16	0.50	0.98	0.01	0.11	1.90	0.16	0.21	0.98	0.01	0.10
CL+P (5–10)	1.45	0.16	0.23	0.98	0.01	0.10	1.60	0.16	0.12	0.98	0.01	0.11
CL+P (2–10)	1.61	0.16	0.48	0.97	0.02	0.24	1.91	0.16	0.20	0.97	0.02	0.24

R2, SEE, and REMS stand for coefficient of determination, estimated standard error, and square root of the residual sum of squares, respectively.

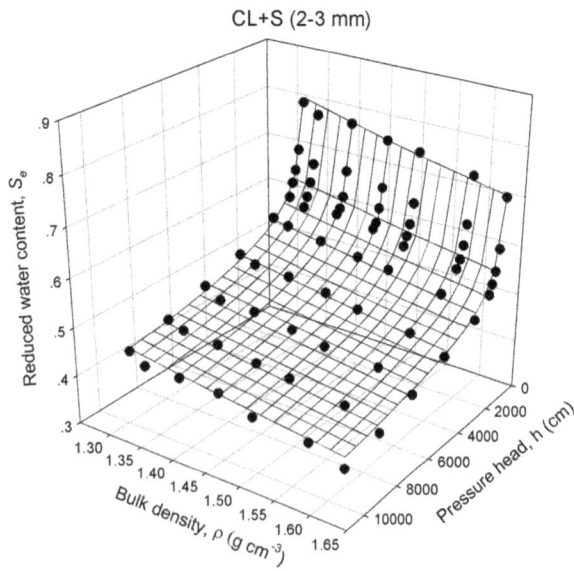

CL+S (2-3 mm)

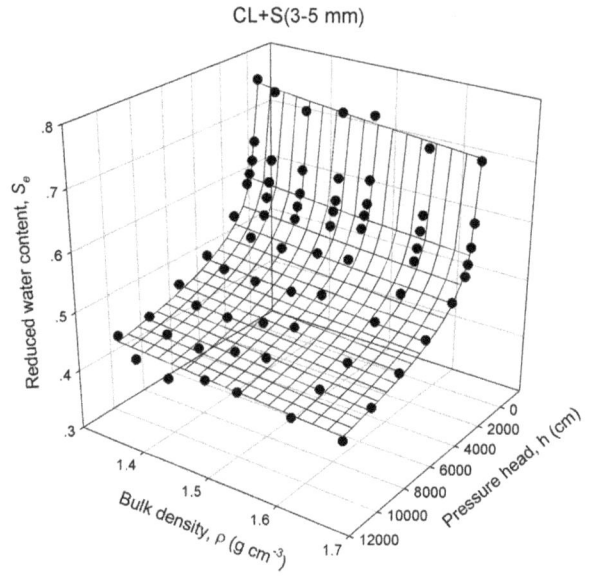

CL+S(3-5 mm)

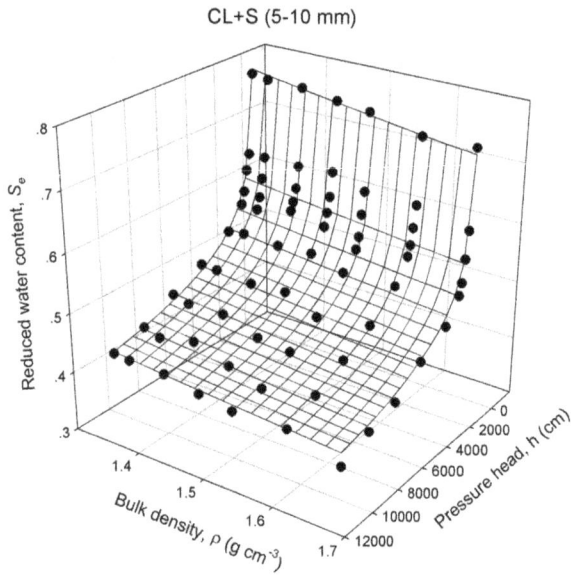

CL+S (5-10 mm)

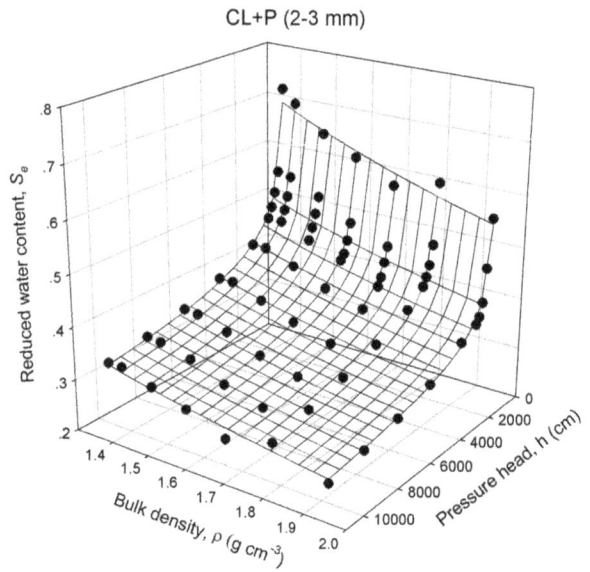

CL+P (2-3 mm)

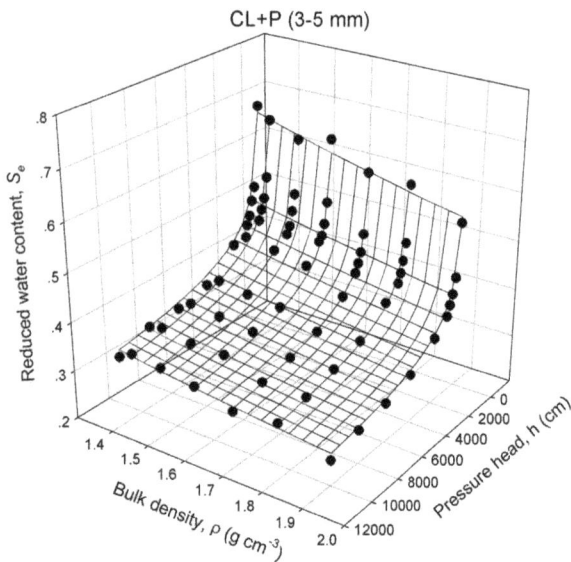

CL+P (3-5 mm)

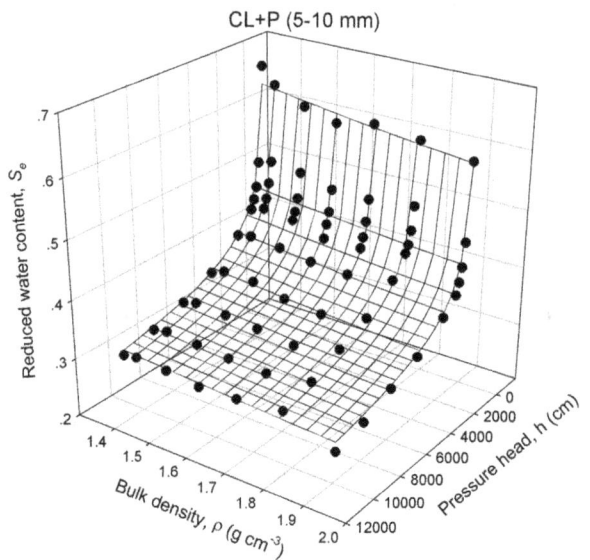

CL+P (5-10 mm)

Figure 3. The measured points and the regression surfaces of power function with variable of samples bulk density for soil-gravel mixtures (•: Measured point; mesh: Regression surface). The pressure head in the axe is the absolute value of the actual pressure head which is negative in the measurement.

and VG-function with the increased suction implied that the water content at $h = 0$ and $h = 100$ cm decided the trend and shape of WRCs in the wet end (h = 0–100 cm), which varied insignificantly when the parameters were changed. Those results indicated that the comparison between WRCs combining extrapolated and measured one for soil-gravel samples was acceptable. It seems that α reflected the order of air-entry values from those α values (lower, media, upper value at 95% confident limits) for various soil-gravel samples due to its high sensitivity to BC-function at $h = 1/\alpha$. One thing should be noted that the measured θ_s and the fitted θ_r in BC- and VG- function were as the fixed value in the analysis of parameter sensitivity.

There were considerable differences between the water retention parameters of soil-gravel mixtures with different gravel contents (Table 4). The air-entry values ($1/\alpha$) of the soil-gravel mixtures decreased gradually with increasing gravel content when it was lower than 55%; the reason may be that the amount of coarse pores increased with gravel content (Although the gravels were packed by hand to increase uniformity, there was still some pores in the soil-gravel mixtures because these gravels possibly overlapped each other and functioned as skeleton). However, when the gravel content reached 65%, especially for the soil-gravel mixtures containing the larger gravels (5–10 mm), the air-entry values increased slightly. It was further noted that the stronger the weathering degree of rock fragments in the soil, the higher the air-entry values, at least in the range of the higher coarse fragment contents for soil-gravel mixtures. Apparently, the shale clasts with smaller sizes had more void characteristics similar to fine soil medium than the pebbles. This observation was confirmed by the fact that the θ_s of clay loam-shale clast mixtures (CL+S) were larger at any gravel content than those of the clay loam-pebble mixtures (CL+P), while this effect was probably aggravated by the side-effect (side conditions in the container) in the experiment. It is reasonable to speculate that the measured θ_s of the soil-gravel mixtures decreased linearly with the increase of rock fragment contents because the majority of water was held by the fine soil of soil-gravel mixtures. The variation of θ_r in the soil-gravel mixtures in this research differed from that reported by Indrawan et al. [40],

according to whom the residual water content decreased with increasingly coarse fragments.

The WRCs for fine soil, gravel, and soil-gravel mixtures with different gavel contents (15%, 35%, and 65%) are presented in Fig. 1. This figure only displays the results of soil-gravel mixtures with 3–5 mm gravel because of the similarity among the different size groups. It can be seen that the WRCs of the soil-gravel mixtures are located between the WRCs of the pure soil and those of the rock fragment media. According to the fitting curves of VG-function, the slopes of the WRCs for the soil-gravel mixtures increased with the coarse grain contents at a low suction range (0–100 cm). That suggested an increase in the rate of the soil water volume changes with respect to the matric suction changes ($\partial\theta/\partial h$). However, the trend to increase in slopes of the WRCs for the soil-gravel mixtures became less significant with increasing rock fragment contents at the high suction range (100–1000 cm), and became opposite at a higher suction range (>1000 cm), even lower than the slope of WRCs of the fine soils, especially for the soil-pebble mixtures (see CL+P in Fig. 1). This indicated that a soil containing a high amount of rock fragments at lower water contents releases less water than the others, which might impair plant growing.

Direct calculation of the saturated water content of soil-gravel mixtures

Knowing the saturated water content is a prerequisite for calculating the effective degree of saturation. At or near saturation, the moisture of the soil-gravel mixtures equals the sum of water amount hold by the fine soil and that by the rock fragments. As a result, the water content of the soil-gravel mixtures could be predicted according to the saturated water content of the fine soil and the rock fragments, respectively, as well as their content in the soil-gravel mixtures. The formula expressing the saturated water content of soil-gravel mixtures is given in Eq. (3):

$$\theta_s^b = \theta_s^f(1 - M_r)\rho_b/\rho_f + \theta_s^r M_r\rho_b/\rho_r \tag{3}$$

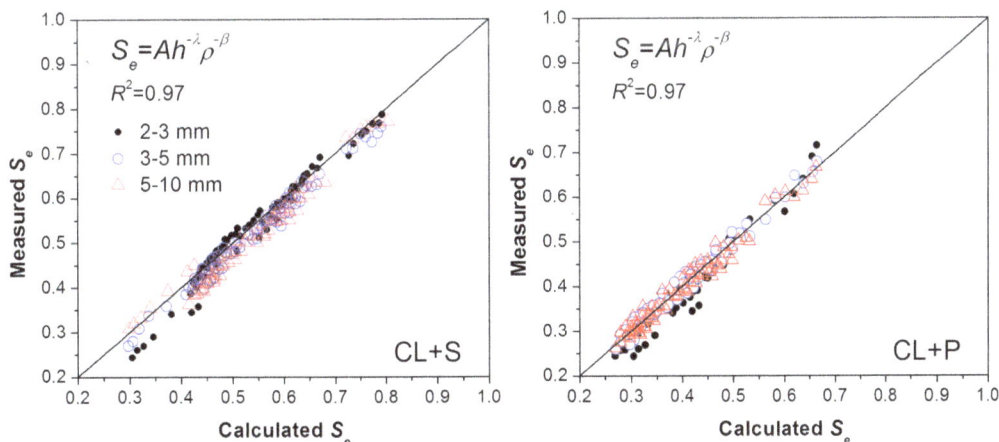

Figure 4. The measured and fitted data calculated by the trade-off parameters of power surface model with the variable of samples bulk density without considering gravel size change.

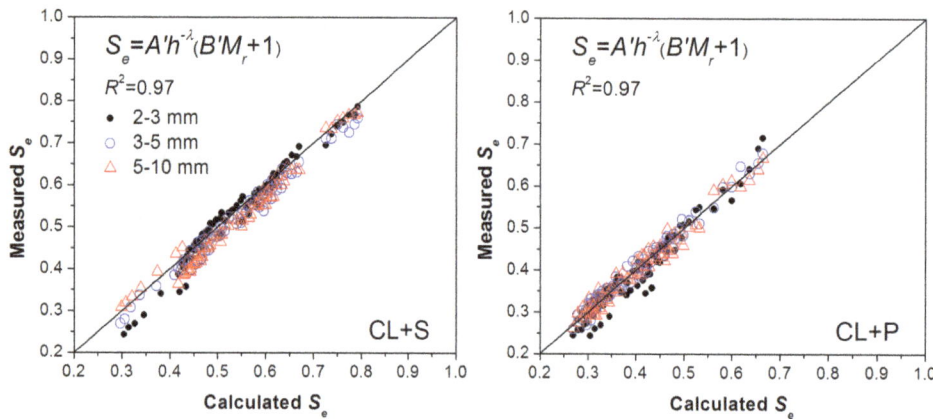

Figure 5. The measured and fitted data calculated by the trade-off parameters of linear-power surface model with the variable of gravel mass content without considering gravel size change.

where θ_s^b, θ_s^f, and θ_s^r are saturated volumetric water content of the soil-gravel mixture, fine soil, and gravel, respectively ($cm^3\ cm^{-3}$); M_r is the gravimetric gravel content ($g\ g^{-1}$); ρ_b, ρ_r, and ρ_f is the bulk density of the soil-gravel mixture, fine soil, and gravel, respectively ($g\ cm^{-3}$); and both superscripts and subscripts refer to bulk sample (soil-gravel mixture) (*b*), fine soil alone (*f*), rock fragments (*r*), and at saturated state (*s*), respectively.

Reasonably good results were obtained using Eq. (3) to predict the saturated water contents of all the samples, even those containing the weathered rock fragments. The values resulting from Eq. (3) and those actually measured were almost identical for the soil-gravel mixtures with all sizes of gravels (see Fig. 2); a small deviation between the measured water contents and the predicted values for CL+P with relative low gravel content due to measured error was recorded (see CL+P in Fig. 2). Of course, when the θ_s^r value of the pebbles was assumed to be zero (i.e., Eq. (3) was the same as Bouwer-Rice equation) [20], the predicted results for CL+P with relative low gravel contents could not be improved.

Revised formulas for water retention processes of soil-gravel mixtures

BC-function with simple power equation, introduces a well-defined air-entry value, which is associated with a largest pore-size

through the relation of Young Laplace, assuming complete wettability [41]. In addition, the parameter α in BC-function could reflect the variation of air-entry value of varied soil-gravel samples even lack data of wet end of WRC. Due to these reasons, BC-function was selected to be the revised model for soil-gravel mixtures in this study. To examine the moisture content of the unsaturated soil-gravel mixtures, we generated water retention curved surfaces (curve surfaces which reflected the water retention variation of soil-gravel mixtures) by adding one variable to BC-function, and then fitted the empirical parameters of the curved surface based on the measured WRCs data of the soil-gravel mixtures. We found that the effective saturation (S_e) of the soil-gravel mixtures at any soil potential had a reasonable power or linear correlation with the gravimetric gravel content (M_r) or bulk density of the soil-gravel mixtures (ρ_b). Thus M_r and ρ_b were added respectively as further variables of BC-function and the revised water retention functions were established. Eq. (4) and (5) express power relations:

$$S_e = (\alpha h)^{-\lambda} M_r^{-\beta} = A_M h^{-\lambda} M_r^{-\beta} \quad (\alpha h > 1) \qquad (4)$$

$$S_e = (\alpha h)^{-\lambda} \rho_b^{-\beta} = A_\rho h^{-\lambda} \rho_b^{-\beta'} \quad (\alpha h > 1) \qquad (5)$$

Table 7. Regression empirical parameters obtained by fitting two measured water retention curves of soil mixtures with 10% and 65% gravel content.

Samples	$S_e = Ah^{-\lambda}\rho^{-\beta}$						$S_e = A'h^{-\lambda}(B'M_r+1)$					
	A	λ	β	R2	SEE	REMS	A'	λ	B'	R2	SEE	REMS
CL+S (2–3)	1.68	0.12	0.74	0.98	0.01	0.04	1.41	0.12	0.26	0.98	0.01	0.04
CL+S (3–5)	1.46	0.12	0.33	0.98	0.01	0.04	1.35	0.12	0.13	0.99	0.01	0.03
CL+S (5–10)	1.65	0.14	0.37	0.98	0.01	0.05	1.51	0.14	0.15	0.98	0.01	0.05
CL+S (2–10)	1.59	0.13	0.48	0.96	0.02	0.18	1.42	0.13	0.18	0.96	0.02	0.18
CL+P (2–3)	1.92	0.17	0.73	0.98	0.01	0.05	1.61	0.17	0.40	0.98	0.01	0.05
CL+P (3–5)	1.63	0.16	0.50	0.98	0.01	0.04	1.44	0.16	0.28	0.98	0.01	0.04
CL+P (5–10)	1.59	0.18	0.30	0.98	0.01	0.10	1.48	0.18	0.18	0.98	0.01	0.04
CL+P (2–10)	1.72	0.17	0.51	0.97	0.02	0.16	1.51	0.17	0.29	0.96	0.02	0.17

R2, SEE, and REMS stand for coefficient of determination, estimated standard error, and square root of the residual sum of squares, respectively.

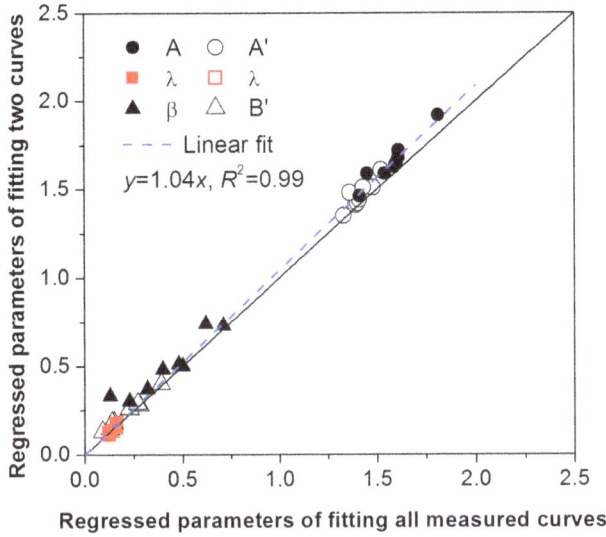

Figure 6. Comparison between the regressed parameters by fitting two and all measured water retention curves of soil-gravel mixtures.

where A_M, A_ρ, β, β', and λ are empirical parameters. Eq. (6) and (7) express power-linear relations.

$$S_e = (\alpha h)^{-\lambda}(BM_r + C) = Ah^{-\lambda}(BM_r + C) \qquad (6)$$

$$S_e = (\alpha h)^{-\lambda}(B\rho_b + C) = Ah^{-\lambda}(B\rho_b + C) \qquad (7)$$

When Eq. (6) and Eq. (7) were simplified to decrease the number of the parameters, the water retention curved surfaces were expressed as Eq. (8) and Eq. (9):

$$S_e = A'_M h^{-\lambda}(1 - B'_M M_r) \qquad (8)$$

$$S_e = A'_\rho h^{-\lambda}(1 - B'_\rho \rho_b) \qquad (9)$$

where A'_M, A'_ρ, B'_M, and B'_ρ are empirical parameters.

The water content of the soil-gravel mixtures can be calculated according to the revised water retention functions and the definition of S_e. By taking both Eq. (5) and Eq. (8) as examples, the water content of the soil-gravel mixtures can be calculated by the formulas listed as follows.

$$\theta^b = Ah^{-\lambda}\rho^{-\beta}(\theta_s^b - \theta_r^b) + \theta_r^b \qquad (10)$$

$$\theta^b = A'h^{-\lambda}(1 - BM_r)(\theta_s^b - \theta_r^b) + \theta_r^b \qquad (11)$$

The saturated water content of soil-gravel mixtures (θ_s^b) in Eq. (10) and Eq. (11) could be obtained by direct measurement in the laboratory, while it can be calculated indirectly using Eq. (3) above if pertinent detailed information is known, such as gravel content,

sample bulk density, gravel density, saturated water content of fine soil, and water content of gravels.

Parameter determination for the proposed revised formulas through nonlinear regression for the soil-gravel mixtures

Two variables (Mr and ρb) which presented a linear or power relation with Se were added to BC-function. In order to test if those revised water retention functions were practical and applicable for obtaining the WRCs parameters of the soil-gravel mixtures, we used the measured data to fit the revised formulas and obtain the parameters of curved surfaces by a nonlinear regression.

Parameter determination for separate classes of the gravel sizes. The fitting results are given in Table 5 and Table 6. It is shown that the values of coefficient of determination ($R2$), estimated standard error (SEE), and square root of the residual sum of squares ($RSME$), which all reflect the nonlinear regression matching level, indicate that the fitting results were in considerable agreement for both the power function and the linear-power function curved surfaces with either selected variable (M_r and ρb). Though the values of $R2$ and SEE for the different curved surfaces of the various soil-gravel mixtures were similar (Table 5 and Table 6), the difference of $Norm$ indicated that the power function with the variable of sample bulk density (Eq. (5)) and linear-power function with the variable of gravimetric gravel content gave a better fitting (Eq. (8)). The regressed curved surfaces of power function with the variable "bulk density" taken as an example here are drawn in Fig. 3 showing the regression results based on the measured data. The two different types of revised formulas, with the variables bulk density or gravel content, provided options for spatial heterogeneous research of regional soil hydrology according to the types of the practical measured data. The above formulas and the empirical parameters listed in Table 5 and Table 6 concur to simplify the determination of variability in water retention properties for soil-gravel mixtures with highly variable gravel content in soilscape.

Model effectiveness and parameter determination for soil-gravel mixtures with distribution of the three classes of gravel sizes. Table 5 and Table 6 show that the empirical parameters for various soil rock fragments mixtures vary with gravel type and size. The shape parameter λ for soil mixtures containing the same type of gravels with three size groups showed little difference, while by comparing the λ values between the two kinds of gravels it was observed that the stronger the weathering degree of the rock fragments in the soil-gravel mixtures, the smaller the λ values. The shape parameter β or B in the power function or linear-power function, respectively, was decreasing with increasing gravel sizes, while the differences in β or B between different size groups were not significant. Therefore, it seemed reasonable to consider obtaining the "trade-off" and appropriate parameters for the curved surfaces to account for the water retention properties of soil-gravel mixture with a large range of gravel size (2–10 mm). The fitting parameters, obtained by fitting the measured data of soil-gravel mixtures within all size groups (2–3 mm, 3–5 mm, and 5–10 mm) for the same kind of gravels, are given in Table 5 and Table 6. The fitting results of the soil-gravel mixtures containing the large range of the gravel sizes from 2 to 10 mm were also rather satisfactory since the $R2$ values of the two typical soil-gravel mixtures were both 0.97. The match effect between measured and calculated data is presented respectively in Fig. 4 and Fig. 5 using the trade-off parameters according to Eq. (5) and Eq. (8). Both Fig. 4 and Fig. 5 show that there was a good match between the measured and simulated data. In addition, the two kinds of the revised WRCs formulas (Eq. (5) and Eq. (8)) had almost the same fitting level. The well fitted results for the soil-gravel mixtures with the gravel sizes of 2–10 mm lead to conclude that the effect of gravel sizes was much lower than that of the gravel contents and types. This conclusion gives support to the effectiveness of revised models for large scale field investigation of the soil-gravel mixtures with highly variable gravel size.

Parameter determination with two data sets of representative gravel contents. As shown above, the water characteristic properties of the soil-gravel mixtures can be well determined based on the seven WRCs data sets corresponding to the seven different gravel mass contents (10%, 15%, 25%, 35%, 45%, 55%, and 65%) by fitting the regression surfaces, but the WRCs measurement work for seven gravel mass contents might be time-consuming. It was explored consequently if two measured water retention curves of soil mixtures with minimum and maximum gravel contents (10% and 65%) could represent well all the measured data within the whole range of gravel contents (10%–65%) in determining the parameter of revised water retention model. To examine that, the measured water retention curves of the soil-gravel mixtures with 10% and 65% gravel contents were used to fit Eq. (5) and Eq. (8), respectively, by the nonlinear regression. The regression results are reported in Table 7 showing that high $R2$ values were achieved. Importantly, the values of the empirical fitting parameters from the two measured curves were rather close to those from the whole measured curves, which indicated that the curved surface obtained based on the two measured curves were acceptable. The comparison between fitting parameters calculated from the two data sets and those from the seven data sets is presented in Fig. 6. The parameters of A and A' for Eq. (5) and Eq. (8) obtained from fitting the two curves were slightly larger than those from fitting the whole measured curves. However, the small deviation did not bring about a considerable difference between the two groups of the curve surface parameters. Therefore the method of determining the parameters of the revised water retention model based on the measured data from two representative gravel contents can be recommended, probably being more practical for the mixtures with embedding gravels in fine soil and for poorly sorted sediments with small size gravels in regional or large scale field investigation.

Conclusion

Based on this investigation it can be seen that the saturated water contents of the soil-gravel mixtures can be directly calculated from the amount and saturated water contents of both fine soil and rock fragments. The experimental results achieved from the different samples containing typical rock fragments with various sizes (2–10 mm) confirmed that the given equation was applicable for computing the water content of the soil-gravel mixtures. Revised formulas for water retention processes of the soil-gravel mixtures were proposed. Revised water retention models were developed by adding one variable (gravimetric gravel content or bulk density of soil-gravel mixtures that reflect the change in rock fragment contents) which had linear or power relation with the effective saturation of soil-gravel mixture to BC-function forming curved surfaces. Furthermore, the laboratory results indicated that the revised water retention models calibrated either by using the experimental data from the soil-gravel mixtures with a small range of gravel sizes or with a large range of gravel sizes, would be acceptable. However, determination and application of the regression surface model based on the data sets from a few typical different gravel sizes with two representative gravel contents would be more practical. Finally, it seems that the revised water retention models, the procedure of calculation parameters and the specific parameter values of the curved surfaces obtained in this study may be applied to different kinds of soil-gravel mixtures, because the whole analyses in this study were based on the measured water retention in soil-gravel mixtures containing two different types of rock fragments. However, their applicability needs to be validated through the water retention data from soils containing different rock fragments.

Acknowledgments

We give our thanks to Prof. Dr. Vito Sardo (University of Catania, Italy) and anonymous reviewer for their invaluable comments on improving the manuscript.

Author Contributions

Conceived and designed the experiments: BX HW MW MS. Performed the experiments: BX HW. Analyzed the data: HW. Contributed reagents/materials/analysis tools: HW. Wrote the paper: BX HW MW.

References

1. Phillips JD, Luckow K, Marion DA, Adams KR (2005) Rock fragment distributions and regolith evolution in the Ouachita Mountains, Arkansas, USA. Earth Surf Proc Land 30: 429–442.

2. Poesen J, Lavee H (1994) Rock fragments in top soils: Significance and processes. Catena 23: 1–28.

3. Cousin I, Nicoullaud B, Coutadeur C (2003) Influence of rock fragments on the water retention and water percolation in a calcareous soil. Catena 53: 97–114.

4. Cuniglio R, Corti G, Agnelli A (2009) Rock fragments evolution and nutrients release in vineyard soils developed on a thinly layered limestone (Tuscany, Italy). Geoderma 148: 375–383.

5. Sauer TJ, Logsdon SD (2002) Hydraulic and physical properties of stony soils in a small watershed. Soil Sci Soc Am J 66: 1947–1956.

6. Scheinost AC, Sinowski W, Auerswald K (1997) Regionalization of soil water retention curves in a highly variable soilscape, I. Developing a new pedotransfer function. Geoderma 78: 129–143.

7. Vereecken H, Kasteel R, Vanderborght J, Harter T (2007) Upscaling hydraulic properties and soil water flow processes in heterogeneous soils. Vadose Zone J 6: 1–28.

8. Patil N, Pal D, Mandal C, Mandal D (2012) Soil water retention characteristics of vertisols and pedotransfer functions based on nearest neighbor and neural networks approaches to estimate AWC. J Irrig and Drain E-ASCE 138: 177–184.

9. Nielsen D, Van Genuchten MT, Biggar J (1986) Water flow and solute transport processes in the unsaturated zone. Water Resour Res 22: 89–108.

10. Pachepsky Y, Rawls WJ, Giménez D (2001) Comparison of soil water retention at field and laboratory scales. Soil Sci Soc Am J 65: 460–462.

11. Margesin R, Schinner F, Wilke B-M (2005) Determination of chemical and physical soil properties, monitoring and assessing soil bioremediation. Springer Berlin Heidelberg. 47–95.

12. Flury M, Bittelli M (2009) Errors in water retention curves determined with pressure plates. Soil Sci Soc Am J 73: 1453–1460.

13. Smagin A (2012) Column-centrifugation method for determining water retention curves of soils and disperse sediments. Eurasian Soil Sci 45: 416–422.

14. Ravina I, Magier J (1984) Hydraulic conductivity and water retention of clay soils containing coarse fragments. Soil Science Society of America Journal 48: 736–740.

15. Ma DH, Shao MA (2008) Simulating infiltration into stony soils with a dual-porosity model. Eur J Soil Sci 59: 950–959.

16. Brouwer J, Anderson H (2000) Water Holding Capacity of Ironstone Gravel in a Typic Plinthoxeralf in Southeast Australia Soil Science Society of America Journal 64: 1603–1608.

17. Hanson CT, Blevins RL (1979) Soil water in coarse fragments. Soil Sci Soc Am J 43: 819–820.

18. Fiès JC, Louvigny NDE, Chanzy A (2002) The role of stones in soil water retention. European Journal of Soil Science 53: 95–104.

19. Gardner WH (1986) Water content. In: A K, editor. Methods of Soil Analysis, Part 1: Physical and Mineralogical Methods. Madison, WI, USA: American Society of Agronomy, Soil Science Society of America. 493–544.

20. Bouwer H, Rice RC (1984) Hydraulic properties of stony vadose zones. Ground Water 22: 696–705.

21. Klute A (1986) Methods of soil analysis. Part 1. Physical and mineralogical methods: American Society of Agronomy, Inc.

22. Dann R, Close M, Flintoft M, Hector R, Barlow H, et al. (2009) Characterization and estimation of hydraulic properties in an alluvial gravel vadose zone. Vadose Zone J 8: 651–663.

23. Khaleel R, Relyea JF (1997) Correcting laboratory-measured moisture retention data for gravels. Water Resour Res 33: 1875–1878.

24. Bouma J (1989) Using soil survey data for quantitative land evaluation. In: Advance in soil science, BA Stewart (ed) Springer-Verlag New York Vol 9 (1989) 225–239.

25. Minasny B, McBratney AB, Bristow KL (1999) Comparison of different approaches to the development of pedotransfer functions for water-retention curves. Geoderma 93: 225–253.

26. Ronsyn J, Van Meirvenne M, Hartmann R, Cornelis WM (2001) Evaluation of pedotransfer functions for predicting the soil moisture retention curve. Soil Sci Soc Am J 65: 638–648.

27. Puhlmann H, von Wilpert K (2012) Pedotransfer functions for water retention and unsaturated hydraulic conductivity of forest soils. J Plant Nutr Soil Sc 175: 221–235.

28. Cornelis WM, Ronsyn J, Van Meirvenne M, Hartmann R (2001) Evaluation of pedotransfer functions for predicting the soil moisture retention curve. Soil Sci Soc Am J 65: 638–648.

29. Brooks RH, Corey AT (1964) Hydraulic properties of porous media. Hydrology Papers, Colorado State University.

30. van Genuchten MT (1980) A closed-form equation for predicting the hydraulic conductivity of unsaturated soils. Soil Sci Soc Am J 44: 892–898.

31. Das BM (2010) Geotechnical Engineering Handbook: J. Ross Publishing.

32. Reatto A, da Silva EM, Bruand A, Martins ES, Lima JEFW (2008) Validity of the centrifuge method for determining the water retention properties of tropical soils. Soil Sci Soc Am J 72: 1547–1553.

33. Li YS (1981) Centrifuge method for determing soil water suction. Soils 4: 143–146 (in Chinese).

34. Khanzode RM, Vanapalli SK, Fredlund DG (2002) Measurement of soil-water characteristic curves for fine-grained soils using a small-scale centrifuge. Can Geotech J 39: 1209–1217.

35. van Genuchten MT (1987) Documented Computer Programs, Part 1 & 2. Riverside, California: U.S. Salinity Laboratory, USDA-ARS.

36. Marquardt DW (1963) An algorithm for least-squares estimation of nonlinear parameters. Journal of the Society for Industrial and Applied Mathematics 11: 431–441.

37. Mace A, Rudolph DL, Kachanoski RG (1998) Suitability of Parametric Models to Describe the Hydraulic Properties of an Unsaturated Coarse Sand and Gravel. Ground Water 36: 465–475.

38. Vogel T, van Genuchten MT, Cislerova M (2000) Effect of the shape of the soil hydraulic functions near saturation on variably-saturated flow predictions. Advances in Water Resources 24: 133–144.

39. Omuto CT (2009) Biexponential model for water retention characteristics. Geoderma 149: 235–242.

40. Indrawan IGB, Rahardjo H, Leong EC (2006) Effects of coarse-grained materials on properties of residual soil. Eng Geol 82: 154–164.

41. Ippisch O, Vogel HJ, Bastian P (2006) Validity limits for the van Genuchten–Mualem model and implications for parameter estimation and numerical simulation. Advances in Water Resources 29: 1780–1789.

23

The Campanian Ignimbrite Eruption: New Data on Volcanic Ash Dispersal and Its Potential Impact on Human Evolution

Kathryn E. Fitzsimmons[1]*, Ulrich Hambach[2], Daniel Veres[3,4], Radu Iovita[1,5]

1 Department of Human Evolution, Max Planck Institute for Evolutionary Anthropology, Leipzig, Germany, **2** Chair of Geomorphology, Laboratory for Palaeo- and Enviro-Magnetism, University of Bayreuth, Bayreuth, Germany, **3** Institute of Speleology, Romanian Academy, Cluj-Napoca, Romania, **4** Faculty of Environmental Sciences, Babes-Bolyai University, Cluj-Napoca, Romania, **5** MONREPOS Archaeological Research Centre and Museum for Human Behavioural Evolution, RGZM, Neuwied, Germany

Abstract

The Campanian Ignimbrite (CI) volcanic eruption was the most explosive in Europe in the last 200,000 years. The event coincided with the onset of an extremely cold climatic phase known as Heinrich Event 4 (HE4) approximately 40,000 years ago. Their combined effect may have exacerbated the severity of the climate through positive feedbacks across Europe and possibly globally. The CI event is of particular interest not only to investigate the role of volcanism on climate forcing and palaeoenvironments, but also because its timing coincides with the arrival into Europe of anatomically modern humans, the demise of Neanderthals, and an associated major shift in lithic technology. At this stage, however, the degree of interaction between these factors is poorly known, based on fragmentary and widely dispersed data points. In this study we provide important new data from Eastern Europe which indicate that the magnitude of the CI eruption and impact of associated distal ash (tephra) deposits may have been substantially greater than existing models suggest. The scale of the eruption is modelled by tephra distribution and thickness, supported by local data points. CI ashfall extends as far as the Russian Plain, Eastern Mediterranean and northern Africa. However, modelling input is limited by very few data points in Eastern Europe. Here we investigate an unexpectedly thick CI tephra deposit in the southeast Romanian loess steppe, positively identified using geochemical and geochronological analyses. We establish the tephra as a widespread primary deposit, which blanketed the topography both thickly and rapidly, with potentially catastrophic impacts on local ecosystems. Our discovery not only highlights the need to reassess models for the magnitude of the eruption and its role in climatic transition, but also suggests that it may have substantially influenced hominin population and subsistence dynamics in a region strategic for human migration into Europe.

Editor: John P. Hart, New York State Museum, United States of America

Funding: This work was funded by the Max Planck Society through the project "Lower Danube Survey for Palaeolithic Sites". D. Veres acknowledges the support of CNCS-UEFISCDI through grant PN-II-RU-TE-2011-3-0062, contract no. 73/05.10.2011. The funders had no role in study design, data collection and analysis, decision to publish, or preparation of the manuscript.

Competing Interests: The authors have declared that no competing interests exist.

* E-mail: kathryn_fitzsimmons@eva.mpg.de

Introduction

The Phlegrean Fields super-eruption and caldera collapse that produced the Campanian Ignimbrite (CI)/Y-5 tephra, which took place 39.28 ± 0.11 ka [1], was one of the most explosive eruptions affecting Europe in the late Pleistocene [2], in terms of both eruption magnitude and volume of volcanic ejecta. The distal ashfall of CI tephra was widely distributed from its source vent near Naples in southern Italy [3,4], eastward over the Balkans [5] and Black Sea [6] to the Russian plain more than 2200 km distant [7–9], and over 1000 km southward to the north African coast [10–12]. The CI tephra thereby provides a powerful chronostrati-graphic marker horizon for palaeoclimatic and archaeological records during Marine Isotope Stage (MIS) 3. Moreover, the timing of this eruption coincides with the onset of the cold, dry climatic phase Heinrich Event 4 (HE4) [13,14]. It has been proposed that the conjunction of these two events may have triggered a positive feedback cycle affecting global, and particularly European, climates for hundreds or even thousands of years

[11,15]. Significantly, the timing of the eruption, and of the extreme environmental conditions during HE4, also coincides with significant changes in the archaeological record in Europe; specifically, the arrival of anatomically modern humans (AMHs) [16–18], a substantial shift in hominin lithic technology [18–20], and the disappearance of Neanderthals from the continent [21,22]. However, the role of the CI super-eruption in the interaction between sudden climatic change, the demise of the Neanderthals, and their replacement by AMHs, remains a matter of hypothesis [10,15,23].

The CI super-eruption took place in the Phlegrean fields of southern Italy [2], and has been dated based on a composite $^{40}Ar/^{39}Ar$ age of 39.28 ± 0.11 ka obtained from proximal ignim-britic deposits [1]. The volcanic cataclysm involved a two-step eruption, consisting of an initial phase with a volcanic column at least 40 km high, followed by collapse and creation of the caldera, a rejuvenated volcanic column, and widespread ignimbritic deposition extending at least 1500 km^2 from the eruption point [9,24] with sufficient force to ascend the surrounding topography to over 1000 m altitude [25]. A peak in sulphate concentration within the GISP2 Greenland ice core, only slightly lower than the Icelandic Z2 ash or Toba eruptions, was correlated with the CI eruption [26]. However, subsequent investigations failed to identify tephra shards at the same depth within the core or a comparable peak within other Greenland ice core records, urging caution when such correlations are proposed based on only geochemical proxies [27,28].

Interaction of the volcanic column with high altitude wind currents transported finer-grained volcanic particles (<250 µm) northeastward and southward, as far as the Russian Plain, eastern Mediterranean, Black Sea and north Africa (Figure 1a) [3,6,12]. Recent modeling of tephra transport [12], based on measured tephra thickness from several sites in the Balkans, Russian Plain and from eastern Mediterranean sea cores, suggested that the magnitude of the eruption was more than twice that of previous estimates. The modeling study also predicted the thickness of ash cover for regions beneath the ash plume for which data previously did not exist. Depending on the model parameters, the thickness of tephra deposited across the Balkans proximal to the Adriatic Sea should average 5–10 cm, decreasing to 2–5 cm in eastern Europe (Romania, Moldova, southern Ukraine), with the plume tapering northeastwards over the Russian Plain (Figure 1a) [12]. However, the model output, while providing a useful estimate of the magnitude of the eruption, contains no data points spanning the 1500 km between the Balkan sites and the Don River on the Russian Plain (Figure 1b).

Significant thick deposits corresponding to the CI tephra have recently been identified in southern Romania [5,29–31], and could provide important information for better constraining the magnitude of the eruption and its likely environmental impact. Moreover, elucidating the nature and spatial extent of the tephra fallout, as well as its likely forcing on regional climates, is also important for assessing the regional impact on hominin populations and their resilience during this critical period of time. In this paper we provide significant new data from the loess steppe environments of Dobrogea in southeastern Romania, a narrow land corridor between the Danube River and Black Sea. Our data not only expand on the critical mass of new data points from a region previously unaccounted for in the models of tephra dispersal, but also indicates an average tephra thickness substantially greater than the models predict (Figure 1b). These observations could present major implications for the magnitude of the eruption and its role within climate and environmental feedbacks.

The timing of the CI super-eruption coincided with the onset of Heinrich Event 4 (HE4) [32,33], a short-lived stadial (Greenland Stadial 9, GS9) associated with enhanced ice-rafting in the North Atlantic Ocean and cold, dry conditions over Europe [34,35]. This was followed by a series of centennial- to millennial-scale warm interstadials (separated by stadials), of which Greenland Interstadial 8 (GIS8) [14] was the longest and most pronounced event, occurring at the transition from Middle to Upper Pleniglacial [36]. Climatic conditions across the northern hemisphere during HE4 were generally cool and dry [35] due to a southward shift of the polar front driven by the collapse of the thermohaline circulation in the northern oceans. Palynological records from the Lago Grande di Monticchio in southern Italy, not far from the Phlegrean Fields, suggest abrupt cooling and more arid conditions during HE4 immediately following CI tephra deposition [32,37,38]. Indications of cooler climates during HE4 within existing eastern European records are not as clearly defined as in the Italian and other Mediterranean palaeoclimate archives [39–41]. In this respect, the sedimentary deposits of the temperate eastern European loess steppe are particularly poorly understood [41,42]. Aside from the limitations of existing records to preserve evidence of HE4 intensity and its connection to the CI-super-eruption, the potential feedback link is difficult to establish beyond mere hypothesis given existing datasets, often of low analytical and consequently low temporal resolution [11]. However, comparable attempts to identify a causal link between volcanic winters and climatic change in the case of the earlier Toba super-eruption in Sumatra have so far proven similarly inconclusive [43–46], although recent chronological constraints have linked the Toba eruption with a substantial cooling event [46]. A better understanding both of the nature and intensity of climate change during HE4 across Europe, the temporal connection with the CI, and potential for establishment of positive climate feedbacks, is clearly necessary to establish the significance of the CI super-eruption within the global climate system.

The timing and impact of the CI super-eruption is of particular interest and relevance to human evolution in Europe, since it coincides with a substantial shift in the archaeological record, associated with the arrival of AMHs to the continent [16–18], and the demise of the Neanderthals [21,22]. Archaeological records within the CI ashfall zone presently provide an inconsistent picture of the eruption's influence on hominin occupation [10,47,48]. In southern Italy, the region closest to the volcanic cataclysm, both open air and rock shelter sites preserve a significant hiatus in occupation during the period immediately following the CI eruption [3,49,50]. This indicates widespread abandonment of habitation sites in the region for some hundreds, if not thousands, of years following the eruption [11,48]. The archaeological record from further afield is contradictory. Several rock shelters in the Balkans and northern Africa preserve CI tephra shards within their stratigraphy, yet interpretations of archaeological records within these sites postulate no significant disruption of the archaeological record and conclude that the CI super-eruption did not affect hominin populations there [10]. Conversely, the open air site of Kostenki 14 on the Russian Plain preserves an archaeological assemblage suggesting sudden catastrophic destruction of a human settlement [51], despite the fact that the tephra layers in that sequence appear to be redeposited [52]. Moreover, thick deposits of tephra within the Montenegran cave site of Crvena Stijena clearly divide the Middle and Upper Palaeolithic [53].

The localised impacts of large volcanic eruptions are not exclusively climatic. Chemical reactions between acidic volatile volcanic gases, and atmospheric and soil moisture, produce acid rain and soil acidification in ashfall zones, contamination of freshwater systems, and fluorosis of herbivores ingesting contaminated vegetation, with associated effects on humans dependent on affected ecosystems [54,55]. Therefore, in considering the impact of the CI super-eruption on human evolution within an ashfall zone, more than just the influence of climate comes into play. In this sense, accurate data on tephra distribution and thickness within the ashfall region becomes critically important.

The lower Danube River valley and its major tributaries in eastern Europe has long been proposed to represent one of the major migration routes for AMHs into Europe [17,56]. Yet this is precisely the region most likely to have been affected by the ecological impacts of the CI ashfall and related impacts on the regional ecosystem. Consequently, at this stage not only is it unclear what impact the CI eruption might have had on hominin occupation; the conjunction between hypothesized migration routes into Europe, potential interaction between hominin species, and tephra deposition may well have intensified this impact.

Map of Lower Danube region showing contrast between modelled (Costa et al. 2012) and observed tephra thickness (this study; Veres et al. 2013). *Inset*: Projected extent of >0.1 cm tephra thickness (Costa et al. 2012).

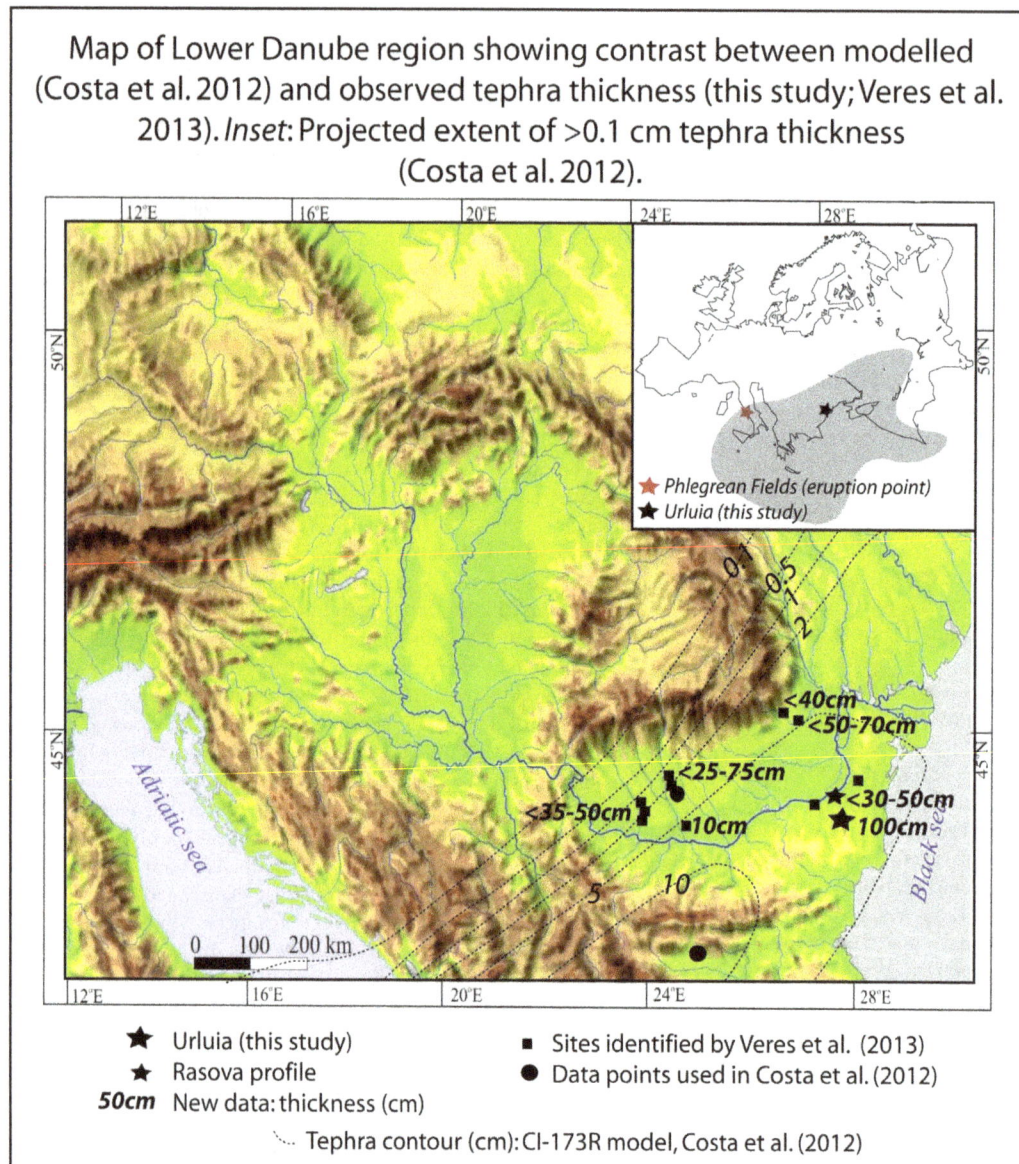

Figure 1. Map of the Lower Danube region showing the contrast between modelled [12] and observed thickness of the CI tephra (this study; [5]). Inset shows the projected extent of >0.1 cm tephra thickness, based on data from [12].

In this paper we present new data from a tephra deposit in the CI distal ashfall region of the lower Danube basin in southeast Romania, consolidated with recently published data confirming widespread CI tephra from other sites nearby [5]. Based on the characteristics of the deposit, we show that the CI tephra was deposited not only rapidly but also more thickly than predicted, 1200 km east of the eruption. We hereby propose that models of the magnitude of the eruption be reassessed using the new data points from this distal region for which data previously did not exist. Although we do not remodel the volcanic event within this paper, we do hypothesise that the eruption was substantially more explosive than previously estimated. We speculate on the potential impact of such rapid, thick deposition of ash on hominin populations in this region, and for the spatial variability of this impact, both climatically and directly on the ecosystem. Our hypotheses not only hold significant implications for human evolution within Europe in general, but also highlight the

complexity and potential vulnerability of hominin dispersals and occupations at varying scales across the continent at a critical time in human evolution.

Results

The Tephra Deposit at Urluia, Southeast Romania

The Quaternary-uplifted Dobrogea plateau, in southeastern Romania, preserves some of the thickest loess deposits in the lower Danube basin [41,57,58]. The loess is derived mostly from aeolian transport of fluvial silts associated with glaciers at the head of the Danube catchment [59–61], with minor components from the Saharan Desert to the south [62] and Russian Plain and Caspian Basin to the east [60,63]. Loess in Dobrogea typically occurs as plateau deposits, characterised by aeolian draping over low angle slopes. Multiple phases of intensified deposition and pedogenesis, typically associated with glacial and interglacial phases respectively, are preserved in the form of loess-paleosol sequences [41,64]. Sedimentary differentiation between paleosols and overlying loess, and in the case of this study, loess and tephra, enables low angle palaeotopography and loess palaeokarst to be clearly distinguished.

In this paper we investigate an especially thick deposit of CI tephra from the Urluia Quarry site, a substantial exposure of loess and limestone basement rocks located on the Dobrogea loess plateau (Figure 2a). Urluia Quarry lies approximately 15 km south of the Danube River, within a zone of particularly thick loess (Figure 2b). At this site, a thick (20+ m) sequence of loess-palaeosol packages overlies the uplifted Cretaceous-Tertiary-age limestone of the Dobrogea plateau [65]. The present land surface is predominantly horizontal, dipping more steeply on the western margins of the quarry into a small tributary of the Danube. Palaeotopography visible from the differentiation of stratigraphic units indicates the land surface to have changed slightly with time, most likely due both to tectonic activity and surface geomorphic processes. Karst phenomena such as dolines, infilled cave conduits and karst springs are also documented on a limited scale.

The tephra at Urluia occurs as a distinct, pale whitish, coarser-grained, fine sand-sized unit within the buff-coloured silty loess of the uppermost loess-paleosol package that represents the last full glacial cycle (Figure 3a). The palaeotopography of the tephra exposure suggests a gentle depression in the central part of the land surface, which was subsequently infilled by tephra and then by loess. Given that the upper limit of the last interglacial paleosol appears largely horizontal, the palaeodepression filled in by the

tephra may have developed after MIS 5. The thickness of the tephra ranges between ca. 0–100 cm, and is thinner on the upper, steeper parts of the palaeotopographic slope, and thickest towards the base of the depression (Figure 3a). The ash was analysed using geochemical and geochronological methods in order to determine its provenance and assess its depositional age. Sedimentological and environmental magnetic studies were used to establish the mode of deposition and the relative stratigraphic context.

Analysis and Provenance of the Tephra at Urluia

The tephra was investigated in detail at two separate locations along the exposed section ("northwest" and "southeast"; Figure 3b). The provenance of the tephra was fingerprinted based on the chemical analyses of two ash samples collected from the base and top of the unit at the northwest profile where the ash is thicker. Analyses of major oxide concentrations from isolated glass shards yield consistent phonolite/trachyte compositions (Table 1, Table S1 in File S1; Figure 4a) consistent with the CI glass shard chemical composition [4,8,66]. The major oxide concentrations measured from both the top and bottom of the tephra (up to 40 grains measured for each sample) are indistinguishable from one another, indicating correspondence with the same eruptive event. The SiO_2, K_2O, Na_2O, CaO and FeO compositions average 58.76–59.95 wt %, 7.10–7.53 wt %, 5.24–6.08 wt %, 1.77–1.83 wt % and 2.75–2.86 wt % respectively (Table 1). The geochemical composition correlates not only with proximal Plinian fall and pyroclastic flow deposits associated with the CI super-eruption in Italy [4,66] but also with distal CI tephra deposits from

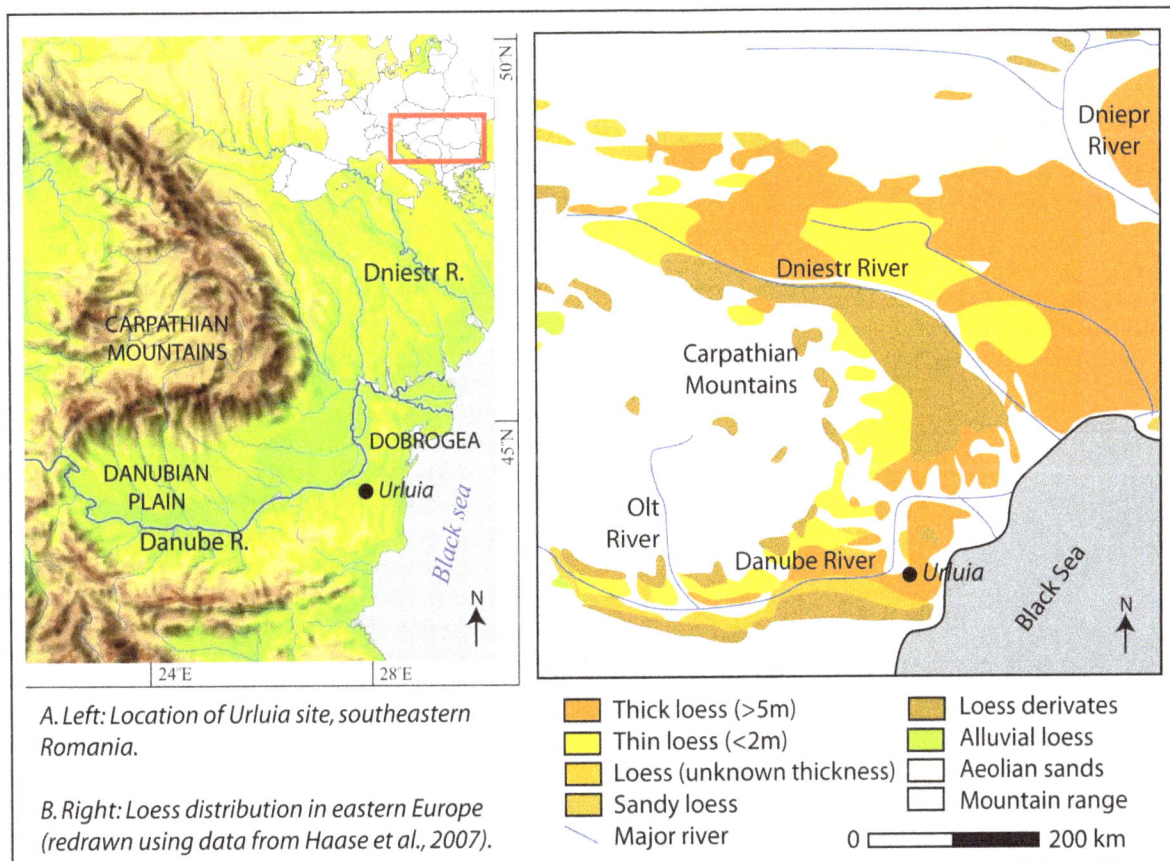

A. Left: Location of Urluia site, southeastern Romania.

B. Right: Loess distribution in eastern Europe (redrawn using data from Haase et al., 2007).

Figure 2. Location of Urluia loess deposit, southeastern Romania (a), and distribution and thickness of loess deposits in the lower Danube region (b; redrawn using data from from [57]).

A. Urluia loess profile, tephra deposit highlighted (looking NE)

B. Stratigraphy and luminescence chronology (looking NE)

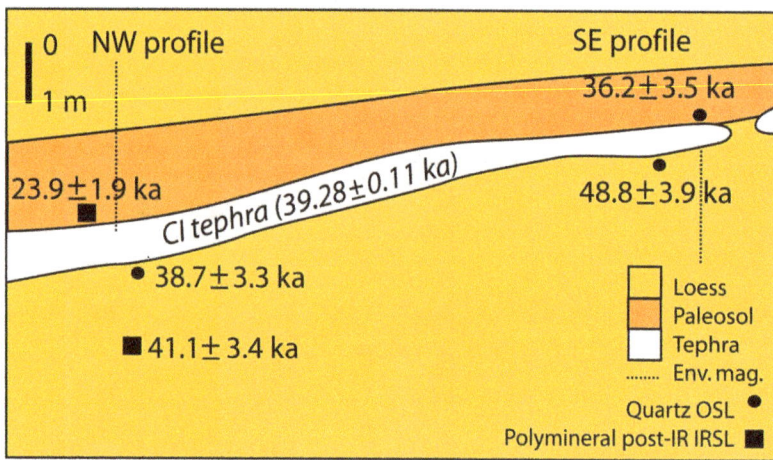

C. Magnetic susceptibility

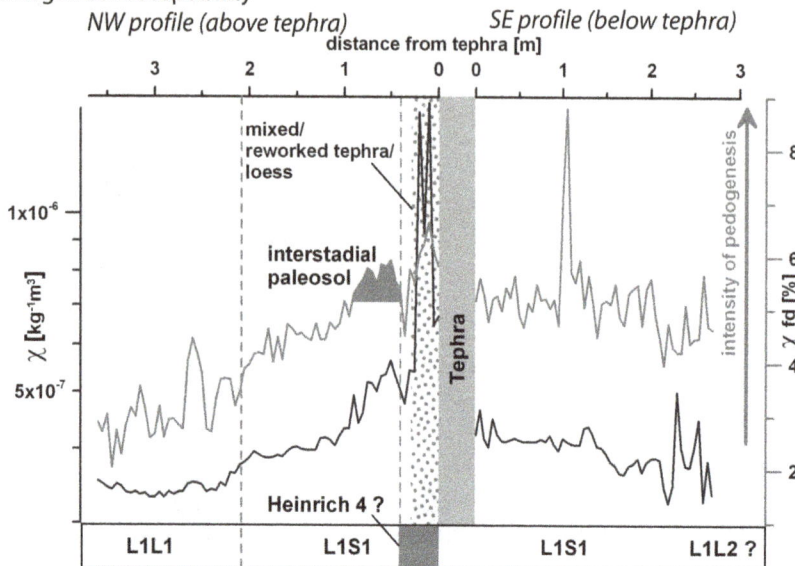

Figure 3. The tephra deposit at Urluia. a. Photograph of the site. b. Stratigraphic section, showing luminescence ages and methods used. c. Magnetic susceptibility of the northwest and southeast loess profiles, showing frequency dependent (χ_{fd}) and magnetic low field susceptibility (χ) as functions of stratigraphy below and above the tephra layer. Bulk χ reflects the concentration of the magnetisable fraction in sediments; χ_{fd} reflects the relative amount of pedogenetic and diagenetic neo-formation of ultrafine Fe-particles.

the Eastern Mediterranean and Russian Plain [8] the Crvena Stijena archaeological site in Montenegro [53], and from other CI tephra occurrences in Romania [5].

The age of the tephra deposit at Urluia was constrained using luminescence dating performed on tephra-bracketing samples collected from the surrounding loess, from the northwest and southeast profiles along the exposed section (Figure 3b). Three samples were collected from below the tephra, and two collected

from the overlying loess. The northwest profile yielded two ages below the tephra of 38.7 ± 3.3 ka and 41.1 ± 3.4 ka, in correct stratigraphic order, and an age from a paleosol horizon overlying the tephra of 23.9 ± 1.9 ka (Table 2). The age of the uppermost sample suggests an unconformable boundary with the underlying tephra, although additional chronological analyses are required to clarify these aspects. The southeast profile yielded an age for the loess underlying the tephra of 48.8 ± 3.9 ka, and for the overlying

Figure 4. Characteristics of the tephra at Urluia. a. Geochemical profile of the Urluia tephra compared with CI data from Italy, eastern Europe and Mediterranean Sea cores [4,8,53,66]. b. Photograph of the contact between the tephra and underlying loess deposits. c. Photomicrograph from scanning electron microscope (SEM) of tephra shards. d. Vegetation imprints at the basal contact of the tephra with the loess (looking upward), collected from the exposure at Urluia.

Table 1. Geochemical analyses of major and minor element concentrations for tephra samples URL1 and URL2.

Oxide	URL1 (base of tephra) Average ± σ (wt %)	URL2 (top of tephra) Average ± σ (wt %)
Na_2O	6.08±0.87	5.24±1.56
SiO_2	59.95±0.99	58.76±3.87
K_2O	7.10±0.61	7.53±2.04
CaO	1.77±0.24	1.83±0.47
FeO	2.86±0.13	2.75±0.73
MgO	0.38±0.10	0.42±0.19
Al_2O_3	18.48±0.21	18.61±2.65
P_2O_5	0.05±0.04	0.09±0.06
TiO_2	0.41±0.04	0.36±0.10
MnO	0.21±0.04	0.17±0.07
Cl^-	0.69±0.13	0.54±0.23

loess of 36.2±3.5 ka (Table 2). The significantly older age of the underlying loess compared with the tephra at this point may be accounted for by erosion of the loess prior to tephra deposition, although equally this age lies within 2σ of the other underlying sample and may be contemporaneous. All underlying ages, accounting for uncertainties, are older than the CI tephra. The two overlying ages are younger. These results corroborate the previously known age of proximal tephra in Italy of 39.28±0.11 ka [1], as well as those obtained for other deposits interbedded with the CI in Romania [30].

The geochemical and geochronological data therefore firmly establish the Urluia tephra as deriving from the CI super-eruption which occurred in the Phlegrean Fields of Italy.

Nature of CI Tephra Deposition

The tephra deposit at Urluia consists of a distinct unit deposited on a gentle palaeoslope, increasing in thickness downslope (Figure 3a). It can be traced for more than 40 m along the quarry wall; colluvial deposits and inaccessible steep slopes resulting from quarrying activities obscure the primary stratigraphy along the rest of the quarry wall. At the thickest point near the depocentre of the palaeotopographic depression, the tephra reaches 100 cm thickness. At the southeastern end of the exposure, the tephra changes from a distinct, thick, fresh pale ash deposit approximately 15 cm

in thickness to an intermittent exposure of blocks of more weathered orange-coloured tephra, the upper surface of which is orange in colour as a result of intensified in situ weathering. To the southeast, the tephra layer pinches out and can only be identified by gravel-sized orange-coloured clasts indicating its stratigraphic position. The distinction between the thicker, fresher tephra and thinner, more weathered deposits further upslope is most likely due to relatively prolonged exposure of the latter at the surface after deposition, and the higher vulnerability of thinner deposits to weathering. Although no screening has yet been performed, it is likely that the CI tephra might be present in cryptotephra form along the profile in sectors where the layer is not visible to the naked eye.

The tephra overlies pale, buff-coloured primary loess, and the contact between the two units is clearly defined (Figure 4b), suggesting a sudden transition. The contact between the tephra and overlying loess is more diffuse, indicating some degree of bioturbation and sediment mixing coeval with post-tephra loess deposition. The variation in thickness of this zone suggests the impact of overlying sedimentation on the comparatively less dense ash which created an irregular surface. At the northwest section, approximately mid-way along the palaeoslope, the tephra is ca. 50 cm thick and grades upwards from pure ash into a 25 cm thick zone comprising mixed tephra and loess. This layer is overlain by carbonate-rich loess, presumably representing HE4, which in turn is overlain by an interstadial weakly developed paleosol. This most intensely developed paleosol, containing humic components and carbonate rootlets, is ca. 1 m thick, and grades upwards through a transitional pedogenic zone into primary, light yellow loess a further 2.3 m above. This sequence may represent the Greenland Interstadials (GI) 8–5. The southeast section, where the sediments below the tephra were investigated, is interspersed with, and overlain by, weakly pedogenic overprinted loess (Figure 3b). The stratigraphic position, relatively weakly developed character and age of the overlying paleosol most likely correlates with the upper more intensively developed part of the L1S1 fossil soil complex identified in other loess profiles across the Danube basin [41,42]. The L1S1 paleosols have been interpreted to correspond to relatively humid MIS 3 interstadial conditions prevailing not only across the Middle to Lower Danube basin [41] but also wider Eastern Europe [42,67]. HE4 is situated within MIS 3 at the end of a series of relatively warm and humid interstadials characterizing early MIS 3 and prior to the Greenland Interstadials (GI) 8–5 representing the so-called Denekamp phase in the terrestrial palaeoclimate stratigraphy [14,68].

The tephra is also clearly distinguished as a discrete stratigraphic unit within the loess by environmental magnetism, along

Table 2. Luminescence dating data and age estimates for the CI tephra at the Urluia site.

Sample code	Depth (m)	D_e (Gy)	σ (%)	Attenuated dose rates (Gy/ka) β	γ	Cosmic	Total dose rate (Gy/ka)	Age (ka)
L-EVA1091	9.5±0.1	114±2	7	2.01±0.20	1.67±0.17	0.07±0.01	4.77±0.37	23.9±1.9
L-EVA1089	9.5±0.1	149±7	17	1.89±0.19	1.67±0.17	0.07±0.01	4.12±0.35	36.2±3.5
CI TEPHRA								
L-EVA1028	9.6±0.1	167±6	14	2.22±0.22	1.58±0.16	0.07±0.01	4.31±0.33	38.7±3.3
L-EVA1029	10.2±0.1	193±4	7	2.05±0.21	1.54±0.15	0.07±0.01	4.70±0.38	41.1±3.4
L-EVA1090	9.5±0.1	199±5	6	1.93±0.19	1.67±0.17	0.07±0.01	4.08±0.31	48.8±3.9

Samples analysed by OSL are shown in plain text; those analysed using the post-IR IRSL protocol are shown in italics.

with varying intensity of pedogenesis in the surrounding loess (Figure 3c). The tephra layer is sandwiched in between loess which shows incipient pedogensis and a weakly developed interstadial paleosol (maximum pedogenesis 0.5–1.0 m) above the tephra. The distinct peak in magnetic susceptibility can be visibly correlated with the overlying reworked tephra in the field; this layer yields χ-values almost one order of magnitude higher than the surrounding sediments. The weakly developed overlying interstadial paleosol can also be identified in the northwest section as a 2 m thick zone of generally increased magnetic susceptibility relative to the surrounding loess (Figure 3c; Table S5 in File S1); these χ-values are generally elevated and most likely mark the more favorable climate during MIS 3 (L1S1). The χ_{fd}-values follow this trend but yield a distinct peak approximately 1 m below the tephra which cannot presently be explained. There is some indication of sediment redeposition during a phase of higher dust input between the tephra and interstadial paleosol which may correlate with HE4, although the resolution of sampling for environmental magnetism unfortunately precludes more detailed palaeoenvironmental reconstruction of this time period. There is no indication either sedimentologically or magnetically of more intense pedogenesis having taken place within the loess prior to tephra deposition.

A number of features indicate that the tephra was deposited rapidly at this site. At the contact between the tephra and underlying loess, imprints of vegetation are preserved (Figure 4d), along with intact ichnofossils of borings made by invertebrates as they attempted to escape the rapid deposition of ash. The low density of the tephra was sufficient not to distort or compact these trace fossils. The tephra itself is relatively poorly sorted, ranging from 10–250 μm diameter, with a high proportion of fine sand-sized clasts and a small component of grains >150 μm (Figure 4c). This grainsize range is broadly finer than the deposit at Crvena Stijena in Montenegro (500 km from the source; [53]) but certainly coarser than at Kostenki on the Russian Plain (2200 km distant from source; [8]). The Urluia site, more than 1200 km from the Phlegrean Fields, lies towards the hypothesized upper limit of distal sand-sized transport [53,69], although without more detailed particle size analysis, further interpretations as to the magnitude of the eruption and associated suspended transport capacity of distal ash cannot be made.

Stratigraphic evidence at both micro and macro scales suggests minimal redeposition of the tephra at Urluia. Microlaminations and fine-scale cross-bedding are present within the unweathered tephra along the length of the exposure. While these features suggest some degree of transport during or just after the ashfall, the integrity of preservation of these sedimentary structures indicates that the bulk of this process occurred rapidly and prior to subsequent loess deposition. Likewise, the preservation of ichnofossils (traces of invertebrate burrowing and vegetation) strongly suggests rapid initial tephra deposition and minimal slumping over a short period of time.

In addition to previously documented exposures of the tephra on the Dobrogea Plateau and Danube Plain further north and west of the Urluia site [5,31], several additional exposures nearby Urluia reinforce our arguments for its previously unrecognised thickness and ubiquity. In particular, an exposure on the Danube River at Rasova to the north of Urluia preserves a horizontal layer of sand-sized tephra up to 50 cm thick (Figure 1b), indicating another significant primary occurrence of CI tephra in the region. Another tephra layer is also exposed in a small road-cut profile on the other side of Urluia village approximately 1 km to the northeast. Although geochemical analyses are not yet available for this ash bed, its visual and microscopic optical appearance is

identical to the Urluia tephra. It moreover occupies the same position within the loess-paleosol stratigraphy, suggesting that it derives from the same source. This latter ash bed is approximately 15 cm thick.

The stratigraphic and sedimentological evidence from the CI tephra deposit at Urluia, combined with emerging data from other deposits in the region ([5]; Figure 1b), strongly argue for widespread, thick, rapid sedimentation more than 1200 km from the source of the super-eruption.

Discussion and Conclusions

Implications for Reconstructing the Scale of the Eruption

The evidence from our study indicates rapid deposition of coarse-grained, distal tephra from the CI super-eruption more than 1200 km from its source.

The unusually thick deposit of CI tephra at Urluia Quarry in the Dobrogea region of southeastern Romania is at its thickest point up to twenty times thicker than is proposed from the computational model for this region (Figure 1; [12]). In addition, the exposure at Urluia corroborates recent observations of 12 additional substantial accumulations across the lower Danube basin (Figure 1b), of which at least four [5], described in detail, are up to five times thicker than proposed by the model. In our view, these data provide a critical mass of information sufficient to alter the computational models for the distal transport of tephra, and will most likely increase the volume of ash and magnitude of the eruption substantially. Accordingly, the climatic impacts of the CI super-eruption, its interaction with the HE4 episode, and implications for Palaeolithic communities living in this region especially, are likely to have been more extreme, and should be carefully reassessed.

Implications for Palaeoclimate and Human Evolution

There is some indication from both the stratigraphic and environmental magnetic profile at Urluia that the CI tephra deposition was immediately followed by a short-lived phase of loess deposition associated with stadial conditions, which is overlain by the interstadial paleosol (Figure 3c). This phase may be correlated with HE4, although at present the resolution of the record is insufficient to extract the intensity of this climate signal. However, if this interpretation is accurate, then it corroborates studies from southern Italy which place the timing of the CI super-eruption at the onset of HE4 [32,33]. Primary loess deposition at Urluia would indicate cooler, more arid conditions, which have been noted in records from central and western Europe at this time [32,37,38], but are as yet poorly defined in eastern Europe [41,42]. At present, however, the potential feedback link between the CI volcanic event and HE4 cannot be established beyond the hypothetical realm [11]. However, since most of southeastern and eastern Europe lies on the "fringe area" of the eruption's impact, should a positive feedback cycle have been established, the region, and the hominins and fauna living within it, would certainly have experienced the climatic deterioration caused by the coupling of HE4 with a volcanic winter [48].

The potential effects of the CI super-eruption and related ashfall on hominin populations are manifold. Our stratigraphic data from the more distal parts of Eastern Europe, combined with a now critical mass of additional tephra occurrences in various sedimentary settings [5,6,70–72], suggest that even such distal regions may have experienced direct and substantial effects of ash deposition. Among the known health hazards associated with volcanic ashes are a variety of respiratory acute (e.g., asthma) and chronic diseases (e.g., silicosis) caused by the inhalation of fine ash particles

[73]. The magnitude of these effects is directly related to the proportion of finer, respirable ash grains, implying that areas situated at different distances from the eruption should experience different localized effects, despite similar ashfall volumes [74]. A further effect of substantial ashfall is fluoride poisoning, which may affect not only people [75] but also fauna, particularly large herbivores [55]. Both modern [76] and archaeological [77] examples of fluoride poisoning have been documented. Fluorosis can lead to death, but it is the associated longer term debilitating bone deformations (e.g., hyperostosis, osteosclerosis, osteomalacia, and osteoporosis) that are most likely to be identifiable in the archaeological record [77].

Archaeological traces of the response of hunter-gatherer populations to the effects of volcanic eruptions have been studied in a variety of contexts [74,78–84], and the patterns differ markedly from case to case. The Toba super-eruption in Sumatra (ca. 74 ka; [46]) seems to have induced few changes in hunter-gatherer adaptations in, for example, the distal Jurreru Valley in India [82,83], even though ashfall there was also substantial. However, despite being one of the largest volcanic events during the late Pleistocene, Toba is perhaps not the most appropriate analogy for the effects of the CI super-eruption [48], since the equatorial location of the eruption result in tephra predominantly falling over the ocean would have induced different environmental impacts [48,85]; furthermore, the temporal resolution of the archaeological record there is low [48]. More relevant to our case is the Laacher See eruption in the Eifel region of Germany, an intraplate volcanic event which occurred at temperate latitudes ca. 12.9 ka ago, approximately two centuries prior to the Younger Dryas abrupt cooling event [86]. The Laacher See eruption appears to correlate with major technological changes, such as the abandonment of the bow and arrow [87]. Technological changes can also result from adaptation to decreased subsistence yields in the environment [88], which may be independent of volcanic eruptions but is nevertheless predicted, possibly on a continental scale, by the magnitude of the combined impact of the CI-HE4 event.

Given the above-mentioned potential effects of super-eruptions on hominin populations, what is the evidence we currently see on the ground? A recent study of several CI-tephra-bearing archaeological sites in the southern and central Balkans interpreted an absence of evidence for a catastrophic effect on human populations, in the form of discontinuities, as evidence for the resilience of populations in the region and the survival of cultural traditions [10]. This contrasts with archaeological sites proximal to the eruption in southern Italy which show a marked hiatus in occupation [3,49,50], with the lithic artifact record at Crvena Stijena in Montenegro east of the Adriatic Sea which suggests that the CI tephra coincides with the boundary between Middle and Upper Palaeolithic technologies [53], and with the Temnata Cave record in Bulgaria which also suggests a transition in lithic industries [70]. The currently known distribution of the thickest ash layers also suggest that it is possible that local population histories may vary even across a small region. Ultimately, however, the region presently lacks systematic and concrete diachronic studies over large areas where the effects of the CI super-eruption on hominin populations can be properly tested. Consequently, we will refrain from speculation regarding the presence or absence of specific hominin species in the region at the time of the eruption, except to say that current data suggest a patchwork structure. The difficulty in pronouncing a judgment on this issue arises mainly from the fact that the number of known archaeological sites in Eastern Europe securely dated to this period is so few that an accurate view of paleodemography is not yet

possible. Nevertheless it is likely that Eastern Europe, and particularly the Danube Basin, was at this time a crossroads between migration routes of AMHs from the south and possibly from the east across the northern Pontic area [56,89], producing a cultural mosaic that is difficult to decipher today. This is further complicated by the fact that most of the landscape under consideration is draped by thick loess and alluvial deposits that might obscure traces of past human presence [41,90]. Nevertheless, AMHs were clearly already present in the Carpathian region at, or slightly before, the time of the CI super-eruption, as demonstrated by the Peştera cu Oase AMH fossil remains [91]. However, because these fossils are not associated with any archaeology, rendering a direct association between stone tools and hominin morphology is currently impossible. In practice, this means that migration routes are proposed by establishing at best tenuous cultural-historical links between local lithic assemblages and more geographically distant techno-complexes (e.g., the Near-Eastern Ahmarian or Emiran) that are more securely associated with hominin species (i.e. AMHs, Neandertals). The process is not helped by complex, locally-defined lithic typologies, assumptions of unilinear progress in time, and, not least, by the frequent use of quartzite or other coarse-grained raw materials which render the typological approach difficult or inconclusive [92].

Speaking probabilistically, focusing on long archaeological sequences (usually in caves or rockshelters) in regions distal to the super-eruption represents our best chance of sampling the time interval of interest. At present, the discrepancy between the archaeological data offered by the few cave sequences in the southern and central Balkans (e.g. Klissoura, Golema Pesht, Franchthi, and Tabula Traiana; [10]), the tephra-bearing long rockshelter sequence at Crvena Stijena [53], and the open-air non-archaeological situations documented in Dobrogea (this study) and southern Romania [5] calls for a more intense and systematic investigation of open-air archives in the region. So far, only a cluster of recently re-excavated archaeological sites in the Romanian Banat (Româneşti-Dumbrăviţa/Coşava/Tincova; [93]) and the Petrovaradin Fortress site in Vojvodina (northern Serbia; [94]) appear to contain the relevant time intervals, but have not yet yielded information regarding tephra deposits, nor is the dating sufficiently precise to be sure that they were occupied at the precise time of the CI super-eruption. Recent systematic survey in Dobrogea has so far documented and dated a number of Palaeolithic sites, constrained to the later Middle Palaeolithic (Cuza Vodă; [95]) and last glacial period [96], but so far no sites have been identified that date to the CI super-eruption or HE4. Given existing data, it remains difficult to evaluate the impact of, and response to, the eruption in Eastern Europe. However, given the high visibility and widespread distribution of the CI tephra across the region, intensive, targeted surveys, making use of modern subsurface surveying techniques, should uncover such sites if they exist. Efforts within our own research programme are currently focused on finding new archaeological sites associated with CI tephra deposits in Southeastern Romania [95].

In conclusion, we have demonstrated that CI ash deposits in the Lower Danube steppe are much thicker than previously modelled. Moreover, the ashfall present at Urluia was deposited quickly and might have constituted a health hazard for mammalian taxa inhabiting the region, including hominins, irrespective of species. Although we have at present no archaeological remains associated with the deposit, the visibility of the tephra in several profiles in the region is encouraging for targeted surveys.

Materials and Methods

The profile at Urluia was cleaned and logged at three sections along the palaeotopographic slope, to cover the range of tephra thickness laterally along the outcrop. The contact between the tephra and the loess immediately underlying and overlying it was also observed with respect to documenting the associated sedimentary features and soil development. Two main sections ("northwest" and "southeast") were selected for detailed investigation (Figure 3b). The tephra layer at the northwest profile is of intermediate thickness (ca. 50 cm) but was more accessible for sampling than the thickest (100 cm) exposure of the tephra. Two samples were collected for geochemical analysis from the base and top of the tephra within the northwestern cleaned section. Block samples of the tephra were also collected from the lower contact with the loess to observe the nature of deposition. Sediment samples for environmental magnetism studies were collected both from the northwest (72 samples) and southeast (55 samples) cleaned sections. Luminescence dating samples were collected in stainless steel, lightproof tubes driven horizontally into the section above and below the tephra at the northwest (3 samples) and southeast (2 samples) cleaned sections. In total, three samples were collected from below the tephra, and two samples were collected from above it. Additional bulk sediment samples were collected from immediately surrounding the tube locations for laboratory analysis of dosimetry.

Geochemistry

Two samples were collected for geochemical analyses from the northwest section, from the base (URL1) and top (URL2) of the tephra. The ash samples were prepared in the laboratory using published protocols [5]. The almost pure tephra was disaggregated by gently pressing the material and then mounted in epoxy resin, ground and polished in preparation for microprobe analysis of 10 major and minor element concentrations. Measurements were made using single-grain, wavelength-dispersive electron microprobe analysis [5] on up to 40 grains from each sample. The measurements were performed at the Bayerisches GeoInstitut (University of Bayreuth) on a Jeol JXA8200 microprobe employing an accelerating voltage of 15 keV. A 6 nA beam current and defocussed beam were used. Peak counting times were 10 s for Na, 30 s for Si, Al, K, Ca, Fe and Mg, 40 s for Ti and Mn, and 60 s for P. Precision is estimated at <1–6% (2σ) and 10–25% (2σ) for major and minor element concentrations respectively.

Geochronology

The five luminescence dating samples were processed in the laboratory using published protocols to extract fine-grained (4–11 µm) polymineral material [97]. A subsample of this material was then etched in hydrofluorosilicic acid to extract fine-grained quartz following published laboratory procedures [98]; three of the five samples prepared with this procedure yielded sufficient quartz extract for equivalent dose measurements. Equivalent dose estimation was undertaken on each sample using 24 aliquots on a Risø TL-DA-20 reader [99,100], using a U340 filter for quartz OSL measurements, and a D410 filter for polymineral post-IR IRSL measurements. Polymineral samples were measured using the post-IR IRSL$_{290}$ protocol [101]. The fine-grained quartz samples were measured using the single aliquot regenerative dose (SAR) protocol [102] using a preheat temperature of 240°C, based on the results of a preheat plateau test for thermal stability on a loess sample overlying the tephra (Figure S1 in File S1). Dose recovery (Figure S2 in File S1 and Table S2 in File S1) and recycling ratios (Table S3 in File S1) lay within 5% and 10% of

unity respectively, indicating suitability for dating. Since the resulting dose distributions arising from each series of aliquots yielded Gaussian populations (Figure S3 in File S1), the equivalent dose for each sample was determined using the central age model (CAM) [103]. Dose rates for the beta and gamma components were derived from high resolution germanium gamma spectrometry measured at the Felsenkeller in Dresden, Germany, compared with in-house beta counting, and corrected using published attenuation factors [104] incorporating the averaged moisture contents from the sample tubes and surrounding bulk sediment samples (Table S4 in File S1). Alpha values of 0.04±0.02 and 0.08±0.02 were used for the quartz and polymineral fine grain samples respectively to account for the lower luminescence efficiency of alpha radiation relative to the beta and gamma components [105,106]. Cosmic dose rates were calculated using published formulae [107]. Further data relating to luminescence dating characteristics can be found in the Supplementary Information section (File S1).

Environmental Magnetism

Environmental magnetism in loessic sediments is applied on the principle enhancement of magnetic minerals derived from silicate weathering, primarily iron oxides and hydroxides, through pedogenesis [108,109], and in this case, also the presence of weathered and fresh volcanic glass and minerals. Since pedogenesis is climatically controlled, variations in magnetic susceptibility down profile can be linked to regional climatic changes such as glacial-interglacial variations, or if the resolution allows, even to stadial-interstadial fluctuations.

Samples for magnetic susceptibility analyses were taken both from the southeast profile (55 samples) and northwest profile (72 samples), from cleaned sections at 5 cm intervals (Table S5 in File S1). Sampling at the northwest profile focused on the loess-palaeosol sequence overlying the tephra, including the contact zone of mixed loess and tephra. At the southeast profile, samples were collected from the primary (L1) loess underlying the tephra. Stratigraphically, the sections are in contact with the tephra as a reference horizon. Samples were collected in air-tight plastic bags and dried at 40°C in the laboratory, then packed in 6.4 cm^3 plastic boxes. Magnetic susceptibility measurements were undertaken using an AGICO KLY-3 3-Spinner-Kappa-Bridge reader (AGICO, Brno, Czech Republic) working at 920 Hz and 300 A m^{-1}, using previously published methods [67,110]. For measurement of the frequency-dependent magnetic susceptibility (χ_{fd}), each specimen was measured twice at two different frequencies (0.3 and 3 kHz; $\chi@_{0.3kHz}$ - $\chi@_{3kHz}/\chi@_{0.3kHz}*100$ in %) in a magnetic AC field of 300 Am^{-1} using a MAGNON VFSM susceptibility bridge (MAGNON, Dassel, Germany). The data are expressed as mass specific magnetic susceptibility (χ in kg^{-1} m^3) and frequency-dependent magnetic susceptibility (χ_{fd} in %), and used as first order proxies for identifying the course of the intensity of pedogenesis with depth within the stratigraphy.

The measurement of χ_{fd} is sensitive to the presence and relative contribution of ultrafine (c. $\leq$30 nm), pedogenetically derived, so called superparamagnetic (SP) particles in a sample. For that reason this parameter is often used as a proxy for pedogenetic formed SP particles and thus for pedogenesis [111].

Supporting Information

File S1   This file contains supporting figures and tables.
Table S1, Geochemical analyses of tephra samples URL1 and URL2. Grains analysed yielding total % concentrations of $<$95% have been removed from the dataset. Table S2, Results of dose

recovery tests for quartz OSL samples L-EVA1089 and L-EVA1090. Both samples were given a calibrated known dose of 42.75 Gy. Table S3, Averaged measured recycling ratios and standard deviations measured for each luminescence dating sample. Samples measured using quartz OSL are shown in plain text; those measured using the post-IR IRSL$_{290}$ protocol are given in italics. Table S4, Moisture contents (and incorporated uncertainties) of the luminescence dating samples. Table S5, Mass specific and frequency dependent magnetic susceptibility of samples from subsections SS1 (NW profile) and SS2 (SE profile). Susceptibilities were measured in a field of 300 Am^{-1} at frequencies of 300 and 3000 Hz, respectively. Figure S1, Results of preheat plateau test for sample L-EVA1027. Figure S2, Recovered dose distributions from dose recovery tests on samples L-EVA1089 and L-EVA1090. Figure S3, Radial plots illustrating the dose distributions of each of the luminescence dating samples.

Acknowledgments

We thank Professor Jean-Jacques Hublin and Dr Shannon McPherron for their support. The work at Urluia was carried out as part of the Lower Danube Survey Project for Paleolithic Sites (LoDanS, http://lodans. wordpress.com/). Thanks to Professor A. Barnea (University of Bucharest) and to Mariana Petrut (National Museum of History, Adamclisi) for logistical assistance. The authors wish to thank Steffi Albert for luminescence dating sample preparation, and Tsenka Tsanova and Tamara Dogandzic for valuable information about archaeological sites. We would also like to express our gratitude to the staff of the Bayerisches GeoInstitut (Bayreuth, Germany) for assistance with the preparations and EPMA analyses, namely Hubert Schulze and Detlef Krauße. Christine Lane of Oxford University is also thanked for guidance with the geochemical analyses. Finally, we wish to thank the two reviewers for their positive and timely reviews.

Author Contributions

Conceived and designed the experiments: KEF UH. Performed the experiments: KEF UH DV. Analyzed the data: KEF UH DV RI. Contributed reagents/materials/analysis tools: KEF UH DV RI. Wrote the paper: KEF. Assisted in writing the manuscript: UH DV RI.

References

1. De Vivo B, Rolandi G, Gans PB, Calvert A, Bohrson WA, et al. (2001) New constraints on the pyroclastic eruptive history of the Campanian volcanic Plain (Italy). Mineralogy and Petrology 73: 47–65.
2. Barberi F, Innocenti F, Lirer L, Munno R, Pescatore TS, et al. (1978) The Campanian Ignimbrite: A major prehistoric eruption in the Neapolitan area (Italy). Bulletin of Volcanology 41: 10–22.
3. Giaccio B, Nomade S, Wulf S, Isaia R, Sottili G, et al. (2008) The Campanian ignimbrite and Codola tephra layers: two temporal/stratigraphic markers for the early upper Palaeolithic in southern Italy and eastern Europe. Journal of Volcanology and Geothermal Research 177: 208–226.
4. Civetta L, Orsi G, Pappalardo L, Fisher RV, Heiken G, et al. (1997) Geochemical zoning, mingling, eruptive dynamics and depositional processes – the Campanian Ignimbrite, Campi Flegrei caldera, Italy. Journal of Volcanology and Geothermal Research 75: 183–219.
5. Veres D, Lane CS, Timar-Gabor A, Hambach U, Constantin D, et al. (2013) The Campanian Ignimbrite/Y5 tephra layer - A regional stratigraphic marker for isotope Stage 3 deposits in the Lower Danube region, Romania. Quat Int 293: 22–33.
6. Nowaczyk NR, Arz HW, Frank U, Kind J, Plessen B (2012) Dynamics of the Laschamp geomagnetic excursion from Black Sea sediments. Earth Planet Sci Lett 351–352: 54–69.
7. Melekestsev IV, Kirianov VY, Praslov ND (1984) Catastrophic eruption in the Phlegrean Fields region (Italy) – possible source for a volcanic ash in late Pleistocene sediments of the European part of the USSR. Vulcanologia i Seismologia 3: 35–44.
8. Pyle DM, Ricketts GD, Margari V, van Andel TH, Sinitsyn AA, et al. (2006) Wide dispersal and deposition of distal tephra during the Pleistocene 'Campanian Ignimbrite/Y5' eruption, Italy. Quat Sci Rev 25: 2713–2728.
9. Rosi M, Vezzoli L, Castelmenzano A, Grieco G (1999) Plinian pumice fall deposit of the Campanian Ignimbrite eruption (Phlegrean Fields, Italy). Journal of Volcanology and Geothermal Research 91: 179–198.
10. Lowe J, Barton N, Blockley S, Ramsey CB, Cullen VL, et al. (2012) Volcanic ash layers illuminate the resilience of Neanderthals and early modern humans to natural hazards. Proc Natl Acad Sci 109: 13532–13537.
11. Fedele FG, Giaccio B, Isaia R, Orsi G (2003) The Campanian Ignimbrite Eruption, Heinrich Event 4, and Palaeolithic Change in Europe: a High-Resolution Investigation. Geophysical Monograph Volcanism and the Earth's Atmosphere. American Geophysical Union 139: 301–325.
12. Costa A, Folch A, Macedonio G, Giaccio B, Isaia R, et al. (2012) Quantifying volcanic ash dispersal and impact of the Campanian Ignimbrite super-eruption. Geophys Res Lett 39: L10310.
13. Wolff EW, Chappellaz J, Blunier T, Rasmussen SO, Svensson A (2010) Millennial-scale variability during the last glacial: The ice core record. Quat Sci Rev 29: 2828–2838.
14. Andersen KK, Svensson A, Johnsen SJ, Rasmussen SO, Bigler M, et al. (2006) The Greenland Ice Core Chronology 2005, 15–42 ka. Part 1: constructing the time scale. Quat Sci Rev 25: 3246–3257.
15. Fedele FG, Giaccio B, Isaia R, Orsi G, Carroll M, et al. (2007) The Campanian Ignimbrite Factor: Towards a Reappraisal of the Middle to Upper Palaeolithic 'Transition'. In: Grattan J, Torrence R, editors. Living Under the Shadow: the Cultural Impacts of Volcanic Eruptions. Left Coast Press: Walnut Creek (CA): 19–41.
16. Zilhão J, Trinkaus E, Constantin S, Milota S, Gherase M, et al. (2007) The Peştera cu Oase people, Europe's earliest modern humans, in Rethinking the human revolution: new behavioural and biological perspectives on the origin and dispersal of modern humans. In: Mellars P, et al., editors. McDonald Institute of Archaeological Research, Cambridge, UK: 249–262.
17. Mellars P (2006) A new radiocarbon revolution and the dispersal of modern humans in Eurasia. Nature 439: 931–935.
18. Gamble C, Davies W, Pettitt P, Richards M (2004) Climate change and evolving human diversity in Europe during the last glacial. Philos Trans R Soc Lond B Biol Sci 359: 243–253.
19. Klein R (1999) The human career. Human biological and cultural origins. 2 nd edition. Chicago: University of Chicago Press.
20. Bar-Yosef O (2002) The Upper Palaeolithic revolution. Annu Rev Anthropol 31: 363–393.
21. Stringer C (2006) The Neanderthal-H. sapiens interface in Eurasia. In: Hublin JJ, Harvati K, Harrison T, editors. Neanderthals revisited: new approaches and perspectives Springer: Netherlands: 315–323.
22. Wood RE, Barroso-Ruiz C, Caparros M, Jorda Pardo JF, Galvan Santos B, et al. (2013) Radiocarbon dating casts doubt on the late chronology of the Middle to Upper Palaeolithic transition in southern Iberia. Proc Natl Acad Sci. DOI:10.1073/pnas.1207656110.
23. Giaccio B, Hajdas I, Peresani M, Fedele FG, Isaia R (2006) The Campanian Ignimbrite tephra and its relevance for the time of the Middle to Upper Palaeolithic shift. In: Conard NJ, editor. When Neanderthals and Modern Humans Met. Kerns Verlag: Tübingen, Germany: 343–375.
24. Pappalardo L, Civetta L, de Vita S, Di Vito M, Orsi G, et al. (2002) Timing of magma extraction during the Campanian Ignimbrite eruption (Campi Flegrei Caldera). Journal of Volcanology and Geothermal Research 114: 479–497.
25. Ort M, Orsi G, Pappalardo L, Fisher RV (2003) Emplacement processes in a far-traveled dilute pyroclastic current: anisotropy of magnetic susceptibility studies of the Campanian Ignimbrite. Bulletin of Volcanology 65: 55–72.
26. Zielinski GA, Mayewski PA, Meeker LD, Whitlow S, Twickler MS (1996) A 110,000-Yr Record of Explosive Volcanism from the GISP2 (Greenland) Ice Core. Quat Res 45: 109–118.
27. Davies SM, Wastegård S, Abbott PM, Barbante C, Bigler M, et al. (2010) Tracing volcanic events in the NGRIP ice-core and synchronising North Atlantic marine records during the last glacial period. Earth Planet Sci Lett 294: 69–79.
28. Abbott PM, S.M Davies (2012) Volcanism and the Greenland ice-cores: the tephra record. Earth Sci Rev 115: 173–191.
29. Upton J, Cole P, Shaw P, Szakacs A, Seghedi I (2002) Correlation of tephra layers found in southern Romania with the Campanian Ignimbrite (~37 ka) eruption. The Quaternary Research Association and First Postgraduate Paleo-environmental Symposium. Universiteit van Amsterdam, Amsterdam.
30. Constantin D, Timar-Gabor A, Veres D, Begy R, Cosma C (2012) SAR-OSL dating of different grain-sized quartz from a sedimentary section in southern Romania interbedding the Campanian Ignimbrite/Y5 ash layer. Quat Geochronol 10: 81–86.
31. Panaiotu CG, Panaiotu EC, Grama A, Necula C (2001) Paleoclimatic record from a loess-paleosol profile in southeastern Romania. Phys Chem Earth, Part A: Solid Earth and Geodesy 26: 893–898.
32. Watts WA, Allen JRM, Huntley B (1996) Vegetation history and palaeoclimate of the last glacial period at Lago Grande di Monticchio, Southern Italy. Quat Sci Rev 15: 133–153.

33. Ton-That T, Singer B, Paterne M (2001) 40 Ar/39 Ar dating of latest Pleistocene (41 ka) marine tephra in the Mediterranean Sea: implications for global climate records. Earth Planet Sci Lett 184: 645–658.

34. Heinrich H (1998) Origin and consequences of cyclic ice rafting in the Northeast Atlantic Ocean during the past 130,000 years. Quat Res 29: 142–152.

35. Hemming SR (2004) Heinrich events: Massive late Pleistocene detritus layers of the North Atlantic and their global climate imprint. Rev Geophys 42: RG1005.

36. Kadereit A, Kind CJ, Wagner GA (2013) The chronological position of the Lohne Soil in the Nussloch loess section – re-evaluation for a European loess-marker horizon. Quat Sci Rev 59: 67–86.

37. Watts WA, Allen JRM, Huntley B (2000) Palaeoecology of three interstadial events during oxygen-isotope Stages 3 and 4: a lacustrine record from Lago Grande di Monticchio, southern Italy. Palaeogeogr Palaeoclimatol Palaeoecol 155: 83–93.

38. Zolitschka B, Negendank JFW (1996) Sedimentology, dating and palaeoclimatic interpretation of A 76.3 ka record from Lago Grande di Monticchio, southern Italy. Quat Sci Rev 15: 101–112.

39. Tzedakis PC, Lawson IT, Frogley MR, Hewitt GM, Preece RC (2002) Buffered Tree Population Changes in a Quaternary Refugium: Evolutionary Implications. Science 297: 2044–2047.

40. Stevens T, Markovic SB, Zech M, Hambach U, Sümegi P (2011) Dust deposition and climate in the Carpathian Basin over an independently dated last glacial-interglacial cycle. Quat Sci Rev 30: 662–681.

41. Fitzsimmons KE, Marković SB, Hambach U (2012) Pleistocene environmental dynamics recorded in the loess of the middle and lower Danube basin. Quat Sci Rev 41: 104–118.

42. Marković SB, Bokhorst MP, Vandenberghe J, McCoy WD, Oches EA (2008) Late Pleistocene loess-palaeosol sequences in the Vojvodina region, north Serbia. J Quat Sci 23: 73–84.

43. van der Kaars S, Williams MAJ, Bassinot F, Guichard F, Moreno E, et al. (2012) The influence of the ~73 ka Toba super-eruption on the ecosystems of northern Sumatra as recorded in marine core BAR94–25. Quat Int 258: 45–53.

44. Timmreck C, Graf HF, Zanchettin D, Hagemann S, Kleinen T, et al. (2012) Climate response to the Toba super-eruption: Regional changes. Quat Int 258: 30–44.

45. Williams M (2012) The ~73 ka Toba super-eruption and its impact: History of a debate. Quat Int 258: 19–29.

46. Storey M, Roberts RG, Saidin M (2012) Astronomically calibrated 40 Ar/39 Ar age for the Toba supereruption and global synchronization of late Quaternary records. Proc Natl Acad Sci 109: 18684–18688.

47. Hoffecker JF, Holliday VT, Anikovich MV, Sinitsyn AA, Popov VV, et al. (2008) From the Bay of Naples to the River Don: the Campanian Ignimbrite eruption and the Middle to Upper Paleolithic transition in Eastern Europe. J Hum Evol 55: 858–870.

48. Fedele FG, Giaccio B, Hajdas I (2008) Timescales and cultural process at 40,000 BP in the light of the Campanian Ignimbrite eruption, Western Eurasia. J Hum Evol 55: 834–857.

49. Accorsi CA, Aiello E, Bartolini C, Castelletti L, Rodolfi G, et al. (1979) Il giacimento Paleolitico di Serino (Avellino): stratigrafia, ambienti e palentologia. Atti della Società Toscana di Scienze Naturali, Memorie A 86: 435–487.

50. Gambassini P (1997) Il Paleolitico di Castelcivita, culture e ambiente. Napoli: Electa Napoli.

51. Sinitsyn AA (2003) A Palaeolithic "Pompeii" at Kostenki, Russia. Antiquity 77: 9–14.52.

52. Sinitsyn AA, Hoffecker JF (2006) Radiocarbon dating and chronology of the Early Upper Paleolithic at Kostenki. Quat Int 152–153: 164–174.

53. Morley MW, Woodward JC (2011) The Campanian Ignimbrite (Y5) tephra at Crvena Stijena Rockshelter, Montenegro. Quat Res 75: 683–696.

54. Frogner Kockum PC, Herbert RB, Gislason SR (2006) A diverse ecosystem response to volcanic aerosols. Chem Geol 231: 57–66.

55. Grattan JP, Gilbertson DD (1994) Acid-loading from Icelandic tephra falling on acidified ecosystems as a key to understanding archaeological and environmental stress in Northern and Western Britain. J Archaeol Sci 21: 851–859.

56. Hoffecker JF (2009) The spread of modern humans in Europe. Proc Natl Acad Sci 106: 16040–16045.

57. Haase D, Fink J, Haase G, Ruske R, Pécsi M, et al. (2007) Loess in Europe–its spatial distribution based on a European Loess Map, scale 1:2,500,000. Quat Sci Rev 26: 1301–1312.

58. Buggle B, Hambach U, Kehl M, Zöller L, Marković SB, et al. (2013) The progressive evolution of a continental climate in SE-Central European lowlands during the Middle Pleistocene recorded in loess paleosol sequences. Geology. DOI: 10.1130/G34198.1. (In press).

59. Smalley IJ, Leach JA (1978) The origin and distribution of the loess in the Danube basin and associated regions of East-Central Europe – A review. Sed Geol 21: 1–26.

60. Buggle B, Hambach U, Glaser B, Marković SB, Glaser I, et al. (2008) Geochemical characterization and origin of Southeastern and Eastern European loesses (Serbia, Romania, Ukraine). Quat Sci Rev 27: 1058–1075.

61. Újvári G, Varga A, Balogh-Brunstad Z (2008) Origin, weathering, and geochemical composition of loess in southwestern Hungary. Quat Res 69: 421–137.

62. Stuut JB, Smalley I, O'Hara-Dhand K (2009) Aeolian dust in Europe: African sources and European deposits. Quat Int 198: 234–245.

63. Kukla GJ (1975) Loess stratigraphy of central Europe. In: Butzer KW, Isaac GL, editors. After the Australopithecines. Mouton Publishers, The Hague: 99–188.

64. Marković SB, Hambach U, Stevens T, Jovanović M, O'Hara-Dhand K, et al. (2012) Loess in the Vojvodina region (Northern Serbia): an essential link between European and Asian Pleistocene environments. Netherlands Journal of Geosciences – Geologie en Mijnbouw 91: 173–188.

65. Munteanu MT, Munteanu E, Stiuca E, Macaleti R, Dumitrascu G (2008) Some aspects concerning the Quaternary deposits in south Dobrogea. Acta Palaeontologica Romaniae 6: 229–236.

66. Signorelli S, Vaggelli G, Francalanci L, Rosi M (1999) Origin of magmas feeding the Plinian phase of the Campanian Ignimbrite eruption, Phlegrean Fields (Italy): constraints based on matrix-glass and glass-inclusion compositions. Journal of Volcanology and Geothermal Research 91: 199–220.

67. Buggle B, Hambach U, Glaser B, Gerasimenko N, Marković SB (2009) Stratigraphy, and spatial and temporal paleoclimatic trends in Southeastern/Eastern European loess-paleosol sequences. Quat Int 196: 86–106.

68. Rousseau DD, Kukla G, McManus J (2006) What is what in the ice and the ocean? Quat Sci Rev 25: 2025–2030.

69. Pyle DM (1989) The thickness, volume and grainsize of tephra fall deposits. Bulletin of Volcanology 51: 1–15.

70. Ferrier C, Laville H (2000) Stratigraphie des depots aurignaciens de la grotte Temnata. In: Ginter B, editor. Temnata Cave: Excavations in Karlokovo Karst Area, Bulgaria 2.1. Jagellonian University Press, Kraków: 13–30.

71. Paterne M (1992) Additional remarks on tephra layers from Temnata Cave. In: Kozlowski JK, Laville H, Ginter B, editors. Temnata Cave: Excavations in Karlukovo Karst Area, Bulgaria. Jagellonian University Press, Kraków: 99–100.

72. Pawlikowskj M (1992) Analysis of tephra layers from TD-II and TD-V excavations. In: Kozlowski JK, Laville H, Ginter B, editors. Temnata Cave: Excavations in Karlukovo Karst Area, Bulgaria. Jagellonian University Press, Kraków: 89–98.

73. Horwell CJ, Baxter PJ (2006) The respiratory health hazards of volcanic ash: a review for volcanic risk mitigation. Bulletin of Volcanology 69: 1–24.

74. Riede F, Bazely O (2009) Testing the "Laacher See hypothesis": a health hazard perspective. J Archaeol Sci 36: 675–683.

75. D'Alessandro W (2006) Human fluorosis related to volcanic activity: a review. In: Kungolos AG, et al., editors. Environmental Toxicology. WIT Press, Southampton: 21–30.

76. Cronin SJ, Neall VE, Lecointre JA, Hedley MJ, Loganathan P (2003) Environmental hazards of fluoride in volcanic ash: a case study from Ruapehu volcano, New Zealand. Journal of Volcanology and Geothermal Research 121: 271–291.

77. Byerly RM (2007) Palaeopathology in late Pleistocene and early Holocene Central Plains bison: dental enamel hypoplasia, fluoride toxicosis and the archaeological record. J Archaeol Sci 34: 1847–1858.

78. Baales M, Jöris O, Street M, Bittmann F, Weninger B, et al. (2002) Impact of the Late Glacial Eruption of the Laacher See Volcano, Central Rhineland, Germany. Quat Int 58: 273–288.

79. Dumond DE (2004) Volcanism and history on the Northern Alaska Peninsula. Arctic Anthropol 41: 112–125.

80. Grattan J (2006) Aspects of Armageddon: An exploration of the role of volcanic eruptions in human history and civilization. Quat Int 151: 10–18.

81. Riede F (2008) The Laacher See-eruption (12,920 BP) and material culture change at the end of the Allerød in Northern Europe. J Archaeol Sci 35: 591–599.

82. Haslam M, Clarkson C, Petraglia M, Korisettar R, Jones S, et al. (2010) The 74 ka Toba super-eruption and southern Indian hominins: archaeology, lithic technology and environments at Jwalapuram Locality 3. J Archaeol Sci 37: 3370–3384.

83. Clarkson C, Jones S, Harris C (2012) Continuity and change in the lithic industries of the Jurreru Valley, India, before and after the Toba eruption. Quat Int 258: 165–179.

84. Jones SC (2012) Local- and regional-scale impacts of the ~74 ka Toba supervolcanic eruption on hominin populations and habitats in India. Quat Int 258: 100–118.

85. Sparks S, Self S, Grattan J, Oppenheimer C, Pyle D, et al. (2005) Super-eruptions: Global effects and future threats. Report of a Geological Society of London Working Group. The Geological Society: London.

86. Lane CS, Blockley SPE, Lotter AF, Finsinger W, Filippi ML, et al. (2012) A regional tephrostratigraphic framework for central and southern European climate archives during the Last Glacial to Interglacial transition: comparisons north and south of the Alps. Quat Sci Rev 36: 50–58.

87. Riede F (2009) The loss and re-introduction of bow-and-arrow technology: a case study from the Northern European Late Paleolithic. Lithic Technology 34: 27–46.

88. Barton L, Brantingham PJ, Ji D (2007) Late Pleistocene climate change and Paleolithic cultural evolution in northern China: implications from the Last Glacial Maximum. Developments in Quaternary Sciences 9: 105–128.

89. Hublin JJ (2012) The earliest modern human colonization of Europe. Proc Natl Acad Sci 109: 13471–13472.

90. Romanowska I (2012) Lower Palaeolithic of central and Eastern Europe: Critical evaluation of the current state of knowledge. In: Ruebens K, Romanowska I, Bynoe R, editors. Unravelling the Palaeolithic. 10 years of Research at the Centre for the Archaeology of Human Origins (British Archaeological Reports). Archaeopress, Oxford: 1–12.

91. Trinkaus E, Moldovan O, Milota S, Bîlgăr A, Sarcina L, et al. (2003) An early modern human from the Peştera cu Oase, Romania. Proc Natl Acad Sci 100: 11231–11236.

92. Anghelinu M, Nita L (2013) What's in a name: The Aurignacian in Romania. Quat Int. DOI:10.1016/j.quaint.2012.03.013. (In press).

93. Sitlivy V, Chabai VP, Anghelinu M, Uthmeier T, Kels H, et al. (in press) Preliminary reassessment of the Aurignacian in Banat (south-western Romania). Quat Int. DOI: 10.1016/j.quaint.2012.07.024.

94. Mihailovic D (2008) New data about the Middle Palaeolithic of Serbia. In: Darlas A, Mihailovic D, editors. The Palaeolithic of the Balkans. British Archaeological Reports, Oxford: 93–100.

95. Iovita RP, Doboş A, Fitzsimmons KE, Probst M, Hambach U, et al. (2013) Geoarchaeological prospection in the loess steppe: preliminary results from the Lower Danube Survey for Palaeolithic sites (LoDanS). Quat Int. (In press).

96. Iovita RP, Fitzsimmons KE, Dobos A, Hambach U, Hilgers A, et al. (2012) Dealul Guran: evidence for Lower Paleolithic occupation of the Lower Danube loess steppe. Antiquity 86: 973–989.

97. Frechen M, Schweitzer U, Zander A (1996) Improvements in sample preparation for the fine grain technique. Ancient TL 14: 15–17.

98. Timar A, Vandenberghe D, Panaiotu EC, Panaiotu CG, Necula C, et al. (2010) Optical dating of Romanian loess using fine-grained quartz. Quat Geochronol 5: 143–148.

99. Bøtter-Jensen L, Bulur E, Duller GAT, Murray AS (2000) Advances in luminescence instrument systems. Rad Meas 32: 523–528.

100. Bøtter-Jensen L, Mejdahl V, Murray AS (1999) New light on OSL. Quat Sci Rev 18: 303–309.

101. Buylaert JP, Jain M, Murray AS, Thomsen KJ, Thiel C, et al. (2012) A robust feldspar luminescence dating method for Middle and Late Pleistocene sediments. Boreas 41: 435–451.

102. Murray AS, Wintle AG (2000) Luminescence dating of quartz using an improved single-aliquot regenerative-dose protocol. Rad Meas 32: 57–73.

103. Galbraith RF, Roberts RG, Laslett GM, Yoshida H, Olley JM (1999) Optical dating of single and multiple grains of quartz from Jinmium rock shelter, northern Australia. Part 1, Experimental design and statistical models. Archaeometry 41: 339–364.

104. Adamiec G, Aitken M (1998) Dose-rate conversion factors: update. Ancient TL 16: 37–50.

105. Rees-Jones J (1995) Dating young sediments using fine grained quartz. Ancient TL 13: 9–14.

106. Rees-Jones J, Tite MS (1997) Optical dating results for British archaeological sediments. Archaeometry 36: 177–187.

107. Prescott JR, Hutton JT (1994) Cosmic ray contributions to dose rates for luminescence and ESR dating: Large depths and long term variations. Rad Meas 23: 497–500.

108. Evans ME, Heller F (2001) Magnetism of loess/palaeosol sequences: recent developments. Earth Sci Rev 54: 129–144.

109. Hambach U, Rolf C, Schnepp E (2008) Magnetic dating of Quaternary sediments, volcanites and archaeological materials: an overview. Eiszeitalter und Gegenwart: Quaternary Science Journal 57: 25–51.

110. Hambach U (2010) Palaeoclimatic and stratigraphic implications of high resolution magnetic susceptibility logging of Würmian loess at the Krems-Wachtberg Upper-Palaeolithic site. In: Neugebauer-Maresch C, Owen LR, editors. New Aspects of the Central and Eastern European Upper Palaeolithic – methods, chronology, technology and subsistence. Proceedings of the Prehistoric Commission of the Austrian Academy of Sciences, Vienna: 295–304.

111. Baumgart P, Hambach U, Meszner S, Faust D (2013) An environmental magnetic fingerprint of periglacial loess: Records of Late Pleistocene loess-palaeosol sequences from Eastern Germany. Quat Int. DOI: 10.1016/j.quaint.2012.12.021 (In press).

24

Validating and Improving Interrill Erosion Equations

Feng-Bao Zhang[1,2], Zhan-Li Wang[1,2]*, Ming-Yi Yang[1,2]

1 State Key Laboratory of Soil Erosion and Dryland Farming on the Loess Plateau, Institute of Soil and Water Conservation, Northwest A&F University, Yangling, P. R. China,
2 Institute of Soil and Water Conservation, Chinese Academy of Science and Ministry of Water Resources, Yangling, P. R. China

Abstract

Existing interrill erosion equations based on mini-plot experiments have largely ignored the effects of slope length and plot size on interrill erosion rate. This paper describes a series of simulated rainfall experiments which were conducted according to a randomized factorial design for five slope lengths (0.4, 0.8, 1.2, 1.6, and 2 m) at a width of 0.4 m, five slope gradients (17%, 27%, 36%, 47%, and 58%), and five rainfall intensities (48, 62.4, 102, 149, and 170 mm h^{-1}) to perform a systematic validation of existing interrill erosion equations based on mini-plots. The results indicated that the existing interrill erosion equations do not adequately describe the relationships between interrill erosion rate and its influencing factors with increasing slope length and rainfall intensity. Univariate analysis of variance showed that runoff rate, rainfall intensity, slope gradient, and slope length had significant effects on interrill erosion rate and that their interactions were significant at p = 0.01. An improved interrill erosion equation was constructed by analyzing the relationships of sediment concentration with rainfall intensity, slope length, and slope gradient. In the improved interrill erosion equation, the runoff rate and slope factor are the same as in the interrill erosion equation in the Water Erosion Prediction Project (WEPP), with the weight of rainfall intensity adjusted by an exponent of 0.22 and a slope length term added with an exponent of −0.25. Using experimental data from WEPP cropland soil field interrill erodibility experiments, it has been shown that the improved interrill erosion equation describes the relationship between interrill erosion rate and runoff rate, rainfall intensity, slope gradient, and slope length reasonably well and better than existing interrill erosion equations.

Editor: Juan A. Añel, University of Oxford, United Kingdom

Funding: Financial support for this research was provided by the National Natural Science Foundation of China (41171227, 40901127, 40971172), Natural Science Foundation of Shannxi Province (S2012JC6612), and the special funds of Northwest A&F University for the operational costs of basic scientific research (QN2011072). The funders had no role in study design, data collection and analysis, decision to publish, or preparation of the manuscript.

Competing Interests: The authors have declared that no competing interests exist.

* E-mail: zwang@ms.iswc.ac.cn

Introduction

Soil erosion on a hill slope consists of two major components: interrill and rill erosion [1]. Interrill erosion is the detachment and transport of soil material from the surface of the soil matrix by raindrop impact and overland flow [2]. More recent process-based models, e.g., WEPP [3], EUROSEM [4], LISEM [5], and PSEM-2D [6], make an explicit distinction between rill and interrill erosion processes and estimate interrill erosion rate using a function of various factors, including interrill erodibility parameter which expresses the soil's resistance to interrill erosion within interrill areas, rainfall erosivity, slope characteristics, hydraulic factors, vegetation, and land use [4–10].

It is difficult to describe and quantify interrill and rill erosion processes simultaneously during a rainfall event, especially to measure the rate of interrill erosion because it is less visible. The mini-plot, which eliminates the effect of rill erosion, has been an important means to study interrill erosion and to establish interrill erosion equations. The effects of rainfall intensity, slope length, slope gradient, runoff rate, and soil type on interrill erosion rate have been described. Power functions were used to describe the effect of rainfall intensity on interrill erosion. The rainfall intensity exponent varied from 1.6 to 2.3 [11] and from 1.36 to 2.54 [12]. The exponent decreased as the clay content of the soil increased [11]. Therefore, interrill erosion rate has been shown to be approximately proportional to the square of the rainfall intensity for a given mini-plot if runoff is not considered [12–15]. To

describe the effect of slope gradient on interrill erosion, several formulas of slope factors have been proposed [16–19]. The role of runoff was initially considered as sediment transport for interrill erosion [20] and was not included in interrill erosion equations [3], but some researchers found that model predictability was improved when a separate runoff parameter was included in interrill erosion equations [21,22]. In general, interrill erosion rate increased gradually with increasing runoff. For the effect of slope length on interrill erosion, there have been inconsistent results, which will be described detailedly in the following discussion.

According to the relationships between interrill erosion rate and its influencing factors, interrill erosion equations have been proposed on the basis of results from mini-plot experiments. In the WEPP model, the following equation was first used to estimate interrill erosion rates [3]:

$$D_i = K_i I^2 S_f \qquad (1)$$

where D_i is interrill erosion rate (kg m^{-2} s^{-1}), K_i is interrill erodibility, I is rainfall intensity (m s^{-1}), and S_f is slope factor that can be calculated using the following equation [16]:

$$S_f = 1.05 - 0.85\exp(-4\sin\theta) \qquad (2)$$

where θ is the slope gradient (°).

As described above, previous studies have shown that model predictability was improved when a separate runoff parameter was included in interrill erosion equations [21,22]. Therefore, Eq. (1) was modified by adding the factor of overland flow [22] and was then used in the WEPP model in 1995 as the following formula [23]:

$$D_i = K_i R I S_f \quad (3)$$

where R is runoff rate (m s^{-1}). In addition, to describe interrill erosion processes more accurately and to develop a simple lumped parameter model, the following equation was proposed by analyzing simulated rainfall data [19]:

$$D_i = K_i q^{1/2} I S^{2/3} = K_i R^{1/2} I S^{2/3} L^{1/2} \quad (4)$$

where q is flow discharge per unit width (m^2 s^{-1}) and S is slope inclination (m m^{-1}). This equation thought that the rainfall intensity term, I, represented the detachment of soil by raindrop impacts and the enhancement of the transport capacity of overland flow and that the term $q^{1/2}S^{2/3}$ represented sediment transport by thin overland flow. Compared to Eq. (3), the effect of

Table 1. Interrill erodibilities (K_i) and coefficients of determination (R^2) for Eqs (1), (3) and (4) for different slope lengths.

L (m)		Eq (1)	Eq (3)	Eq (4)
0.4	K_i	1,773,995	3,734,418	43,430
	R^2	0.90	0.90	0.96
0.8	K_i	2,026,011	3,420,684	31,387
	R^2	0.77	0.78	0.88
1.2	K_i	1,815,288	2,927,228	22,287
	R^2	0.79	0.80	0.84
1.6	K_i	1,531,958	2,435,898	16,294
	R^2	0.71	0.70	0.82
2.0	K_i	1,571,555	2,437,679	15,029
	R^2	0.64	0.64	0.77
Mean	K_i	1,743,761	2,991,181	25,686
CV		11%	19%	46%

runoff on interrill erosion rate was weakened, and the positive effect of slope length on interrill erosion rate was added to Eq. (4).

Parameters for soil type (K_i), slope gradient, rainfall intensity, and overland flow are included in Eqs. (1), (3), and (4). The parameter for slope length is included only in Eq. (4). However, inconsistent results have been reported for the effect of slope length on interrill erosion. Running the similar experiments which simulated slope length using inflow rates, some experiments indicated consistently greater interrill erosion from increased overland flow [24], but other experiments showed that greater discharge rates resulted in decreased soil loss [25,26]. The authors thought that interrill erosion at a particular downslope distance is dictated by the soil detachment or sediment transport capacity characteristics existing at that particular location [25,26]. The effects of row-sideslope length within a range of 0.15 m to 0.6 m on sideslope erosion for four soils at several rain intensities demonstrated that slope length affected erosion very little until the soils began to rill, but had a major effect once rilling occurred [27]. However, the inconsistent viewpoint, which was that the effect of slope length on interrill erosion might have been due to a change from erosion which is dominated by raindrop-induced flow transport (RIFT) to erosion which is dominated by raindrop detachment-flow transport (RD-FT) except for the effect of slope length on rilling, was concluded by using the same data [28]. Other experiments indicated that soil losses were significantly correlated with rainfall intensity ($r = 0.89$; P<0.001) and slope length ($r = 0.43$; P<0.001), and suggested that careless scale transfer of erosion data may lead to erroneous conclusions [29]. In addition, some researchers have attributed the effect of slope length on interrill erosion to rainfall intensity and soil properties [27,30] and have assumed that slope length has little or no effect on interrill erosion per unit of area [31,32]. In short, although numerous studies were reported, the effect of slope length on interrill erosion has not been clearly established.

In addition, Eqs. (1), (3), and (4) were developed for smaller plot sizes (less than 0.5 m^2), shorter slope lengths (less than 0.8 m), and small slope gradients, and the different researchers who developed these equations used different plot sizes and slope lengths [3,19,22,27]. They suggested that the applicability of their model to longer slopes required further testing [19]. To date, no

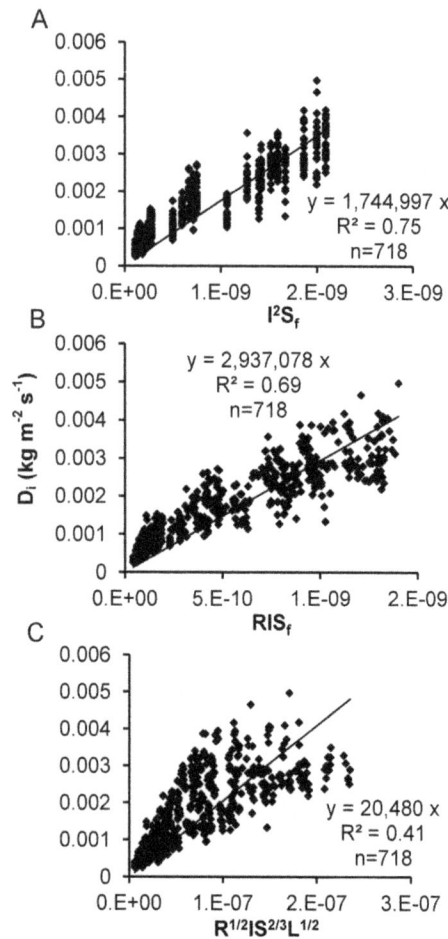

Figure 1. Linear regressions with zero intercept. Using linear regression with zero intercept between interrill erosion rate and the scaling factors I^{2}S$_f$ (A), RIS$_f$ (B), and IR$^{1/2}$S$^{2/3}$L$^{1/2}$ (C) to assess the previous interrill equations based on the mini plots.

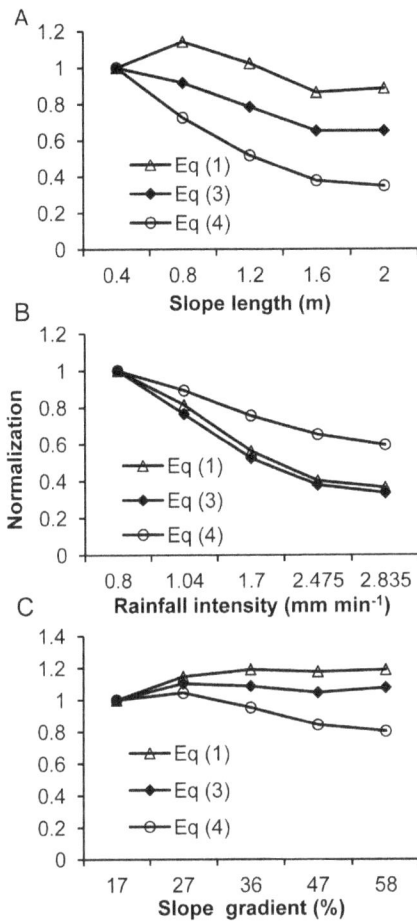

Figure 2. Normalized interrill erodibilities. Variations in interrill erodibility for the three equations for different slope lengths (A), rainfall intensities (B), and slope gradients (C), normalized to the lowest values of the independent factors.

Table 2. Interrill erodibilities (K_i) and coefficients of determination (R^2) for Eqs. (1), (3), and (4) for different rainfall intensities.

I (mm min^{-1})		Eq (1)	Eq (3)	Eq (4)
0.8	K_i	4,386,185	7,967,808	29,897
	R^2	0.26	0.49	0[a]
1.04	K_i	3,572,441	6,097,944	27,063
	R^2	0.39	0.45	0
1.7	K_i	2,465,436	4,171,479	23,615
	R^2	0.4	0.37	0
2.475	K_i	1,761,220	3,009,673	20,424
	R^2	0.6	0.39	0
2.835	K_i	1,596,803	2,662,866	19,345
	R^2	0.46	0.17	0
Mean	K_i	2,756,417	4,781,954	24,069
CV		43%	47%	18%

[a]R^2 was negative for the linear regression with zero intercept between interrill erosion rate and $Iq^{1/2}S^{2/3}$ in Eq (4); therefore, it was replaced by zero.

were required for the field studies described. The soil contained 38.72% sand, 45.59% silt, 15.69% clay, and 0.53% organic matter. The soil was collected from the top 0.25 m layer in cultivated land, air-dried, crushed to pass through a 4-mm sieve, and thoroughly mixed. The soil moisture content was held constant at 14% for all experiments.

Perforated metal flumes with different lengths (0.4, 0.8, 1.2, 1.6, and 2 m), 0.4 m in width and 0.25 m in depth, were manufactured for use in the tests. The slope gradient of each flume could easily be adjusted from 0% to 84%. A trough was placed at the lower edge of the flume to collect runoff and sediment samples. The soil was packed uniformly into the flumes in four 5.5-cm layers to a total depth of 22 cm with a bulk density of 1.3 g cm^{-3}. To reduce discontinuities between layers, the surface of each soil

systematic experiments based on mini-plot experiments have been conducted to validate and compare the existing interrill erosion equations for different slope lengths or plot sizes and a wider range of slope gradients and rainfall intensities. The objective of the present study is to validate the existing interrill erosion equations under different controlled experimental conditions and to describe interrill erosion rate accurately by adding or adjusting factors in the interrill erosion equations.

Materials and Methods

Erosion tests were conducted at the State Key Laboratory of Soil Erosion and Dryland Farming on the Loess Plateau in Yangling, China. Artificial rainfall was produced using a rain-making machine with an effective rainfall area of 2 m×3 m at a height of 8.67 m, which produced a simulated rainstorm at a controllable intensity with distribution uniformity greater than 85% and a median raindrop diameter of 2.2 mm, comparable to natural rainfall.

The soil used in the tests was cultivated Huangmian soil (Calcaric Cambisols, FAO) from Ansai County in Shaanxi Province, located in the northern part of the Loess Plateau. The cultivated land sampled in this study is publicly owned and managed by the Ansai Research Station of Soil and Water Conservation, Chinese Academy of Science. No specific permits

Table 3. Interrill erodibilities (K_i) and coefficients of determination (R^2) for Eqs. (1), (3), and (4) for different slope gradients.

S (%)		Eq (1)	Eq (3)	Eq (4)
17	K_i	1,503,203	2,753,304	23,327
	R^2	0.71	0.7	0.47
27	K_i	1,722,260	3,036,752	23,789
	R^2	0.74	0.68	0.44
36	K_i	1,788,967	2,989,530	21,976
	R^2	0.72	0.65	0.37
47	K_i	1,766,816	2,875,945	19,512
	R^2	0.71	0.64	0.34
58	K_i	1,784,369	2,958,936	18,592
	R^2	0.75	0.67	0.34
Mean	K_i	1,713,123	2,922,893	21,439
CV		7%	4%	11%

Table 4. Univariate Analysis of Variance for selected parameters with interrill erosion rate as a dependent variable using all effective steady-state values.

Source	Type III Sum of Squares	df	Mean Square	F Value	Sig.
Corrected Model	.001[a]	125	5.405E-06	115.591	.000
Intercept	2.797E-06	1	2.797E-06	59.824	.000
Runoff (R)	2.747E-07	1	2.747E-07	5.874	.016
Intensity (I)	3.518E-06	4	8.794E-07	18.808	.000
Slope length (L)	1.939E-05	4	4.847E-06	103.660	.000
Slope gradient (S)	3.017E-05	4	7.542E-06	161.305	.000
I * L	1.104E-05	16	6.901E-07	14.760	.000
I * S	1.507E-05	16	9.419E-07	20.144	.000
L * S	8.892E-06	16	5.558E-07	11.886	.000
I * L * S	8.576E-06	64	1.340E-07	2.866	.000
Error	2.768E-05	592	4.676E-08		
Total	.003	718			
Corrected Total	.001	717			

[a]$R^2 = 0.961$ (Adjusted $R^2 = 0.952$).

layer was gently scored. The surface of the top layer was smoothed to minimize microtopographic effects.

Two replications of all combinations of one soil type, five levels of slope length (0.4, 0.8, 1.2, 1.6, and 2 m), five levels of slope gradient (17%, 27%, 36%, 47%, and 58%), and five levels of rainfall intensity (48, 62.4, 102, 149, and 170 mm h^{-1}) were tested using a randomized factorial design. A total of 250 runs were conducted. Replicate means were used in the data analyses.

Simulated rainfall was generally applied for 60 min for each rainfall event and was terminated if rill erosion occurred during the 60 min. Sediment and runoff samples were collected continuously throughout the rainfall event. Samples were collected at 1, 2, 3, 4, and 5 min within the first 15 minutes after runoff production and then every 5 min thereafter. Sediment and runoff were measured gravimetrically. In general, steady states of soil loss and runoff rate were reached within 30 min after runoff commenced. To minimize the effect of surface condition, steady-state values were used in the analysis [3,19,27]. Values identified as outliers were not included [3]. In total, 36 of 1740 experimental values and 22 of 740 steady-state values were rejected. Statistical analyses were performed using the SPSS PASW Statistics (Version 18.0) software.

Results and Discussion

Validation of existing interrill erosion equations

Interrill erodibility based on mini-plot experiments for a given soil type was defined as a constant for each equation. This constant was determined from linear regression between the interrill erosion rate D_i and the scaling factors I^2S_f in Eq. (1), RIS_f in Eq. (3), and $IR^{1/2}S^{2/3}L^{1/2}$ in Eq. (4) for all effective steady-state values. A zero intercept was used in the regression, and the slope of the regression line was assumed to be the interrill erodibility (K_i). Figure 1 presents the results of linear regression with zero intercept for Eqs. (1), (3), and (4). Good linear regressions could not be gained for the three interrill erosion equations for all effective steady-state values, which indicated that in some cases the existing interrill erosion

equations did not adequately describe the relationship between interrill erosion rate and its influencing factors.

To find further key factors influencing the equation's predictive power, interrill erodibility was also calculated using linear regression with a zero intercept under various controlled experimental conditions. Table 1 presents interrill erodibility (K_i) and coefficient of determination (R^2) for different slope lengths for the three existing equations. When rainfall intensity and slope gradient were varied and slope length was increased from 0.4 m to 2 m, the interrill erodibility (K_i) from Eq. (1) varied slightly, with a coefficient of variation (CV) of 11%, whereas that from Eq. (3) decreased by approximately 1.5 times, with a CV of 19%, and that from Eq. (4) varied widely and decreased by approximately 2.9 times, with a CV of 46% (Fig. 2a). These results indicated that slope length plays a key role in Eqs. (3) and (4) and should be included in these equations. Compared with Eq. (1), the wide variations in interrill erodibility for Eqs. (3) and (4) might imply interactions between runoff and slope length. The change in the coefficient of determination (R^2) corresponding to the change in slope length from 0.4 m to 2 m demonstrated that the three existing interrill equations could describe interrill erosion processes effectively for plots with shorter slope lengths, especially those experiments of approximately 0.4 m. This finding is consistent with existing interrill erosion equations established on the basis of data from mini-plots with shorter slope lengths in previous works [3,12,27].

Table 2 presents the interrill erodibility (K_i) values and coefficients of determination (R^2) obtained for different rainfall intensities using the three existing equations. Interrill erodibility (K_i) for the three equations decreased as rainfall intensity increased from 0.8 to 2.835 mm min^{-1} (Fig. 2b). Interrill erodibility (K_i) varied obviously for Eqs. (1) and (3), with a CV of approximately 45%, but varied relatively little for Eq. (4), with a CV of 18% because of the decrease in the weight of runoff rate in the equation. These results indicated that the effects of rainfall intensity or runoff rate are overestimated by the three existing interrill erosion equations. When the weight of runoff or rainfall intensity in the interrill erosion equations was decreased, the equations' predictive power was improved. Therefore, an adjustment to these weights should be considered.

Table 3 presents the interrill erodibility (K_i) values and coefficients of determination (R^2) obtained for different slope gradients for the three existing equations. Only for Eq. (4) did interrill erodibility decrease with an increase in slope gradient (Fig. 2c). In general, interrill erodibility from these three equations varied slightly with slope gradient. These results suggested that the effect of slope gradient on interrill erosion rate is almost completely expressed by the existing equations, but that use of the S_f term in Eqs. (1) and (3) yields better results than use of the $S^{2/3}$ term in Eq. (4). The lower coefficient of determination (R^2) values shown in Tables 2 and 3 also implied that slope length is the most important factor affecting interrill erosion processes and results in lower coefficients of determination (R^2).

In summary, the results from the analysis described above suggested that the three existing interrill erosion equations considered do not completely describe the relationship between interrill erosion rate and its influencing factors, especially slope length and rainfall intensity or runoff. In addition, univariate analysis of variance was conducted on the assumption that runoff rate was the covariant and that rainfall intensity, slope gradient, and slope length were the fixed factors. The results of this analysis showed that runoff rate, rainfall intensity, slope gradient, and slope length had significant effects on interrill erosion rate and that their interactions were significant at p<0.01 (Table 4). These results

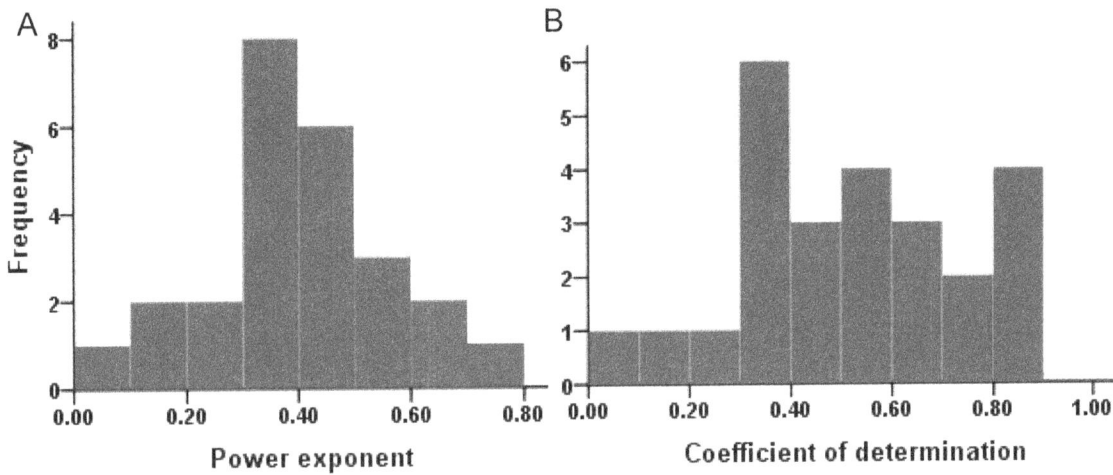

Figure 3. Regression parameters of power function. Power exponents (A) and coefficients of determination (B) for the relationship between sediment concentration and slope gradient for all tests.

also implied that the effect of slope length on interrill erosion rate should be considered in the interrill erosion equations. Therefore, an interrill erosion equation is needed that includes a factor for slope length and weight adjustments for the other factors.

Improved interrill erosion equation

According to Eqs. (1), (3), and (4) and the results of the analysis described above, the interrill erosion rate for a given soil type can be described by a function of the following general form:

$$D_i = f(R, I, S, L) \qquad (5)$$

In this function, slope gradient (S), rainfall intensity (I), and slope length (L) are three independent variables that directly affect runoff rate (R) and directly or indirectly affect interrill erosion rate (D_i). Runoff rate also affects interrill erosion rate. In fact, these factors are interdependent and interactive. Multivariate regression does obtain the best-fit expression, but does not reflect the physical interpretation of the interaction between interrill erosion rate and its influencing factors. It is difficult to eliminate the interdependence and interaction among these factors and to express accurately the relationship between interrill erosion rate and its influencing factors using multivariate regression. Therefore, before structuring the interrill erosion equation, it is necessary to analyze individually the relationships between interrill erosion rate and each influencing factor and to understand their physical interpretation under controlled conditions.

Sediment concentration (C) is a composite indicator whose variations reflect the dynamic variations of runoff and soil loss. Transport by interrill overland flow is the predominant process in interrill erosion [2], and the detachment process can be considered negligible [20]. Based on the theory of erosion by rainfall-disturbed flow [33,34], sediment concentration was determined by soil factors, rainfall intensity, kinetic energy per unit quantity of rain, slope length, and slope gradient [28]. Because the kinetic energy of rain is an expression of rainfall intensity [35], the sediment concentration for a given soil type can also be expressed by a function of the following general form:

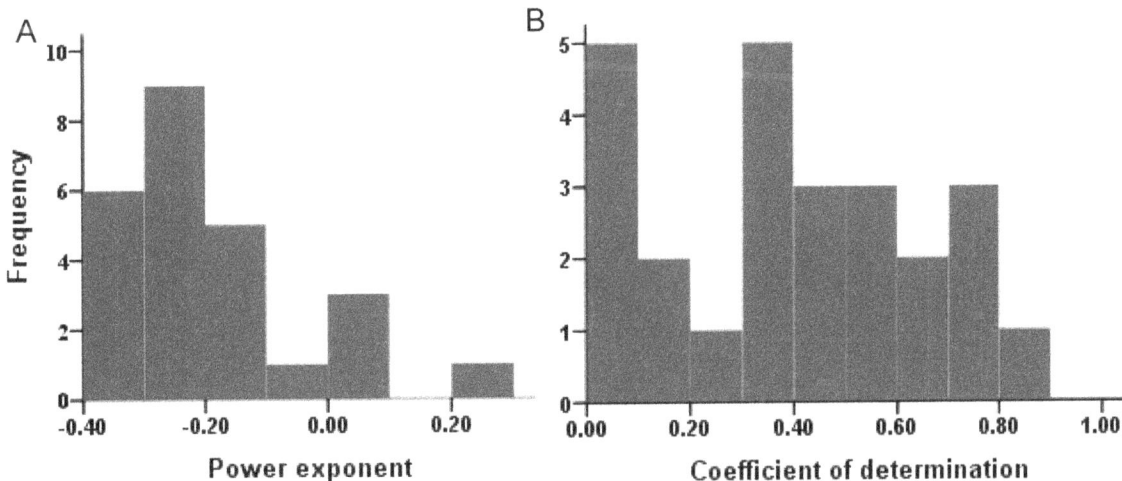

Figure 4. Regression parameters of power function. Power exponents (A) and coefficients of determination (B) for the relationship between sediment concentration and slope length for all tests.

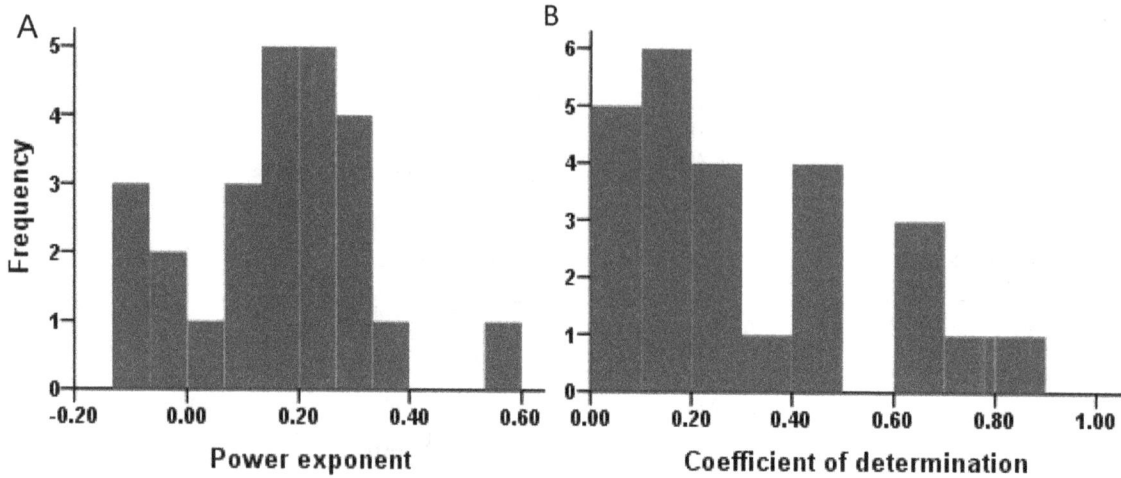

Figure 5. Regression parameters of power function. Power exponents (A) and coefficients of determination (B) for the relationship between sediment concentration and rainfall intensity for all tests.

$$C = f(I, S, L) \tag{6}$$

Compared with Eq. (5), there is one dependent parameter (C) in Eq. (6), and the effects of the interdependence between runoff rate and rainfall intensity, slope gradient, and slope length in Eq. (5) are not considered in Eq. (6), while interactions among rainfall intensity, slope gradient, and slope length may exist in Eq. (6).

Sediment concentration (C) can also be calculated by the following formula:

$$C = D_i / R \tag{7}$$

Therefore, if the relationships between sediment concentration and slope gradient, rainfall intensity, and slope length can be determined, the relationships between interrill erosion rate and runoff rate, slope gradient, rainfall intensity, and slope length can also be determined. The relationships between sediment concentrations and each influencing factor are quantified in the following discussion.

For the same slope length and rainfall intensity, highly significant power-function relationships ($p<0.05$) between sediment concentration and slope gradient were found except for one combination test, although the power exponents were different for different combinations of slope length and rainfall intensity (Fig. 3). In general, the change in sediment concentration was less than linearly related to the change in slope gradient, with a mean power exponent of 0.4 and a CV of 0.41. The power exponents and the coefficients of determination (R^2) decreased with increasing slope length. This suggested that the effect of slope gradient on sediment concentration decreases with increasing slope length. The variability of the power exponent was less for higher rainfall intensities than for lower rainfall intensities.

For the same slope gradient and rainfall intensity, significant negative power-function relationships ($p<0.05$) between sediment concentration and slope length were detected in 19 out of 25 combination tests, and one had a negative power-function relationship ($p>0.05$). Figure 4 presents the frequencies of the power exponents and the coefficients of determination of the relationship between sediment concentration and slope length for all tests. For five combination tests with positive exponents, four tests were insignificant ($p>0.05$), and one had a significantly positive correlation ($p<0.05$). The positive power exponents

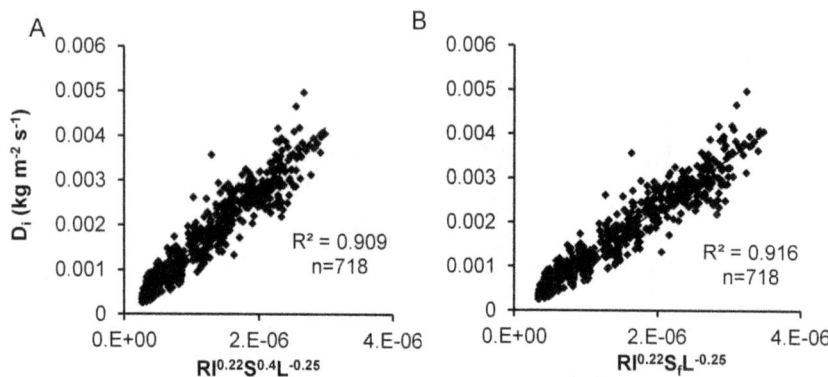

Figure 6. Linear regressions with zero intercept. Linear regression with zero intercept between the interrill erosion rate and the scaling factors $R\,I^{0.22}S^{0.4}L^{-0.25}$ (A) and $R\,I^{0.22}\,S_f\,L^{-0.25}$ (B) to assess the slope factor of $S^{0.4}$ and S_f.

Table 5. Slopes of the linear regression (K_i) and coefficients of determination (R^2) from Eq (10) for different experimental conditions.

L (m)		Eq (10)	I (mm min^{-1})		Eq (10)	S (%)		Eq (10)
0.4	K_i	1,132	0.8	K_i	1,271	17	K_i	1,082
	R^2	0.95		R^2	0.5		R^2	0.86
0.8	K_i	1,234	1.04	K_i	1,190	27	K_i	1,197
	R^2	0.95		R^2	0.49		R^2	0.89
1.2	K_i	1,163	1.7	K_i	1,192	36	K_i	1,182
	R^2	0.93		R^2	0.57		R^2	0.92
1.6	K_i	1,050	2.475	K_i	1,148	47	K_i	1,134
	R^2	0.90		R^2	0.72		R^2	0.9
2.0	K_i	1,153	2.835	K_i	1,134	58	K_i	1,140
	R^2	0.86		R^2	0.67		R^2	0.94
Mean	K_i	1,146	Mean	K_i	1,187	Mean	K_i	1,147
CV		5.7%	CV		4.5%	CV		3.9%

between sediment concentration and slope length might have resulted from poor control of the experimental conditions or from other random factors. The mean power exponent of the negative correlation was -0.25, with a CV of 0.33. In general, the power exponents decreased as slope gradient and rainfall intensity increased. The negative correlation between sediment concentration and slope length might indicate that the role of raindrop detachment on detached soil particles from the soil surface weakens with increasing slope length and that transport-limited erosion processes transfer to detachment-limited erosion with increasing slope length.

A power-function relationship between sediment concentration and rainfall intensity was derived for controlled conditions of fixed slope length and gradient. Among the 25 combination tests, 18 were significant positive power-function relationships ($p<0.05$), 2 were insignificant positive power-function relationships ($p>0.05$), and 5 were insignificant negative power-function relationships ($p>0.05$). These results indicate that positive power-function relationships exist between sediment concentration and rainfall intensity, while for a few tests, an insignificant negative correlation was detected, and that the power exponents are different for different tests (Fig. 5). The mean power exponent of the positive correlations was 0.22, with a CV of 0.55. In general, the power exponents decreased as slope length increased, which indicated that the effect of rainfall intensity on sediment concentration weakens due to the increase of flow depth and the reduction in raindrop detachment with increasing slope length and plot size.

Based on the results of the analysis described above, power-function relationships were derived between sediment concentration (C) and slope gradient (S), rainfall intensity (I), and slope length (L) for all combination tests. The averaged power exponents were used as a basis for describing the relationship between C and I, S, and L in Eq. (6). As a result, Eq. (6) can be expressed as:

$$C = AI^{0.22}S^{0.4}L^{-0.25} \qquad (8)$$

In Eq. (8), A is a constant for a given soil type and is equivalent to the interrill erodibility (K_i) of that soil type.

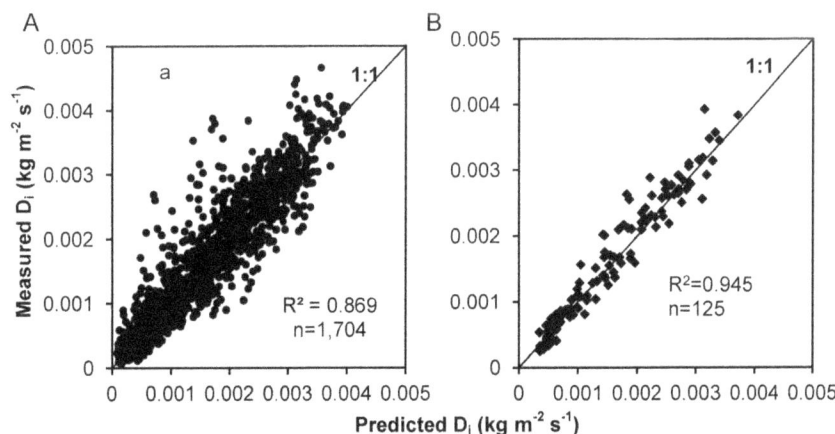

Figure 7. Measured vs. predicted interrill erosion rates. Measured and predicted interrill erosion rates for Eq. (10) for the erosion processes during a rainfall event (A) and the average of each rainfall event (B).

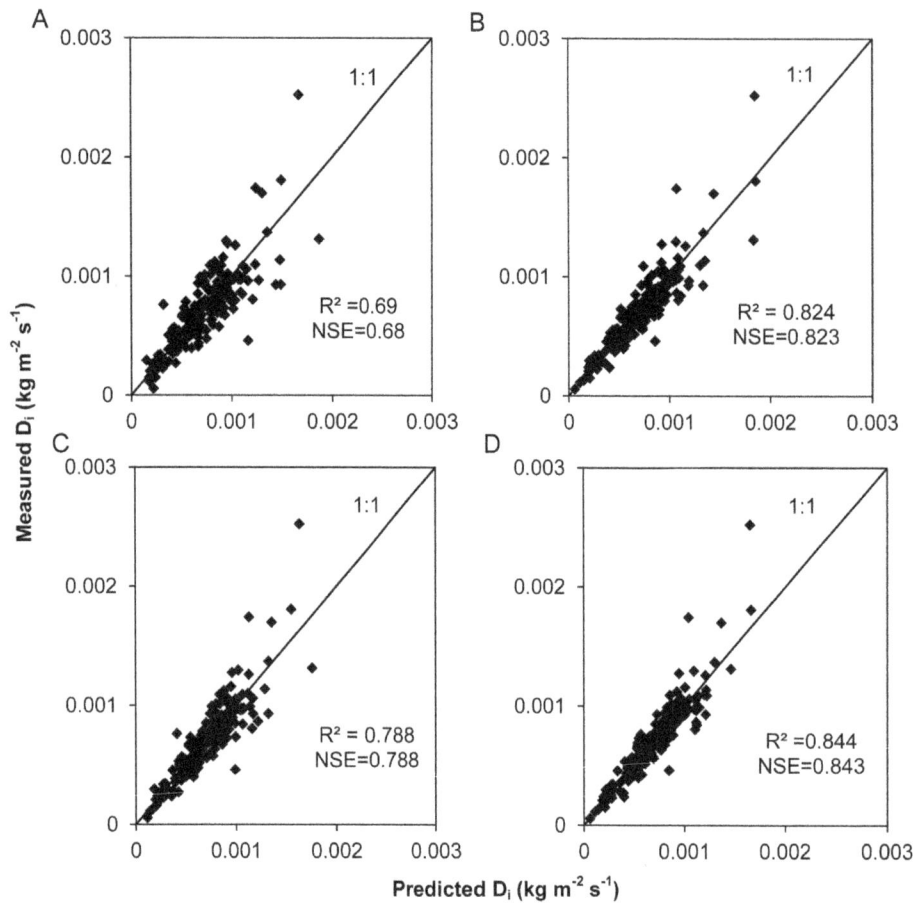

Figure 8. Measured vs. predicted interrill erosion rates. Measured and predicted interrill erosion rates for Eq. (1) (A), Eq. (3) (B), Eq. (4) (C), and Eq. (10) (D) using experimental data from WEPP cropland soil field interrill erodibility experiments.

Combining Eqs. (7) and (8), the interrill erosion rate, D_i (kg m^{-2} s^{-1}), can be expressed by the following equation:

$$D_i = K_i RI^{0.22} S^{0.4} L^{-0.25} \qquad (9)$$

However, as the previous work noted [19], the exponents of rainfall intensity, slope gradient, and slope length can be slightly altered to account for interactions among soil type, gradient, slope length, and rainfall intensity.

Figure 6 shows linear regressions with zero intercept between the interrill erosion rate (D_i) and the scaling factor $RI^{0.22} S^{0.4} L^{-0.25}$ and the scaling factor $RI^{0.22} S_f L^{-0.25}$. Although the coefficient of determination (R^2) values using $S^{0.4}$ and S_f were not significantly different, using S_f to describe the effect of gradient on interrill erosion yielded slightly better results than using $S^{0.4}$. Moreover, using S_f was consistent with results obtained by other researchers [16]. Equation (9) could therefore be transformed as follows:

$$D_i = K_i RI^{0.22} S_f L^{-0.25} \qquad (10)$$

To assess the performance of Eq. (10), table 5 presents the linear regression slopes (K_i) and coefficients of determination (R^2) obtained from linear regressions with zero intercept between the

interrill erosion rate (D_i) and the scaling factor $RI^{0.22} S_f L^{-0.25}$ for different experimental conditions. The higher coefficients of determination (R^2) and the lower CVs of interrill erodibility (K_i) indicated that Eq. (10) is better than previously developed equations for interrill erosion and adequately predicts interrill erosion for a wide range of rainfall intensities, slope gradients, and slope lengths.

As stated previously, Eq. (10) was developed for steady-state overland flow and interrill erosion rate. Equation (10) must be validated for non-steady-state conditions. Equation (10) was therefore used to predict interrill erosion rates during a rainfall event and the averaged interrill erosion rate of each rainfall event for all rainfall events in this study. The measured and predicted interrill erosion rates are presented in Fig. 7. Correlation coefficients of 0.932 and 0.972 between the measured and predicted interrill erosion rates were obtained, and Nash-Sutcliffe coefficients of 0.865 and 0.942 were obtained for the interrill erosion rates of erosion processes and the average of each rainfall event respectively. These results suggested that Eq. (10) could predict interrill erosion rates reasonably well, despite a few predicted values that were considerably different from the corresponding measured values.

Validation of the improved interrill erosion equation

To evaluate further the effectiveness of the improved interrill erosion equation, experimental data from WEPP cropland soil field interrill erodibility experiments were used for validation.

Table 6. Interrill erodibility and its CV from the different interrill erosion equations for the experimental data from the WEPP cropland soil field interrill erodibility experiments.

Soil type	No. of plots	$D_i = K_i I^2 S_f$		$D_i = K_i RIS_f$		$D_i = K_i q^{1/2} I S^{2/3}$		$D_i = K_i R I^{0.22} S_f L^{-0.25}$	
		K_i	CV (%)	K_i	CV (%)	K_i	CV (%)	K_i	CV (%)
Sharpsburg	6	1,917,372	24.05	2,457,363	13.22	23,850	17.18	323	13.03
Hersh	6	4,153,649	37.64	6,801,244	31.43	58,530	31.28	920	26.69
Keith	6	3,403,141	23.66	4,229,851	24.27	44,048	21.04	597	19.94
Amarillo	6	4,213,438	31.33	6,235,421	33.63	65,118	31.16	865	34.25
Woodward	6	4,592,143	27.85	7,568,362	20.60	74,617	28.95	1,049	27.73
Heiden	6	1,336,998	24.65	1,638,827	30.53	17,529	25.49	211	30.66
Whitney	6	3,101,189	20.41	4,367,166	16.02	48,903	18.38	590	14.56
Academy	6	2,994,017	25.59	3,932,019	11.12	40,099	16.97	514	9.04
Los Banos	6	2,405,840	14.00	3,262,543	17.95	30,191	10.92	414	10.58
Portneuf	6	1,382,542	16.90	2,314,773	14.54	20,975	13.61	280	12.42
Nansene	6	3,206,879	22.89	4,936,611	16.09	44,005	15.82	682	11.03
Palouse	6	4,003,363	18.95	5,201,600	12.97	48,338	13.65	682	9.57
Zahl	6	3,267,617	29.70	3,840,678	20.37	42,288	21.38	512	14.65
Pierre	6	2,300,664	17.06	2,719,824	7.11	27,829	11.27	367	6.60
Williams	6	2,684,238	24.36	3,548,762	11.90	38,403	10.74	508	10.17
Barnes - ND	6	1,876,045	10.85	3,056,669	10.09	31,502	8.41	474	8.12
Sverdrup	6	2,362,425	17.69	3,958,183	14.65	34,180	13.69	533	14.54
Barnes - MN	6	1,503,947	21.63	2,576,945	20.84	24,530	18.80	409	17.40
Mexico	6	2,970,456	16.24	4,153,863	15.04	41,848	13.23	570	12.10
Grenada	6	2,633,592	19.66	3,372,939	10.67	35,349	13.24	462	10.24
Tifton	6	652,793	54.22	1,603,183	27.15	12,053	36.29	221	23.88
Bonifay	6	871,497	73.08	4,226,263	15.32	22,179	47.29	581	21.05
Cecil	6	1,858,964	25.03	2,282,079	18.68	24,807	21.10	319	19.42
Hiwassee	6	1,873,194	18.16	2,243,785	17.27	25,030	15.54	320	13.94
Gaston	6	2,036,993	13.15	2,205,358	16.98	24,961	14.49	297	16.56
Opequon	6	3,195,673	16.05	3,398,535	9.63	39,643	11.97	474	7.20
Frederick	6	2,477,747	29.91	3,331,759	15.01	35,043	21.67	476	14.22
Manor	6	2,694,182	23.21	3,563,361	12.63	38,111	17.67	515	13.31
Caribou	6	1,551,246	11.13	1,845,618	7.78	20,635	8.21	263	7.82
Collamer	6	3,456,598	16.26	4,121,058	9.61	44,942	10.17	566	6.60
Miamian	6	1,591,767	22.99	2284814	18.64	22548	20.61	310	19.44
Lewisburg	6	2,257,824	20.43	2,803,097	15.93	29,807	16.97	381	14.62
Miami	6	1,970,749	26.84	2,579,164	18.97	26,927	20.07	356	15.01

These experiments were part of the USDA-ARS WEPP project, which included thirty-three soils from all areas of the United States in ridge plots. Flat plots were excluded from the validation because their slope gradients were much less than the lowest slope gradient in these experiments and some soil types were not represented in flat plots. The detailed information about the experimental materials and procedures were described [3]. Based on their data analysis techniques, the means of the last four erosion rates and flow discharges were selected as the erosion rate and flow discharge for a given plot in this validation. If there was an outlier value within the last four, then it was not considered, and the four most consistent of the last five values were used [3]. Although the fixed slope length was an obstacle to validating the improved interrill erosion equation, these experimental data could be used to judge the quality of different interrill erosion equations.

The "intrinsic" interrill erodibility for a given soil is independent of runoff rate, rainfall intensity, infiltration, slope length, and gradient. If an interrill erosion equation adequately describes the relationship between interrill erosion rate and its influencing factors, the interrill erodibility, as estimated by an interrill erosion equation for a given soil, should change little over a wide range of conditions. Therefore, the CV of interrill erodibility estimated from an interrill erosion equation for a given soil under a wide range of conditions can be used to judge the quality of the interrill erosion equation. Interrill erodibility values, as estimated from the different interrill erosion equations discussed in this paper, along with their CV for 33 soils in the United States, are given in Table 6.

In general, there were common trends in the interrill erodibilities predicted by the various interrill erosion equations. For all soils, very strong positive correlations ($p<0.05$) were found between the interrill erodibility values predicted from the various interrill erosion equations. Although the trends showed that interrill erodibility values predicted from the different interrill erosion equations reflected the strength of soil resistance to interrill erosion, the precision of the predictions was different among the equations. The CV of interrill erodibility from Eq. (10) was less than that from Eq. (1) for 30 soils, that from Eq. (10) was less than that from Eq. (3) for 25 soils, and that from Eq. (10) was less than that from Eq. (4) for 29 soils. For 23 soils, the CV of interrill erosion from Eq. (10) was the lowest among those from the four equations. Figure 8 shows the observed and predicted values for the various equations. The correlation coefficients and Nash-Sutcliffe coefficients between the measured and predicted values for Eq. (10) were the highest among all the equations, which suggested that Eq. (10) described the relationship between interrill erosion rate and its influencing factors better than previously proposed interrill erosion equations, even though the slope length of the plots in the experiments used for validation was fixed.

The CV of interrill erodibility from Eqs. (3) and (10) was generally less than that from Eq. (1), indicating that rainfall intensity or flow discharge alone cannot completely express interrill erosion processes. Compared with Eq. (3), the reduced proportion of flow discharge in Eq. (4) resulted in an increase in the CV of interrill erodibility, but the reduced proportion of rainfall intensity in Eq. (10) resulted in a decrease in the CV of interrill erodibility. These results indicated that flow discharge is still the most important transport agent for interrill erosion processes, even though rainfall intensity has two effects, i.e., detachment of soil and enhancement of the transport capacity of thin overland flow. The previously proposed interrill erosion equations overemphasized the effect of rainfall intensity on interrill

erosion when slope length was relatively longer. The CVs of interrill erodibility from Eq. (3) and Eq. (10) were closer, perhaps because slope length was fixed in the WEPP cropland soil field interrill erodibility experiments.

Conclusions

Previously proposed interrill erosion equations based on mini-plot experiments largely ignored the effect of slope length and plot size on interrill erosion rate. A series of simulated rainfall experiments was conducted to validate the previously proposed interrill erosion equations. The results indicated that slope gradient almost adequately predicted interrill erosion in the previously proposed equations, but that the interrill erodibilities calculated from previously proposed equations generally decreased with increasing slope length and rainfall intensity. This result suggested that adding a factor for slope length and adjusting the factor for rainfall intensity are necessary to improve prediction of interrill erosion rate. An improved interrill erosion equation, $D_i = K_i RI^{0.22} S_f L^{-0.25}$, was developed by analyzing the relationships between sediment concentration and rainfall intensity, slope length, and slope gradient. To evaluate the effectiveness of the improved interrill erosion equation, part experimental data from WEPP cropland soil field interrill erodibility experiments were used for validation. The results indicated that the improved interrill erosion equation predicted interrill erosion rate better than previously proposed interrill erosion equations, even though the slope length of the plots in the WEPP experiments was fixed.

Author Contributions

Conceived and designed the experiments: ZW FZ MY. Performed the experiments: ZW FZ. Analyzed the data: FZ ZW MY. Contributed reagents/materials/analysis tools: ZW. Wrote the paper: FZ ZW MY.

References

1. Meyer LD, Foster GR, Romkens MJM (1975) Source of soil eroded by water from upland slopes. In: Present and prospective technology for predicting sediment yields and sources, Washington, USDA-Agricultural Research: 177–189.
2. Kinnell PIA (1991) The effect of flow depth on sediment transport induced by raindrops impacting shallow flows. Transactions of the ASAE 34: 161–168.
3. Elliot WJ, Liebenow AM, Laflen JM, Kohl KD (1989) A Compendium of Soil Erodibility Data from WEPP Cropland Soil Field Erodibility Experiments 1987 and 1988. NSERL Report No 3, The Ohio State University, and USDA Agricultural Research Service.
4. Morgan RPC, Quinton JN, Smith RE, Govers G, Poesen JWA, et al. (1998) The European Soil Erosion Model (EUROSEM): a dynamic approach for predicting sediment transport from fields and small catchments. Earth Surf Proc Land 23: 527–544.
5. De Roo APJ, Wesseling CG, Ritsema CJ (1996) LISEM: A single-event physically based hydrological and soil erosion model for drainage basins I: theory, input and output. Hydrol Process 10: 1107–1117.
6. Nord G, Esteves M (2005) PSEM_2D: A physically based model of erosion processes at the plot scale. Water Resour Res 41: W08407.
7. Valmis S, Dimoyiannis D, Danalatos NG (2005) Assessing interrill erosion rate from soil aggregate instability index, rainfall intensity and slope angle on cultivated soils in central Greece. Soil Till Res 80: 139–147.
8. Yan FL, Shi ZH, Li ZX, Cai CF (2008) Estimating interrill soil erosion from aggregate stability of Ultisols in subtropical China. Soil Till Res 100: 34–41.
9. Ascough JC, Baffaut C, Nearing MA, Liu BY (1997) The WEPP watershed model: I. Hydrology and erosion. Transactions of the ASAE 40: 921–933.
10. Wei H, Nearing MA, Stone JJ, Guertin DP, Spaeth KE, et al. (2009) A New Splash and Sheet Erosion Equation for Rangelands. Soil Sci Soc Am J 73: 1754–1754.
11. Meyer LD (1981) How Rain Intensity Affects Interrill Erosion. Transactions of the ASAE 24: 1472–1475.
12. Watson DA, Laflen JM (1986) Soil strength, slope, and rainfall intensity effects on interrill erosion. Transactions of the ASAE 29: 98–102.
13. Foster GR, Meyer LD, Onstad CA (1977) An Erosion Equation Derived from Basic Erosion Principles. Transactions of the ASAE 20: 678–682.

14. Meyer LD, Harmon WC (1984) Susceptibility of agricultural soils to interrill erosion. Soil Sci Soc Am J 48: 1152–1157.
15. Guy BT, Dickinson WT, Rudra RP (1987) The Roles of Rainfall and Runoff in the sediment transport capacity of interrill flow. Transactions of the ASAE 30: 1378–1386.
16. Elliot WJ, Laflen JM, Kohl KD (1989) Effect of soil properties on soil erodibility. Paper no 89-2150 Transactions of the ASAE, St Joseph, MI.
17. Mccool DK, Brown LC, Foster GR, Mutchler CK, Meyer LD (1987) Revised slope steepness factor for the universal soil loss equation. Transactions of the ASAE 30: 1387–1396.
18. Neal JH (1938) The Effect of the Degree of Slope and Rainfall Characteristics on Runoff and Soil Erosion. Soil Sci Soc Am J 2: 525–532.
19. Zhang XC, Nearing MA, Miller WP, Norton LD, West LT (1998) Modelling interrill sediment delivery. Soil Sci Soc Am J 62: 438–444.
20. Young RA, Wiersma JL (1973) The role of rainfall impact in soil detachment and transport. Water Resour Res 9: 1629–1636.
21. Truman CC, Bradford JM (1995) LABORATORY DETERMINATION OF INTERRILL SOIL ERODIBILITY. Soil Sci Soc Am J 59: 519–526.
22. Kinnell PIA (1993) Runoff as factor influencing experimentally determined interrill erodiblities. Aust J Soil Res 31: 333–342.
23. Flanagan DC, Nearing MA (1995) USDA-Water Erosion Prediction Project: Technical documentation. NSERL Rep no 10 Natl Soil Erosion Res Lab, West Lafayette, IN.
24. Monke EJ, Marelli HJ, Meyer LD, DeJong JF (1977) Runoff, erosion, and nutrient movement from interrill areas. Transactions of the ASAE 20: 58–61.
25. Gilley JE, Woolhiser DA, McWhorter DB (1985) Interrill soil erosion - Part II: Testing and use of model equations. Transactions of the ASAE 28: 154–159.
26. Gilley JE, Woolhiser DA, McWhorter DB (1985) Interrill soil erosion-Part I. Development of model equations. Transactions of the ASAE 28: 147–153, 159.
27. Meyer LD, Harmon WC (1989) How Row-Sideslope Length and Steepness Affect sideslope erosion. Transactions of the ASAE 32: 639–644.
28. Kinnell PIA (2000) The effect of slope length on sediment concentrations associated with side-slope erosion. Soil Sci Soc Am J 64: 1004–1008.
29. Chaplot VAM, Le Bissonnais Y (2003) Runoff features for interrill erosion at different rainfall intensities, slope lengths, and gradients in an agricultural loessial hillslope. Soil Sci Soc Am J 67: 844–851.

30. Gabriels D (1999) The effect of slope length on the amount and size distribution of eroded silt loam soils: short slope laboratory experiments on interrill erosion. Geomorphology 28: 169–172.
31. Foster GR (1982) Modeling the erosion process. In Hydrologic modeling of small watersheds C T Haan et al, eds Chapter 8: 297–380. ASAE Monograph No. 5, St. Joseph, MI: ASAE.
32. Renard K, Foster GR, Weesies G, McCool DK, Yoder D (1997) Predicting Soil Erosion by Water; Guide to Conservation Planning with the Revised Universal Soil Loss Equation (RUSLE). ARS, USDA, Agricultural Handbook No 703.
33. Kinnell PIA (1993) Sediment concentrations resulting flow depth drop size interactions in shallow overland flow. Transactions of the ASAE 36: 1099–1103.
34. Kinnell P (1994) The effect of pre-detached particles on soil erodibilities associated with erosion by rain-impacted flows. Soil Res 32: 127–142.
35. Wischmeier WH (1958) Rainfall energy and its relationship to soil loss. Trans AGU 39: 285–291.

Permissions

All chapters in this book were first published in PLOS ONE, by The Public Library of Science; hereby published with permission under the Creative Commons Attribution License or equivalent. Every chapter published in this book has been scrutinized by our experts. Their significance has been extensively debated. The topics covered herein carry significant findings which will fuel the growth of the discipline. They may even be implemented as practical applications or may be referred to as a beginning point for another development.

The contributors of this book come from diverse backgrounds, making this book a truly international effort. This book will bring forth new frontiers with its revolutionizing research information and detailed analysis of the nascent developments around the world.

We would like to thank all the contributing authors for lending their expertise to make the book truly unique. They have played a crucial role in the development of this book. Without their invaluable contributions this book wouldn't have been possible. They have made vital efforts to compile up to date information on the varied aspects of this subject to make this book a valuable addition to the collection of many professionals and students.

This book was conceptualized with the vision of imparting up-to-date information and advanced data in this field. To ensure the same, a matchless editorial board was set up. Every individual on the board went through rigorous rounds of assessment to prove their worth. After which they invested a large part of their time researching and compiling the most relevant data for our readers.

The editorial board has been involved in producing this book since its inception. They have spent rigorous hours researching and exploring the diverse topics which have resulted in the successful publishing of this book. They have passed on their knowledge of decades through this book. To expedite this challenging task, the publisher supported the team at every step. A small team of assistant editors was also appointed to further simplify the editing procedure and attain best results for the readers.

Apart from the editorial board, the designing team has also invested a significant amount of their time in understanding the subject and creating the most relevant covers. They scrutinized every image to scout for the most suitable representation of the subject and create an appropriate cover for the book.

The publishing team has been an ardent support to the editorial, designing and production team. Their endless efforts to recruit the best for this project, has resulted in the accomplishment of this book. They are a veteran in the field of academics and their pool of knowledge is as vast as their experience in printing. Their expertise and guidance has proved useful at every step. Their uncompromising quality standards have made this book an exceptional effort. Their encouragement from time to time has been an inspiration for everyone.

The publisher and the editorial board hope that this book will prove to be a valuable piece of knowledge for researchers, students, practitioners and scholars across the globe.

List of Contributors

Yunbin Qin, Zhongbao Xin, Xinxiao Yu and Yuling Xiao
Institute of Soil and Water Conservation, Beijing Forestry University, Beijing, China

Bing Wang, Fenxiang Wen, Jiangtao Wu and Xiaojun Wang
College of Environmental Science and Resources, Shanxi University, Taiyuan, China

Yani Hu
Library, Hebei University of Science and Technology, Shijiazhuang, China

Xiaobo Yi and Li Wang
College of Resources and Environment, Northwest A&F University, Yangling, Shaanxi, China
State Key Laboratory of Soil Erosion and Dryland Farming on the Loess Plateau, Northwest A&F University, Yangling, Shannxi, China

Yaai Dang
State Key Laboratory of Soil Erosion and Dryland Farming on the Loess Plateau, Institute of Soil and Water Conservation, Northwest A&F University, Yangling, Shaanxi, China
International Center for Climate and Global Change Research, School of Forestry & Wildlife Sciences, Auburn University, Auburn, Alabama, United States of America
College of Science, Northwest A&F University, Yangling, Shaanxi, China

Wei Ren, Bo Tao, Guangsheng Chen, Chaoqun Lu, Jia Yang, Shufen Pan and Hanqin Tian
International Center for Climate and Global Change Research, School of Forestry & Wildlife Sciences, Auburn University, Auburn, Alabama, United States of America,

Guodong Wang
College of Science, Northwest A&F University, Yangling, Shaanxi, China

Shiqing Li
State Key Laboratory of Soil Erosion and Dryland Farming on the Loess Plateau, Institute of Soil and Water Conservation, Northwest A&F University, Yangling, Shaanxi, China

Xiaoying Wang and Yanan Tong
College of Natural Resources and Environment, Northwest A&F University, Yangling, China

Key Laboratory of Plant Nutrition and the Agri-environment in Northwest China, Ministry of Agriculture, Yangling, China

Yimin Gao, Pengcheng Gao, Fen Liu, Zuoping Zhao and Yan Pang
College of Natural Resources and Environment, Northwest A&F University, Yangling, China

Fulin Yang, Qiang Zhang and Runyuan Wang
Key Laboratory of Arid Climatic Change and Reducing Disaster of Gansu Province, Key Open Laboratory of Arid Climatic Change and Disaster Reduction of China Meteorological Administration (CMA), Institute of Arid Meteorology, CMA, Lanzhou, China

Jing Zhou
State Key Laboratory of Grassland Agro-ecosystems, College of Pastoral Agriculture Science and Technology, Lanzhou University, Lanzhou, China

Lubo Gao, Huasen Xu, Xiaoyan Wang, Biao Bao, Chao Bi and Yifang Chang
College of Water and Soil Conservation, Beijing Forestry University, Beijing, P.R. China

Huaxing Bi
Key Laboratory of Soil and Water Conservation, Ministry of Education, Beijing, P.R. China,
College of Water and Soil Conservation, Beijing Forestry University, Beijing, P.R. China

Weimin Xi
Department of Biological and Health Sciences, Texas A&M University-Kingsville, Kingsville, Texas, United States of America

Lei Deng and Zhou-Ping Shangguan
State Key Laboratory of Soil Erosion and Dryland Farming on the Loess Plateau, Northwest A&F University, Yangling, Shaanxi, China

Sandra Sweeney
Institute of Environmental Sciences, University of the Bosphorus, Istanbul, Turkey
Fazhu Zhao, Gaihe Yang, Xinhui Han, Yongzhong Feng and Guangxin Ren
College of Agronomy, Northwest A&F University, Yangling, Shaanxi, China; and The Research Center of Recycle Agricultural Engineering and Technology of Shaanxi Province, Yangling, Shaanxi, China

Mingguo Zheng
Key Laboratory of Water Cycle and Related Land Surface Processes, Institute of Geographic Sciences & Natural Resources Research, Chinese Academy of Sciences, Beijing, China,

Yishan Liao
Guangdong Institute of Eco-environment and Soil Sciences, Guangzhou, China

Jijun He
Base of the State Laboratory of Urban Environmental Processes and Digital Modelling, Capital Normal University, Beijing, China

Yihe Lu¨, Bojie Fu, Xiaoming Feng, Yu Liu and Ruiying Chang
State Key Laboratory of Urban and Regional Ecology, Research Center for Eco-Environmental Sciences, Chinese Academy of Sciences, Beijing, China,

Yuan Zeng and Bingfang Wu
Institute of Remote Sensing Applications, Chinese Academy of Sciences, Beijing, China

Ge Sun
USDA-Forest Service, Southern Research Station, Raleigh, North Carolina, United States of America

Hong Wang
Institute of Soil and Water Conservation, Chinese Academy of Sciences and Ministry of Water Resources, Yangling, Shaanxi Province, China
University of Chinese Academy of Sciences, Beijing, China

Jian-en Gao
Institute of Soil and Water Conservation, Chinese Academy of Sciences and Ministry of Water Resources, Yangling, Shaanxi Province, China

Institute of Soil and Water Conservation, Northwest A&F University, Yangling, Shaanxi Province, China
College of Water Resources and Architectural Engineering, Northwest A&F University, Yangling, Shaanxi Province, China
College of Natural Resources and Environment, Northwest A&F University, Yangling, Shaanxi Province, China

Shao-long Zhang and Xing-hua Li
College of Water Resources and Architectural Engineering, Northwest A&F University, Yangling, Shaanxi Province, China

Meng-jie Zhang
College of Natural Resources and Environment, Northwest A&F University, Yangling, Shaanxi Province, China

Xueling Yao, Bojie Fu, Yihe Lu¨, Feixiang Sun, Shuai Wang and Min Liu
State Key Laboratory of Urban and Regional Ecology, Research Center for Eco-Environmental Sciences, Chinese Academy of Sciences, Beijing, P. R. China

Ruiying Chang
State Key Laboratory of Urban and Regional Ecology, Research Center for Eco-Environmental Sciences, Chinese Academy of Sciences, Beijing, China,
Institute of Mountain Hazards and Environment, Chinese Academy of Sciences, Chengdu, Sichuan, China

Tiantian Jin
State Key Laboratory of Urban and Regional Ecology, Research Center for Eco-Environmental Sciences, Chinese Academy of Sciences, Beijing, China
China Institute of Water Resources and Hydropower Research, Beijing, China

Yihe Lu¨, Guohua Liu and Bojie Fu
State Key Laboratory of Urban and Regional Ecology, Research Center for Eco-Environmental Sciences, Chinese Academy of Sciences, Beijing, China

Liguang Sun, Xin Zhou, Wenhan Cheng and Nan Jia
Institute of Polar Environment and School of Earth and Space Sciences, University of Science and Technology of China, Hefei, China

Yuhong Wang
Advanced Management Research Center, Ningbo University, Ningbo, China

Mingming Li and Qing Zhen
Institute of Soil and Water Conservation, Chinese Academy of Sciences and Ministry of Water Resources, Yangling, PR China
University of Chinese Academy of Sciences, Beijing, PR China

Xingchang Zhang and Fengpeng Han
University of Chinese Academy of Sciences, Beijing, PR China
State Key Laboratory of Soil Erosion and Dryland Farming on the Loess Plateau, Institute of Soil and Water Conservation, Northwest A & F University, Yangling, PR China

Liguang Sun, Xin Zhou, Wenhan Cheng and Nan Jia
Institute of Polar Environment and School of Earth and Space Sciences, University of Science and Technology of China, Hefei, China

Yuhong Wang
Advanced Management Research Center, Ningbo University, Ningbo, China

Mingming Li and Qing Zhen
Institute of Soil and Water Conservation, Chinese Academy of Sciences and Ministry of Water Resources, Yangling, PR China
University of Chinese Academy of Sciences, Beijing, PR China,

Xingchang Zhang and Fengpeng Han
University of Chinese Academy of Sciences, Beijing, PR China
State Key Laboratory of Soil Erosion and Dryland Farming on the Loess Plateau, Institute of Soil and Water Conservation,Northwest A & F University, Yangling, PR China

Lan Mu, Yinli Liang and Ruilian Han
Institute of Soil and Water Conservation, Northwest A&F University, Yangling Shaanxi Province, China
Institute of Soil and Water Conservation, Chinese Academy of Sciences and Ministry of Water Resources, Yangling, Shaanxi Province, China

Huifang Wang
Beijing Research & Development Center for Grass and Environment, Beijing Academy of Agriculture and Forestry Sciences, Beijing, China
Center for Water System Security, Graduate University of Chinese Academy of Sciences, Beijing, China
State Key Laboratory of Soil Erosion and Dryland Farming on the Loess Plateau, Institute of Soil and Water Conservation, Chinese Academy of Sciences, Yangling, Shaanxi, China

Bo Xiao
Beijing Research & Development Center for Grass and Environment, Beijing Academy of Agriculture and Forestry Sciences, Beijing, China
State Key Laboratory of Soil Erosion and Dryland Farming on the Loess Plateau, Institute of Soil and Water Conservation, Chinese Academy of Sciences, Yangling, Shaanxi, China

Mingyu Wang
Center for Water System Security, Graduate University of Chinese Academy of Sciences, Beijing, China

Ming'an Shao
State Key Laboratory of Soil Erosion and Dryland Farming on the Loess Plateau, Institute of Soil and Water Conservation, Chinese Academy of Sciences, Yangling, Shaanxi, China

Kathryn E. Fitzsimmons
Department of Human Evolution, Max Planck Institute for Evolutionary Anthropology, Leipzig, Germany

Ulrich Hambach
Chair of Geomorphology, Laboratory for Palaeo- and Enviro-Magnetism, University of Bayreuth, Bayreuth, Germany

Daniel Veres
Institute of Speleology, Romanian Academy, Cluj-Napoca, Romania
Faculty of Environmental Sciences, Babes- Bolyai University, Cluj-Napoca, Romania

Radu Iovita
MONREPOS Archaeological Research Centre and Museum for Human Behavioural Evolution, RGZM, Neuwied, Germany

Feng-Bao Zhang, Zhan-Li Wang and Ming-Yi Yang
State Key Laboratory of Soil Erosion and Dryland Farming on the Loess Plateau, Institute of Soil and Water Conservation, Northwest A&F University, Yangling, P. R. China
Institute of Soil and Water Conservation, Chinese Academy of Science and Ministry of Water Resources, Yangling, P. R. China

Index